Lecture Notes on Data Engineering and Communications Technologies

Volume 52

Series Editor

Fatos Xhafa, Technical University of Catalonia, Barcelona, Spain

The aim of the book series is to present cutting edge engineering approaches to data technologies and communications. It will publish latest advances on the engineering task of building and deploying distributed, scalable and reliable data infrastructures and communication systems.

The series will have a prominent applied focus on data technologies and communications with aim to promote the bridging from fundamental research on data science and networking to data engineering and communications that lead to industry products, business knowledge and standardisation.

**** Indexing: The books of this series are submitted to SCOPUS, ISI Proceedings, MetaPress, Springerlink and DBLP ****

More information about this series at http://www.springer.com/series/15362

Ketan Kotecha · Vincenzo Piuri ·
Hetalkumar N. Shah · Rajan Patel
Editors

Data Science and Intelligent Applications

Proceedings of ICDSIA 2020

Springer

Editors
Ketan Kotecha
Faculty of Engineering
Symbiosis Institute of Technology
Pune, India

Vincenzo Piuri
Department of Computer Science
Università degli Studi di Milano
Milan, Italy

Hetalkumar N. Shah
Gandhinagar Institute of Technology
Gandhinagar, Gujarat, India

Rajan Patel
Department of Computer Engineering
Gandhinagar Institute of Technology
Gandhinagar, Gujarat, India

ISSN 2367-4512 ISSN 2367-4520 (electronic)
Lecture Notes on Data Engineering and Communications Technologies
ISBN 978-981-15-4473-6 ISBN 978-981-15-4474-3 (eBook)
https://doi.org/10.1007/978-981-15-4474-3

This Springer imprint is published by the registered company Springer Nature Singapore Pte Ltd.
The registered company address is: 152 Beach Road, #21-01/04 Gateway East, Singapore 189721, Singapore

Preface

The International Conference on Data Science and Intelligent Applications (ICDSIA-2020) was organized by Gandhinagar Institute of Technology (GIT) during 24–25 January 2020. It was sponsored by Gujarat Council on Science and Technology (GUJCOST), Gandhinagar and Platinum Foundation, Ahmedabad, in association with Gujarat Technological University (GTU), Ahmedabad, and Indian Society for Technical Education (ISTE), New Delhi.

The focus of the papers of the conference is on original, significant and quality research by researchers on the theories and practices of emerging technologies in the main areas of data science, intelligent applications and communication technologies. It provided the scope for opportunities to researchers and practitioners from academia and industry. The conference was projected to invite a large number of quality submissions and stimulate the cutting-edge research discussions among many academic pioneering researchers, scientists, industrial engineers and students from all around the world and provide a forum to researchers. The idea of the conference was drawn by the requirements and impact of data science and AI in society. It has a big and positive impact on the society and the economy. Robots and AI will help people to perform their tasks better than now. The combination of man and machine will be unstoppable. AI and data science also significantly reduce the probability of human error and study historical data to cut costs. More than 129 research submissions were received and out of them 79 presentations were scheduled, which will give benefits to more than 240 authors.

The conference was chaired by Dr. Navin Sheth (Hon. VC, GTU), followed by Dr. G. T. Pandya (IAS—Gujarat Higher Education), Dr. Narottam Sahoo (GUJCOST), Dr. N. M. Bhatt (LDCE), Dr. H. N. Shah (Director, GIT and convener of ICDSIA-2020) and conference secretaries: Dr. Rajan Patel and Prof. Archana Singh. The two-day programme includes keynote speech by Dr. Sheng-Lung Peng, Professor, National Dong Hwa University, Taiwan; Dr. Ketan Kotecha, Director, Symbiosis International University, Pune, India; Dr. Nilanjan Dey, Techno India College of Technology, Kolkata, India; Dr. Y. P. Kosta, Provost, Marwadi University, Rajkot, India; and Mr. Deepak Pareek, Digi Agri, Ahmedabad, India.

We would like to express our sincere appreciation to all active authors for their contributions to this book. We would like to extend our thanks to all the reviewers and session chairs for their constructive comments and reviews on all the papers. We are thankful to keynote speakers for their technical contribution to the conference. We are also very thankful to the Trustees of GIT and Director Dr. H. N. Shah for providing all types of excellent infrastructure to organize this international conference. We are also thankful to dedicated TPC team members Prof. Sejal Bhavsar and Prof. Margil Shah, all faculty members of the computer engineering department and all staff members of GIT. Especially, we would like to thank the dedicated and motivated organizing team of the conference for their hard work. We appreciate the initiative and support from Mr. Aninda Bose and his colleagues in Springer Nature for their strong support towards publishing this volume in the Lecture Notes on Data Engineering and Communications Technologies (LNDECT) series of Springer Nature.

Pune, India — Ketan Kotecha
Milan, Italy — Vincenzo Piuri
Gandhinagar, India — Hetalkumar N. Shah
Gandhinagar, India — Rajan Patel
Gandhinagar, India — Archana Singh

Contents

About the Editors

Dr. Ketan Kotecha is a Director at Symbiosis Institute of Technology, Pune. In the course of a 25-year career, he has served in technical education leadership positions at some of the finest engineering institutions in India. Holding a Ph.D. (CE) from the IIT, Bombay, he has published more than 100 papers and filed three patents. He is a member of the Governing Council or Academic Council at various universities.

Dr. Vincenzo Piuri is a Professor at the University of Milan, Italy. He has published innovative results in more than 400 papers in international journals and conference proceedings. He has acted as IEEE Director, IEEE Fellow, IEEE-HNK member, ACM Scientist and INNS member. In the course of 20 years, he has served on the technical program committees of various international conferences and organized more than 100 conferences and workshops. He is an Honorary Professor at Obuda University, Budapest, Hungary, Guangdong University of Petrochemical Technology, China, Muroran Institute of Technology, Japan, and Amity University, India.

Dr. Hetalkumar N. Shah is a Director of Gandhinagar Institute of Technology, India. With more than 22 years of academic, administrative and industry experience, he has published more than 40 research papers in prominent national and international journals and conference proceedings. A member of e.g. the ISTE, AMM, IIPE and IEI, he is a recipient of a three-year Research Scholar Fellowship Award from the AICTE, New Delhi, and an International Travel Grant from the DST to Taiwan and Hong Kong.

Dr. Rajan Patel is an Associate Professor at Gandhinagar Institute of Technology, Gandhinagar. He has more than 15 years of teaching experience in the field of Computer Engineering. A member of e.g. the CSI, ISTE and UACEE, he has authored more than 40 publications. He has also received numerous awards, honors and certificate of excellence.

Archive System Using Big Data for Health care: Analysis, Architecture, and Implementation

Suraj Tekchandani, Jigar Shah, and Archana Singh

Abstract In the era of digitalization, since the last few decades, we have seen the advancements of technology in all the domains, and health care does not remain in situ. With the growing medical science field, the growth of the EMRs/EHRs skyrockets the growth of data. Hence, storing the patients' data and reports has become an ache this day. This is often too large and heterogeneous, and changes are often to be stored, processed, and transformed into the required form, thus resulting in a lack of memory storage, processing power and bandwidth for transmission of the data. So hospitals are now lapsing into a dilemma for buying more storage and more processing power for preserving the patients' data. The patients' data not only includes textual data but also images, videos, and different files. This data can vary from patient to patient and differs from one hospital to another. In this paper, we will be talking about the room for the big data technology and data compression in the fields of medicine and healthcare by reduction in storage consumption, cleaning, and formatting the patients' data. We can furthermore use this clean data for research and analytical purposes. Therefore, we will be proposing an architecture for the Patient Data Archiving System including Big data technologies and data compression technique through this paper.

Keywords Big data · Cassandra · ETL · PACS · Storage techniques · PostgreSQL · CQL · Clustering · Network topology · Image processing · Image compression · DICOM · Feature extraction · Analytics · Health care · Framework · Methodology

S. Tekchandani · J. Shah (✉)
Birla Institute of Technology and Science, Pilani, India
e-mail: jigarshah1994@gmail.com

S. Tekchandani
e-mail: surajteckchandani@gmail.com

A. Singh (✉)
Gandhinagar Institute of Technology, Kalol, India
e-mail: archana.singh@git.org.in

K. Kotecha et al. (eds.), *Data Science and Intelligent Applications*, Lecture Notes on Data Engineering and Communications Technologies 52,
https://doi.org/10.1007/978-981-15-4474-3_1

1 Introduction

Every day the number of internet users increases with the advancements in the technologies, which consecutively increase the volume of data. This is true for all domains, and health care is no in situ. With the growing medical science field, the growth of the electronic medical records/electronic health records has skyrocketed the growth of data. Hence, storing the patients' data and reports has become an ache this day. This is often too large and heterogeneous, and changes are often to be stored, processed, and transformed into the required form, thus resulting in a lack of memory storage, processing power, and bandwidth for transmission of the data. Therefore, hospitals are now lapsing into a dilemma for buying more storage and more processing power for preserving the patients' data. The patients' data not only includes textual data but also images, videos, and different files. This data can vary from patient to patient and differs from one hospital to another.

2 Problem Definition

2.1 Problem Statement

With the increasing amounts of the patients, the patient-related information is also increasing. These data consist of 80–90% images data and 10–20% textual and other data. The medical images are being generated in the form of DICOM. Therefore, we identified the need of the hybrid architecture for archiving system including big data technologies and image compression to reduce the storage requirements and requirement for fault tolerance system as well as zero data loss.

The hospitals are struggling to store the data, the current system has capacity of 2–3 years to store patients' data and their medical images, and it can be changed based on policy of hospital. To manage patients' data and their medical images, hospitals have to spend a huge amount of money.

2.2 Problem Definition

As discussed, medical data comprises 80% of the medical images and 20% of the textual patients' data. The hospitals need to bear extensive cost of the storage and management of the images and patients' data. Since the data is huge, we can apply compression along with big data techniques to reduce the amount of storage required for storing this enormous amount of data.

Table 1 Sample storage calculation [3]

	Devices	Number of studies per day	Study size MB	1 year GB
CR	1	150	30	1080
CT	1	40	60	576
CT	1	40	375	3600
DX	1	150	30	1080
MG	1	20	240	1152
MR	1	20	26	124.8
MR	1	20	250	1200
US	1	50	150	1800
XA	1	15	60	216
XA	1	15	250	900
Total data in GB per year	11,728.8			

With the help of traditional methods, it is very difficult to perform data processing that includes collecting, analyzing, and leveraging consumer, patient, and clinical data as the data is complex, huge, and scattered. As value-based care has increased, it has motivated the enterprise to take future decisions that use predictive analytics.

In addition to all this, the volume of data will exponentially increase as smart wearable and IoT devices are getting more popular. These devices will help in the constant monitoring of the patient; this will add huge amounts of data to big data stores. With this data, healthcare marketers to find and maintain patients with the highest service propensity can embed a big amount of healthcare ideas [1, 2].

Table 1 shows the data generated by the different modalities in a year and amount of the storage required to store. Currently, hospitals are not able to store all the data related to patient for more than 2–3 years. If we can store this data with lesser storage requirements, we can use this stored data for the research and further medical field inventions and enhancements.

3 Literature Review

See Table 2.

Table 2 Literature reviews

References	Outcome
[4]	Impact of big data in the healthcare industry and its effects and importance of big data technologies that are used to deal with the growing data
[5]	Adoption of big data framework in storage and processing technologies to attain high-performance potential. Once this is respected, descriptive, predictive, or prescriptive analytics can be incorporated easily to get valid significant insights from the EHRs data
[6]	Employing healthcare analytics with efficient organization, streamlining and analysis of big data, will ensure prompt and accurate diagnosis, reduction in preventable mistakes and appropriate treatment
[7]	The solution for efficient data collection, healthcare delivery integrated with analytics
[8]	Requirement of the compression in PACS and lossless compression using Huffman coder
[9]	Extraction of the model image and residual image and quad tree compression
[10]	ROI-based compression and lossy ROI and lossless ROI and compression of the MRI files and ratio of 25%

3.1 Fieldwork

To understand the details of how the hospitals are dealing with medical data and medical imaging data, a survey is done for two multispecialty hospitals in Ahmedabad. Following are the discussions with the hospitals:

Dr. Arvind Sharma, Neurophysician, Zydus Hospital, Ahmedabad
According to Dr. Sharma, hospitals are dealing with lots of images on a daily basis. Depending upon the kind of modality required, the images generated per patient vary. For instance, if it is an MRI scan, it may generate 500–800 images for an individual. The hospital has given the task to the vendor to store and handle the data, it has to go for license and as the storage space assigned goes full, and they have to purchase new space. Moreover, due to the cost involved they do not prefer to store the data for a longer duration. The other option with them is to go on the cloud, but again the major issue was cost-effectiveness. Therefore, according to them if a system is developed to overcome this, it is going to be welcomed.

Dr. Gargey Sutaria, Head of the Department, Shalby Hospital Ahmedabad
Discussion with Dr. Gargey was a detailed understanding of the ways the hospital is working with images. The survey introduces us to the PACS system used by the hospital to store and manage the data. Here images are stored in a format called DICOM. These images are rich images that allow doctors to create 3D views and measure the height and width of the wounds, density of the affected area, and many more. As the size of DICOM images is still a bottleneck, the doctors want it to be processed. According to the doctors in the near future, the number of images

generated is going to increase exponentially and will be difficult to handle. From the amount of big data stored and managed by hospital, 80% of the space and processing is done for images. PACS is a platform-dependent storage system, and sometimes their retrieval is a hurdle. In addition, as vendors manage the PACS, the cost here is a big factor to be considered. Therefore, they say if a solution to this could be provided and that too in-house will be a great contribution.

4 Proposed Work

In order to solve the above-mentioned problem, here a large patients' dataset was considered to simulate the real ecosystem of the patients' data store. Figure 1 shows few table structures from the dataset that we will be working.

This dataset has been converted to PostgreSQL database to depict any real EHR/EMR System. After considering the pros and cons of the various big data technologies in our use case, we came up with the following architecture proposal as shown in Table 3.

The data is read from the PostgreSQL through the AWS Lambda and is pushed to the SQS queues. The Apache spark continues pool these queues for the data. Whenever the data is pushed to SQS is read by Apache Spark, which consecutively writes the data to Cassandra. The files, i.e., medical images, are read from the PACS drive and are fed to Apache Spark for the preprocessing and compression, and then the compression file is pushed to Cassandra through Apache Spark. We have divided the whole execution into two different flows based on the migration performed, i.e., text migration and files migration. Figure 2 depicts the text migration or the text data migration.

The text migration starts by reading the data of the patients and pushing it to SQS queues through the AWS Lambda. Then, we the Apache Spark running in the Docker that continuously pool the SQS queues. When the Apache Spark gets the data from the SQS, then it processes the data and is then pushed to Cassandra. This is how the textual data is migrated from the PostgreSQL to Cassandra. The next flow in the proposed architecture deals with migrating the files from PACS drive to Cassandra. Figure 3 shows the flow for the same.

In this flow, the DICOM images are read from the PACS drive and are pushed to SQS using the Lambda for the preprocessing and compression through Apache Spark. The DICOM images read, their features are extracted, Huffman encoding is applied on the extracted Pixel Array to compress the pixel array, this compressed pixel array is written back to the original DICOM file, and file is pushed to Cassandra. The flowchart for the compression process is as shown in Fig. 4.

The compression is part of the processing of the data by Apache Spark, which starts by reading the files from the queue and extracting the feature and compression using Huffman encoder and writing the files to Cassandra. This process will loop until the queue is empty.

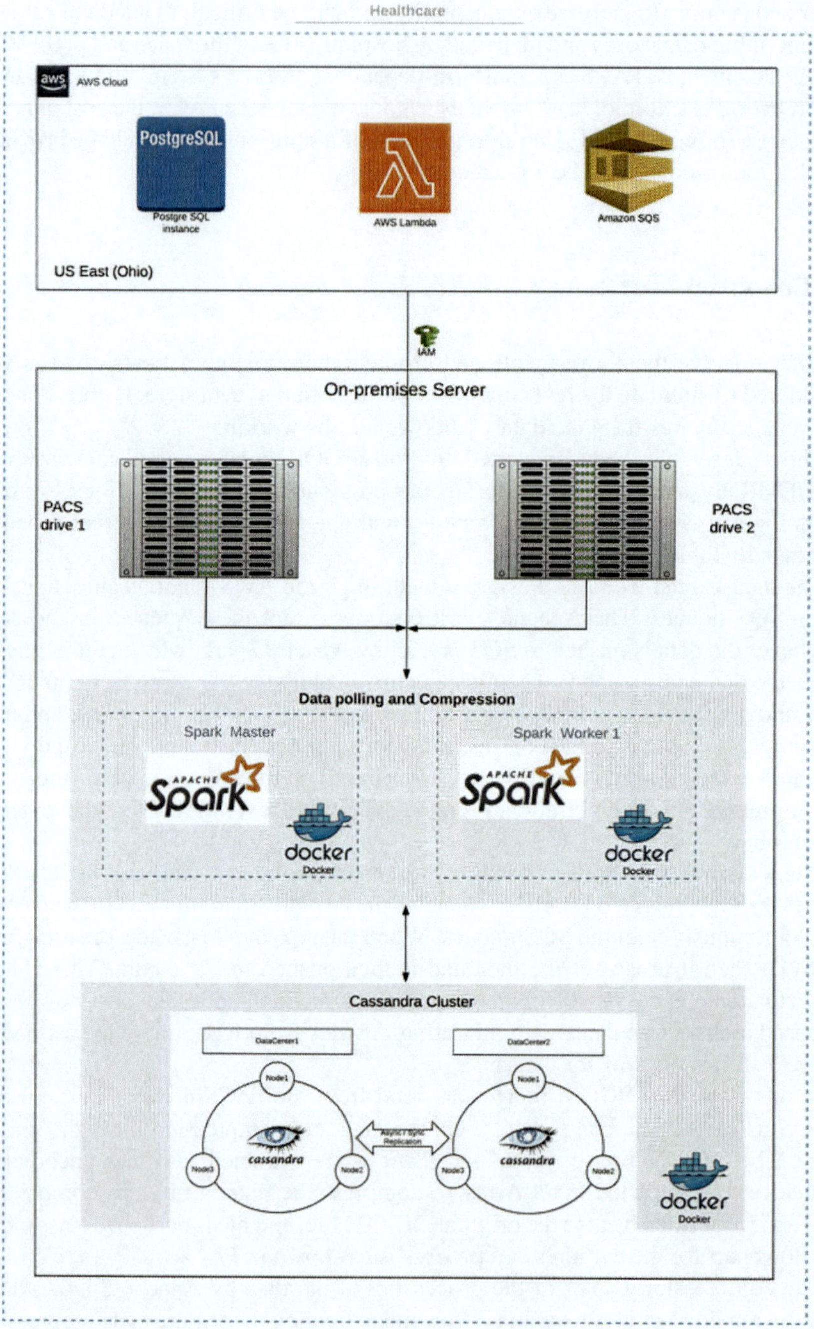

Fig. 1 Migration architecture

Table 3 Dataset [11]

	Table name	Total count
1	Observations	79,674
2	Encounters	20,524
3	Claims	20,523
4	Immunizations	13,189
5	Careplans	12,125
6	Procedures	10,184
7	Conditions	7040
8	Medications	6048
9	Patients	1462
10	Allergies	572

The cost reduction is achieved with the help of the containers, the fault tolerant by the gossip protocol in Cassandra, where the data is replicated and so even if any node goes down, there is no loss of the data and by using the SQS, there is no loss of the data even while migrating it.

5 Conclusion

We saw the advancements in the big data field. Different tools and technologies were introduced with the time to reduce the problem of storage. We had done survey and research on the current system hospitals are using for storing the patients' data that include textual as well as medical imaging and the problem they are facing with the current technique. We saw that this data consists of 20% of textual data and 80% of medical images. Hospitals are struggling to store the historical data of patient, using their current system they can only store 2–3 years of data and at the same time maintaining the privacy of the data is additional and important issue to be taken care [12]. Currently, they are using two different systems: one for textual data and one for storing medical images. Here, big data comes into picture, and we can solve this problem with the help of big data. We are working on replacing their current storage system with Cassandra; In Cassandra, we can store both textual data and medical images so that we can remove overhead to maintaining two different systems.

With our current implementation, we can solve one of the major problems of storage faced by the hospitals, and this architecture can also help to reduce the maintenance cost of the system. It targets to solve the problems of data security such as, what if the on-premises server fails, without raid, we will lose the data. As in the case of the big data, we have data replication, so even if any node goes down, we do not lose data. By doing so, we achieve a centralized data store, so even if a new facility of any hospital gets opened we can easily accommodate it's data though overall data remains synchronized. On the other hand with the traditional system,

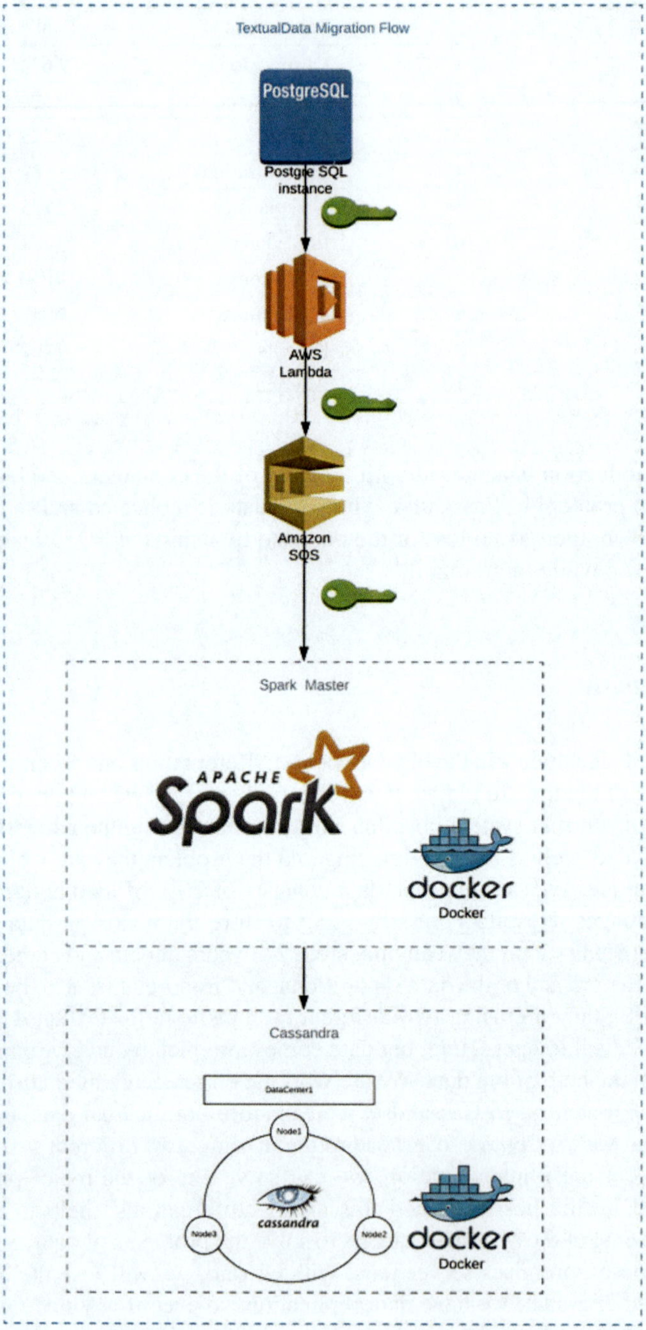

Fig. 2 Text data migration

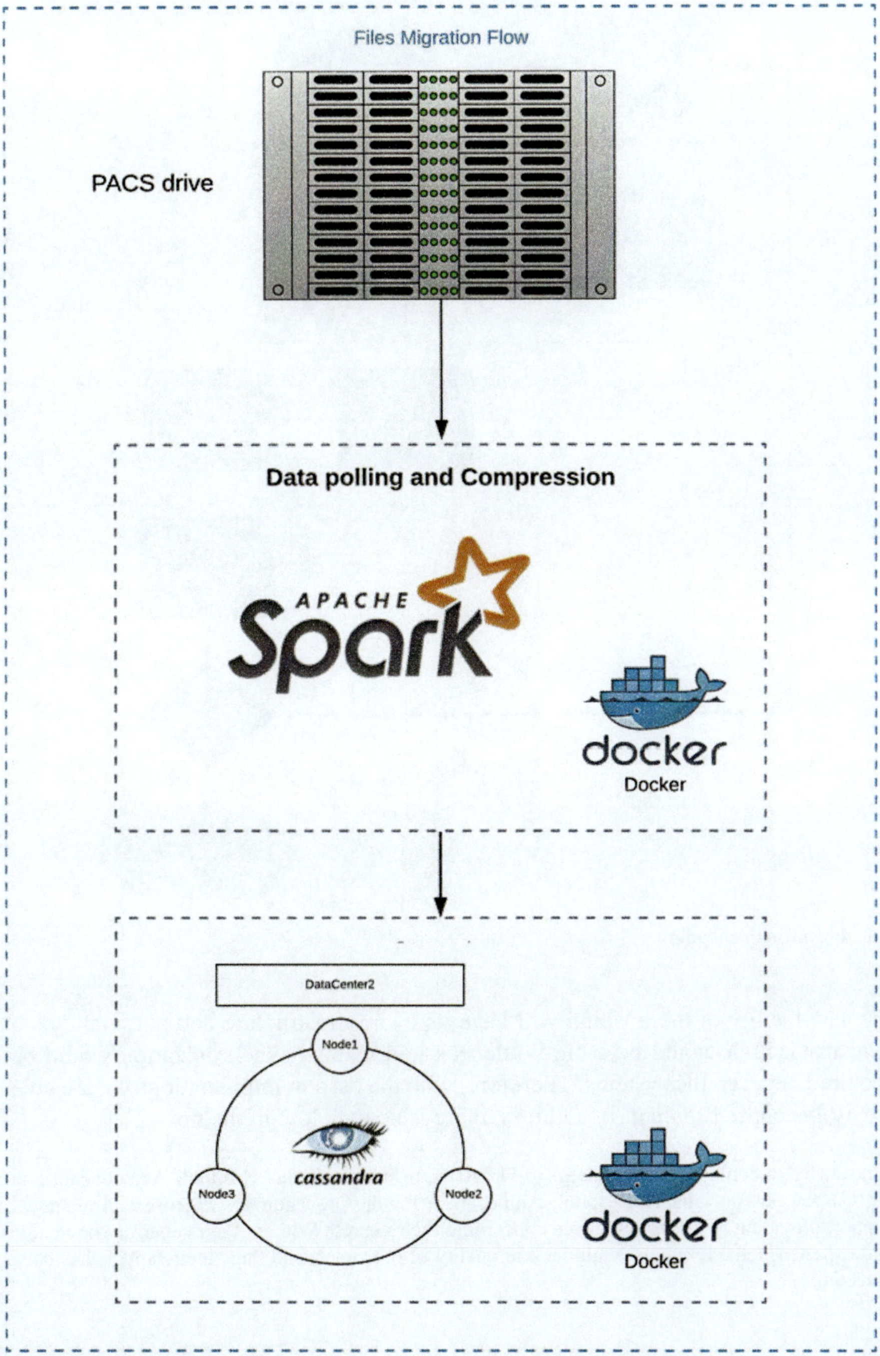

Fig. 3 Files migration

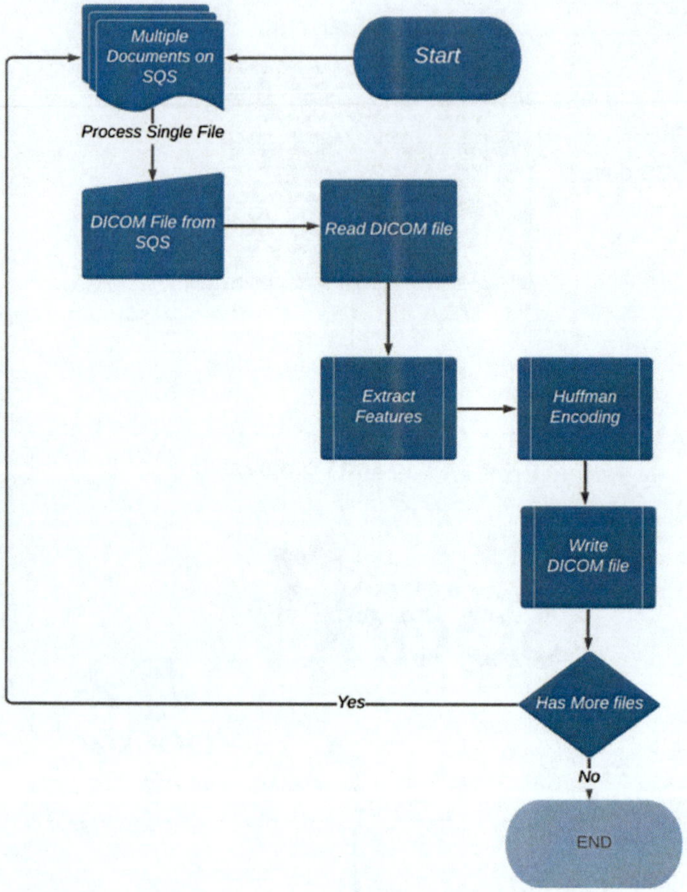

Fig. 4 Huffman encoding

the new facility of the hospital will have its own infrastructure and automatic data synchronization would be a big bottleneck and manual synchronization would be required between the systems. Therefore, with the current implantations we are able to resolve entire problem. In addition, there is no data loss in the flow.

Acknowledgements We are grateful to Dr. Arvind Sharma, Zydus Hospital, Ahmedabad, and Dr. Gargey Sutaria, Shably Hospital, Ahmedabad, for the provision of expertise and technical support providing necessary guidance concerning the research. Without their superior knowledge and experience, the research would lack in quality of outcomes, and thus, their support has been essential.

References

1. Slobogean GP (2015) Bigger data, bigger problems. J Orthop Trauma 29:43–46
2. Scruggs SB (2015) Harnessing the heart of big data. Circ Res 116(7):1115–1119
3. PACS Storage Calculator - https://www.dicomlibrary.com/dicom/pacs-storagecalculator/
4. Kumar S, Singh M (2019) Big data analytics for healthcare industry: impact, applications, and tools. Big Data Min Anal 2(1):48–57
5. Dinov ID (2016) Methodological challenges and analytic opportunities for modeling and interpreting big healthcare data. Gigascience 5(1)
6. Wang W, Krishnan E (2014) Big data and clinicians: a review on the state of the science. JMIR Med Inform 2(1)
7. Panda M, Ali SM, Panda SK (2017) Big data in health care: a mobile-based solution. In: International conference on big data analytics and computational intelligence (ICBDAC). IEEE
8. Wilson DL (1995) Image compression requirements and standards in PACS. In: Proceedings of SPIE 2435, medical imaging: PACS design and evaluation: engineering and clinical issues
9. https://www.inderscienceonline.com/doi/pdf/10.1504/IJMEI.2011.039075
10. Rumsfeld JS, Joynt KE, Maddox TM (2016) Big data analytics to improve cardiovascular care: promise and challenges. Nat Rev Cardiol 13(6)
11. Dataset - https://synthea.mitre.org/downloads
12. Al Hamid HA (2017) A security model for preserving the privacy of medical big data in a healthcare cloud using a fog computing facility with pairing-based cryptography. IEEE Access 5:22313–22328; Krishnan SM (2016) Application of analytics to big data in healthcare. In: 2016 32nd Southern biomedical engineering conference (SBEC). IEEE

Data Science Team Roles and Need of Data Science: A Review of Different Cases

Tejashri Patil and Archana K. Bhavsar

Abstract The paper first looks at the benefits of well-known roles and then discusses the relative lack of structured roles within the data science community, possibly because of the field's novelty. The paper reports extensively on five case studies which discuss five separate attempts to establish a standard set of roles. The paper then leverages the findings of these case studies to discuss the use in online job posts for data science positions. While some positions often appeared, such as data scientist and software engineer, no role in all five case studies was regularly used. The paper concludes, however, by acknowledging the need to build a structure for data science workforce that students, employers, and academic institutions can use. This framework would allow organizations to more accurately employ their data science teams with the desired skills.

Keywords Big data · Data science · Project management · Data science roles

1 Introduction

Big data represents a significant shift in the information-intensive computational methods and technologies used. The switch to parallelism has added complexity to big data approaches and created the need for a range of specialized skills. In conjunction with this, the word "information science" has become omnipresent and used to describe any event that affects data. This is compounded by the fact that the profession progresses from an individual's job conducting data science to a data science team [1]. We lack a vocabulary in this new context to define the responsibilities and skills needed for an active data science/big data team. This lack of vocabulary creates many problems (e.g., identifying the appropriate person to be hired in a data science

T. Patil (✉) · A. K. Bhavsar (✉)
SSBT's College of Engineering and Technology, Jalgaon, Maharashtra, India
e-mail: Ptejashri2293@gmail.com

A. K. Bhavsar
e-mail: asgujar@rediffmail.com

K. Kotecha et al. (eds.), *Data Science and Intelligent Applications*,
Lecture Notes on Data Engineering and Communications Technologies 52,
https://doi.org/10.1007/978-981-15-4474-3_2

team for a specific role). To address this challenge, this paper provides examples and definitions of the data science workforce.

Section 2 provides some background and motivation to develop standard categories and definitions of skills. Section 3 provides a number of case studies in the use of various job titles. Section 4 offers a contrast between the case studies as well as our study of online job postings using these positions. Section 5 describes future trends that will affect the data science workforce. Finally, Sect. 6 presents our conclusions and next steps.

2 Background

With the advent of new techniques for large data file storage (Hadoop Distributed File System [HDFS]), physically distributed logical datasets (Hadoop), parallel processing of distributed data (MapReduce), and the paradigm shift known as big data occurred in the mid-2000s.

The Hadoop ecosystem relies on horizontal scaling to distribute data across independent nodes, with scalability arising from the addition of nodes. The result is the use of parallelism for scalable, data-intensive application development. While it has been given a number of differing conceptual definitions [2], big data is not "bigger data" than common techniques that can handle but rather data that requires parallel processing to meet the time constraints for end-to-end analysis performance at an affordable cost.

2.1 Workforce Descriptions

The main motivation for job definitions is the need to identify, recruit, train, develop, and maintain an adequately skilled workforce by having a common vocabulary to categorize and clarify the type of data science work that needs to be done. Although the functions and terminology are unique to data science, the need for explanations of the workforce is not limited to data science.

In other words, data science is not the only discipline which needs to explain responsibilities and skills. Cyber-security, for example, is another domain where there has been this need. For data security, the US National Institute for Science and Technology (NIST) developed the National Initiative for Cyber-security Education (NICE) Cyber-security Workforce Framework [4] that clarifies the categories, specialty areas, and work roles for cyber-security practitioners. In addition, they provided lists of tasks, knowledge, skills, and ability descriptions mapping them to work roles.

Another effort to provide workforce definitions was the US Department of Defense Cyber Workforce Framework [5]. This work is ongoing, including revisions to a companion document a role-based model for federal information technology/cybersecurity training [6]. The benefits listed in the NICE report apply equally well to the domain of data science and include the following:

- Employers—track staff skills, training, and qualifications; improve position descriptions; develop career paths; and analyze proficiency
- Educators—develop curriculum and conduct training for programs, courses, and seminars for specific roles
- Technology Providers—identify work roles, tasks, and knowledge, skills, and abilities associated with their products.

This work would also be of interest to learners (in learning how to map their training to different possible tasks) and workers (in recognizing positions where their talents can be most efficiently leveraged). Therefore, including job titles and job descriptions that describe roles, expertise, abilities, and skills more clearly would help the data science community and eliminate the data scientist word overload.

2.2 *Skills and Roles*

Although there are some common skills in the data science across different types of roles, some skills may be unique to a specific role. Just as the NICE workforce framework has knowledge, skills, and abilities that can be applied to multiple work roles, it will be important to ensure that each function of data science work is described in a similar way. More generalist practitioners can fit into a number of roles but it is important not to overlap role descriptions to ensure consistency. Depending on the type of data science project, however, skills can vary, such as projects involving more or less experimentation in the analysis [7].

2.3 *Challenge Due to Lack of Process Model*

The development of a framework for data science workers poses a number of challenges, but perhaps the most critical is that there is no agreed data science process template [1, 8]: the model Cross-Industry Standard Process for Data Mining Framework (CRISP-DM) in the late 1990s [9]. This model remains that the method practiced by the largest number of practitioners [10], but it has predated Internet, big data, machine learning, agile, "Internet of things," and so on and has not tackled system development or management processes [8].

2.4 *In Relation to Software Development Lifecycles*

Most analytical system development results in situational awareness through reporting or business intelligence. Software development lifecycles (SDLCs) are geared to this type of development for requirement-driven analytics systems. Nonetheless, advanced analytics systems are results-driven and involve creativity in selecting data and software features, building models, and testing and optimizing models. Many skillets in data science would overlap those used in SDLCs, but tasks would need to be explicitly defined to be complete. Care should be taken, however, to align the data science and SDLC models, in particular lining them up with agile and develop standard methodologies.

3 Methodology

Therefore, providing job titles and job descriptions that more clearly identify tasks, knowledge, skills, and abilities would benefit the data science community and remove the overloading of the term data scientist. A cross section of organizations has been explored to help ensure that a representative view of current thinking and usage across the field of data science has been captured. Explicitly, our case studies were selected from two standard bodies, two industry organizations, and one consultant/advisory firm.

The analysis of the defined roles was based on written documentation from each organization for each case study, where documentation was not as robust, and discussions were also held with individuals from the identified organizations.

4 Case Studies

In the section Case Studies, different cases in data science technology are to be discussed.

4.1 *NIST*

The NIST Big Data Public Working Group (NBD-PWG) aims to build consensus on relevant, fundamental concepts related to big data in order to promote progress in big data. The findings were recorded in the volume series [11] of the NIST Big Data Interoperability System. One of the main activities of the NBD-PWG is the creation of big data reference architecture (RA) which categorizes big data system components. It also describes the roles of those whose tasks are included in that

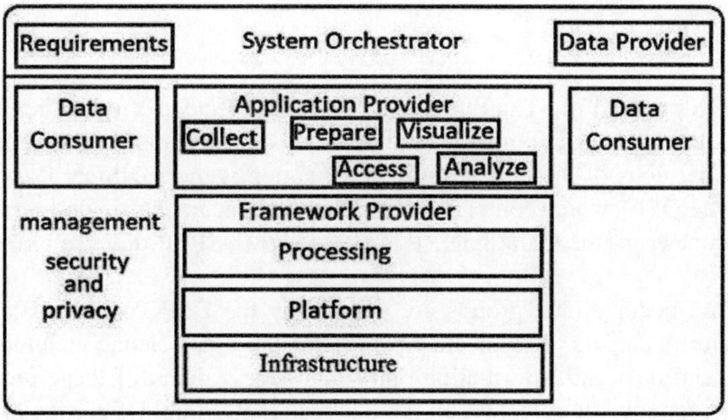

Fig. 1 NIST big data reference architecture

component at a high level. The RA consists of five components and identifies their respective roles, as shown in Fig. 1 and as follows:

- System Orchestrator—defines and integrates the required data application activities into an operational vertical system.
- Data Provider—introduces new data or information feeds into the big data system.
- Big Data Application Provider—executes a life cycle to meet security and privacy requirements as well as system orchestrator-defined requirements.
- Big Data Framework Provider—establishes a computing framework in which to execute certain transformation applications while protecting the privacy and integrity of data.
- Data Consumer—includes end-users or other systems who use the results of the big data application provider.
- Security and Privacy—interacts with the system orchestrator for policy, requirements, and auditing and with both the big data application provider and the big data framework provider for development, deployment, and operation.
- Management—management of big data systems should handle both system- and data-related aspects of the big data environment, that is system management and big data lifecycle management. System management includes activities such as provisioning, configuration, package management, software management, backup management, capability management, resources management, and performance management. Big data lifecycle management involves activities surrounding the data lifecycle of collection, preparation, analytics, visualization, and access.

4.2 EDISON

The EDISON project [12] is an European Union (EU)-funded effort to "speed up the increase in the number of competent and qualified data scientists across Europe and beyond." The focus of the collection of information is the EDISON Data Science Framework (EDSF), which comprises several related documents, including the Competence Framework, the Data Science Professional (DSP) Profiles, and the Model Curriculum.

Four main occupational groups are defined by the DSP Profiles: data science infrastructure managers, data science professionals, data science technology professionals, and data and information entry and access. Each of these profiles has descriptions for specific roles within the occupational group. Of particular focus is the data science professional, which has the following roles:

- Data Scientist—data scientists find and interpret rich data sources, manage large amounts of data, merge data sources, ensure consistency of datasets, and create visualizations to aid in understanding data. Build mathematical models, present, and communicate data insights and findings to specialists and scientists and recommend ways to apply the data.
- Data Science Researcher—It applies scientific discovery research/process, including hypothesis and hypothesis testing, to obtain actionable knowledge related to a scientific problem, business process, or reveal hidden relations between multiple processes.
- Data Science Architect—It designs and maintains the architecture of data science applications and facilities. It creates relevant data models and processes workflows.
- Data Science Programmer—It designs, develops, and codes large data (science) analytics applications to support scientific or enterprise/business processes.
- Data/Business Analyst—It analyzes large variety of data to extract information about system, service, or organization performance and present them in usable/actionable form.

4.3 Springboard

Springboard, an online data science education startup, defines three roles: data engineer, data scientist, and data analyst. As one can see from Fig. 2, all of these roles require software engineering, math/stats, and data communication skills. Below we describe some possible roles, based on springboard's definitions:

Data Engineer—It relies mostly on his or her software engineering experience to handle large amounts of data at scale: typically focuses on coding, cleaning up datasets, and implementing requests that come from data scientists, and typically known a broad variety of programming languages, from Python to Java. When

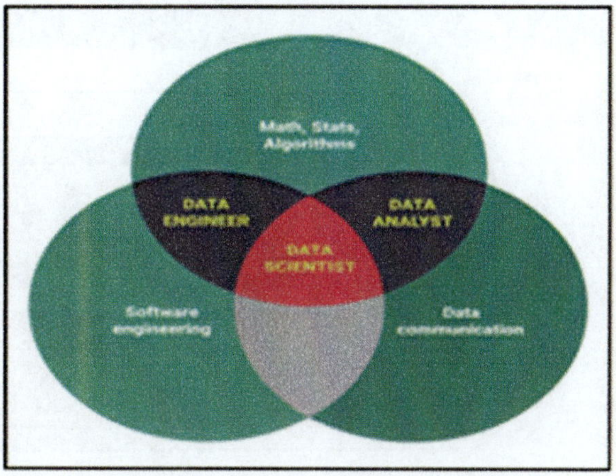

Fig. 2 Springboard overview in software engineering

somebody takes the predictive model from the data scientist and implements it in code, they are typically playing the role of a data engineer.

Data Scientist—It bridges the gap between the programming and implementation of data science, the theory of data science, and the business implications of data. It can take a business problem and translate it to a data question, create predictive models to answer the question, and tell a story about the findings.

Data Analyst—It looks through the data and provides reports and visualizations to explain what insights the data are hiding. When somebody helps people from across the company understand specific queries with charts, they are filling the data analyst role. One role not shown in the diagram, but mentioned by springboard is a data architect, who focuses on structuring the technology that manages data models.

4.4 SAIC

SAIC is a systems integrator that works primarily for the federal government, including civilian, defense, and intelligence customers. To increase efficiency in developing and deploying BDA systems, SAIC developed an internal process model known as Data Science Edge TM [8], as shown in Fig. 3; the model extends the earlier limited data mining process of CRISP-DM to add in big data, systems development, and data-driven decision-making considerations, including the alignment with agile process models.

This overarching process model aligns with the general data science roles SAIC uses. In addition to the traditional roles for software and systems development, SAIC has specific roles for data science, big data platforms, and data management:

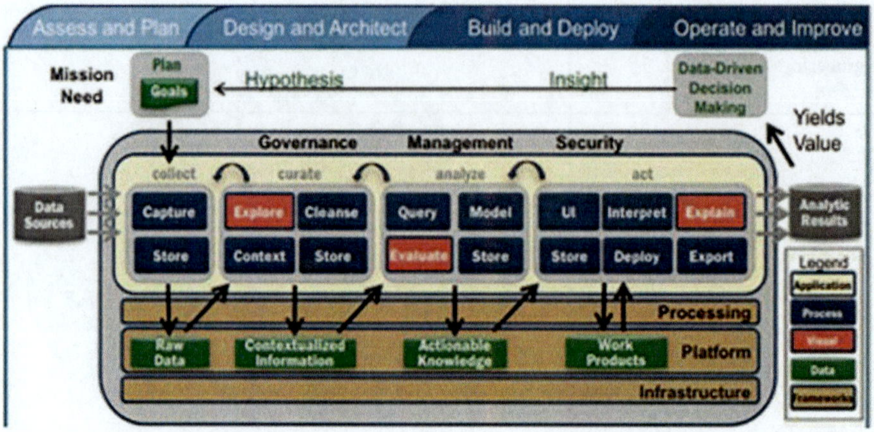

Fig. 3 SAIC's data science edge

Information Architect: It designs shared information environments involving models or concepts of information. It develops data models for optimal performance in databases. It designs data structures for data interchange and develops data standards and converts data to controlled vocabularies.

Data Scientist: It works in cross-functional teams with data at all stages of the analysis lifecycle to derive actionable insight and follows a scientific approach to generate value from data, verifying results at each step.

Metrics and Data: It develops, inspects, mines, transforms, and models data to raise productivity, improve decision-making, and gain competitive advantage. It conducts statistical analysis of data and information to ensure correct predictive forecasting or classification and manages all aspects of end-to-end data processing.

Knowledge and Collaboration Engineer: It designs and implements tools and technologies to promote knowledge management and collaboration within the enterprise.

Big Data Engineer: Using big data technology, it builds parallel Information-intensive systems and works with Hadoop's full open-source stack from cluster management to data warehouses and to scheduler analytics technology.

In addition, SAIC has management roles for each of these positions. Note that while DSE calls out visualization explicitly in three distinct types, there are no specific roles for visualization or business intelligence specialists, which are usually found among software engineers or data scientists SAIC recognizes the importance of understanding the provenance of data in a given domain. BDA systems are developed by teams that collectively contain all the skills needed for agile BDA system development [13].

Table 1 Comparative analysis of all the cases mentioned in the paper

	Data science researcher	Data scientist	Data architect	Data analyst	Data science programmer	Data engineer
NIST	No	No	No	No	No	No
EDISON	Yes	Yes	Yes	Yes	Yes	No
SAIC	No	Yes	No	No	No	Yes
SPRINGBOARD	No	Yes	Yes	Yes	No	Yes

5 Discussion

Table 1 indicates the positions that more than one of our case studies have used. As you can see, the positions used are not used regularly. Certain positions listed ranged from commonly used roles like information architects and software engineers to rare roles like data providers and origin system experts.

6 Conclusion

Providing consistency and greater precision for the word data scientist will help with the expertise needed to train and hire specialists. One way to achieve a comprehensive description of the data science workforce would be through a consortium of stakeholders from government, industry, and academia. The first task of the consortium would be to develop a detailed data science process model that reflects all participants' consensus on what defines a data science activity before beginning to identify the skills and job titles required by the data scientists. This work then succeeded in following the NICE model to provide categories, specializations, work roles, and tasks to clarify the differences in roles.

One possible difficulty will be that conventional analytics systems that require simple summary statistics, analysis, or business intelligence are built almost exclusively by computer and process engineers along with traditional roles of data modelers, database analysts, and database administrators. Given the ongoing evolution of big data and data science, we note that current use may not show how the industry is evolving, so another, complementary next step may be to resume a future (perhaps every six months) analysis of role usage in the industry in order to identify trends over time.

References

1. Saltz JS, Shamshurin I (2016) Big data team process methodologies: a literature review and the identification of key factors for a project's success. In: 2016 IEEE international conference on Big Data, pp 2872–2879
2. Chang WL, Grady N (2015) NIST big data interoperability framework: volume 1, big data definitions. No. special publication (NIST SP)-1500-1
3. Newhouse WD (2017) Nice cybersecurity workforce framework: national initiative for cybersecurity education. No. special publication (NIST SP)-800-181
4. Newhouse W, Keith S, Scribner B, Witte G (2017) National initiative for cybersecurity education (NICE) cybersecurity workforce framework. NIST special publication 800:181
5. Toth P, Klein P (2013) A role-based model for federal information technology/cyber security training. NIST special publication 800-16:1–152
6. Saltz J, Shamshurin I, Connors C (2017) Predicting data science sociotechnical execution challenges by categorizing data science projects. J Assoc Inf Sci Technol 68(12):2720–2728
7. Shearer C (2000) The CRISP-DM: the new blueprint for data mining. J Data Wareh 5(4)
8. Saltz JS (2015) The need for new processes, methodologies and tools to support big data teams and improve big data project effectiveness. In: IEEE international conference on big data, pp 2066–2071
9. Framework (2015) DRAFT NIST big data interoperability framework, volume 7, standards roadmap. NIST special publication 1500-7
10. Saltz JS, Grady NW (2017) The ambiguity of data science team roles and the need for a data science workforce framework. In: IEEE international conference on big data, pp 2355–2361
11. Saltz J, Crowston K (2017) Comparing data science project management methodologies via a controlled experiment. In: Proceedings of the 50th Hawaii international conference on system sciences

Performance Analysis of Indian Stock Market via Sentiment Analysis and Historical Data

Amit Bardhan and Dinesh Vaghela

Abstract Stock market has always generated interest among the people since the time of its initiation. It is a very complex and challenging system, where people invest to gain money in order to attain higher gains. But the scenario may be negative if the investment is made in the stocks without any proper analysis. The performance of any stock depends on many parameters and factors like historical prices, social media data, news, country economics, production of the company, etc. In our research, we consider two major factors like historical prices and social media data and will let the investors give an idea about the stock performance in the nearby future. Therefore, we combine the sentiments of the different stakeholders across the Internet with historic prices of the stock to predict the stock recital. For combining the above approaches, we are using the decision tree approach of machine learning for classification and prediction for more accurate prophecy. The proposed algorithm gives above 70% accuracy for the given data.

Keywords Stock market analysis · Sentiment analysis · Machine learning · Stock prediction

1 Introduction

Stock market has always remained the topmost inclination from the date of its initiation when it comes to investment. Here, the people invest because they generally get more profit with respect to the other sectors of investment. Indian stock market is considered to be very volatile in nature, so before investors place their hard-earned

A. Bardhan
Som-Lalit Institute of Computer Applications, Navrangpura, Ahmedabad, Gujarat, India
e-mail: amitsbardhan@gmail.com

D. Vaghela (✉)
Shantilal Shah Engineering College, Sidsar Campus, Vartej, Bhavnagar, Gujarat, India
e-mail: dineshvaghela28@gmail.com

K. Kotecha et al. (eds.), *Data Science and Intelligent Applications*, Lecture Notes on Data Engineering and Communications Technologies 52,
https://doi.org/10.1007/978-981-15-4474-3_3

money into the investment vehicles like stock markets and mutual funds they should always analyze and then invest as the risk component is considered to be very high.

On the other hand, companies also need investors for their firm expansions and other activities. So companies always try to create its good image in the market by spreading the news related to company profit, growth, projects it has achieved over news websites or even on social media like twitter [1]. Because nowadays social media and news websites have become a major platform to reach maximum people in minimum time, so many organizations are using these tools to keep the investors informed with the recent updates of any company [2].

Out of the above-mentioned factors the investors generally use three factors extensively to invest in the stock market, i.e., historic prices, social media data and data from the news websites [3]. In our research work, we are going to use these parameters and its components of it, in order to help investors to analyze and predict the Indian stock market [4].

In BSE, which is the oldest stock exchange in the country, nearly 5000 companies are listed out of which approx 2000 are actively traded. Indian stock market BSE/NSE is very much sentiment/opinion driven which comes from different social networking sites and news websites, so there should be some automation while analyzing people's sentiments correctly, because 70% of the data from these online engines remains text, i.e., unstructured data. So sentiment analysis (SA) or opinion mining (OM) will help us to classify the opinions related to the company and will directly help the investors to predict the trend of the company stock [5]. But only sentiment analysis (SA) won't be that much helpful until and unless we compare it with the analysis of the historic prices. Therefore in our research work we integrate both sentiment analysis (SA) and historic stock prices using natural language processing (NLP) and machine learning (ML) algorithms on a single platform so as to analyze and predict the stock prices before the investment has been made by the investors [1].

In our proposed work, we are using decision tree classifier (DTC) for combining the output of trend and sentiment score in order to analyze the given stock.

2 Background Theory

2.1 Natural Language Processing

The natural language we mean the words that human being uses for the day-to-day communication like English, Hindi, Portuguese, etc. These words are generally in the form of unstructured text in distinction to the synthetic programming languages and numerical notations [6]. NLP can be used in data mining, modeling and information detection to analyze the natural language. NLP can also be used with linguistic algorithms and facts structures in vigorous speech processing information systems. Challenges in the NLP include linking words and machine insight.

2.2 Machine Learning

Machine learning (ML) teaches computers to do what comes logically to any individual [7]. Machine learning algorithms employ computational methods to learn information unswerving from the premeditated data without relying on the predefined paradigm. ML has basically two types of approaches: supervised learning and unsupervised learning. In supervised learning, a model is on accepted I/O data so that it can forecast the potential output. It uses classification or regression techniques for the formation of predictive models. In classification, the given I/P data into categories and is used in medicinal imaging, language identification, etc. Regression is used to calculate constant responses for, e.g., changes in the stock prices, changes in the temperature or fluctuations in power demands, etc.

In unsupervised learning, the model searches secreted prototype in input data and is generally used in object detection, series examination, etc. ML can be used in variety of purposes like computational finance and biology, image processing, power creation for electric energy demand predicting, NLP, etc.

2.3 Sentiment Analysis

Sentiment analysis/opinion mining/emotion mining uses text mining, NLP and machine learning algorithms in order to identify the skewed content from the given text. Here, the term opinion means a person's outlook about an object or issue which can be positive, negative or neutral [5] (Fig. 1).

Sentiment analysis tries to conclude whether a specified wording is subjective or objective [8]. Analyzing sentiment is a very challenging job because a specific word used in a statement can be positive or negative depending upon the emotions attached to it [9]. Sentiment classification methods can be chiefly alienated into the machine learning approach and Lexicon-based procedure [10].

3 Related Work

In the latest study by Nayak et al. [11] on prediction models for the Indian stock market reveals 70% accuracy using supervised machine learning approaches for everyday forecasting. While studying the monthly trend, it is observed that the trend of one month is having a close correlation with the trend of its previous month. In a different study by Bhardwaj et al. [5] used three models like Naïve Bayes, support vector machine (SVM), random forest and the results prove that accuracy of SVM is high. Also, observations reveal that for a specific time interval retrieved values of Sensex and Nifty relics unvarying.

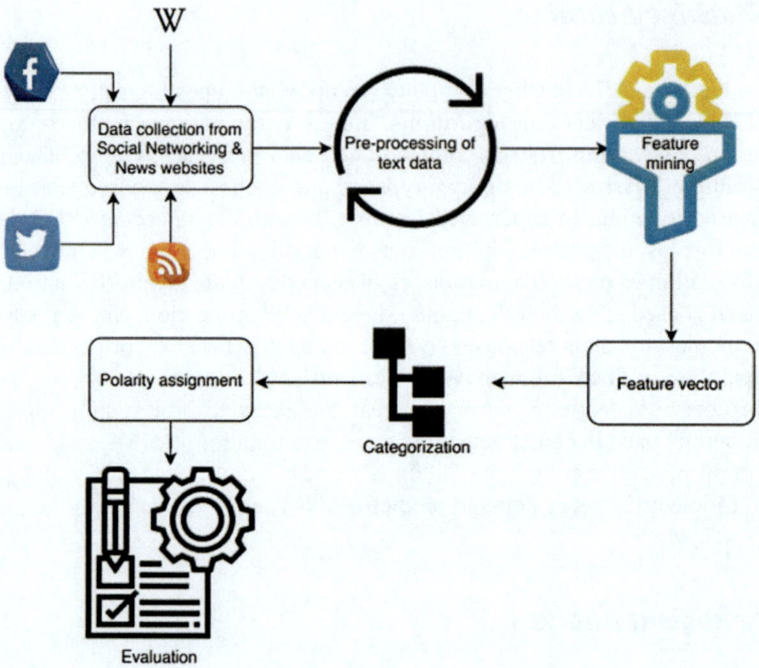

Fig. 1 Process of sentiment analysis

Bing et al. [1] have collected 15 million tweets and processed through NLP and used Naïve Bayes classifier and SVM and the results show SVM had a better average predictive accuracy. Stock price of some companies is predicted with an average accuracy of 76.12%. Ahuja et al. [12] presented a study in which the data set used is from BSE (June 2013–Dec 2013) from Yahoo Finance and Publically available Twitter data. The Polarity distribution is prepared by means of OF (Opinion Finder) and GPOMS (Google profile of moods states) and the classification methods used are SVM, linear regression, logistic regression, SOFNN (Self-organizing Fuzzy Neural Networks) and the results of SOFNN are better. The results give 75.56% accuracy when Fuzzy Neural Networks were used.

Zhang et al. [13] have collected the historical data from Dow Jones, S&P 500 and NASDAQ (March 30, 2009, to September 7, 2009) and twitter data from 8100 to 43,040 per day (2.5 million tweets). Polarity of a text was classified in terms of (Hope, Happy, Fear, Worry, Nervous, Anxious and Upset). Here, the authors are focusing on the number of retweets per day and the time taken to react on the buzz.

4 Proposed Method

The proposed approach for predicting the performance of a given stock is shown in Fig. 2.

Historic records of stock process are fetched from Yahoo Finance available in CSV, Excel or in R-format. Twitter API is used to fetch tweets for the given time from the handles of moneycontrol, NDTV Profit, Zee Business and ET Markets and also different data sets are available on data.world website. Next is pre-processing of the historic data and extracting the adjusted price of a particular stock for the quarter specified, its net NPA, net profit/loss and setting the investment price.

Next in pre-processing is creating bag of words which is used to pre-define some words like bullish, bearish, buy, sell, etc. so that sentiment polarity can be assigned to the given utterance accordingly. It is created from the data set after removing the noisy data. After creating bag of words, we assign and calculate the total sentiment score according to lexicons in the bag of words and calculate the total sentiment. Once the total sentiment score is calculated for the particular stock it is added in the data set. Lastly, we assign class labels like accumulate; book profit, stop loss and hold and the rules are stated in Table 1.

Once the training data set is ready, following attributes are considered like dates of the quarter, adjusted price, sentiment and analysis (class Label). According to the training data set given to various ML classification algorithms like KNN, SVM, Naïve Bayes and decision tree, we calculate the percentage accuracy of correctly classified instances of the given the set. The classifier which gives the highest accuracy is selected and the model is trained accordingly. After the prediction model gets ready, ML prediction algorithms implemented to test data set (without class label) in the model that is already built. Finally, prediction is carried out and result's accuracy was checked.

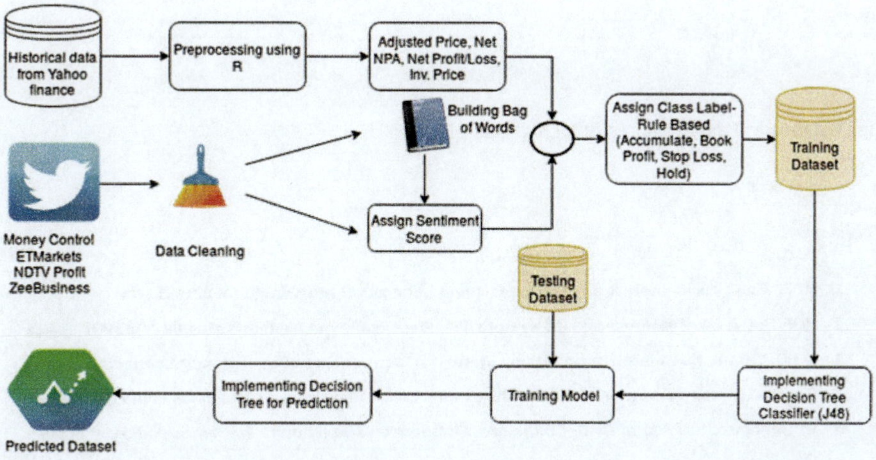

Fig. 2 Proposed approach

5 Implementation

For the implementation, we are using RStudio and Weka 3. The data set sample used in this experiment is obtained using RStudio. Here, we first obtain the stock data for HDFCBANK from Yahoo Finance for a period for all the four quarters of the year 2017. The data set that is fetched contains the attributes as date, open, high, low, close, volume traded and adjusted amount of the range specified. The proposed model analyzes the data of the first three quarters data and tries to predict the values of the fourth quarter using different machine learning algorithms. But before that, it fetches the sentiments of the people across the web and classifies the sentiments in terms of positive, negative and neutral. Figure 3 shows the historical values of the HDFC bank for the first quarter, and similarly, for the next quarter, it will be fetched.

Similarly, tweets for the given stock script for the particular day from various twitter handles are like moneycontrol, ET Markets, NDTV Profit and Zee Business and merge the entire corpus into a single object in R. The corpus obtained is unstructured and requires converting it into separate frames staining all its attributes separately in order to classify whether the text was tweet or retweet, its screen name, location, language, source, id, etc. Figure 4 shows the outcome of the corpus after converting them into data frames.

The next step is to prepare a bag of positive and negative lexicons in order to assign sentiment polarity to a given text and also to perform fetch the particular attributes from the data frames and perform the cleansing operations of the given corpus. This operation includes cleansing of punctuation, control characters, URL, repeated words which are not necessary, the script names, the handle name of tweets, etc. and removing the stop words (English) from the given corpus. Now, we split the

Filter

Date	HDFCBANK.NS.Open	HDFCBANK.NS.High	HDFCBANK.NS.Low	HDFCBANK.NS.Close	HDFCBANK.NS.Volume	HDFCBANK.NS.Adjusted
1/2/2017	1209.45	1210.10	1192.05	1197.10	927164	1189.19
1/3/2017	1200.80	1201.95	1185.75	1189.90	961875	1182.04
1/4/2017	1189.85	1192.00	1183.80	1185.50	1741780	1177.67
1/5/2017	1189.00	1194.50	1183.10	1186.20	1694096	1178.37
1/6/2017	1190.10	1198.40	1188.20	1193.65	1381315	1185.77

Fig. 3 Historical data acquired of HDFC Bank

Filter

	text
1	HDFC Bank hikes interest on FDs; loans could get costlier https://t.co/mCcZmVLoUz
2	HDFC bank hikes interest on FDs by upto 1%, loans could get costlier- https://t.co/a3vHMupAez
3	▶ HDFC Bank hikes interest on FDs by up to 1%; loans could get ... - https://t.co/itlacUZbDk:... ...
4	RT @moneycontrolcom: #HDFCBank hikes interest on FDs by up to 1%; loans could get costlie...
5	RT @moneycontrolcom: #HDFCBank hikes interest on FDs by up to 1%; loans could get costlie...

Fig. 4 Corpus converted into data frame

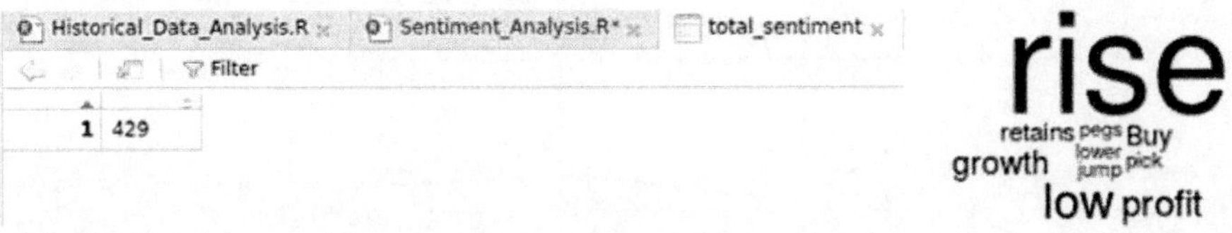

Fig. 5 Total sentiment score and word cloud

tweets into words and match each word with lexicon generated before, which will give the sentiment of each word in terms of positive and negative and calculate the score for every positive and negative word like assigning score of 1 to every positive word and −1 to every negative word. Depending upon the score, we generate a word cloud that will give us a glimpse about the sentiment of the particular stock in the current scenario. In Fig. 5, the size of the word "rise" is largest which gives a sight that the overall sentiment is going to be positive. Also, we calculate the total sentiment score of the given script for that particular day as shown below which is positive.

Lastly, we combine the sentiment score calculated for daily with the adjusted price and also net NAV, net profit/loss, investment price and assign the class label according to the rules specified in the algorithm (Fig. 6).

Now we apply the training data to the WEKA software for pre-processing and the attributes used are date, adjusted price, sentiment score and analysis. Once training data set is loaded, next we have used different classifiers like Naïve Bayes, SVM, KNN and decision tree in which the accuracy of the decision tree is highest. Here, the decision tree classifier gives the maximum accuracy for classification as shown in the table given below (Fig. 7).

Next, we perform cross-validation of tenfold which gives the following output which contains the class-wise output of the parameters like TP (true positive), FP (false positive), Precision, Recall, F-Measure, MCC, ROC area and PRC area (Fig. 8).

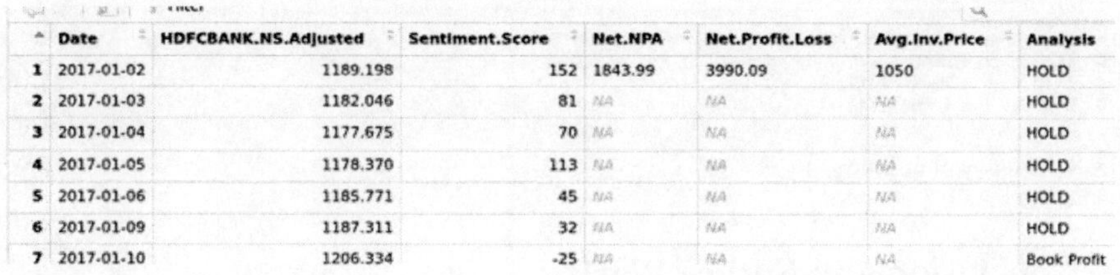

	Date	HDFCBANK.NS.Adjusted	Sentiment.Score	Net.NPA	Net.Profit.Loss	Avg.Inv.Price	Analysis
1	2017-01-02	1189.198	152	1843.99	3990.09	1050	HOLD
2	2017-01-03	1182.046	81	NA	NA	NA	HOLD
3	2017-01-04	1177.675	70	NA	NA	NA	HOLD
4	2017-01-05	1178.370	113	NA	NA	NA	HOLD
5	2017-01-06	1185.771	45	NA	NA	NA	HOLD
6	2017-01-09	1187.311	32	NA	NA	NA	HOLD
7	2017-01-10	1206.334	-25	NA	NA	NA	Book Profit

Fig. 6 Final training data set

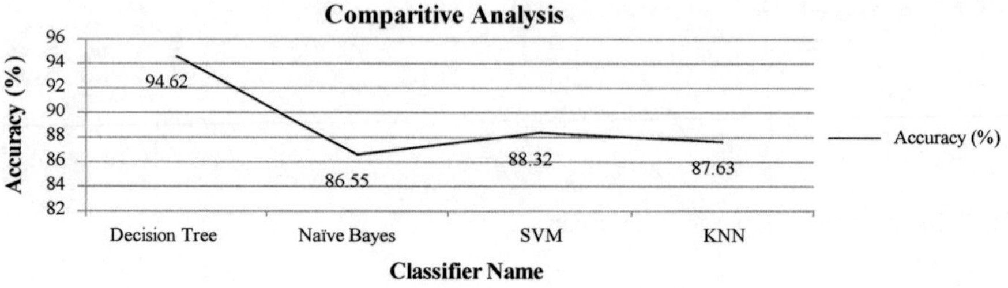

Fig. 7 Comparative analysis of results obtained using different ML classifiers

Table 1 Class label and their rules

Class label	Rule
Accumulate	Net profit > Net NPA Trading price < Investment price, sentiment is +ve
	(Trading price − Investment price) > (% returns last year) Net profit > Net NPA, sentiment is +ve
Book profit	(Trading price − Investment price) > (% returns last year) Sentiment is −ve
Stop loss	(Trading price − Investment price) < (% returns last year) Net profit < Net NPA, sentiment is −ve
Hold	(Trading price − Investment price) < (% returns last year) Net profit > Net NPA, sentiment is +ve

```
=== Detailed Accuracy By Class ===

                TP Rate  FP Rate  Precision  Recall  F-Measure  MCC    ROC Area  PRC Area  Class
                0.879    0.126    0.870      0.879   0.874      0.753  0.882     0.810     HOLD
                0.818    0.007    0.964      0.818   0.885      0.867  0.922     0.835     Book Profit
                0.903    0.081    0.848      0.903   0.875      0.810  0.925     0.846     Accumulate
Weighted Avg.   0.876    0.090    0.879      0.876   0.876      0.792  0.903     0.827

=== Confusion Matrix ===

  a  b  c   <-- classified as
 80  1 10 |  a = HOLD
  6 27  0 |  b = Book Profit
  6  0 56 |  c = Accumulate
```

Fig. 8 Cross-validation result

6 Conclusion and Future Work

Analyzing the stock market is a very typical as there are many factors that are required to be considered. But the study reveals that majority of the market is sentiment-driven and now in this digital age where a person is using digital devices and sharing their sentiments online, those sentiments can be used for analyzing the stock market with its historical data. Our proposed method is considering both historic data like adjusted price, net NPA, net profit/loss and the sentiment for the particular share to forecast the performance of the given stock in nearby future. For this, we have used a supervised machine learning approach in which the classifier is a decision tree, which provides improved analysis rather than only performing the analysis in the historical data. In the future work, we will consider more parameters for deciding the rule-based class label for forecasting the stock market prices. Automatic generation of bag of words for assigning the polarity.

References

1. Bing L, Chan KCC, Ou C (2014) Public sentiment analysis in twitter data for prediction of a company's stock price movements. In: 2014 IEEE 11th IEEE international conference on e-bus engineering, pp 232–239
2. Trends SP, Chowdhury SG, Routh S, Chakrabarti S (2014) News analytics and sentiment analysis to predict. Int J Comput Sci Inf Technol 5(3):3595–3604
3. Li Q, Wang T, Li P, Liu L, Gong Q, Chen Y (2014) The effect of news and public mood on stock movements. Inf Sci (NY) 278:826–840
4. Smailović J, Grčar M, Lavrač N, Žnidaršič M (2014) Stream-based active learning for sentiment analysis in the financial domain. Inf Sci (NY) 285(1):181–203
5. Bhardwaj A, Narayan Y, Vanraj, Pawan, Dutta M (2015) Sentiment analysis for indian stock market prediction using sensex and nifty. Procedia Comput Sci 70:85–91
6. Wu DD, Zheng L, Olson DL (2014) A decision support approach for online stock forum sentiment analysis. IEEE Trans Syst Man Cybern Syst 44(8):1077–1087
7. Nassirtoussi AK, Aghabozorgi S, Ying Wah T, Ngo DCL (2015) Text mining of news-headlines for FOREX market prediction: a Multi-layer Dimension Reduction Algorithm with semantics and sentiment. Expert Syst Appl (1):306–324
8. Haddi E, Liu X, Shi Y (2013) The role of text pre-processing in sentiment analysis. Procedia Comput Sci 17:26–32
9. Li X, Xie H, Chen L, Wang J, Deng X (2014) News impact on stock price return via sentiment analysis. Knowl-Based Syst 69(1):14–23
10. Medhat W, Hassan A, Korashy H (2014) Sentiment analysis algorithms and applications: a survey. Ain Shams Eng J 5(4):1093–1113
11. Nayak A, Pai MMM, Pai RM (2016) Prediction models for indian stock market. Procedia Comput Sci 89:441–449
12. Ahuja R, Rastogi H, Choudhuri A, Garg B (2015) Stock market forecast using sentiment analysis. In: 2nd international conference on computing for sustainable global development, INDIACom, pp 1008–1010
13. Zhang X, Fuehres H, Gloor PA (2011) Predicting stock market indicators through twitter 'I hope it is not as bad as I fear'. Procedia Soc Behav Sci 26(2007):55–62

D-Lotto: The Lottery DApp with Verifiable Randomness

Kunal Sahitya and Bhavesh Borisaniya

Abstract The true ingredient of any lottery system to be successful is randomness of its underlying algorithm. Since the introduction of gambling and lottery industry into cryptocurrency market, trust and security have been less of a concern for organizations, unlike randomness and verifiability. Many of the existing lottery designs use dynamic attributes of either game or blockchain to introduce randomness in its algorithm and only a handful of them are verifiable. In this paper, we introduce a lottery system design having novel random function, which uses combination of game and blockchain state to produce randomness in underlying algorithm and is verifiable. Proposed lottery DApp, built on Ethereum platform, includes smart contracts that help system to achieve properties like decentralization, transparency, and immutability. These properties combined with randomness and verifiability lead to this significant lottery design to be unique of its kind.

Keywords Lottery · Ethereum · DApp · Randomness · Verifiability

1 Introduction

Cryptocurrency and lottery shares similarities at Indian subcontinent of being topics for discussion and partial acceptance rather than having wholehearted technological embrace. While on the other hand, many of the European countries already have blockchain-based fortune lottery companies like FireLotto [1] and Kibo [2] making trillion dollar profit-count annually [3]. Lottery is legalized in less than half states of India, but surprisingly, it estimates to generate per annum revenue of around fifty thousand crores alone for states and companies despite taking hit from tax increment by government [4]. Online lotteries have seen its fair share of failures like

K. Sahitya (✉) · B. Borisaniya
Shantilal Shah Engineering College, Bhavnagar, Gujarat, India
e-mail: kunal2sahitya@gmail.com

B. Borisaniya
e-mail: borisaniyabhavesh@gmail.com

K. Kotecha et al. (eds.), *Data Science and Intelligent Applications*, Lecture Notes on Data Engineering and Communications Technologies 52,
https://doi.org/10.1007/978-981-15-4474-3_4

HotLotto scandal [5], which included theft of 14.3 Million US dollars by deployment of self-destructing malware to affect randomness of underlying algorithm and many more.

A typical lottery system includes three phases: announcement of lottery, purchase of lottery and lottery drawing, and distribution of winnings. A traditional process is physical, tiresome, and error-prone. On the other hand, e-lotteries have improved in terms of speed and security; however, still the system lacks transparency and is vulnerable to single point of failure. Now has come an era of distributed, fully transparent, and immutable lottery systems which work without a slightest intervention of third-party organizers. Formal-mentioned systems are either independent distributed peer-to-peer lotteries or they are built on blockchain platforms like Ethereum, Neo, Cardano, Zilliqa, etc. Highlights of such systems include fully decentralized, distributed, peer-to-peer, transparent, immutable, based on fiat as well as cryptocurrencies, and operable worldwide. To solve the disbelief in participants toward any electronic lottery system, it must ensure qualities like, unpredictability, intrigue-resistivity, immutability, open membership, verifiability, and auditability [6]. Hence, it is hard to achieve transparency while ensuring the trust in electronic lotteries. When blockchain platforms combined with lottery system design can help with decentralization and immutability, and it is hard to achieve trustworthy randomness in that case. The design of lottery system discussed in this paper safeguards all of the above criteria that make it novel.

The rest of the paper is organized as follows: Section 2 discusses the background of lottery and related work in terms of blockchain-based lotteries. Section 3 focuses on actual system design and its underlying algorithm description. Properties of proposed system as whole and related advantages are discussed in Sect. 4 with conclusion and references at the end.

2 Background Theory and Related Works

Ethereum is an open source, distributed, and decentralized platform, which supports smart contracts. It is mainly known for its adoption of detailed scripting languages in cryptocurrency domain which can be used to write smart contracts. Cyptocurrency of Ethereum is known as ether. Smart contract is a tamper-proof, immutable program executing the terms mentioned in its transactional protocol. Smart contracts run on virtual machine (VM) supported by every node running on the network in a distributed manner. Ethereum supports its own higher level programming language called Solidity to write smart contracts. DApp is nothing but combination of one or more smart contracts to develop a piece of decentralized software. Inspired from these ideas, proposed system is designed.

The proof of first lottery drawn by human kind can be found in fifteenth century. Modern lotteries run by state government began around 60s in New Hampshire, USA, to generate revenue without incrementing taxes [7]. Hence, the lottery has been around since 600 years without changing a lot other than its forms. Computer has surely taken place of traditional lotteries as in storing the data of participants and drawing winners to transferring funds, but at the heart of it, basics remain the same. Many new forms of lotteries can be found in gambling ranging from instantaneous reward, lotto, number games, and scratch lottery, etc. [7].

Blockchain-based lottery systems are still finding its way in the maze of this new combination of cryptocurrencies and lottery scheme. Liao and Wang [8] have described blockchain-based lottery scheme as smart city application. They used a cryptographic model called hawk, to hide sensitive information from the participants on blockchain. Authors assert this model as lightweight enough to be implemented using IoT devices in new smart cities applications. Jia et al. [9] has proposed a lottery model built using smart contracts in Solidity language and deployed on Ethereum blockchain network. It uses existing RANDAO algorithm [6] to generate random numbers for lottery drawings. This model asserts resistant against Sybil and Node Attacks, and runs without any third-party intervention. Chen et al. [10] define a lottery DApp which determines its randomness not only from game state and blockchain state, but also using a specific set of people called 'committee' from the list of joined participants. Other good electronic lottery examples include [11, 12], and [13] that show properties like multi-level hash chain result drawing, verifiability, and distributed architecture, respectively.

Cryptocurrency and blockchain technology together can surely transform the era of modern gambling. Though, there have been a handful of attempts to develop lottery DApp, proposed system is unique due to its verifiability and novel randomness algorithm.

3 D-lotto System Design

D-lotto is a lottery DApp design that ensures lottery procedure to be fair, transparent, and verifiable. It is important to stick as close as possible to basic lottery system design while improving it. Hence, D-lotto has kept overall lottery design untouched as depicted in Fig. 1. While dealing with D-lotto, any dealer can initiate the new lottery using D-lotto interface using their electronic device. Any interaction medium including WebApp, DesktopApp, or an android application can be used as D-lotto interface. Participants have separate interface instance for joining the same lottery and they have that particular smart lottery contract in common. Also, it is not mandatory for dealer to not participate in lottery. He can behave same as other participants once the lottery is initiated, and that shows the real strength of immutability in smart contracts. Because not even creator of smart contracts can tamper them once they are deployed on network. Hence, if an individual wants to participate, we can treat even the dealer/organizer of lottery as any other participant after deployment of D-lotto. Below are the phases of D-lotto system design.

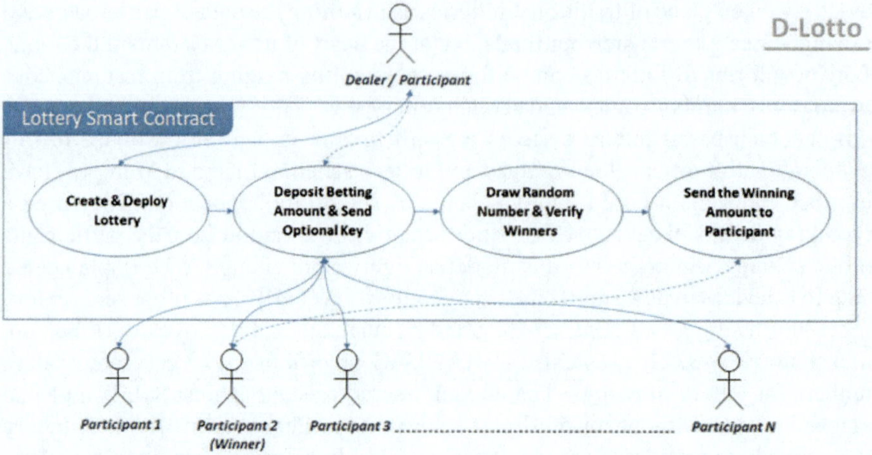

Fig. 1 D-lotto usecase diagram

3.1 Announcement of Lottery

This phase resembles with core lottery design facing slight variation. Dealer declares the own lottery scheme using D-lotto interface. Smart contract of that particular lottery instance is deployed on network by dealer after paying required transaction fee for same. Dealer defines and discloses everything from betting amount to number of maximum participants, pooling amount, winning amount, maintenance/dealer's share, lottery starting and ending time, etc., in detail for participants joining. Now, ideal betting share of each participant can be decided by following equation.

$$B_s = \frac{W_a + D_p + D_c}{T_g} \times 10^{1/\ln(S_f)} \tag{1}$$

Here, B_s is betting share, W_a is winning amount, D_p is dealer profit (Optional), D_c is lottery DApp deployment charges, and B_s, W_a, D_p, $D_c \in R+$. T_g defines number of tickets generated and S_f is a security factor, where $T_g \in N$ and $S_f \in [1, 2]$.

Security factor is used to keep betting share high, so that counterfeiting using dummy nodes and addiction to game can be avoided. Generated extra revenue can be used to declare extra jackpot prices or a consolation price for all the participants. After deciding details of lottery, dealer pays the transaction charges to deploy contract on Ethereum network and lottery goes live, provided having its operative interface. Individuals can have its own strategy to promote particular lottery scheme.

3.2 *Purchase of Lottery*

Depositing Betting Amount: After deployment of lottery by its dealer, system starts accepting the participant requests at a dealer specified start time through smart contract. Participants can simply join the lottery by paying the mentioned betting share calculated using Eq. 1. Single participant can buy more than one lottery tickets. If a participant is buying Ts tickets, his payment can be managed as per following equation:

$$T_s = \frac{P_a - R_a}{B_s} \tag{2}$$

Here, T_s is tickets share, R_a is returned amount, and P_a is paid amount. If a participant is not paying in whole multiples of B_s, he will not be awarded a full ticket and his remaining amount R_a will be sent back to his account by smart contract itself. For example, if a participant pays 110 ETH for 50 ETH B_s in lottery, he will be awarded two tickets as per Eq. 2 and his remaining amount of 10 ETH will be sent back to his address.

Sending an Optional Key: To participate in randomness drawn by D-lotto drawing algorithm 1, every participant can submit an optional random key of k hexadecimal bits or $k \times 4$ binary bits of his choice. Value of k can be arbitrary or can be best determined based on the factors in D-lotto drawing algorithm. Despite of buying more than one tickets, every participant can send only one random key to D-lotto using an interface. Keys will be stored and processed by smart contracts. Here, if any participant or all the participants choose not to send the key, it would not stop algorithm proceedings and will not affect its fair share of randomness.

3.3 *Lottery Drawing and Distribution of Winnings*

Generating Random Numbers for Lottery Drawing: Random number generation algorithm for D-lotto is mentioned in algorithm 1. As shown in the algorithm, every key submitted by participant is XORed by algorithm to produce participant key P_{key}. As discussed earlier, either value of key can be kept arbitrary with cap of 40 hexadecimal digits (160 binary bits) or it can be kept of a predetermined value for all participants. Participants failing to submit key of described parameters, will not be participating in randomness of D-lotto drawing algorithm. Another job is to produce block key B_{key}, which is nothing but the XORing of every block hash produced before $E_t + \delta$ time. Here, δ is a small amount of time compared to lottery buffer time, which is used to incorporate natural randomness underlying Ethereum blockchain. Blocks found during δ time also contribute to generation of random number, which is not known by any individual in advance. This feature gives D-lotto an edge over other lottery DApps for providing randomness.

After calculating P_{key} and B_{key}, evaluate f_{SHA} (.) for producing seed. Given function represents hashing using any algorithm from family of Secure Hash Algorithms (SHAs). We suggest using Keccak-256 [14] algorithm from SHA-3 standards, as it is used by Ethereum in Proof-of-Work (PoW) consensus for calculation of Nonce based on random hash addresses. Now, produced natural random seed is the only requirement of any pseudorandom number generator (PRNG). f_{PRNG} (.) function is used to calculate a set of W pseudo random numbers, where W is the set of winning tickets determined in the lottery. We recommend using ISAAC [15], ChaCha20 [16] or HC-256 [17] for PRNG, as they produce considerable pure randomness and are cryptographically secure to meet the application requirements. These random numbers can be capped by performing modulo operation with number of tickets being sold. This random number generation stage will be drawing out specified number of winners for lottery scheme. Every parameter in the described algorithm is public and/or traceable after all, because smart contracts work in fair and transparent manner. This makes algorithm verifiable after lottery completion and gives participants fair and transparent lottery proceedings without intervention of any third party including dealer itself.

Algorithm 1 D-lotto Drawing Algorithm

Require:
    Participant state $\boldsymbol{P_{ki}}$.
    Blockchain state $\boldsymbol{B_{kh}}$.
Ensure:
    A set of random numbers W (W = $\boldsymbol{w_1}, \boldsymbol{w_2}, \ldots, \boldsymbol{w_n}$)

```
Initialize P_key with 40 bits of zeroes or ones.
Initialize B_key with block hash including deployed lottery DApp.
Initialize h with latest block height at lottery start time.
if C_t = E_t + δ then
        while (P_ki ≠ NULL) && (i < no. of participants) do
                P_key ← P_key ⊕ P_ki ;  i ← i + 1
        end while
        while h < Latest block height at C_t do
                B_key ← B_key ⊕ B_kh ;  h ← h + 1
        end while
end if
Seed ← f_SHA (P_key ⊕ B_key)
W = f_PRNG(Seed)
return W
```

Distribution of Shares: After drawing out winners using RNG algorithm specified in previous state, smart contract will send the share of their money to winning participants on earlier verified receiving addresses. If a dealer has participated in lottery as a player and has been listed among winners, he will get his fair share from it. As mentioned earlier, it will be pretty clear to every participant about dealer's share/profit in the lottery since beginning, so remaining money will be transferred to dealer on his mentioned receiving address in smart contract. Even if the dealer chooses to avoid taking dealer's share, he will be paid back with lottery deployment charges bared by him earlier.

4 Comparison and Advantages

D-lotto, being a decentralized, Peer-to-Peer (P2P) application has its own advantages over traditional lottery systems and any electronic lottery system as described in Table 1. Along with low operational cost and quick response rate, it has no third-party intervention in any way. D-lotto specific additional unique features can be mentioned in terms of auditability and verifiability. Because of its underlying blockchain technology, D-lotto also incorporates blockchain technology advantages along with its architectural design benefits. Few of them are discussed below.

Transparency and Trust: Leaning toward decentralization and distributed computing has its own set of advantages. D-lotto, being built on public decentralized ledger, is totally transparent system. As it is based on distributed computing and uses smart contracts, need of third party and/or control organization has been entirely wiped out. This phenomenon builds trust among its users despite of no connection whatsoever with each other.

Auditability and Verifiability: Auditability in D-lotto is, being able to see every penny flowing through system without any privacy barriers. In public smart contracts, D-lotto can be traced down for its fund transfers and users can verify it with declarations of fund distribution at the time of lottery announcement. Verifiability

Table 1 Comparison of D-lotto with other lottery approaches

Property	Traditional lottery	E-lottery	D-lotto
Unpredictable	Yes	Yes	Yes
Membership	Private	Private/Public	Public
Third-party intrigue	High	Moderate	None
Operational cost	High	Moderate	Low
Response rate	Slow	Moderate	Quick
Auditability	No	Yes/No	Yes
Verifiability	No	No	Yes

can be described as any public procedure with proof of no conspiracy while drawing out lottery numbers. As described earlier, D-lotto uses game state and blockchain state to construct verifiable RNG algorithm. Hence, verifiability makes D-lotto is unique among other DApp lottery designs.

Reduced Fraud and Increased Accessibility: Blockchain technology possesses features like immutability and distributed computing. Hence, it automatically gifts D-lotto system with reduction in fraud, saving it from significant amount of loss. D-lotto uses cryptocurrency as medium of exchange in secure financial transactions, which helps it to remove geographical boundaries along with some legal aspects to participate in a lottery. Anyone is good to operate just by using good Internet connection.

5 Conclusion

Lottery can be used as a tool of entertainment and fund-raising apart from gambling. Hence, a novel lottery DApp design is proposed here on Ethereum platform using smart contracts. Ethereum as a platform ensures the availability of live blockchain parameters needed to build randomness in D-lotto drawing algorithm proposed here. D-lotto has an edge over other existing blockchain-based lottery systems in a manner that it uses game sate and blockchain state both to produce its verifiable randomness. Former property is a major contribution in trust and transparency of the system. D-lotto can be claimed as the novel system design in Ethereum-based lottery research due to its verifiable randomness.

References

1. Firelotto—white paper. https://firelotto.io/whitepaper_en.pdf. Last accessed 2019/11/30
2. Kibo—ethereum smart contracts based lottery. https://kiboplatform.net/en/landing.html. Last accessed 2019/11/30
3. Takya RA (2019) Blockchain lottery platform transforming lottery industry—bringing fairness to the lottery ecosystem. https://www.leewayhertz.com/blockchain-lottery-revolutionize-lottery-industry/. Last accessed 2019/11/30
4. Sharma M (2019) Impact of a streamlined lottery industry on the job market & development in india. https://www.stoodnt.com/blog/impact-of-a-streamlined-lottery-industry-on-the-job-market-development-in-india/. Last accessed 2019/11/30
5. Rodgers G (2015) Guilty verdict in hot lotto trial. https://www.desmoinesregister.com/story/news/crime-and-courts/2015/07/20/hot-lotto-verdict/30411901. Last accessed 2019/11/30
6. Randao: Verifiable random number generation. https://www.randao.org/whitepaper/Randao_v0.85_en.pdf. Last accessed 2019/11/30
7. Ariyabuddhiphongs V (2011) Lottery gambling: a review. J Gambl Stud 27(1):15–33
8. Liao D, Wang X (2017) Design of a blockchain-based lottery system for smart cities applications. In: IEEE 3rd international conference on collaboration and internet computing (CIC), pp 275–282

9. Jia Z, Chen R, Li J (2019) Delottery: a novel decentralized lottery system based on blockchain technology
10. Chen Y, Hsu S, Chang T, Wu T (2019) Lottery DApp from multi-randomness extraction. In: IEEE international conference on blockchain and cryptocurrency, pp 78–80
11. Liu Y, Liu H, Hu L, Tian J (2006) A new efficient e-lottery scheme using multi-level hash chain. In: 2006 international conference on communication technology, pp 1–4
12. Kuacharoen P (2012) Design and implementation of a secure online lottery system. In: Advances in information technology. Springer, Heidelberg, Berlin, pp 94–105
13. Grumbach S, Riemann R (2017) Distributed random process for a large-scale peer-to-peer lottery. In: Distributed applications and interoperable systems. Springer, Cham, Berlin pp 34–48
14. Bertoni G, Daemen J, Peeters M, Van Assche G (2013) Keccak In: Annual international conference on the theory and applications of cryptographic techniques. Springer, Berlin, pp 313–314
15. ISAAC: a fast cryptographic random number generator. http://burtleburtle.net/bob/rand/isaacafa.html. Last accessed 2019/11/30
16. Bernstein DJ (2008) Chacha, a variant of salsa20. In: Workshop record of SASC, vol 8, pp 3–5
17. Wu H (2004) A new stream cipher hc-256. In: Fast software encryption. Springer, Berlin, pp 226–244

Review of Machine Learning and Data Mining Methods to Predict Different Cyberattacks

Narendrakumar Mangilal Chayal and Nimisha P. Patel

Abstract Cybersecurity deals with various types of cybercrimes, but it is essential to identify the similarities in existing cybercrimes using data mining and machine learning technologies. This review paper reveals various data mining algorithms and machine learning algorithms, which can be help to create some specific schema of different cyberattacks. Machine learning algorithms can be helpful to train system to identify anomaly, specific patterns to predict the cyberattacks. Data mining plays a critical role to provide a predictive solution to rectify possible cybercrime and modus operandi and explore defense system against them. This is the era of big data, so it is very difficult to analyze and investigate the irregular activity on cyberspace. Data mining methods, while allowing the system to analyze hidden knowledge and to train expert system to alert and decision-making process. This review paper explores various data mining method like classification, association, and Clustering, while machine learning includes different methods like supervised, semi-supervised, and unsupervised learning methods.

Keywords Cybersecurity · Machine learning · Data mining · Cybercrime · Crime prediction

1 Introduction

Cybercrime is one of the biggest challenges in the twenty-first century globally [1–4]. This paper reveals exiting data mining and machine learning (DM/ML) methodology to predict the different cyberattacks. Different methods and techniques of DM/ML are described concerning different cyberattacks [5–8]. This paper focuses on how existing methods can be helpful to combat the existing and future cybercrime by

N. M. Chayal (✉) · N. P. Patel
Department of Computer Engineering, Sankalchand Patel University, Visnagar, India
e-mail: narendrachayal@gmail.com

K. Kotecha et al. (eds.), *Data Science and Intelligent Applications*, Lecture Notes on Data Engineering and Communications Technologies 52,
https://doi.org/10.1007/978-981-15-4474-3_5

providing the crime prediction methodology by using DM/ML algorithms and techniques. Cyber includes everything related to computer and communication technologies connected to Internet world [9–14]. Cybersecurity includes collection of technologies, methods, process, protocols to protect computer, network, server, mobile devices, critical software infrastructure, programs, etc. [15, 16]. Data mining deals with mining an important information from existing data to identify the threats, risk, or crime pattern to create an external repository to train the expert system by using some artificial and machine learning methodologies to predict crime pattern automatically. According to Veena et al. [17–21], DM/ML methodology can be helpful to perform the automatic forensics process to identify the important artefacts during cyberattack. References [22, 23] describes Web mining (Data mining) methods by using the classification algorithm to identify the Web-based cyberattacks. Crime mining can be identified by applying different DM algorithms and techniques to create the datasets and clusters containing its behavior, impact, pattern, target information, and its modus operandi [24–27]. Cyberattack could take place from either inside the network or outside the network which is known as internal and external attacks. It is not necessary that security mechanism like antivirus, IDS/IPS, firewall, or UTM can identify the attack and respond to attack [28–30]. Sometimes, new types of attack cannot be identified by the security mechanism. Malware is one of the biggest threat or risk which spread via different methods. Wannacry, Loki, and Prism are the examples of ransom-ware, which has affected millions of system in the past and destroyed valuable information [31]. This review paper will focus on various machine learning and data mining algorithms used in the field of cybersecurity to prevent and predict the various cyberattacks. There are various algorithms available in machine learning and data mining like support vector machine, k-means, k-nearest neighbor, Naïve Baysian, ID3, etc. [32, 33]. It depends on the cyberattacks what kind of algorithms will be applicable for forecasting and predicting the alert or countermeasure about attack. Rest of part of this review paper will discuss various types of cyberattacks and their identification mechanism with help of DM/ML algorithms [34].

Data Mining Methodologies: Data mining is similar to conventional treasure mining from a cave. It could be a mathematical or statistical process which uses computational power to mine information from huge amount of data. By utilizing statistical formulas and algorithms, information in the form of hidden knowledge can be mined. These hidden knowledge include some interesting pattern which helps to understand the importance of data for decision making, crime identification, and prevention for forecasting [35]. In the field of cybersecurity, data mining could be very important to identify the malicious activity on social media, malware analysis, blocking spam messages and emails, intrusion detection and prevention, phishing and for network, system and server log analysis to identify the hidden patterns [36, 37]. Apart from this, data mining approach is very useful in the field of cyberforensics which includes investigation part of cybercrimes. Data mining algorithms are generally used in the field of providing intelligence to business and organization by analyzing customer data to predict and forecast the sales and profit of the organization [38] DM algorithms are very useful in field health care, marketing, sales, stock exchange, business process re-engineering, etc. But these methodologies could be very useful for identifying and countering the various cyberattacks in cyberspace. Next section of this paper will explore various DM methodologies and algorithm which can be helpful in identifying the various cybercrimes [39].

Malware Analysis: Data mining plays very important roles in understanding the behavior–r of malware in windows and Linux environment. Normally, it utilizes association, classification, and clustering methodology to analyze the behavior of malware family in a sandbox environment.

Association rule of data mining in malware analysis can be useful to identify the file signature which malware affects. Like sometimes, malware tries to look for some specific DLL files but some executable files are also associated with that DLL, so applying association-based algorithm of DM can be helpful to understand behavior and impact of specific malware [40]. Generally, data mining methodologies help in following ways to understand impact and behavior of malware.

- Anomaly detection can be detected by comparing the behavior of the infected system with uninfected system. Data mining algorithm identifies hidden pattern during idle system behavior which includes running services and process, registry information, network connection, port status, Web traffic, etc. [41]. Data mining algorithm understands and compares the behavior of malware-infected system with above parameters to identify the presence of malware and its impact on the system. But sometimes, these systems can detect real user behavior as a malicious and report it as a false positive. But supervised learning algorithm of machine learning can be useful to train a legitimate user behavior of system for further guidance.
- Signature-based detection: This kind of detection of data mining algorithms looks for some specific signature or clue in the infected machine which is known to the machine. So, the machine will look for some specific pattern or information that DM algorithm already contains. For this methodology, classification and clustering methods of DM are used.
- Hybrid detection: This detection system uses both above approaches to understand the behavior of data. In short, this method employs association, classification, clustering, and regression methodology of data mining to identify the nature of cyberattack and intrusion. It generally does not use a single method but mostly utilized a combination of algorithms like Apropri, support vector mechanism, linear regression, etc.

All the cybersecurity mechanisms are generating logs of each and every event occurring in a cyberspace. So data mining could be very useful to analyze the log data in a way to identify and predict the next attack. For example, attacker is trying a brute force attack using a specific tool and cybersecurity mechanism is generating logs for that so by analyzing log association rule can predict the next possible attack by some other tool which is more efficient than other tool which fails to crack a password like if Hydra tool fails then maybe attacker uses burp suite to crack a password [42]. Below are some data mining methods which can be useful for cyberattack prevention and forecasting.

- Association rules: This method applies where one event can show a probability of the next event happening. For example, if customer will buy milk then it may be possible next product will be bread. So milk and bread are associated with each other. Same way, in the field of cybersecurity, if one event takes place which gives trigger to other event which takes place after finishing the first event. Example, if the attacker finds vulnerability in one website, so to exploit that vulnerability he will try all the payloads associated with that particular vulnerability. This algorithm identifies strong relationship and pattern between vulnerability and possible attacks and provides alert and prevention strategy in real time [43].
- Classification rule: This method is very useful because it classifies and builds a model based on the classification of the existing dataset. This method classifies known cyberattack with known pattern in that datasets. And apply these patterns to new datasets or live datasets for comparison and to identify the cyberattack. For example, during denial of service attack, Web server will get request from multiple (hundreds) IP address simultaneously and same pattern is identified in the existing dataset, so this attack can be easily identified by log analysis quickly [44, 45].
- Clustering: This method groups the attacks with similar data objects and patterns into multiple data groups containing similar hidden pattern of evidence of cyberattack. These similarity and dissimilarity are decided by the data mining algorithm which uses statistical techniques to study and analyze multiple attributes of collected datasets. This clustering analysis includes partitioning methods, hierarchical methods, density-based methods, etc. [46].

2 Phishing

It is an art of collecting confidential user information by applying social engineering techniques. Phishing can be placed by various methods but intension of attacker or cybercriminal is same to collect the confidential information by sending a phishing link by SMS, social media, e-mail, and in form of image [14]. Data mining and machine learning strategies are very helpful to identify and rectify the phishing link and message from your email and message box. Supervised learning with identified pattern along with classification and clustering methods can be used to compare the nature of phishing mail with a legitimate e-mail [47].

3 Machine Learning

It is a field of artificial intelligence which provides a framework to learn and train a data without explicitly programming. It is completely based on the mathematical techniques which help to analysis the data. ML and DM are inter-related to each other where data mining is used to generate the pattern and hidden knowledge and machine learning is used to utilize that DM knowledge to train the existing system and apply it to new dataset [47]. Machine learning technologies are useful in various field like health care, aerospace, cybersecurity, business, stock exchange, etc. But these techniques are very effective in the field of cybersecurity to train the security mechanism to prevent the cyberattack.

Machine learning consists of supervised and unsupervised learning. Supervised learning deals with the existing learned data or already known dataset which consists of pattern or knowledge. These methods are useful to identify the distributed denial of service attack and denial of service attack with the help of IP address and associated information [48]. It is useful to predict or identify whether this IP address was involved in malicious activity or not. Generally, these techniques use regression methods like linear, logistic regression, SVM, and decision tree algorithm.

Unsupervised learning consists to train a machine without data which is not actually labelled or classified. In short, the machine is learning by itself without any guidance or user input. Here, machine will analyze the dataset deeply to identify the hidden similarity or feature from each other attributes of dataset [49]. Machine groups data according to their similarity, dissimilarity, or extracted features without interference of anyone. This method uses an association rule and data clustering techniques.

3.1 Importance of Machine Learning in CyberSecurity

Cyberattacks are increasing day by day, so researchers have given various theory and automated solutions based on machine learning to prevent various cybercrime and cyberattacks. Researchers and cyberexperts have utilized these techniques to create various tools to rectify and stop cyberattack and cybercrime at some extent. Like data mining, machine learning is very useful for malware analysis, intrusion detection, and prevention system [50]. It leverages the capabilities of cybersecurity tools with automatic routine auditing risk and threats in the cyberenvironment which makes assistance to cybersecurity experts. Following are some machine learning applications in the field of cybersecurity.

4 Cyberthreat Identification, Detection, and Prevention

Machine learning-based tools and methods are designed to identify the cyberattack and respond quickly before they start affecting. Supervised learning models are made on the basis of previously identified cyberattack patterns and indicators which actually help to identify real-time attack to respond. Clustering and classification techniques are used to analyze the behavior of various malware and their attack vector which is helpful to identify the behavior of new malware. Specifically, machine learning algorithms are very helpful in identifying the distributed denial of service attack made using a botnet. By analyzing the behavior of malware, these expert systems can be helpful to prevent from the ransom-ware attack like Wannacry and Prism.

5 Trojan Horse Detection

Trojan is one kind of malware which behaves like a legitimate computer program but it is responsible for data stealing and information gathering maliciously. Most of times, antivirus program cannot detect them. Android-based malware 'Agent Smith' is targeting android devices with android version of 6.0 and 5.0. This malware was installed on the legitimate mobile application downloaded from the Google play store. Detection of this kind of malware is very difficult sometimes, but machine learning algorithms are very helpful in identifying its malicious activity.

6 Social Media Analysis

Use of social networking technologies are increasing nowadays. Social media applications like, Facebook, Twitter, Instagram, etc., are the easiest way to communicate. It creates social networking graph between profile owner, friends, friends of friends, organization, and business. Social media network provides a common platform to share their views, activities, interests, audio, video, images, and news also. All the social media applications are Web-based and use to interact with each other. But cybercriminals use these platforms to execute the cybercrimes. Cybercriminal uses social engineering methods to conspire the cybercrime like malware injection, key logger injection, phishing, fake profile, fake news, financial attack, etc. Cybercriminals upload links, images, gif media containing Trojan horse, key logger, spyware, or backdoor to collect the user credentials like user name and passwords [50]. Sometimes, criminal uses fake profile to perform cyberterrorism activities like in past ISIS uses the Facebook platform to recruit terrorism by spreading explicit materials on Facebook accounts. Here, data mining and machine learning techniques and methods are useful to monitor and analyze the social media activities of users.

Extract, transfer, and loading (ETL) methods of data warehousing can be used to collect the data from different social media profiles. Due to security policies of social media websites, ETL process can be used to collect data which is available for public [51]. After collection of data, transformation methods are applied to filter the user profile. These filters include keyword search, regular expression, media object detection, link analysis, hyperlink analysis, etc. Data mining algorithms like association, classification, and clustering are useful to train the data to predict the possible cybercrime on social media websites.

Machine learning algorithm and methods like supervised and unsupervised techniques are used to provide real-time alert to identify the anomaly of particular user account if he/she posted something malicious on the public domain of social [52]. Machine learning and data mining methods can be useful to identify the attackers or criminals profile and details.

7 Conclusion

There are lots of important information available on Internet and criminals utilize these information to execute cybercrime. Machine learning and data mining provide such capability to identify different cybercrimes using classification and clustering algorithms. These algorithms provide automated facility to identify specific cybercrimes, analyzing behavior of different malware. These algorithms are also capable to identify and predict various cybercrime like phishing, distributed denial of service and denial of service attack. To minimize the risk and threat of cybercrime, machine learning and data mining algorithms play a vital role to analyze and identify the behavior of criminals. Machine learning provides automated crime detection and prevention methods to combat the cybercrimes.

References

1. Fischer EA (2014) Cybersecurity issues and challenges: in brief
2. Han J, Kamber M, Pei J (2012) Data mining concepts and techniques. In: Han J (ed) Introduction, 3rd edn. Elsevier, USA
3. John S, Sara G, Gillian M (2013) Comprehensive study on cybercrime. In: John S (ed) Connectitivity and cybercrime, 1st edn. United Nations Office on Drugs and Crime, Vienna, USA
4. Artur A (2014) Legal aspects of cybersecurity. In: Artur A (ed) Cybersecurity as an umbrella concept, 1st edn. University of Copenhagen, Denmark
5. Azene Z, Mufaro S, Andrei C et al (2019) Cyber threat discovery from dark web. EPIC Ser Com 64:174–183
6. Benjamin V, Li W, Holt T et al (2015) Exploring threats and vulnerabilities in hacker web: forums, IRC and carding shops. Paper presented at USA, University of Maryland, Baltimore County
7. Mavroeidis V, Bromander S (2017) Cyber threat intelligence model: an evaluation of taxonomies, sharing standards, and ontologies within cyber threat intelligence. Paper presented at Greece, pp 11–13
8. Nunes E, Diab A, Gunn A (2016) Darknet and deepnet mining for proactive cybersecurity threat intelligence paper presented at USA, pp 28–30
9. Anshu S, Shilpa S (2008) An intelligent analysis of web crime data using data mining. Int J Eng Innovative Technol 2(3)
10. Hsinchun C, Wingyan C, Yi Q (2003) Crime data mining: an overview and case studies. In: Paper presented at proceeding of the annual national conference on digital government research, Boston, MA

11. Abraham T, De Vel O (2006) Investigating profiling with computer forensic log data and association rules. In: Paper presented at proceedings of the IEEE international Conference on data mining
12. Lin HF, Liang JM (2005) Liang event based ontology design for retrieving digital archieves on human religious self-help consulting. In: Paper presented at proceedings of IEEE international conference on e-technology, e-commerce and e-service
13. Zarri GP (2002) Semantic web and knowledge representation. In: Paper presented at proceedings of the 13th international workshop on database and expert system applications
14. Malathi A, Babboo SS, Nbarasi A (2011) An intelligent analysis of a city crime data using data mining. In: International conference information electronic engineering, pp 130–134
15. Veena HB, Prasanth GR, Deepa PS (2011) Data mining approach for data generation and analysis for digital forensic application. Int J Web Eng Tech 2(3): 313–319. https://doi.org/10.7763/ijet.2010.V2.140
16. Robert R (2004) A ten step approach for forensic readiness. Int J Dig Evi 2(3):313–319
17. Kara N, Brian H, Matt B (2009) Digital forensics: defining a research agenda. In: Paper presented at proceedings of the forty second Hawaii international conference on system sciences
18. Pollitt M (1995) Computer forensics: an approach to evidence in cyberspace. In: Paper presented at proceedings of the national information systems security conference, USA
19. Reith M, Carr C, Gunsch G (2002) An examination of digital forensic models. Int J Dig Evi 1(3):01–12
20. Kohn M, Eloff J, Oliver M (2006) Framework for a digital forensic investigation. In: Paper presented at proceedings of information security from insight to foresight conference, South Africa
21. Freiling FC, Schwittay B (2007) A common process model for incident response and computer forensics. In: Paper presented at proceedings of conference on incident management and IT forensics, Germany
22. Brian C, Eugene H (2003) Spafford: getting physical with digital investigation process. Int J Dig Evi 3(2):1–20
23. Siti RS, Robiah Y, Shahrin S (2006) Mapping process of digital forensic investigation framework. Int J Comput Sci Netw Secur 8(10):163–169
24. Azah AN, Suraya H, Maw MH, Suraya IT (2017) Security threats and techniques in social networking sites: a systematic literature review. In: Paper presented at future technologies conference, Vancouver, Canada, pp 29–30
25. Ellison NB (2007) Social network sites: definition, history, and scholarship. J Comput mediated commun 13(1):210–230
26. Hydara I, Sultan ABM, Zulzalil H, Admodisastro N (2015) Current state of research on cross-site scripting (XSS): a systematic literature review. Inf Sof Tec 58:170–186
27. Devmane M, Rana N (2013) Security issues of online social networks. Adv Comput Commun Control 14:740–746
28. Faghani MR, Matrawy A, Lung CH (2012) A study of trojan propagation in online social networks. In: Paper presented at the international conference on new technologies, mobility and security, IEEE
29. Ahmed F, Abulaish M (2013) A generic statistical approach for spam detection in online social networks. Comput Commun 36(10–11):1120–1129
30. Lee S, Kim J (2014) Early filtering of ephemeral malicious accounts on twitter. Comput Commun 54:48–57
31. Soumajyoti S, Mohammad A, Jana S (2018) Predicting enterprise cyber incidents using social network analysis on the darkweb hacker forums. Cornell University. Available via DIALOG. https://arxiv.org/abs/1811.06537
32. Almukaynizi M (2017) Predicting cyber threats through the dynamics of user connectivity in darkweb and deepweb forums. ACM Comput Soc Sci 01:1–9
33. Leo B (2001) Random forests machine learning. available vai DIALOG. https://www.stat.berkeley.edu/~breiman/randomforest2001.pdftitel

34. Anna S, Alessandro B, Saranya D, Paulo S (2017) Early warnings of cyber threats in online discussions. Cornell University. Available via DIALOG https://arxiv.org/abs/1801.09781
35. Herley C, Dinei F (2010) Nobody sells gold for the price of silver: dishonesty, uncertainty and the underground economy of information security and privacy. Springer 01:33–53
36. Meier L, Van De Geer S, Bühlmann P (2008) The group lasso for logistic regression. J R Stat Soc: Ser B (Stat Methodol) 70(1):53–71
37. Allodi L, Corradin M, Massacci F (2016) Then and now: on the maturity of the cybercrime markets the lesson that blackhat marketeers learned. IEEE Trans Emerg Top Comput 4(1):35–46
38. Palash G, Tozammel HKSM, Ashok D, Nazgol T (2018) Discovering signals from web sources to predict cyber-attacks. IEEE Sys 10(10):1–11
39. Eric N (2016) Darknet and deepnet mining for proactive cybersecurity threat intelligence. In: Paper presented at IEEE conference on intelligence and security informatics, Tucson, AZ, USA, pp 28–30
40. Adgaonkar A, Shaikh H (2015) Privacy in online social networks. Int J Adv Res Com Sci Sof Engg 05(03):01–09
41. Egele M, Stringhini G, Kruegel C, Vigna G (2017) Towards detecting compromised accounts on social networks. IEEE Trans 14:447–460
42. Vishwanath A (2017) Getting phished on social media. Decis Support Syst 103:70–81
43. Ali S, Rauf A, Islam N, Farman H (2017) User profiling: a privacy issue in online public network. Sindh Univ Res J (Sci Seri) 49:125–128
44. Global journal of computer science and technology: C software & data engineering. Online ISSN: 0975-4172
45. Sapienza A (2017) Early warnings of cyber threats in online discussions. In: Data mining workshops (ICDMW)
46. Adgaonkar A, Shaikh H (2015) Privacy in online social networks. Int J Adv Res Comput Sci Softw Eng 5(3)
47. Nunes E (2016) Darknet and deepnet mining for proactive cybersecurity threat intelligence. In: IEEE ISI
48. Norman AA, Hamid S, Hanifa MM, Tamrin SI (2017) Security threats and techniques in social networking sites: a systematic literature review. In: Future technologies conference, Vancouver, Canada, pp 29–30
49. Egele M, Stringhini G, Kruegel C, Vigna G (2017) Towards detecting compromised accounts on social networks. IEEE Trans Dependable Secure Comput 14:447–460
50. Sharma A, Sharma S (2012) An intelligent analysis of web crime data using data Mining. Certif Int J Eng Innov Technol 2(3). ISSN: 2277-3754 ISO 9001:2008

Sentiment Analysis—An Evaluation of the Sentiment of the People: A Survey

Parita Vishal Shah and Priya Swaminarayan

Abstract Online archives have received a great deal of attention in recent years from a person's view and thoughts as primary platform. Set of circumstances gives rise to increasing interest in methods for automatically collecting and evaluating individual opinions from online documents such as customer reviews, Weblogs and comments on electronically accessible media, emphasis of current studies are mainly on attitude analysis. Interest of human beings is on designing a structure which can categorize feelings of individuals in the form of automated letter. Retrieving and determining beliefs from Web require appropriate mechanism that can be used to acquire and estimate thoughts of the desires of online consumers, which could be useful for economic or marketing research. An aspect of natural language processing (NLP), sentiment analysis (SA) has experienced a growing interest in the past decade. The difficulties and chances of this rising field are likewise talked about prompting our postulation that the analysis of multimodal sentiment has a significant untapped potential.

Keywords Opinion mining · Sentiment analysis · Sentiment classification

1 Introduction

Analysis of sentiment is the automated process of analysis of text data and classification of opinions as negative, positive or neutral. Usually, notwithstanding distinguishing the conclusion, these frameworks remove properties from the articulation, for example, Extremity: if the speaker communicates a constructive or pessimistic

P. V. Shah (✉)
Parul University, Vadodara 391760, India
e-mail: paritaponkiya@gmail.com

P. Swaminarayan
Faculty of IT & CS, Parul University, Vadodara 391760, India
e-mail: priya.swaminarayan@paruluniversity.ac.in

K. Kotecha et al. (eds.), *Data Science and Intelligent Applications*,
Lecture Notes on Data Engineering and Communications Technologies 52,
https://doi.org/10.1007/978-981-15-4474-3_6

sentiment, Subject: the thing being talked about, Opinion holder: the individual or element communicating the conclusion [1].

Since it has many practical applications, sentiment analysis is currently a topic of great interest and development. Organizations use sentiment analysis to evaluate survey responses, product reviews, social media comments and the like automatically to gain valuable insights into their brands, goods and services. Examination of supposition is a kind of content examination otherwise known as mining. It applies a blend of insights, common language handling (NLP) and AI to distinguish and remove abstract data from content records, for example, the sentiment, considerations, decisions or assessments of an analyst about a specific subject, occasion or an organization and its exercises as depicted previously. Otherwise called supposition mining or full of feeling rating is this kind of examination.

You can examine content at different degrees of detail, and your objectives rely upon the degree of detail. For example, you can characterize a normal passionate tone of a gathering of surveys to recognize what level of clients your new assortment of garments enjoyed. On the off chance that you have to recognize what guests like or aversion about a specific article of clothing and why, or if different brands contrast it with comparable things, each survey sentence should be examined with an emphasis on explicit perspectives and use or explicit watchwords. Two kinds of estimation can be utilized, contingent upon the scale: coarse-grained and fine-grained. Coarse-grained investigation enables an inclination to be depicted at the degree of an archive or sentence. Furthermore, with fine-grained examination, in every one of the sentence parts, you can remove a sentiment.

2 Previous Work

Sentiment analysis or opinion mining is the method by which subjective data is defined and detected using natural language processing, text analysis and computational linguistics. In short, the purpose of sentiment analysis is to extract information on the writer's or speaker's attitude toward a specific topic or a document's total polarity. The first articles with their keywords that used sentiment analysis were published around a decade ago, but the discipline may trace its roots back to the mid-nineteenth century. The general inquirer is one of the leading tools for examining sentiments [2].

Opinion distinguishing proof is an extremely intricate issue, and consequently, much exertion has been placed into breaking down and attempting to comprehend its various perspectives, see for example. Regular wellsprings of obstinate writings have been film and item surveys, Web journals and Twitter posts [3]. As news stories have generally been viewed as nonpartisan and free from suppositions, little spotlight has been on them. In any case, the enthusiasm for this area is developing as computerized exchanging calculations represent a consistently expanding piece of the exchange [4]. A quick and straightforward strategy for deciding the feeling of a book is utilizing a pre-characterized assortment of assessment bearing words and basically accumulating the assumptions found. Further developed techniques do not treat all words similarly yet allot more weight to significant words relying upon their situation in the sentence [1].

Unfortunately, most domains are very unique, which means that in another domain, one collection of words that is most likely appropriate in one domain will not work as well. For example, efforts were made to resolve this shortcoming [5]. Another branch of sentiment analysis used a more linguistic approach and focused on extracting opinion holders and quotes from texts. While natural language processing technologies continue to improve and computational power continues to become cheaper, further resources are likely to be put into sophisticated automated methods of text processing [6].

3 Data Source

The opinion of the user is a major criterion for improving the quality of the rendered services and improving the deliverables. Blogs, review sites, blogs and microblogs provide a good understanding of the product and service reception rate. Operating on various datasets and experimenting with different methods are the key part of mastering sentiment analysis [7].

3.1 Blogs

Blogging is growing rapidly with increasing Internet usage. Blog pages have become the most popular means of expressing one's personal views. Bloggers document their daily life activities and express their views, opinions and emotions in a blog. Many of these forums have coverage of a lot of products, problems, etc. In huge numbers of the examinations identified with supposition investigation, online journals are utilized as a wellspring of conclusion [7].

3.2 Review Sites

The opinions of others may be an important factor for any user to make a purchasing decision. On the Internet, there is a large and growing body of user-generated feedback. Product or service reviews are usually based on the views expressed in a highly unstructured format. The data used by the reviewers in most of the grouping considers is gathered from the Web-based business sites [7].

3.3 Micro-Blogging

Twitter is a popular micro-blogging service for users to create "tweets" status messages. Sometimes, these tweets express views on different topics. Twitter messages are also used to classify sentiment as a data source [7].

4 Applications [3]

4.1 Social Media Monitoring

Dissect tweets or potentially posts from Facebook over some stretch of time to see a particular crowd's inclination. Investigate your feelings on every single social medium references to your image and order them naturally by direness

4.2 Brand Monitoring

Break down news stories, blog entries, gathering exchanges and different messages on the Web over some stretch of time to see assumption of a specific crowd. Naturally, sort criticalness of every online notice to your image by means of supposition examination.

4.3 Customer Feedback

Break down news stories, blog entries, exchange discussions and other online messages over some stretch of time to see a particular crowd's inclination. Classify the significance of every single online reference to your item naturally by notion investigation.

4.4 Customer Service

Automate applications on all incoming customer support requests to conduct sentiment analysis. Detect disgruntled customers quickly and top these tickets. Connect questions to the best qualified to respond to particular team member. Using analytics to gain in-depth insight into what is going on through your customer support.

4.5 *Market Research*

Analysis of sentiment enables market research and competitive analysis of all kinds. Whether you are exploring a new market, anticipating future trends or maintaining a competitive edge, an analysis of feelings can make all the difference.

5 Sentiment Analysis Importance

Some of the advantages of sentiment analysis include the following [8].

5.1 *Scalability*

Handling client assistance discussions or client audits physically figuring out a great many tweets, there is simply an excessive amount of information for manual preparing. Affectability investigation empowers productive and savvy handling of information on a scale.

5.2 *Real-Time Examination*

We can utilize estimation investigation to order basic data that permits ongoing situational mindfulness in explicit situations. Is there going to break a marketing emergency in online life? A furious customer nearly stirring? A notion investigation framework can assist you with characterizing and make a move rapidly on these sorts of circumstances.

5.3 *Consistent Criteria*

For assessing the sentiment of a bit of content, people do not watch clear criteria. Various individuals are evaluated to concur just around 60–65% of when making a decision about the inclination for a specific bit of content. It is an abstract action vigorously affected by close to home encounters, sentiments and feelings. Associations may apply similar norms to every one of their information by utilizing a brought together conclusion examination system. This diminishes blunders and improves consistency of information.

6 Sentiment Analysis Algorithms [9]

There are many methods and algorithms to implement sentiment analysis systems, which can be classified as [9].

6.1 Rule-Based Systems

It conducts an evaluation of emotions based on a set of rules that are manually designed. In some kind of scripting language, rule-based strategies as a rule characterize a lot of decides that group subjectivity, extremity or the subject of a conclusion. The rules can utilize an assortment of sources of info, for example, great NLP strategies, for example, stemming, tokenization, voice labeling and parsing.

Different assets, for example, dictionaries (e.g., word records and expressions) can be extremely hard to keep up as new standards might be expected to include support for new expressions and jargon. Furthermore, because of collaboration with past principles, including new guidelines may have bothersome outcomes. These frameworks in this way require critical interest in manual tuning and rules upkeep.

6.2 Automatic Systems

To learn from information, it relies on machine learning techniques. Like rule-based systems, automated methods do not rely on manually designed rules but on techniques of machine learning. Usually, the task of slant examination is displayed as an arrangement issue where a classifier is nourished a content and returns relating classification, for example, positive, negative or impartial (on account of an extremity investigation).

7 Sentiment Analysis Types [10]

There are different methods to analyze the "sentiment data."

7.1 Document Level of Sentiment Analysis

That report focuses on a single entity or occurrence in document-level sentiment analysis and includes opinion from a single owner of opinion. The opinion here can be classified as either positive or negative (probably neutral) in two simple classes.

For instance: a summary of the product: "I bought a new phone a few days ago. It's a nice phone, but it's a bit big. It's a good touch screen. The strength of the voice is stronger. I just love the phone." The subjective opinion is said to be optimistic, taking into account the words or phrases used in the analysis (nice, fine, great, love). Objective opinions are measured using the star or poll system, with positive 4 or 5 stars and negative 1 or 2 stars.

7.2 *Sentence Level of Sentiment Analysis*

We should move to the sentence level in order to have a more refined view of the different opinions expressed in the document about the entities. This level of analysis of sentiment filters out those sentences that do not contain an opinion and determines whether the entity's opinion is positive or negative.

7.3 *Aspect-Based Sentiment Analysis*

Examination of the emotion level of the text and the sentence level works well when referring to a single entity. In many cases, however, people are talking about entities with many aspects or attributes. They will have different views on different aspects as well. In product reviews and discussion forums, it often happens. For instance: "I am a lover of the Smartphone. I like the phone's feel. The screen is clear and big. The camera is amazing. But there are also a few downsides; the life of the battery is not up to the mark and it is difficult to access Whatsapp." Categorizing the positive and negative aspects of this review hides valuable product information. The aspect-based sentiment analysis therefore focuses on the identification of all emotional words within a particular document and the issues to which the opinions related.

8 Evaluation of Sentiment Classification [5]

Precision, recall and precision are standard metrics used to assess a classifier's performance. Precision measures how many texts from every one of the writings that were anticipated (effectively and erroneously) as having a place with the class were accurately anticipated to have a place with a given classification. Recall measures how many texts from all the texts that should have been predicted to belong to the category were correctly predicted as belonging to a given category. We also know that the better it will be to remember the more data we feed our classifiers. Accuracy measures how many texts out of all the texts in the corpus were correctly predicted (both as belonging to a category and not belonging to the category).

9 Sentiment Analysis Challenges [9]

By addressing a portion of the primary difficulties and restrictions in the field, the vast majority of the work in opinion investigation as of late has been around growing increasingly exact conclusion classifiers.

9.1 *Subjectivity and Tone*

It is just as important to detect subjective and objective texts as to analyze their tone. In fact, there are no explicit feelings in so-called objective texts.

9.2 *Setting and Polarity*

Eventually in time, every one of the articulations is spoken by and to certain individuals, you get the point. All announcements are communicated in setting. It turns out to be very hard to examine emotions without setting. Machines, in any case, cannot think about settings except if they are expressly referenced. One of the issues emerging from the setting is the extremity changes.

10 Conclusion

Analysis of sentiment has many applications in information systems, including classification of reviews, summarization of reviews, extraction of synonyms and antonyms, tracking of opinions in online discussions and so on. This paper attempts to introduce the problem of sentiment classification at different levels, i.e., document level, sentence level, word level and aspect level. Also, some techniques were introduced that were used to solve these problems. In this paper, we presented a summary of the theory and goals of sentiment analysis, examined the state of the art and explored the field-related issues and perspectives.

References

1. Rani S, Kumar P (2019) A journey of Indian languages over sentiment analysis a systematic review. Artif Intell Rev 52:1415–1462. https://doi.org/10.1007/s10462-018-9670-y
2. Hussein DME-DM (2018) A survey on sentiment analysis challenges. J King Saud Univ Eng Sci 30(4):330–338. https://doi.org/10.1016/j.jksues.2016.04.002

3. Iti C, Erik C, Roy W, Francisco H (2017) Distinguishing between facts and opinions for sentiment analysis: survey and challenges, Elsevier
4. Mäntylä M, Graziotin D, Kuutila M (2018) The evolution of sentiment analysis—a review of research topics venues, and top cited papers, Elsevier
5. Erik C, Soujanya P, Alexander G, Mike T (2017) Sentiment analysis is a big suitcase. IEEE
6. Schouten K, Frasincar F (2015) Survey on aspect-level sentiment analysis. IEEE Trans Knowl Data Eng 28(3):813–830
7. Amandeep K, Vishal G (2013) A survey on sentiment analysis and opinion mining techniques. JETW
8. Moshe K, Jonathan S (2016) The importance of neutral examples for learning sentiment
9. Pradhan V, Vala J, Balani P (2016) A survey on sentiment analysis algorithms for opinion mining. Int J Comput Appl
10. Arti B, Chandak M, Akshay Z (2013) Opinion mining and analysis: a survey. IJNLC

A Comprehensive Review on Content-Based Image Retrieval System: Features and Challenges

Hardik H. Bhatt and Anand P. Mankodia

Abstract Over a couple of years, huge attention is being paid by the researchers on the content-based image retrieval (CBIR) in order to successfully retrieve the contents from large-scale multimedia databases. Typically, each day gigabytes of multimedia contents are being generated by the digital camera, cell phone, and PC and they are available in the form of multimedia database. It is critical to find out the desired data from this vast collection of database. CBIR is not only efficient in performing the image retrieval, but also organizes the common contents of a digital library in the indented database. In this work, totally 25 research works are reviewed under CBIR techniques with respect to certain analytical views. On the basis of different algorithmic models, they are categorized into transform-based CBIR technique, metaheuristic-based CBIR technique, learning-based CBIR technique, fuzzy-learning-based CBIR technique, and other CBIR techniques. The analytical representations are defined by means of graphs and tabular columns. Finally, a detailed description of research gaps and challenges is also presented under this scenario.

Keywords Image retrieval · CBIR technique · Algorithmic analysis · Performance evaluation · Research gaps · Challenges

1 Introduction

Image retrieval is one among the recent hot topics concerned with the searching of digital images in the database as well as the retrieval of the images. Huge count of researchers is undertaken in this topic as it is primarily employed in various fields of image processing, RS, multimedia, database applications, digital libraries,

H. H. Bhatt (✉)
Research Scholar, Ganpat University, Mehsana, Gujarat 384012, India
e-mail: hbhatt992@gmail.com

A. P. Mankodia
Department of Electronics and Communication, Ganpat University, Mehsana, Gujarat, India
e-mail: anand.mankodia@ganpatuniversity.ac.in

K. Kotecha et al. (eds.), *Data Science and Intelligent Applications*,
Lecture Notes on Data Engineering and Communications Technologies 52,
https://doi.org/10.1007/978-981-15-4474-3_7

and also in the other related area. On the basis of the query image, the relevant images are retrieved in an efficient image retrieval system and the closeness of the human perception in the image retrieval is confirmed by the QI [1, 2]. Typically, two major research communities are namely computer vision and database management retrace images from various perspectives like TBIR technique and VBIR technique. In the TBIR techniques, the contents of the image are described in the textual format, whereas the contents in the form of images are retrieved as visual features in CBIR or VBIR [3, 4]. Due to the advancement of the multimedia and Internet in various fields like satellite data, surveillance system, still images repositories, medical treatment, and digital forensics, CBIR is utilized as an alternative to TBIR [2, 5].

In CBIR, the retrieval of the images is categorized on the basis of the features as high-level features and low-level features. In the early researches, single features among the color, shape, and texture features were utilized and this produced in satisfactory retrieval results due to the availability of various visual characteristics. Thus, the recent researches are focusing on the combination of these three features. The image retrieval is performed, when RGB-based QI enters the retrieval system and these RGB query images are transformed into HSB color image [6, 7]. Then, from the color, texture, and shape features, the color, text, and shape feature vectors are extracted. The similarity between the query images and the combined feature vectors are calculated to search for the most similar target images. Moreover, the low-level features enclosing the spatial relations, texture, shape, and color are used in different applications and many sophisticated algorithms were developed to represent these approaches (color, shape, and texture features) [8–10]. But, these algorithms were not able to satisfy the comfort of human perception due to the lack of low-level image features in forming the high-level concepts in a picture-perfect manner.

Moreover, the machine learning techniques are also employed, and they are a good degree of efficiency in extracting the low-level features automatically from images. Apart from this, the high-level concepts in users' mind are not described by the low-level image features, and it is essential to bridge the semantic gap between these features and it is one among the challenging problems in the image retrieval [11, 12]. Yet, significant efforts have been carried out in the image retrieval research, and there is still a huge gap between the mappings of low-level concepts to high-level concepts. Therefore, the recent researches are being conducted in the formation of intelligent image retrieval that has the capability of abstracting the high-level concepts of the image accurately, by extracting the semantic feature automatically and has the ability to interpret the user query [13, 14]. Under this line, this paper aims to make a survey on CBIR system, and the major contribution of this paper is as follows: This survey makes a compact review of the existing CBIR systems by reviewing 25 papers. The analytical review is carried out with respect to the performance measures of reviewed papers. The review results are represented by means of diagrammatic representation, bar charts, and tabulation.

The organization of this paper is as follows: Sect. 2 depicts the literature review under the research area of CBIR systems, and Sect. 3 portrays the analytical review on content-based image retrieval system. The research gaps and challenges of CBIR are represented in Sect. 4, and the clear conclusion to this review work is given in Sect. 5.

2 Literature Works

2.1 Related Works

In 2019, Shamna et al. [15] have proffered T&L model for the automated medical image retrieval system. The guided LDA method was utilized for the formulation of the topic information, and the spatial information was incorporated by utilizing the location model. The rank of the positional matrix was determined with the aid of a position weighted precision approach. In 2019, Mezzoudj et al. [16] have proposed fast CBIR-S with the intention of diminishing the long execution time of large-scale images. The initial contribution of this research was to enhance the speed associated with the indexation process by MapReduce distributed model in the indexation step. Moreover, the write operation was enhanced with the help of memory-centric distributed storage system. Then, in the second contribution, the speed of image retrieval was increased with k-NN search method.

In 2018, Tzelepi et al. [17] formulated an innovative model retraining method with the aid of deep CNN in CBIR for gathering knowledge on the convolutional representations. On the basis of the available information, three approaches were formulated. In case of non-availability of the information, the fully unsupervised retraining approach was used and retraining with relevance information approach will be employed in the presence of labels. Moreover, relevance feedback-based retraining was utilized in the presence of the users' feedback. In 2018, Raza et al. [18] introduced CPV-THF as a novel feature descriptor for CBIR. The correlations existing between the texture orientation, LSSI, color as well as the intensity of an image were evaluated in order to integrate the semantic information as well as the visual content of the image. Moreover, the box-shaped structural elements were constructed in the image texture analysis on the basis of the texton theory.

In 2018, Dai et al. [19] designed the content-based RS image retrieval system, in which the spectral information, as well as the spatial information content of RS images, was characterized with image description method. In RS image, the sparsity was exploited using supervised retrieval method and the spatial content was modeled using the extended bag of spectral values descriptors. In 2018, Shamna et al. [20] developed an unsupervised CBMIR framework in CBIR systems on the basis of the visual words in spatial matching approach. Then, with SSI, the spatial similarity available between the visual words is computed. In 2018, Mistry et al. [21] formulated a hybrid feature-based efficient CBIR system by utilizing different distance measures as well as BSIF feature descriptors, spatial descriptors, CEDD descriptors, and frequency descriptors. The precision was enhanced using the Gabor wavelet transform features.

In 2018, Jin et al. [22] suggested an innovative CBIR technique by fusing both CSL and DML method. The shortcoming of class imbalance was resolved by arranging the samples into minority classes and majority class on the basis of the weight. Further, to make the proposed model more sufficient for CBIR, three major metrics like margin variance, misclassification penalty cost, and margin mean were considered. In 2018, Unar et al. [23] proposed a fully automatic CBIR method with the intention of retrieving similar images with the visual and textual characters. With the assistance acquired from the global descriptors, the image features were extracted from the image in the global-based image retrieval approach and in the region-based image retrieval approach, the based match similarity between the images was evaluated by means of splitting the image into smaller pieces. Then, SURF descriptor extracted the salient visual features from the image and the embedded text in the image was identified with the MSER algorithm. The stop-filters were employed to distinguish the non-text objects and the text objects in the image. The strings, as well as keywords, were formulated using the neural probabilistic language model.

In 2018, Alsmadi [24] formulated an efficient CBIR with MA to retrieve the images from the datasets. The shape, color as well as color signature features are extracted from the QI by CBIR. The retrieval of the image was accomplished with MA and the quality of solution was enhanced with the ILS algorithm and GA algorithm. Meanwhile, the different images query assessed the CBIR system in retrieving similar images. In 2017, Pedronette et al. [25] introduced the rank diffusion approach as a hybrid method that had utilized the diffusion process with concern to the ranking information. The low-complexity re-ranking algorithm utilized the diffusion strategy, in which the rank information was considered only once. The proposed model was evaluated using different detectors on six public image datasets, and the resultant of the analysis was higher effectiveness gains in the proposed mode over the existing model. In 2017, Islam et al. [26] proffered a novel CBIR system with different fuzzy-rough feature selection methods. The retrial, as well as ranking, was made on the basis of the prominent feature subset. The upper approximation was computed on the basis of the fuzzy-rough framework in order to override the shortcoming of fuzzy-indiscernibility relation.

In 2017, Zhu et al. [27] introduced a novel approach for enhancing the performance of CBIR via the swarmed particles mechanism. Moreover, RF of user was utilized with the intention of interpreting the user-provided feedback. The major advantage of the content features, as well as the relevance feedback, was users' feedback interpretation and independence. In 2017, Alsmadi [28] developed GA as well as great deluge algorithm for effective CBIR and here the retrieval of images was accomplished using MA algorithm. The retrieval efficiency of QI with limited features was extracted and stored in the feature repository. The outcomes of the extensive experiments exhibited a higher precision–recall value while compared to the extant CBIR systems. In 2017, Amira et al. [29] proposed a novel CNN-based learning model in CBIR, in which two parallel CNNs were utilized with the aim of extracting the image features from semantic metadata of images in the convolutional layers. The local feature correlations were modeled in the image via the extracted features. The pre-trained deep CNN models were utilized for unsupervised fine-tuning of CBIR parameters.

In 2017, Giveki et al. [30] formulated a novel approach for CBIR by utilizing an image descriptor that had the capability of working in HOG, LTP, LBP, SIFT, and LDP. Moreover, the matches and the similarities between the images were obtained using LDP and SIFT. The efficiency of the pixel-based descriptors was enhanced by means of capturing the higher level of semantic segments or patches. In 2017, Fadaei et al. [31] developed LDRP as a novel LPD to represent the texture in CBIR and it was based on the gray-level difference of pixels. The multilevel coding was utilized for determining the difference between the referenced pixel and the adjacent pixel. The proposed model was compared with the existing models like LBP, LVP, LTP, and LTrP in terms of average precision and the resultant of the analysis exhibited superiority in the proposed model. In 2014, Yasmin et al. [32] introduced a novel CBIR system with EI pixels classification. Initially, the image decomposition, as well as the EI classification, was performed on the images to extract the features and the extracted features were grouped together via clustering. A comparative analysis was made between the proposed and the traditional models and the resultant of the analysis exhibited an enhancement in the performance of the proposed model in terms of precision and recall.

In 2017, Srivastava and Khare [33] projected multiresolution analysis framework in CBIR for image retrieval and WT was utilized for exploiting the multiresolution analysis. Moreover, LBP of the image was combined at multiple resolutions along with the Legendre moments for performing the task of wavelet decomposition in the image. The texture features were determined from the image by using the LBP codes of DWT coefficients. In addition, the feature vectors of LBP codes were constructed using the Legendre moments. In 2017, Tang et al. [34] developed a content-based SAR image retrieval method with the intention of determining both initial as well as later refined outcomes of SAR. The similarity that occurred between SAR images were extracted using the RFM. Initially, at the superpixel level, the SAR image patches were segmented into brightness-texture regions in order to diminish the negative influence of speckle noise. Then, the issues associated with multiscale property were resolved in SAR images using the multiscale edge detector. The original ranked list was refined using MRF scheme.

In 2017, Fadaei et al. [35] projected a uniform partitioning scheme for CBIR in HSV domain using both textures as well as color feature sets in order to extract DCD features. During the image translation, the curvelet features and the wavelet features were defined to ignore the noise. Then, the PSO algorithm was employed to combine both the texture features as well as the color features optimally. In 2016, Mohamadzadeh and Farsi [36] formulated an innovative CBIR technique to interpret the image concept and not the manual texture with vision features via the sparse representation The features corresponding to IDWT were extracted from the image using the HSI color spaces as well as CIE-L*a*b* color spaces. The images were retrieved using PCA as well as DCT. In 2016, de Ves et al. [37] proposed an improved PCA in order to deal with the high dimension corresponding to the low-level feature vector. The low-level feature vectors in the image were utilized for storing the visual content and the semantic gap existing between the low-level features were bridged using LL-LM. Moreover, the dimension of the feature vector was diminished by utilizing PCA and in PCA; the non-overlapped groups were adjusted via dynamic local logistic regression models.

In 2013, Mukhopadhyay et al. [38] proposed CMR technique for CBIR in order to overcome the limitations of the conventional classifier-based retrieval approaches as well as the conventional distance approaches. The statistical texture features were extracted from the image using WT. Initially, in the neural network, the fuzzy class membership and the class label of QI were computed. Then, in the complete search space, the retrieval operation was performed by utilizing the simple and weighted distance metric. In 2015, Dash et al. [39] proposed an innovative CBIR approach referred to as class and CM-CCR. They from the query image, and the fuzzy output class memberships were gathered by utilizing the artificial neural network. The degree of classification confidence was recorded on the basis of the threshold values by using second label classifier and extensive experiments were conducted to securitize the threshold values.

2.2 *Chronological Review*

The chronological review and the percentage of contribution on each of the concerned technique in peculiar years are shown in Fig. 1. In the year 2018–2019, the overall contribution of CBIR technique in this paper is 40% and there is 48% of contribution in the year 2017–2016 and the percentage contribution of CBIR in the year 2015–2013 is 12%, respectively.

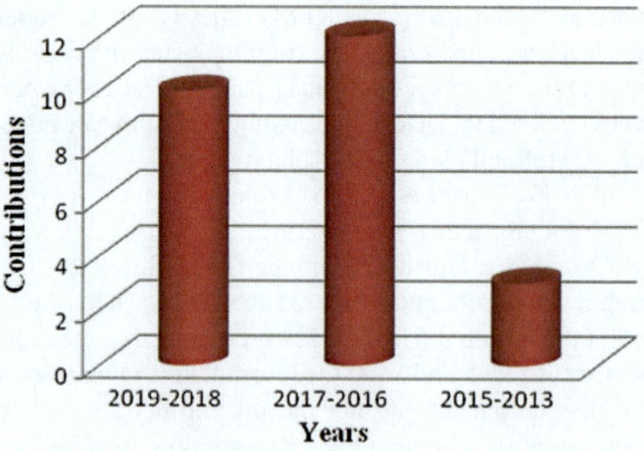

Fig. 1 Bar chart representation of chronological review

3 Analytical Review on Content-Based Image Retrieval System

3.1 Performance Evaluation on Different Research Work

Table 1 represents the best values recorded in each of the research work. The highest precision is achieved in [20] is 95% and the highest value of mean average precision is 94.7% in [30]. The count of the feature dimension is 24 in [15] and the lowest computational time is achieved as 0.16 s in [32]. The cosine distance score and the average memory of single image have the best value as 2.8 and 0.06 Kb in [16, 29], respectively. The best value of Euclidean distance score is found to be in [16] and the corresponding value is 2.75. The highest accuracy and f-score are available in [19, 23] with values as 61.7% and 0.71, respectively. Moreover, the rank diffusion, average recall, and average time of image vectorization have their best values in [25, 28, 29] with 86.17%, 0.146, and 221 ms, respectively. The feature vector length and feature computation time have their best values as 20 and 0.007 s in [38, 39]. The best value of 33 ms and 0.0081 s is recorded as average time of query search and

Table 1 Best performance evaluation of various techniques of CBIR

S. No.	Performance measures	Best value	Citation
1	Precision	95%	[20]
2	Mean average precision	94.7	[20]
3	Feature dimension	24	[15]
4	Computing time	0.16 s	[32]
5	Cosine distance score	2.8	[16]
6	Average memory of single image	0.06 Kb	[29]
7	Euclidean distance score	2.75	[16]
8	Accuracy	61.7%	[19]
9	f-score	0.71	[23]
10	Rank diffusion	86.17%	[25]
11	Average recall	0.146	[28]
12	Average time of image vectorization	221 ms	[29]
13	Feature vector length	20	[38]
14	Feature computation time	0.007 s	[39]
15	The average time of query search	33 ms	[29]
16	Feature search time	0.0081 s	[39]
17	Computational complexity	$0\,(N\ \log N)$	[38]
18	Feature extraction time	0.007 s	[38]
19	Searching time for features	0.0123 s	[38]
20	Recall	79%	[32]

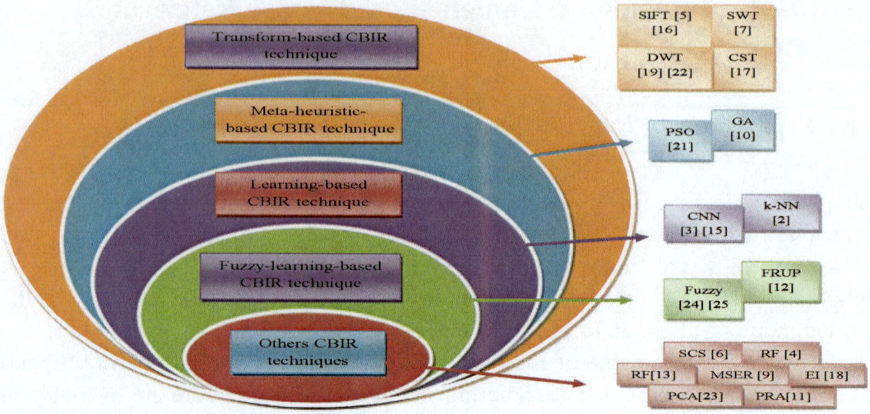

Fig. 2 Bar chart representation of chronological review

feature search time in [29, 39], respectively. The computational complexity achieves its best value 0 ($N \log N$) in [38], the best feature extraction time is 0.007 s in [38], the best searching time for features is 0.0123 s in [38], and the best recall percentage is 79%.

3.2 Algorithmic Analysis

Figure 2 represents the different techniques utilized in different approaches to CBIR reviewed in this paper. On the basis of different algorithmic models, they are categorized into transform-based CBIR technique, metaheuristic-based CBIR technique, learning-based CBIR technique, fuzzy-learning-based CBIR technique, and other CBIR techniques. The transform-based CBIR technique encloses SIFT algorithm in [19, 30], SWT in [21, 22], CST algorithm in [31], and DWT in [33, 36]. The GA in [24] and PSO in [35] fall under the category of metaheuristic-based CBIR technique. Moreover, k-NN algorithm in [16] and CNN in [17] as well as [29] belongs to learning-based CBIR technique. The fuzzy-learning-based CBIR technique comprises of FRUP in [26] and fuzzy-based retrieval in [38, 39]. The topic modeling algorithm in [15] RF in [18], SCS algorithm in [20], MSER algorithm in [23], page rank algorithm in [25], RF in [27], mimetic algorithm in [28], EI pixels classification in [32], and PCA [37] belongs to other CBIR techniques.

4 Research Gaps and Challenges

In recent years, huge research is being conducted in the field of image retrieval to make the retrieval of images as efficient as possible. But, till now no perfect and efficient algorithm, since it is complex to retrieve the images from huge and varied collection of database. In CBIR, it is not easier to search for the spatial relationship of QI on the basis of the spatial relationship between QI and the vector feature image. It is not feasible to segment the regions of the image in a few applications and the selection of the features in CBIR is also a complex task. The semantic gap refers to the divergence taking place in between the user's query and the outcome image of the retrieval system. More algorithms were developed to reduce this semantic gap between the images. RF algorithm has the ability to accept or reject the images on the basis of the similarity between the output images [40]. The statistical-based RF methods using delta mean algorithm have the ability to determine the most relevant features. But, the small size features cannot be calculated to find the exact variance between the datasets. Further, the standard deviation and variance-based RF methods have the capability of exhibiting the specific features of the image and it is impossible to accurately predict the inverse proportional between the relevant image set. QPM-based RF methods calculate the perfect query point for retrieval of the ideal relevant images. The major drawback of this mode is, QPM is unable to make use of the irrelevant samples efficiently when there exist no unimodal images [40]. In kernel-based RF methods using Bayesian framework, the pros of this approach lie in the efficient user interaction during image retrieval and the cons of this approach are its individual approach toward the extraction of color, texture, and shape features. Support vector machine has the capability of providing better pattern identification for image retrieval and the major shortcoming of this model is its sensitivity to smaller data sizes [40]. Biased discriminant analysis is efficient in computing the linear transformation between the positive images as well as the scattered negative image. The major flaw of this model is its Gaussian distribution of the image for the relevant dataset. The major advantage of conventional color histogram is its low-computational complexity and its simplicity in usage. Still, the major limitation of this model is its sensitivity to noise and its low encoding efficiency to handle rotation and translation global spatial information [41]. In the images, the issues related to the translation and rotation are solved by invariant color histograms and it is invariant under any geometry of the surface is its limitation. On the basis of the fuzzy-set relationship function, the degree of encoding similarity between the pixel colors is accomplished in FCH. Here, the global color properties can be described in the image only when there is equal dimensionality with FCH features [41]. Moreover, a higher retrieval result with spatial information is achieved in color coherence vector. Apart from this, the feature space has low dimensionality and higher computations. A lower-computational complexity is achieved with color correlogram, and it suffers from the problem of lower discrimination power. CBIR-based color–shape approach encodes both the shape of the object as well as the size. However, this mode is more sensitive to noise and contrast. CBIR-based color and texture feature have higher retrieval accuracy as the advantage and the insufficient feature set are its major shortcoming. Moreover, CBIR using color, shape, and texture features has the advantage of robust feature set and its drawback lies in the higher semantic gap. The semantic gap is diminished in the CBIR using RF and feature selection and the major disadvantage is high time complexity [41]. This clearly exhibits that during the transformation from the existing CBIR to intelligent CBIR system, there is few advantages as well as shortcomings. Thus, to design an efficient CBIR, it is necessary to hold the upcoming key points in mind. (a) The needs, as well as information-seeking behavior of the image users, need to be identified, (b) suitable approaches for extracting the features from raw images, (c) optimal selection of the features from images to diminish the storage space, and (d) the similarity between the images need to be identified in addition to reduce the semantic gaps between the images. Finally, it is determined that much scope is there in the future to have more researches on CBIR to make the image retrieval as efficient and accurate as possible.

5 Conclusion

This paper has provided a descriptive review of different CBIR techniques and had exhibited the advantages as well as the disadvantages of each of the technique. These CBIR techniques were categorized into transform-based CBIR technique, metaheuristic-based CBIR technique, learning-based CBIR technique, fuzzy-learning-based CBIR technique, and other CBIR techniques in order to grab a clear knowledge of different CBIR techniques. In conclusion, this research work has reviewed 25 research papers concerning different CBIR techniques and has portrayed the benefits and the drawbacks of each technique. The gathered CBIR techniques were grouped under different techniques to understand these approaches in a better way. The performance analysis of each of the approach is collected and the best performance is highlighted in the table. Finally, to have further researches on this area, the research gaps and the challenges of the various existing techniques were portrayed.

References

1. Liu D, Hua KA, Vu K, Yu N (2009) Fast query point movement techniques for large CBIR systems. IEEE Trans Knowl Data Eng 21(5):729–743
2. Feng Y, Ren J, Jiang J (2011) Generic framework for content-based stereo image/video retrieval. Electron Lett 47(2):97–98
3. Lai C, Chen Y (2011) A user-oriented image retrieval system based on interactive genetic algorithm. IEEE Trans Instrum Meas 60(10):3318–3325
4. Iakovidis DK, Pelekis N, Kotsifakos EE, Kopanakis I, Karanikas H, Theodoridis Y (2009) A pattern similarity scheme for medical image retrieval. IEEE Trans Inf Technol Biomed 13(4):442–450

5. Su J, Huang W, Yu PS, Tseng VS (2011) Efficient relevance feedback for content-based image retrieval by mining user navigation patterns. IEEE Trans Knowl Data Eng 23(3):360–372
6. Murala S, Maheshwari RP, Balasubramanian R (2012) Local tetra patterns: a new feature descriptor for content-based image retrieval. IEEE Trans Image Process 21(5):2874–2886
7. Akakin HC, Gurcan MN (2012) Content-based microscopic image retrieval system for multi-image queries. IEEE Trans Inf Technol Biomed 16(4):758–769
8. Chen J, Su C, Grimson WEL, Liu J, Shiue D (2012) Object segmentation of database images by dual multiscale morphological reconstructions and retrieval applications. IEEE Trans Image Process 21(2):828–843
9. Quellec G, Lamard M, Cazuguel G, Cochener B, Roux C (2010) Adaptive nonseparable wavelet transform via lifting and its application to content-based image retrieval. IEEE Trans Image Process 19(1):25–35
10. Rahman MM, Antani SK, Thoma GR (2011) A learning-based similarity fusion and filtering approach for biomedical image retrieval using SVM classification and relevance feedback. IEEE Trans Inf Technol Biomed 15(4):640–646
11. Zhang J, Ye L (2009) Content based image retrieval using unclean positive examples. IEEE Trans Image Process 18(10):2370–2375
12. Zhang L, Wang L, Lin W (2012) Generalized biased discriminant analysis for content-based image retrieval. IEEE Trans Syst Man Cybern Part B (Cybern) 42(1):282–290
13. Chen R, Cao YF, Sun H (2011) Active sample-selecting and manifold learning-based relevance feedback method for synthetic aperture radar image retrieval. IET Radar Sonar Navig 5(2):118–127
14. Quellec G, Lamard M, Cazuguel G, Cochener B, Roux C (2012) Fast wavelet-based image characterization for highly adaptive image retrieval. IEEE Trans Image Process 21(4):1613–1623
15. Shamna P, Govindan VK, Abdul Nazeer KA (2019) Content based medical image retrieval using topic and location model. J Biomed Inform 91:103112
16. Mezzoudj S, Behloul A, Seghir R, Saadna Y (2019) A parallel content-based image retrieval system using spark and tachyon frameworks. J King Saud Univ Comput Inf Sci (In press, available online)
17. Tzelepi M, Tefas A (2018) Deep convolutional learning for content based image retrieval. Neurocomputing 275:2467–2478
18. Raza A, Dawood H, Dawood H, Shabbir S, Mehboob R, Banjar A (2018) Correlated primary visual texton histogram features for content base image retrieval. IEEE Access 6:46595–46616
19. Dai OE, Demir B, Sankur B, Bruzzone L (2018) A novel system for content-based retrieval of single and multi-label high-dimensional remote sensing images. IEEE J Sel Top Appl Earth Obs Remote Sens 11(7):2473–2490
20. Shamna P, Govindan VK, Abdul Nazeer KA (2018) Content-based medical image retrieval by spatial matching of visual words. J King Saud Univ Comput Inf Sci
21. Mistry Y, Ingole DT, Ingole MD (2018) Content based image retrieval using hybrid features and various distance metric. J Electr Syst Inf Technol 5(3):874–888
22. Jin C, Jin S-W (2018) Content-based image retrieval model based on cost sensitive learning. J Vis Commun Image Represent 55:720–728
23. Unar S, Wang X, Zhang C (2018) Visual and textual information fusion using Kernel method for content based image retrieval. Inf Fusion 44:176–187
24. Alsmadi MK (2018) Query-sensitive similarity measure for content-based image retrieval using meta-heuristic algorithm. J King Saud Univ Comput Inf Sci 30(3):373–381
25. Pedronette DCG, Torres RS (2017) Unsupervised rank diffusion for content-based image retrieval. Neurocomputing 260:478–489
26. Islam SM, Banerjee M, Bhattacharyya S, Chakraborty S (2017) Content-based image retrieval based on multiple extended fuzzy-rough framework. Appl Soft Comput 57:102–117
27. Zhu Y, Jiang J, Han W, Ding Y, Tian Q (2017) Interpretation of users' feedback via swarmed particles for content-based image retrieval. Inf Sci 375:246–257

28. Mutasem K (2017) Alsmadi: an efficient similarity measure for content based image retrieval using memetic algorithm. Egypt J Basic Appl Sci 4(2):112–122
29. Alzu'bi A, Amira A, Ramzan N (2017) Content-based image retrieval with compact deep convolutional features. Neurocomputing 249:95–105
30. Giveki D, Soltanshahi MA, Montazer GA (2017) A new image feature descriptor for content based image retrieval using scale invariant feature transform and local derivative pattern. Optik 131:242–254
31. Fadaei S, Amirfattahi R, Ahmadzadeh MR (2017) Local derivative radial patterns: A new texture descriptor for content-based image retrieval. Sig Process 137:274–286
32. Yasmin M, Sharif M, Irum I, Mohsin S (2014) An efficient content based image retrieval using EI classification and color features. J Appl Res Technol 12(5):877–885
33. Srivastava P, Khare A (2017) Integration of wavelet transform, Local Binary Patterns and moments for content-based image retrieval. J Vis Commun Image Represent 42:78–103
34. Tang X, Jiao L, Emery WJ (2017) SAR image content retrieval based on fuzzy similarity and relevance feedback. IEEE J Sel Top Appl Earth Obs Remote Sens 10(5):1824–1842
35. Fadaei S, Amirfattahi R, Ahmadzadeh MR (2017) New content-based image retrieval system based on optimised integration of DCD, wavelet and curvelet features. IET Image Proc 11(2):89–98
36. Mohamadzadeh S, Farsi H (2016) Content-based image retrieval system via sparse representation. IET Comput Vision 10(1):95–102
37. de Ves E, Benavent X, Coma I, Ayala G (2016) A novel dynamic multi-model relevance feedback procedure for content-based image retrieval. Neurocomputing 208:99–107
38. Mukhopadhyay S, Dash JK, Gupta RD (2013) Content-based texture image retrieval using fuzzy class membership. Pattern Recogn Lett 34(6):646–654
39. Dash JK, Mukhopadhyay S, Gupta RD (2015) Content-based image retrieval using fuzzy class membership and rules based on classifier confidence. IET Image Proc 9(9):836–848
40. Shubhankar Reddy K, Sreedhar K (2016) Image retrieval techniques: a survey. Int J Electron Commun Eng 9(1):19–27
41. Wadhai SA, Kawathekar SS (2017) Techniques of content based image retrieval: a review. IOSR J Comput Eng (IOSR-JCE) 75–79

A Comparative Study of Classification Techniques in Context of Microblogs Posted During Natural Disaster

Harshadkumar Prajapati, Hitesh Raval, and Hardik Joshi

Abstract Microblogs generally contribute to statements that are made in public by the users. Tweets made on the Twitter platform fall in the microblog category. Microblogging sites have become important source in disaster events. In this paper, we compare various algorithms to study the effectiveness of retrieval and classification of tweets that were collected during disasters. The overall goal is to identify the relief work and retrieve it efficiently from the microblog tweets. The evaluation metrics that were used are precision, recall and F-score. We have observed that support vector machine (SVM) has the highest accuracy in classification of tweets based on pre-defined retrieval criteria.

Keywords Tweets · Information retrieval · Classification · Microblog · Natural disasters

1 Introduction

Information retrieval (IR) is gaining importance due to penetration of World Wide Web. Moreover, the mobile devices have enabled a lot of people to place their footprints over the Internet. This has resulted into a lot of data generation. Searching material of an unstructured nature which satisfies the information need from within large collections is the need in current scenarios. User generated content in microblogging sites like Twitter [1] is known to be an important source of real-time information

H. Prajapati (✉)
Faculty of Computer Science, Sankalchand Patel University, Visnagar, Gujarat, India
e-mail: harshadnprajapati@gmail.com

H. Raval
Shri C. J Patel College of Computer Studies, Sankalchand Patel University, Visnagar, Gujarat, India

H. Joshi
Department of Computer Science, Gujarat University, Ahmedabad, Gujarat, India
e-mail: hardikjjoshi@gmail.com

K. Kotecha et al. (eds.), *Data Science and Intelligent Applications*, Lecture Notes on Data Engineering and Communications Technologies 52,
https://doi.org/10.1007/978-981-15-4474-3_8

on various natural disaster events like floods, earthquakes and man-made disaster terrorist attacks, etc., that were posted during disasters. Users post information like personal views and situational information related to disasters.

Situational information helps the disaster authorities to gain a high-level understanding of the situation during disaster and plan relief efforts accordingly. During disaster of Nepal earthquake in April 2015, approximately 50,000 microblogs were collected by the FIRE community [2]. There were challenges similar to information retrieval from microblogs during disasters (IRMiDis) [3] known as SMERP [4] where the tweets during earthquake of Italy were collected and rolled out as a student challenge to retrieve the matching tweets using IR systems. In this paper, we have focused on the tweets provided by FIRE community and have evaluated few techniques for the Nepal earthquake tweets.

2 The Microblog Retrieval Experiment

The Text Retrieval Conference—Microblog Track [5]—supports for the microblog retrieval evaluation, and it is one of the pioneers among IR community to present the standard evaluation of information retrieval systems. Microblogging sites like Twitter are increasingly used for supporting relief operations during disaster events. IRMiDis track in FIRE provides dataset for various experiments. To help disaster work, annotators are given tweets to classify into various categories like need tweets, availability tweets, fact-checkable and non-fact-checkable. The researchers participate and submit their evaluation results known as runs.

2.1 Test Collection for Evaluation

Test collections are required to evaluate the IR systems. A good collection of tweets with relevance judgements are required to build an IR system and evaluate different techniques on the collection. The disaster management agencies can contribute to build such collections since it may be helpful to them for efficient relief operations. The collection for this research was provided by FIRE [2]. The team who created test collection consulted members of two different non-governmental organizations (NGOs) who regularly engage themselves in post-disaster work like relief operations. They identified typical information needs during a disaster for relief operation [6]. The challenge was to identify information needs related to seven different topics that were critical in nature from the collection of microblogs that can be helpful in efficient running of the relief operations. Table 1 states in the format of Text Retrieval Conference topics. Each topic contained an identifier "(num)" that denotes numeric value and three fields "(title, desc, narr)" for title, description and narration, describing the types of document (microblogs) to be deemed relevant to the topic [7]. These identified follow the conventions as used in most of the TREC forums [8].

Table 1 Sample queries (topics)

<num> Number: 1 <title> What resources were available <desc> Identify the messages which describe the availability of some resources <narr> A relevant message must mention the availability of some resource like food, drinking water, shelter, clothes, blankets, blood, human resources like volunteers, resources to build or support infrastructure, like tents, water filter, etc
<num> Number: T2 <title> What resources were required <desc> Identify the messages which describe the requirement or need of some resources <narr> A relevant message must mention the requirement or need of some resource like food, drinking water, shelter, clothes, blankets, blood, human resources like volunteers, resources to build or support infrastructure, like tents, water filter, etc
<num> Number: T3 <titles> What medical resources were available <desc> Identify the messages which give information about availability of some medical resource like medicines, medical equipment, blood, supplementary food items (e.g., milk for infants) human resources like doctor etc
<num> Number: T4 <title> What medical resources were required <desc> Identify the messages which give information about requirement of some medicine or other medical resources <narr> A relevant message must mention the requirement of some medical resources like medicines, medical equipments, supplementary foods, blood, human resources like doctor/staff and resources to build or support medical infrastructure like tents. water filter, power supply etc
<num> Number: T5 <title> What were the requirements/availability of resources at specific locations <desc> Identify the messages which describe the requirement or availability of resources at some particular geographical location. <narr> A relevant message must mention both the requirement or availability of some resource, (e.g., human resources like volunteers/medical staff, food, water, shelter, medical resources, tents, power supply) as well as a particular geographical location. Messages containing only the requirement/availability of some resource, without mentioning a geographical location would not he relevant.
<num> Number: T6 <title> What were the activities of various NGOs/Government organisations <desc> Identify the messages which describe on-ground activities of different NGOs and Government organizations <narr> A relevant message must contain information about relief-related activities of different NGOs and Government organisations in rescue and relief operation. Messages that contain information about the volunteers visiting different geographical locations would also be relevant. However, messages that do nol contain the name of any NGO/Government organisation Would not be relevant.
<num> Number: T7 <title> What infrastructure damage and restoration were being reported <desc> Identify the messages which contain information related to infrastructure damage or restoration <narr> A relevant message must mention the damage or restoration of some specific infrastructure resources, such as structures (e.g.. dams, houses, mobile tower), communication infrastructure (e.g., roads, runways, railway), electricity, mobile or Internet connectivity, etc. Generalised statements without reference to infrastructure resources would not be relevant

2.2 *Tweet Dataset*

The microblog dataset provided by FIRE community was a collection of tweets of the Nepal earthquake that happened on 25 April 2015. It was noticed that many users on Twitter started sharing the tweets during the earthquake and post-earthquake. The tweets were about the happenings during the earthquake and relief operations. However, many tweets had the same information since they were being retweeted [7]. While building the microblog collection, the most important part involved was removal of duplicate tweets, since they can lead to overestimation of the IR system performances and may also create more overhead for the human annotators in the form of information overload [7].

3 Microblog Retrieval and Ranking

Our approach for retrieval of relevant tweets and classifying them as per the requirement of FIRE community followed the following four steps. Each step was performed in the following sequence:

- Pre-processing of microblogs/tweets
- Applying expansion of tweets using word embeddings
- Indexing and retrieval of the tweets
- Classification of tweets using machine learning algorithms.

3.1 *Pre-processing*

Since the tweets contained a lot of noise in the form of punctuation marks, URLs, special symbols, emoticons, etc., we applied a rule-based pre-processor to remove unwanted symbols and stopwords. Later, the pre-processing phase also involved applying rule-based stemmer using standard Porter stemmer to stem the tweets.

3.2 *Applying Word Embeddings*

Word embedding is a technique that can help identify the related words for any given term. Implementations like Word2Vec and GloVe are widely used. The Word2Vec model generates a vector for each term in the corpus, and we can identify the relevant terms using vectors that fall in the bracket of cosine similarity. We used the Word2Vec technique for tweets. A model was trained using Word2Vec over the set of 10,000 tweets which were already pre-processed [9]. Continuous bag of words (cbow) model was used with the hierarchical softmax. A query for given retrieving tweets calculates the cosine relation among given query vector and each vector and ranks the tweets in lower levels.

3.3 Indexing and Retrieval

We used Indri system [7] for information retrieval. It implements different retrieval models [10], and the TF-IDF model had been used as the baseline. All the tweets from the collection were indexed using Indri, and later the query topics were used to retrieve and rank the tweets against the query topics.

4 Evaluation Metrics

In this section, we describe various evaluation measures for the given dataset that performs accuracy measurement. IR evaluation metrics are used to evaluate how well the IR system meets the information needs of its users. Different user interprets the same result differently by individual users. The metrics that we used in our experiments were precision (P), recall (R) and F-measure.

Precision is the most widely used measures. It measures the ability to retrieve top ranked documents that are mostly relevant. Precision is the proportion of the documents that were retrieved from the collection and are relevant to the information need [11]. Recall is one of the important measures. It searches to find all the relevant items in the corpus. In other words, recall is the proportion of the documents that are relevant to the query from the collection and were successfully retrieved [11]. F-score or F-measure is used in the statistical analysis of binary classification to measure the test's accuracy. F-measure is the weighted average of the recall and precision, and its score reaches its best value at 1 and worst at 0. Alternatively, we can say that it is the harmonic mean of precision and recall. We have also tested various algorithms against the accuracy score. Since there were tasks related to classification of microblogs into different sets, we got two distinct experimental results. The results obtained are discussed in next section.

5 Experimental Results

In this section, we discuss the performance of the IR methods for the proposed supervised approach. To evaluate the performance measurement, precision, recall and F1-score were used. The macro average is the simply the harmonic mean of these two figures. It used to know how the system performs overall across the sets of data. Micro average can be useful, and the dataset varies in size. For the given

Table 2 Comparison of the all the algorithms in context of accuracy measurement precision, recall and F1

Algorithm	Precision		Recall		F1	
	Macro	Micro	Macro	Micro	Macro	Micro
AdaBoost	0.66	0.70	0.55	0.59	0.60	0.64
Decision tree	0.61	0.64	0.57	0.60	0.59	0.62
Naive_Bayes	0.47	0.50	0.46	0.53	0.46	0.51
SVM_Linear	0.73	0.75	0.60	0.66	0.65	0.59
Random forest	0.76	0.78	0.37	0.42	0.47	0.55

Table 3 Accuracy score of the machine learning techniques

Algorithm	Accuracy score
AdaBoost	0.40
Decision tree	0.48
Naive_Bayes	0.24
SVM_Linear	0.49
Random forest	0.37

task, various machine learning techniques like Naive_Bayes, support vector machine (SVM), decision tree, random forest and AdaBoost were used. A summary of our experiments performed using different algorithms and their results are shown in Table 2.

We have observed that in the classification task, random forest algorithm outperforms the rest of the techniques for precision, while the SVM technique outperforms in recall values. However, SVM performs reasonably well in terms of both precision and recall. The SVM linear model gives better performance rather than using other models. The Naive_Bayes algorithm fails in terms of precision and recall values. We had also evaluated the accuracy of the above listed techniques which is shown in Table 3. As depicted in Table 3, techniques like SVM and decision tree perform equally well in terms of accuracy scores. The Naive_Bayes technique fails to obtain better accuracy scores.

6 Conclusion

We performed two different experiments for the microblog retrieval task. We have observed that the SVM technique with the linear model performs reasonably well in the classification task, and we have also obtained a good accuracy score using this technique. Few other SVM models did not perform as expected, and we have not included in our results. The other good performers are random forest technique and decision tree approach. Random forest performs well in precision values, while the overall accuracy score obtained using decision tree approach is appreciable. In future work, we aim to analyse various pre-processing techniques effectively and apply more features to obtain better accuracy. We wish to propose additional accuracy measures like time stamping and build a system that works in real-time mode.

Acknowledgements We would like to extend a token of gratitude to Dr. Vikram Kaushik for being the mentor and giving continuous guidance, support and motivation in this research work. We also acknowledge organizers of Microblog Track of FIRE-2016 to provide the dataset for research. We would also like to acknowledge the Ministry of Information Technology (MeitY), New Delhi, for partial funding towards this project and Sankalchand Patel University, Visnagar, for providing infrastructure resources.

References

1. Twitter Homepage. https://twitter.com/home. Last Accessed on 2019 Dec 5
2. FIRE Microblog Track. https://sites.google.com/site/fire2016microblogtrack. Last Accessed 2019 Dec 5
3. IRMiDis homepage, Information Retrieval from Microblogs during Disasters (IRMiDis). https://sites.google.com/site/irmidisfire2017/. Last Accessed on 2019 Dec 5
4. TREC Homepage. https://trec.nist.gov/. Last Accessed on 2019 Dec 5
5. Lin J, Efron M, Wang Y, Sherman G, Voorhees E (2018) Overview of the TREC-2015 microblog track
6. Basu M, Roy A, Ghosh K, Bandyopadhyay S, Ghosh S (2017) Micro-blog retrieval in a disaster situation: a new test collection for evaluation. In: SMERP, ECIR
7. Strohman T, Metzler D, Turtle H, Croft WB (2004) Indri: a language model-based search engine for complex queries. In: Proceedings ICIA
8. Ghosh S et al (2017) First international workshop on exploitation of social media for emergency relief and preparedness (SMERP). In: Advances in information retrieval, ECIR 2017, vol 10193
9. Mikonos T, Yih W, Zweig G (2013) Linguistic regularities in continuous space word representation. In: NAACL HLT
10. Ponte JM, Croft WB (1998) A language modeling approach to information retrieval. In: Proc. ACM SIGIR, pp 275–281
11. Baeza-Yates R, Ribeiro Neto B (2011) Modern Information Retrieval, 2nd edn. Pearson Publication

Feature Selection in Big Data: Trends and Challenges

Suman R. Tiwari and Kaushik K. Rana

Abstract Big data is a term used to represent data that is big in volume, speed, and variety. With inflammation, these characteristics are also inflated to 42 V's. We have focused our survey for feature selection in big data, as feature selection is one of the most used methods for dimensionality reduction techniques. Feature selection is used for elimination of irrelevant and redundant features from dataset to improve the classification performance. This paper includes big data characteristics, different feature selection method, and current research challenges of feature selection. We observed that swarm intelligence techniques are the most popular methods among researchers for feature selection in big data. Further, we conclude that gray wolf optimization and particle swarm optimization are the most preferred algorithms by researchers.

Keywords ABC · IG · CFS · CBF · GWO · ZB · YB

1 Introduction

Big data is used to represent large and complex datasets, which makes processing more difficult using traditional data processing applications [1]. It has been estimated that data is growing by 40% compound interest rate and may reach 45 Zetabyte in 2020 [2]. Big data has the potential to help organizations to improve their operations and helps to take faster and intelligent decisions [3].

Big data is categorized into three different types, named structured, unstructured, and semi-structured data. Structured data can be stored in a SQL database in forms of rows and columns. They have a pre-defined data structure and relational keys which

S. R. Tiwari (✉)
Computer Department, R.C. Technical Institute, Ahmedabad, Gujarat, India
e-mail: rsumantiwari@gmail.com

K. K. Rana
Vishwakarma Government Engineering College, Ahmedabad, Gujarat, India
e-mail: kkr@vgecg.ac.in

K. Kotecha et al. (eds.), *Data Science and Intelligent Applications*, Lecture Notes on Data Engineering and Communications Technologies 52,
https://doi.org/10.1007/978-981-15-4474-3_9

can be stored easily into tabular form, queried, and analyzed [4]. Customer, sales, or patients data can be considered as a structured data. The management of unstructured data is recognized as one of the major unsolved problems in the information technology (IT) industry; the main reason being that the tools and techniques that have proved so successful for structured data, and they does not work when it comes to unstructured data. Unstructured data means data is not in a structured format or they do not have any defined structure [4]. Hence, they do not fit into RDBMS for processing and analyzing, so special tools are needed to analyze them. PDF, Natural language text, Images or Video can be considered as unstructured data. Semi-structured data does not reside in a relational database but they have some organizational properties that make it easier to analyze. It can be stored in a relational database with some processing. XML files or call centers log can be considered as semi-structured data.

There are various characteristics that differentiate big data with simple data. Initially, in 2001, Doung Lanely has characterized big data by 3 V's known as big data spectrum [5–7], namely volume, velocity, and variety. As inflammation increases, big data characteristics also inflated and with time, new characteristics value is added in 2012–13, known as 4 V's of big data. In 2013–14, big data characteristic was increased to 7–8 V's with the inclusion of veracity and visualization which reached 42 V's of big data in 2017 [8]. Volume indicates a large scale or size of data. It can be measured in form of Terabyte, Petabyte, and Exabyte (10^18), ZB (10^21) and YB (10^24) and soon this will reach to range known as Padma (10^32). Hence, acquisition, curation, storage, processing, and visualization from the sea of data offer challenges to the current world. We have a data in huge amount; either we can throw it or we can dive inside the data and find hidden treasures from it, and we can use derived information for business organization, medical, science, and technology and for the betterment of human being. Velocity indicates the speed of incoming data. According to forbs, 2.5 quintillion bytes of data created each day. Over the last two years, 90% of the data in the world was generated [9]. Stream data which comes continuously from twitter, Google map, Facebook, etc., are an example of big data which comes with high velocity. Variety suggests the different formats of data. Big data comes from various sources, which provides the data in a different format. Data comes from homogeneous as well as heterogeneous resources, and this causes many problems due to the heterogeneity of data resources, data format, and infrastructure. Data from the sensor network, PDF, X-ray, Video, and Audio are in a different format and they need to be integrated properly for learning. The value indicates the impact and usefulness of derived information from big data. There is no meaning of storing large and complex dataset and process it when it does not return valuable information. Having a large amount of data is one thing, but how much it is useful is another thing. Veracity indicates the quality of data or data uncertainty. It indicates the trustworthiness of data to be used for classification, while visualization is used to represent the data in the understanding format. There are various tools that can be used for big data visualization, for example, Google chart, D3, Fusion chart, etc.

Big data is now rapidly expanding in all science and engineering domains in large volume. According to IDC, 23% of the information in the digital universe would be useful for big data if it were tagged and analyzed properly [2]. As only 23% of data found useful, big data may have noise with different degrees, heterogeneous, imbalanced or they may have low-value density, which makes feature selection difficult in ever-growing data; hence, there is a need for efficient feature selection method for big data [2]. Some of the issues related to the volume of big data are class imbalance, curse of dimensionality [5, 10, 11], feature engineering, nonlinearity, and processing performance [12]. Hence, it requires a new way of thinking for the processing of big data. We have focused our survey related to feature selection problems in big data. Section 2 describes the existing feature selection mechanism. Section 3 covers the research challenges in this area and Sect. 4 gives details about the research work done in this area, followed by conclusion.

2 Feature Selection Method

Feature selection is the most important data processing technique and used to delete redundant, irrelevant, and co-related features from a dataset. Random, noisy, and irrelevant features decrease efficiency of classifier and increase the complexity. The feature selection method can be used as a dimensionality reduction strategy. Feature selection maintains the original feature; hence, it is more useful when knowledge interpretation and knowledge extraction are important [12].

The following are different traditional methods used for feature selection.

1. Filter-based approach [10, 13–17]
2. Wrapper-based approach [10, 13, 14, 16–18]
3. Embedded-based approach [13, 14]
4. Swarm Intelligence-based approach [1, 6, 10, 17, 19–21].

2.1 Filter-Based Approach

The filter method measures the quality of features based on relevancy scores and Top 'X' features are selected manually for classification. Here, Top 'X' feature is different based on classification accuracy and the required number of features. Filter-based feature selection method is independent from a specific classifier hence it is faster than the wrapper method. As they are independent, they can be used for any learning algorithm. The main issue of using filter-based feature selection method is manual selection of top 'X' features and independently selecting the features without consideration of classifier performance; hence, the same features give a different result for the different classifiers. [5] has tried to implement a filter-based Feature selection approach based on data complexity measures. Data complexity is a recent proposal which represents data in terms of particularities that adds complexity to the classification task.

2.2 *Wrapper-Based Approach*

Wrapper-based feature selection method uses feedback from the classifier; hence, it is more accurate but slower than the filter-based method. Wrapper-based method evaluates the selected features with a machine learning algorithm and selects only if it meets the specified criteria, otherwise new feature subset is selected. Recursive feature elimination is an example of a wrapper-based method. As the wrapper-based feature selection method uses the loop, it is a time-consuming mechanism for a big dataset.

2.3 *Embedded Feature Selection Method*

Embedded feature selection method is part of the training or learning of classifiers. They embed feature selection strategy within a machine learning algorithm. This method has the advantage of filter and wrapper method, but they are specific to the machine learning algorithm. The decision tree is an example of an embedded feature selection method, which uses features entropy to split the node.

2.4 *Swarm Intelligence*

Swarm intelligence has been proved as a technique that can solve NP-hard problems [6]. Swarm-based feature selection returns a group of features, which satisfy the desired fitness function condition and used for classification or clustering. Swarm-based technique searches the features in search space based on search strategy like global or local search method.

Each swarm intelligence technique has some basic phases defined as below

- Initialize population and parameters
- Define stopping criteria
- Evaluate fitness function
- Update agent position
- Return global best solution.

There are different swarm-based feature optimization techniques like particle swarm optimization, gray wolf optimization, ant colony optimization, artificial bee colony optimization, grass-hopper optimization, etc.

3 Research Challenges

Feature selection has shown its effectiveness in many data mining applications, but the unique characteristics of big data presents challenges for feature selection [22]. When the data size increases beyond the storage capacity, then it is better to keep only important features related to the application and discard the unimportant features [5]. Existing feature selection does not perform well with big data and sometimes, they become inapplicable with large amount data [5]. The following section describes the area where we need to concentrate more to fill the gap in research for big data feature selection.

3.1 Decentralized Environment

Volume is the primary characteristic in big data, which gives great challenges to machine learning, because learning from a huge amount of data and finding relevant value are time-consuming and existing machine learning algorithms performance degrades as the size of data becomes larger. Big data comes in large volume so it is infeasible to move them at the central location due to security, legal issue, and cost factors. Machine learning algorithm's efficiency degrades and does not perform well when a portion of data is stored at a different location, because they work better on a centralized mechanism. Hence, there is a need of feature selection mechanism that can work on the decentralized environment.

3.2 Curse of Dimensionality

Irrelevant and redundant feature degrades the classifier performance and leads toward "The Curse of Dimensionality" [5, 11, 23] during classification. Curse of dimensionality is a problem where existing machine learning algorithm fails to work and their performance is degraded due to large dimension. Hence, we need to work on a mechanism to reduce the dimension of big data during classification and need more research on compression, and feature selection techniques.

3.3 Class Imbalance

Class imbalance is a problem where the total number of a class of data is far less than the total number of class of another data [15, 24], which degrades the performance of machine learning classifiers because ML algorithm works and learns better in the case where both class samples are roughly same. Class imbalance problem is common in case of big data as its volume is too large and always it is not possible that collected information has an equal share from all kinds of samples.

3.4 Feature Engineering

Feature engineering is the process that uses knowledge of data to construct variables [25], features that can be used to train a predictive model. According to lussa massaron, "The features you use influence more than everything else the result. No algorithm alone, to my knowledge, can supplement the information gain given by correct feature engineering" [26]. Finding appropriate features for machine learning is difficult, time consuming, and needs experts' knowledge in case of big data and can be considered as a big research issue because traditional available methods such as binning, log transformation, feature split, etc., are not able to give the desired result in big data.

The process of extracting features from raw dataset is known as feature engineering or representation learning. A poor data representation may reduce the performance of advanced, complex machine learner, while a good data representation can provide a high performance for relatively a simpler machine learning algorithm [27]. Thus, feature engineering is an important element of machine learning. For heterogeneous data, data integration, which combines data located at a different location in a different format and then applying representation learning is useful [3]. There are three subtopics of representation learning: feature selection, feature extraction, and distance metric learning, but to include multidomain learning ability, different representation learning algorithms like automatic, biased, cross-domain representation learning have been proposed by [27], although this area need concentration from researchers.

3.5 Stream Data

Velocity indicates that speed of incoming data. It is data which is generated continuously in huge amount. In a real-world application, which generates a large amount of data like credit card transaction, sensor network, stock market exchange, social media produce a tremendous amount of traffic.

In-stream data is generated continuously, and we need to perform online feature selection upon arrivals of new features [28]. Reference [29] has proposed an approach for the frequent pattern mining algorithm to improve the classification accuracy using a levy flight bat algorithm along with online feature selection, which is used to filter the low-quality features from the big data in an online manner. They have proposed entropy frequent pattern mining to reduce the computation time which is more in case of stream data. Reference [20] has proposed an innovative online feature selection method accelerated bat algorithm (ABA) and a framework for applications that streams the features in advance with indefinite knowledge of the feature space.

In the worst case, stream data size and features are unknown or may be infinite in size and coming continuously; thus, it is not a practical solution to wait to perform feature selection. Hence, it is more preferable to perform streaming feature selection which can find new features as per changes.

3.6 Structured Data

IDC estimated that unstructured data will be 95% of the global data in 2020 with an estimated 65% annual growth rate [20]. Industries estimate that 20% of the data is in structured form while the remaining 80% of data is in the semi-structured form [11], this is challenging for data processing. Traditional feature selection works on assumptions that data does not have any explicit correlations and ignores the implicit structure but in case of big data, available data is generally a combination of structured, unstructured, and semi-structured format. Data captured from human behavior or user-generated data like tweets and re-tweets has a complex internal co-relation with one another which makes feature selection complex and time consuming. Data can be in different structures like link data where data is linked with different links, group, tree, graph which requires special attention for big data feature selection because classification and clustering performance can be improved with the use of prior knowledge of the structure of data and feature selection can be applied more efficiently.

3.7 Heterogeneous Data

Heterogeneous data are any data with high variability of data types and with different formats [30]. These collected data are ambiguous and data with low quality which can not directly used for classification due to large number of missing values, high data redundancy, and untruthfulness [30]. Reference [24] has proposed one class-oriented feature selection and used mutual information and cluster space representation for classification of each kind of target class in turn. Reference [31] has proposed feature selection method to extract the most relevant features from heterogeneous urban open data. According to [32], no single selection algorithm is capable of optimal results for heterogeneous data in terms of predictive performance and stability, hence they thinks toward "ensemble" approaches involving the combination of different selectors. As we have stated earlier that big data may be in a combination of structured, unstructured, and semi-structured format as well as they arrive from different sources [15], hence merging heterogeneous dataset together and extract most relevant features to meet the business information demand is challenging issue for big data.

3.8 Data Cleansing

Data cleansing is one among all big issues in big data analysis. It is laborious to clean dirty data before it can be used for accurate data analysis as 50–80% of a data scientist's time is used in data cleansing [33]. Un-cleaned, low-quality data is not just time-consuming to work with but it is also uneconomic. Machine Learning is far more effective when it is provided sanitized data as per the trend micro survey. It is not good to process a large amount of data without getting any significant results. Hence we need to do research for proper feature selection methods which can be effective in case of missing and noisy data.

4 Related Work

To deal with the large volume of data distributed feature selection approach is discussed by researchers and seen as a new guideline to solve the big data feature selection. [5] has tried to overcome the problem of centralized mechanism and deciding threshold value to select optimum number of features of existing feature selection methods. They have portioned their dataset using horizontal and vertical portioning method and proposed a mechanism to select number of features dynamically using data complexity measures. They have decided threshold value for feature selection using D-F1, Firsher discriminant ratio, D-F2 length of overlapping region and D-N2 nearest neighbor distance at runtime based on complexity of data. [13] has proposed fast-mRMR based on filter-based feature selection using mutual information. To overcome the limitation of mutual information, author have proposed greedy approach for feature selection. They have transformed original computation complexity of mutual information into linear order for big data. Reference [15] has proposed feature selection method for microarray data. They have implemented traditional feature selection method in distributed manner and selected top- ranked features from all distributed nodes.

Number of authors has worked on swarm intelligence technique for feature selection. Sudhakar Ilango et al. [1] proposed the Artificial Bee colony based clustering approach for feature selection, which combines the local search method carried out by employed and onlooker bees and global search methods by onlookers and scouts. They have designed map/reduce programming for ABC mechanisms and implemented in a single node and multi-node Hadoop environment. They found that proposed ABC is more efficient than PSO and DE. Tripathi et al. [22] has proposed a novel Gray wolf optimizer for feature selection using map-reduce for big data. They have introduced two suggestions in gray wolf optimizer namely levy flight and binomial crossover to improve the exploration and exploitation capabilities. They have implemented their proposed work on Hadoop with map/reduce programming for the distributed node but suggested to use spark for a large database to reduce the computation time. Faris et al. [34] has analyzed recent variants and applications of GWO which can be used for feature selection. They have reviewed that there is a positive future of GWO in Machine learning and feature selection algorithm. They have given various research directions for gray wolf optimization which is still needed to be explored. Emary et al. [17] has proposed Multi-objective gray wolf optimization for attribute reduction. They have used a swarm-based optimization method which can optimize the feature set with minimum redundancy and keeps classification performance better. Gupta et al. [11] have proposed scale-free binary particle swarm optimization for feature selection in big data. They have used multi-class SVM for the classification and found the proposed method is better than existing particle swarm optimization. Hodge et al. [35] proposed a Hadoop neural network for parallel and distributed feature selection methods. They have implemented five different feature selection algorithm constructed using an artificial neural network framework embedded in Hadoop YARN. They have identified similarities among the feature selection algorithm and implemented parallel in the Hadoop framework, which allows the best feature selector and the actual feature to be selected from a large and high dimensional data set.

Zhang et al. [10] proposed multi-objective PSO for a cost-based feature selection problem, which uses probability-based encoding technology and an effective hybrid operator, with the idea of crowding distance, the external archive, and the Pareto domination relationship. The proposed method is compared against another multi-objective feature selection algorithm on five benchmark datasets where the proposed method is more effective for solving cost-based feature selection problems. Emary et al. [14] has proposed feature subset selection approach by gray wolf optimization with K-nearest neighbor using forward feature selection method to select the optimum feature set, which gives them better results in terms of feature reduction and classification accuracy as compared with PSO and GA. They have evaluated their proposed work in four different scenarios where proposed GWO is robust against the initialization of parameters as compared to PSO and GA. Emary et al. [36] has proposed Binary Gray wolf optimization approaches for feature selection using two different methods. In the first approach individual steps toward the first three best solutions are binarized and then stochastic crossover is performed among the three basic moves to find the updated binary gray wolf optimization. In the second approach, a sigmoidal function is used to squash the continuously updated position and then threshold these values to find the updated gray wolf optimization position. Jundong Li et al. [12] has reviewed the various challenges of feature selection for big data analytics. They said that feature selection is effective in many application but big data features may be structured, unstructured features with linked data or multi-view or multi-source data. To tackle the challenges of feature selection for big data they proposed an open-source feature selection repository called scikit-feature. Qiu et al. [3] has done a detailed survey of machine learning for big data processing for distributed and parallel, transfer and active learning. They have reviewed various critical issues of big data such as learning for large scale, different types, high speed, uncertain and incomplete and low-value density data. Fong et al. [37] has proposed accelerated PSO swarm search feature selection for mining of stream data on the fly. An incremental approach is used for various classification algorithms, which gives them higher accuracy as compared to the traditional approach. Weng et al. [23] has reviewed various dimension reduction strategies that can be used for big data for volume reduction. According to authors PCA and sufficient dimension reduction are being increasingly used in the era of big data. Gu et al. [20] has used competitive swarm optimizer for feature selection in high dimensional classification. They have grouped entire particle into two groups and compared both the groups. A group with the highest fitness goes for the next iteration (Table 1).

Feature selection and swarm intelligence techniques are used as a dimensionality reduction technique. Traditional feature selection is used mostly as a dimensionality reduction technique in data mining, but sometimes they become inefficient and sometimes inapplicable in large volumes. Figure 1 shows that big data researchers are using swarm intelligence techniques more as compared to the feature selection method (Fig. 2).

Apart from this, the feature selection method suffers limitations, which makes them unpopular for big data feature selection. Firstly, for n number of features, total 2^n possible solution can exist, which makes traditional feature selection time consuming, complex, and in-appropriate in terms of process in case of large number of features, while swarm intelligence is proved as a method which can be used to solve NP-Hard problems, hence swarm intelligence is more preferable for big data feature selection [6]. Secondly, the traditional feature selection method needs to know the optimum number of features to be selected in advance [11, 21], Hence providing a threshold value for feature selection can be considered as a major drawback, where in case of swarm intelligence technique optimum set of features are selected based on fitness function. Third, performance of selected features differs with different classifiers, hence it is difficult to design any generalized model [11] because the feature selection method are custom formulated for specific classifier and optimizer [21], where the swarm intelligence is flexible in integrating with any classifier just like plug and play. Fourth, Feature selection method ranked the feature based on their usefulness like information gain and entropy, irrelevant features are deleted from bottom based on specified criteria or number of features, and due to non-linearity relations of the features and concept targets, sometimes removing an attribute which has littler co-relation with another features may lead toward performance degradation [21]. In the case of swarm intelligence, features are not ranked based on entropy or information gain; they search the features which is close to target value and selected based on their fitness.

Computational intelligence (CI) is a subfield of artificial intelligence, and swarm intelligence is another form of CI used to solve the optimization problem [34]. Swarm intelligence is considered as the fastest growing algorithm used for optimization as per our survey. Hence, in future, we will focus on swarm intelligence technique for feature selection method and do the detailed study of different swarm intelligence technique like particle swarm optimization, genetic algorithm, artificial bee colony, ant colony optimization, gray wolf optimization, grass-hopper optimization, etc. We will find recent changes done in swarm optimization technique for feature selection and develop our new swarm-based feature selection algorithm.

5 Conclusion

We have concentrated our research survey on current research challenges for feature selection in big data. We end up with reason why the swarm-based feature selection

Table 1 Summary of the literature survey

Article	Proposed algorithm	Feature selection technique	Research objective	Remarks
[14]	GWO based on Knn based on forward feature selection	Swarm intelligence	Used Gray wolf optimization with K-nearest neighbor using forward feature selection method to select the optimum feature set	The proposed GWO is robust against the initial initialization of parameters and performs better than PSO and GA
[36]	BGWO: They have proposed stochastic crossover and sigmoidal function in GWO	Swarm intelligence	Stochastic crossover and sigmoid function is used in GWO	Better than PSO and GA
[35]	Proposed a Hadoop neural network for parallel and distributed feature selection methods	Framework	Proposed a Hadoop neural network for parallel and distributed feature selection method and implemented five different feature selection algorithm constructed using ANN framework embedded in Hadoop YARN	We need to investigate Apache spark, in-memory data analytics, and cluster computing framework
[13]	Fast mRMR to overcome the problem of mRMR	Filter method	Fast mRMR Optimization, which overcomes the computation drawback of MI	Spark version of implementation performed better in a distributed environment, in case of a large dataset The computation of MI has been improved

(continued)

Table 1 (continued)

Article	Proposed algorithm	Feature selection technique	Research objective	Remarks
[5]	IG, CFS, CBF, Relief	Filter approach	Implemented filter feature selection approach and calculated threshold value for feature selection using D-F1 Fisher discriminant ratio, D-F2 Length of the overlapping region and D-N2 Nearest neighbor distance	Data complexity measure is used to calculate threshold in the existing filter-based feature selection method
[22]	GWO	Swarm intelligence	Introduced GWO with binomials crossover and levy flight for Feature selection and optimization	Enhanced GWO outperforms than K-means, PSO, GSA, BA Poor accuracy in case of distributed implementation
[1]	Artificial Bee colony	Swarm intelligence	Combines the local and global search for feature selection and designed map/reduce programming model for single and multimode environment	The proposed method performs better than PSO and DE in terms of time and efficiency Selected top 10 features manually for clustering
[15]	Information Gain, Relief	Filter approach	They have implemented a distributed feature selection method for microarray data	Successfully distributed feature selection process
[11]	PSO	Swarm intelligence	Implemented Binary scale-free topology controlled PSO for feature selection	The proposed algorithm gives better result than conventional PSO
[10]	PSO	Swarm intelligence	Multi-objective PSO for coast-based feature selection	They have combined the hybrid operator, crowding distance and Pareto dominance relations

(continued)

Table 1 (continued)

Article	Proposed algorithm	Feature selection technique	Research objective	Remarks
[17]	GWO	Swarm intelligence	Proposed multi-objective GWO for feature reduction	Performs better than PSO and GA
[20]	Canonical PSO	Swarm intelligence	Changed the calculation method of Pbest and Gbest. They found the group of swarm and group with the highest fitness is allowed to go for the next iteration	Performs better than PSO in feature selection and classification performance
[37]	PSO	Swarm intelligence	They have proposed accelerated PSO for mining stream data based on incremental learning algorithms	

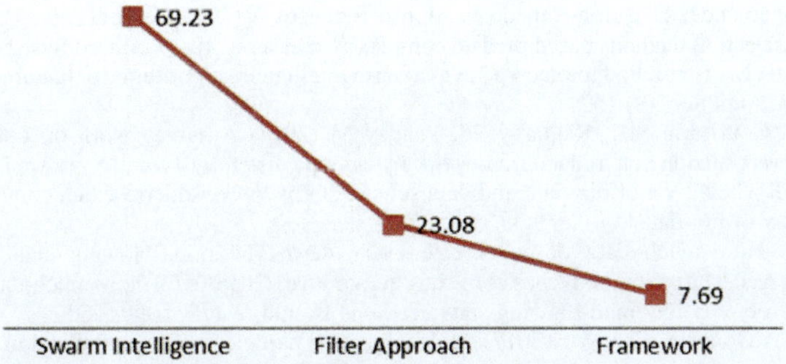

Fig. 1 The popularity of feature selection method

method is most popular method for finding relevant features as compared to traditional feature selection methods. According to survey, gray wolf optimization and particle swarm optimization are most used technique for feature selection. In the future, we will develop the swarm-based feature selection algorithm for big data to overcome the limitation of the traditional feature selection algorithm.

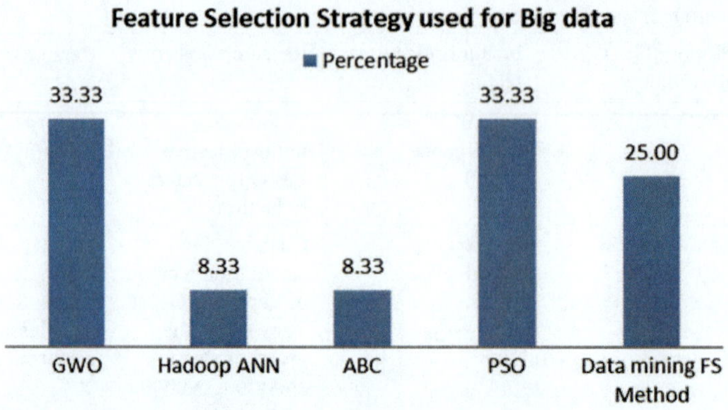

Fig. 2 Comparison among different approach used for feature selection

References

1. Sudhakar Ilango S, Vimal S, Kaliappan M, Subbulakshmi P (2018) Optimization using artificial bee colony based clustering approach for big data. Cluster Comput 1–9
2. Devi DR, Sasikala SJ (2019) J Big Data 103. https://doi.org/10.1186/s40537-019-0267-3
3. Qiu J, Wu Q, Ding G, Xu Y, Feng S (2016) A survey of machine learning for big data processing. EURASIP J Adv Sig Process 1687–6180 (2016)
4. Enterprise Big Data Framework. https://www.bigdataframework.org/data-types-structured-vs-unstructured-data/
5. Morán-Fernández L, Bolón-Canedo V, Alonso-Betanzos A (2017) Centralized vs distributed feature selection methods based on data complexity measures. Knowl Based Syst 117:27–45
6. Brezočnik L, Fister I, Podgorelec V (2018) Swarm intelligence algorithms for feature selection: a review. Appl Sci 8(9):1521
7. Jena B, Gourisaria MK, Rautaray SS, Pandey M (2017) A survey work on optimization techniques utilizing map reduce framework in hadoop cluster. Int J Intell Syst Appl 9(4):61
8. Shafer T. The 42 v's of big data and data science. https://www.elderresearch.com/company/blog/42-v-of-big-data
9. Mar B. How much data do we create every day? The mind-blowing stats everyone should read. https://www.forbes.com/sites/bernardmarr/2018/05/21/how-much-data-do-we-create-every-day-the-mind-blowing-stats-everyone-should-read/#24ee8aaa60ba
10. Zhang Y, Gong D, Cheng J (2017) Multi-objective particle swarm optimization approach for cost-based feature selection in classification. IEEE/ACM Trans Comput Biol Bioinform (TCBB) 14:64–75
11. Gupta SL, Baghel A, Iqbal A (2019) Big data classification using scale-free binary particle swarm optimization. In: Harmony search and nature inspired optimization algorithms. Springer, Singapore, pp 1177–1187
12. Li J, Liu H (2017) Challenges of feature selection for big data analytics. IEEE Intell Syst 32:9–15
13. Ramírez-Gallego S, Lastra I, Martínez-Rego D, Bolón-Canedo V, Benítez JM, Herrera F, Alonso-Betanzos A (2017) Fast-mRMR: fast minimum redundancy maximum relevance algorithm for high-dimensional big data. Int J Intell Syst 32:134–152
14. Emary E, Zawbaa HM, Grosan C, Hassenian AE (2015) Feature subset selection approach by gray-wolf optimization. In: Afro-European conference for industrial advancement. Springer, Cham, pp 1–13
15. Bolón-Canedo V, Sánchez-Maroño N, Alonso-Betanzos A (2015) Distributed feature selection: an application to microarray data classification. Appl Soft Comput 30:136–150

16. Kawamura A, Chakraborty B (2017) A hybrid approach for optimal feature subset selection with evolutionary algorithms. In: IEEE 8th international conference on awareness science and technology (iCAST), pp 564–568
17. Emary E, Yamany W, Hassanien AE, Snasel V (2015) Multi-objective gray-wolf optimization for attribute reduction. Procedia Comput Sci 65:623–632
18. Majdi MM, Mirjalili S (2017) Hybrid whale optimization algorithm with simulated annealing for feature selection. Neurocomputing 260:302–312
19. Mirjalili SZ, Mirjalili S, Saremi S, Faris H, Aljarah I (2018) Grasshopper optimization algorithm for multi-objective optimization problems. Appl Intell 48(4):805–820
20. Gu S, Cheng R, Jin Y (2018) Feature selection for high-dimensional classification using a competitive swarm optimizer. Soft Comput 22:811–822
21. Fong S, Yang X-S, Deb S (2013) Swarm search for feature selection in classification. In: 2013 IEEE 16th international conference on computational science and engineering, pp 902–909
22. Tripathi AK, Sharma K, Bala M (2018) A novel clustering method using enhanced grey wolf optimizer and mapreduce. Big Data Res 14:93–100
23. Weng J, Young D (2017) Some dimension reduction strategies for the analysis of survey data. J Big Data 4(1):1–19
24. Boyle T. Dealing with imbalanced data. https://towardsdatascience.com/methods-for-dealing-with-imbalanced-data-5b761be45a18
25. L'Heureux A, Grolinger K, Elyamany HF, Capretz MAM (2017) Machine learning with big data: challenges and approaches. IEEE Access 5(5):777–797
26. Rencberoglu E. Fundamental techniques of feature engineering for machine learning. https://towardsdatascience.com/feature-engineering-for-machine-learning-3a5e293a5114
27. Wang L (2017) Heterogeneous data and big data analytics. Autom Control Inf Sci 3(1):8–15
28. Devi SG, Sabrigiriraj M (2017) Swarm intelligent based online feature selection (OFS) and weighted entropy frequent pattern mining (WEFPM) algorithm for big data analysis. Cluster Comput 1–13
29. Rong M, Gong D, Gao X (2019) Feature selection and its use in big data: challenges, methods, and trends. IEEE Access 7:19709–19725
30. Seijo-Pardo B, Porto-Díaz I, Bolón-Canedo V, Alonso-Betanzos A (2017) Ensemble feature selection: homogeneous and heterogeneous approaches. Knowl Based Syst 118:124–139
31. Hossain MA, Jia X, Benediktsson JA (2016) One-class oriented feature selection and classification of heterogeneous remote sensing images. IEEE J Sel Top Appl Earth Obs Remote Sens 9(4):1606–1612
32. Chen L, Zhang D, Pan G, Ma X, Yang D, Kushlev K, Zhang W, Li S (2015) Bike sharing station placement leveraging heterogeneous urban open data. In: Proceedings of the 2015 ACM international joint conference on pervasive and ubiquitous computing. ACM, pp 571–575
33. Oliver J. Is big data enough for machine learning in cyber security? https://www.trendmicro.com/vinfo/us/security/news/security-technology/is-big-data-big-enough-for-machine-learning-in-cybersecurity
34. Faris H, Aljarah I, Al-Betar MA, Mirjalili S (2018) Grey wolf optimizer: a review of recent variants and applications. Neural Comput Appl 1–23
35. Hodge VJ, O'Keefe S, Austin J (2016) Hadoop neural network for parallel and distributed feature selection. Neural Netw 78:24–35
36. Emary E, Zawbaa HM, Hassanien AE (2016) Binary grey wolf optimization approaches for feature selection. Neurocomputing 172:371–381
37. Fong S, Wong R, Vasilakos A (2015) Accelerated PSO swarm search feature selection for data stream mining big data. IEEE Trans Serv Comput 9:33–45

38. Rezek IA, Roberts SJ (1998) Stochastic complexity measures for physiological signal analysis. IEEE Trans Biomed Eng 45(9):1186–1191
39. Cornejo FM, Zunino A, Murazzo M (2018) Job schedulers for machine learning and data mining algorithms distributed in hadoop, In: VI Jornadas de cloud computing & big data (JCC&BD), La Plata

Big Data Mining on Rainfall Data

Keshani Vyas

Abstract India is a country with leading economy driven by agriculture, which is majorly depended on the climatic changes. Rainfall events in India have diversified irregular pattern across the country, which makes it more complex and challenging to mine. Rainfall data comes under the meteorological data which is spatiotemporal in nature. The meteorological data is increasing day by day, to handle that huge amount of data imposing a wide range of challenges for storage and analysis. Big data technologies, like Hadoop Distributed File System (HDFS), Hbase, and many query processing tools for processing data like Hive and Pig are popularly known to handle such type of data. But these tools and techniques individually are inadequate for mining data efficiently. We have reviewed the research in the area of mining rainfall data and identified the gaps in the existing approaches.

Keywords Spatiotemporal data · Rainfall · Big data mining

1 Introduction

The climate of India follows some seasonal patterns like winter, summer and monsoon even the weather of India varies from one region to another region like tropical wet areas, tropical dry areas, hill stations, desert areas, near river basin have different amounts of rainfall and cold and snowfall. So, the prediction of weather in India with this diversity is quite complex and challenging task [1]. There are many factors that affect the climate. The data that used for weather prediction with these factors is known as meteorological data. There are many parameters or factors of meteorological data which contains the value for rainfall, atmospheric temperature, atmospheric pressure, wind speed, humidity, and other factors such as visibility and thunderstorm. These factors affect the climate of India. There are wide variety of data generation

K. Vyas (✉)
L. D. College of Engineering, Ahmedabad, Gujarat, India
e-mail: khvyas1997@gmail.com

K. Kotecha et al. (eds.), *Data Science and Intelligent Applications*, Lecture Notes on Data Engineering and Communications Technologies 52, https://doi.org/10.1007/978-981-15-4474-3_10

like it can be generated by satellite, radar, weather stations with many different sensors and many other ways. These data generation may vary from day to hour basis as some satellite may generate data at every hour some generate at the end of the day so basically it depends on the purpose behind selecting data. Raw data formats usually used for rainfall data is Geographic Tagged Image File Format (Geotiff), Network Common Data Form (netCDF), Hierarchical Data Format (HDF5) and Binary data [2].

Meteorlogical data have both spatial and temporal components in it. Spatial component in terms of latitude and longitude and as temporal component it gives time and date range [3]. Analysis on these data can be helpful for not only predicting droughts, floods, tropical cyclones events but also in for consecutive effects that focus on water resources, agriculture land, soil moisture and other events. There are many Web sites and applications and news channels which are predicting the weather condition and tracking cyclone events so accurately so these can be possible by applying data mining algorithm on the meteorological data and retrieving useful knowledge out of it which can be more useful for many applications like prediction of rainfall, trend of rainfall, outlier detection, heavy and low rainfall region classification and many more [4]. We can apply the following data mining techniques on these data (i) classification, (ii) regression, (iii) clustering, (iv) forecasting, (v) outlier detection, (vi) pattern tracking and (vii) association rule mining.

2 Review of Various Big Data Mining Approaches

2.1 Spatiotemporal Analysis of Rainfall Trends over a Maritime State (Kerala) of India During the Last 100 Years

In this paper [5], the aim is to find trend of rainfall in Kerala region by using data from the last 100 years and find variation of rainfall in different regions. Data is collected from Ministry of Earth and Science (MoES) and Indian Meteorological Department (IMD). The authors divided the collected data of Kerala state into 14 districts and grouped by three main regions like north, central and south Kerala. In order to find variations of rainfall seasonality index is used for each district in monthly basis. Mann-Kendall test statistic method is used to estimate trend in rainfall.

The Mann-Kendall S is given as,

$$S = \sum_{i=1}^{n-1} \sum_{j=i+1}^{n} \operatorname{sgn}(x_j - x_i) \tag{1}$$

Here, x refers to time series point, sgn function stands for sigmoid function where it gives 1 if $(x_j - x_i) > 0$, it gives 0 if $(x_j - x_i) = 0$ and -1 if $(x_j - x_i) < 0$. The variance is,

$$\text{Var}(S) = \frac{n(n-1)(2n+5) - \sum_{i=1}^{n} t_i(i)(i-1)(2i+5)}{18} \tag{2}$$

And finally test statistic Zc is calculated as:

$$\text{Zc} = \begin{Bmatrix} \frac{S-1}{\sqrt{\text{Var}(S)}}, & S > 0 \\ 0, & S = 0 \\ \frac{S+1}{\sqrt{\text{Var}(S)}}, & S < 0 \end{Bmatrix} \tag{3}$$

Positive value of Zc gives upward trend and negative value of Zc gives downward trend. They have also computed seasonality index which shows variability of monthly rainfall through year.

$$\text{SI} = \frac{1}{R}\sum_{n=1}^{12}\left|X_n - \frac{R}{12}\right| \tag{4}$$

Here, X_n stands for mean rainfall and R stands for annual rainfall. Lower the value of SI, the better is the distribution of monthly rainfall among the months of a year. For different months, the authors have given trend and mean rainfall. Mann-Kendall statistical test can be replaced with other approaches like data mining and machine learning as one of the limitations of Mann-Kendall test is, it is not suited for seasonal effects. The test is applied on for predefined regions or districts only. Visualization of the trend is not depicting proper and accurate rainfall trend.

2.2 A Hybrid Approach to Rainfall Classification and Prediction for Crop Sustainability

This paper [6] aims at predicting rainfall by considering affecting factors and applies regression analysis on rainfall data which is helpful for predicting crops and classify the states based on crop suitability. Rainfall data is divided based on seasons as winter and monsoon; and monsoon is further divided into pre- and post-monsoon. Data is from 1901 to 2002 from the Indian Meteorological Department (IMD) and Open Government Data (OGD). The average temperature, cloud cover, vapor pressure, wet day frequency, potential evapotranspiration and precipitation factors are considered and average is performed for all these factors for experiment. Regression analysis is applied on the factors that affect rainfall and also correlationship is considered between those factors. Additionally, the authors also classify data based on crop ranges for different states of India using decision trees. Correlation and covariance can be calculated for the average temperature, cloud cover, vapor pressure, wet day frequency, potential evapotranspiration and precipitation factors, in which if we consider X and Y as set of n observation, then covariance is positive, if X is greater, then expected value of X and Y is greater than expected value of Y. If any of the factors X or Y is lesser than their expected value, then covariance becomes negative. Simple linear and multiple linear regression analysis are applied where coefficients are calculated by least squares method and then labeling of data is performed where decision tree is applied to predict the suitable season for particular crop.

The authors have not consider the variation of rainfall from one state to another considering it may help them to get better results also The authors have not perform preprocessing on the data so removing noisy data and resolving missing values also helps.

2.3 Modeling Rainfall Prediction Using Data Mining Method—A Bayesian Approach

In this paper [7], the authors have applied Bayesian theory for rainfall prediction. The dataset is provided by Indian Meteorological Department (IMD), Pune. The attributes considered are station level pressure, mean sea level pressure, relative humidity, temperature, vapor pressure, wind speed and rainfall. The preprocessing and transformation are applied on missing and noisy data. The Bayesian formula classifies the rainfall (yes or no). The data is analyzed for three cities Pune, Mumbai and Delhi with accuracy of 90%. The authors have not preprocessed the data for missing values of classification. In case of missing values, the classifier takes it as zero and leads to wrong prediction.

Train-test splitting performed by 70:30 ratio where training data is used for model generation and testing data is applied on the generated model.

$$p(C|F_1, \ldots F_n) = \frac{p(C)p(F_1, \ldots, F_n|C)}{p(F_1, \ldots, F_n)} \tag{5}$$

Here, F_1 and F_2 stand for features or attributes and C stands for classifying the rainfall as yes or no. If $P(\text{yes}|t) > P(\text{No}|t)$ (where t stands for different features in dataset), then we can classify data as Yes category otherwise No. Accuracy will increase with amount of data.

2.4 Novel Weather Data Analysis Using Hadoop and MapReduce—A Case Study

In this research paper [8], the objective is to develop system that works on historical data of weather and apply MapReduce and Hadoop methods to analyze it. Data is collected from weather sensors which generate data for every one hour and datasets from NCDC with size of 20 GB. Hadoop contains HDFS which is used for storing huge amount of weather data in distributed file system environment. MapReduce is framework which is used here for managing and processing these huge amounts of data. MapReduce works in two phases. In map phase, the mapper will split the data and convert data into key value pairs. After applying mapper on the dataset, the name of the station and date will work as a key while precipitation, temperature and wind speed will works as a value. Reduce phase will shuffle the output of map phase data and merge them and give it as reduce phase output, so it will apply operations like min, max, and average based on key join. MapReduce technique is used here for temperature analysis. The main goal of this research is to find out the best-suited tools for query on the datasets whether it is Pig or Hive. When data is huge Hive and Pig tools take more time for processing and analyzing. Most of meteorological data have spatiotemporal characteristics and performing query on them is quite complex task using Pig and Hive.

2.5 Comparative Analysis of Data Mining Techniques for Malaysian Rainfall Prediction

In this paper [9], the authors have applied different classifiers like Naive Bayes, decision tree, neural network, support vector machine and random forest for rainfall prediction on the data and give best technique for classifying rainfall data. Data is collected from multiple weather stations in Malaysia, Malaysia Meteorological department and Malaysia Drainage and Irrigation Department, data is collected between January 2010 and April 2014 year. These dataset contains temperature, relative humidity, rainfall, water level and flow as features. In order to remove noisy data and for filling missing values, preprocessing tasks have been performed. The mean average mechanism is used for filling missing values and normalization is used for converting data values into specific data range, after all cleaning and preprocessing tenfold cross-validation was performed on data with the different ratio of training and testing split data. In order to find performance matrix recall, precision and F-measure are used. By using 10% training and 90% testing data, random forest correctly identified 1043 instances out of 1581 instances and other classifiers correctly identified instances are for Naive Bayes has 1015, decision tree has 1039, ANN has 1015 and SVM has 1034. So, the random forest is most suited classifier for their data. Here, the result of best-suited technique will vary with ratio of train-test split. The authors are not considering any correlation between features which affect rainfall so considering relationship between the features may help in increasing precision and recall values.

3 Conclusion

In this paper, we have reviewed different approaches that can be applied for analysis of meteorological data. The first challenge is storing and processing of huge amount of data. These papers used HDFS and Hive, Hbase, Pig as query processing tool which is inefficient for spatiotemporal data. Applying data mining techniques can give more reliable results rather than going for statistical approaches. As applying statistical methods on large data result into more complex and time-consuming process while data mining works well with large amount of data. Boundary line analysis is a major challenge. All papers have analyzed fixed regions depicting states or districts. There does not exist any framework that supports scalable solution for mining rainfall data.

References

1. https://www.toppr.com/guides/geography/climate/climate-of-india/
2. https://climatedataguide.ucar.edu/climate-data-tools-and-analysis/common-climate-data-formats-overview
3. https://www.omnisci.com/learn/resources/technical-glossary/spatial-temporal
4. http://www.statsoft.com/textbook/data-mining-techniques
5. Nair A, Ajith K, Nair KS (2014) Spatio-temporal analysis of rainfall trends over a maritime state (Kerala) of India during the last 100 years. Atmos Environ 88:123–132
6. Rao P, Sachdev R, Pradhan TA (2016) Hybrid approach to rainfall classification and prediction for crop sustainability. In: Thampi S, Bandyopadhyay S, Krishnan S, Li KC, Mosin S, Ma M (eds) Advances in Signal processing and intelligent recognition systems. Advances in intelligent systems and computing, vol 425. Springer, Cham
7. Nikam V, Meshram BB (2013) Modeling rainfall prediction using data mining method: a bayesian approach. In: Fifth international conference on computational intelligence, modelling and simulation, pp 132–136
8. Suryanarayana V, Sathish B, Ranganayakulu A, Ganesan P (2019) Novel weather data analysis using Hadoop and MapReduce. In: 5th International conference on advanced computing and communication systems (ICACCS), Coimbatore, India
9. Zainudin S, Jasim D, Abu BA (2016) Comparative analysis of data mining techniques for malaysian rainfall prediction

Disease Prediction in Plants: An Application of Machine Learning in Agriculture Sector

Zankhana Shah, Ravi Vania, and Sudhir Vegad

Abstract Agriculture is the mainstay of the Indian economy. Almost 70% people depend on it and share major part of the GDP. Diseases in crops are mostly on the leaves which affect on the reduction of both quality and quantity of agricultural products. Perception of human eye is not so much stronger so as to observe minute variation in the infected part of leaf. It needs lengthy process involving knowledge regarding plants as well as considerable processing time. Hence, machine learning can be used for the detection of plant diseases. Disease detection involves the steps like image acquisition, image pre-processing, image segmentation, feature extraction, and classification. In this paper, we have tried to provide solution to automatically detect and classify cotton plant leaf diseases which in turn will enhance productivity of crops.

Keywords Machine learning feature extraction · Classification · Disease detection

1 Introduction

Agriculture is the backbone of every economy. India is an agricultural country, and crop making is having much importance. Due to increasing population, the demand for food is increasing in days to come. The agriculture sector needs to adopt advance techniques to survive the changing conditions of the economy and to meet the need

Z. Shah (✉) · R. Vania
Information Technology Department, B V M Engineering College, Vallabh Vidyanagar, Anand, Gujarat, India
e-mail: zankhana.shah@bvmengineering.ac.in

R. Vania
e-mail: ravivaniya4911@gmail.com

S. Vegad
Information Technology Department, A D Patel Institute of Technology, Vallabh Vidyanagar, Anand, Gujarat, India
e-mail: svegad@gmail.com

K. Kotecha et al. (eds.), *Data Science and Intelligent Applications*, Lecture Notes on Data Engineering and Communications Technologies 52,
https://doi.org/10.1007/978-981-15-4474-3_11

of the country. For optimum yield, the crop should be healthy for finest harvesting. Hence, automated techniques are must for periodic monitoring of crop. Crop disease is one of the major factors which indirectly influence the significant reduction of both the quality and quantity of agricultural products. The diseases in plant can be controlled with the help of many pesticides to be applied which in turn will increase the crop production. But finding the most current disease, appropriate and effective pesticide to control the infected disease is difficult and requires experts advise which is time-consuming and expensive. As the diseases can be predicted by examining the leaves, an automatic, accurate, and less expensive machine learning system needs to be devised.

The work presented here is based on disease prediction in the cotton plant. The paper is organized in this way: Chap. 2 discusses the related work done by different researchers. Literature review is presented in Chap. 3. Chapter 4 gives the details of implementations, and result analysis is presented in Chap. 5 which is followed by conclusion in Chap. 6.

2 Related Work

The proposed system by Mainkar et al. [1] performs leaf disease detection and classification using image processing techniques. It includes feature extraction using GLCM and image segmentation using K-means clustering. It also provides graphical user interface to detect disease, but GUI can only be accessed via desktop application. Premalatha et al. [2] proposed system to identify diseases in cotton plant using spatial FCM and PNN classification. The proposed system also helps in classifying the pest in cotton plant. In pre-processing noise removal is performed using median filter. The system did not provide any diagnosis measure to control disease. Bagdage et al. [3] proposed an approach to detect disease in crops like wheat and cotton. Here, it uses machine learning techniques to perform classification. Classification is performed using canny edge detection algorithm. Although, the proposed system did not provide any method to perform pre-process and feature extraction method. The method did not include any result analysis.

Tijare et al. [4] proposed system include two diseases alternaria and fungi. Cotton feature like HSV features extracted from the capture of segment and ANN. Experimental accuracy is around 80%. But, this system did not provide a user interface to interact a user with the system. Khirade and Patil [5] use image processing techniques for detection of disease in any plant. Here, image is captured through only mobile camera, classification by k-means clustering, and ANN used for classification. Here, feature extraction is performed using color co-occurrence method where RGB image converted into HSI translation.

3 Literature Review

In India, there is a drastic change in agriculture sector in line to technology advancement. But in very few cases, these technologies reach to farmers who are the people responsible for growing the crops. It is a time to aware the farmers about technology. An automated system is needed to facilitate farmer with early detection of disease in plants or crops. This kind of system includes machine learning approach incorporating various sub-steps like pre-processing, feature extraction, training, and classification.

Pre-processing: To make the image smooth and uniform, various preprocessing techniques are available, and one of them is Gaussian Blur. Gaussian function is used to calculate the transformation which is applied to every pixel of the image. The Gaussian function in two dimensions is,

$$G(x) = \frac{1}{2\pi\sigma^2} \cdot e^{-\frac{x^2+y^2}{2\sigma^2}} \tag{1}$$

where x and y are the distance from the origin in the horizontal and vertical axis, respectively, and σ is the standard deviation of the Gaussian distribution. Each pixel's new value is set to a weighted average of that pixel's neighborhood. The original pixel's value gets replaced with the heaviest weight, and adjacent pixels receive smaller weights as their distance to the original pixel increases. This results in a blur effect [6].

Feature extraction: Feature extraction is a dimensionality reduction process, where an initial set of raw variables is reduced to more manageable groups (features) for processing, while still accurately and completely describing the original data set. A GLCM holds pixels' positions having similar gray level values. A co-occurrence matrix is a two-dimensional array that represents a set of possible image values. A GLCM $C\mu(i, j)$ is defined as

$$C\mu(i, j) = n_{ij} \tag{2}$$

where n_{ij} is the number of occurrences of the pixel values (i, j) lying at distance μ in the image. The co-occurrence matrix $C\mu$ has dimension $n \cdot n$, where n is the number of gray levels in the image.

Training of model: This is a supervised machine learning model where the class label is already known. An epoch is a hyper-parameter which is defined before training a model. One epoch is an iteration and passed in neural network only once for one cycle of an entire dataset.

Classification model: In machine learning, the perceptron is an algorithm for supervised learning of binary classifiers. It is a type of linear classifier, i.e., it predicts the class based on a set of weights combined with the feature vector. Linear classifier defined that the training data should be classified into corresponding categories, i.e., if we are applying classification for the two categories, then all the training data must be lie in these two categories. Binary classifier defines that there should be only two categories for classification [7].

4 Implementation

To cope up with the challenges of meeting demand of population for food, it is needed to have an automated system which improves the quality of crop by early detection of disease in leaves. This work proposes AI-based model that help in early detection of disease in cotton plant. Figure 1 shows the proposed methodology.

An image captured by mobile may not be always of good quality, so pre-processing of the image is required which will improve quality of image. Here, for de-noising image, we applied Gaussian Blur algorithm which internally converts image from RGB to gray scale. Normalization of values is performed, so that any outlier present in dataset can be analyzed. Then to improve contrast in image, histogram equalization [8] technique is applied.

Gray-level co-occurrence matrix (GLCM) is used as feature extraction technique which basically extracts texture-based features. How frequently a pixel with the particular intensity value i occurs in a specified spatial relationship to a pixel with the value j is calculated. Total eight texture-based features are extracted as discussed below:

Contrast: It is a measure of the intensity difference between a pixel and its adjacent pixel over the entire image. Contrast is 0 for a constant image. $C(i, j)$ is pixel value of location(i, j).

$$\text{Contrast} = \sum_{i,j=0}^{n-1} c(i, j)(i, j)^2 \tag{3}$$

Energy: Returns the sum of squared elements in the SGDM. Range = [0 1]. Energy is 1 for a constant image.

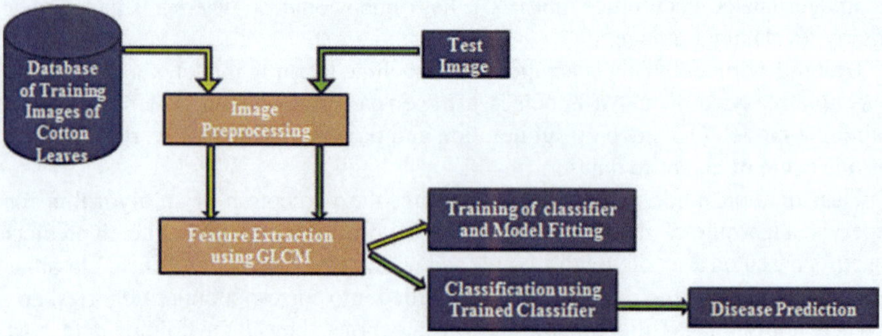

Fig. 1 Proposed methodology

$$\text{Energy} = \sum_{i,j=0}^{n-1} c(i, j)^2 \tag{4}$$

Homogeneity: It measures image homogeneity as it assumes larger values for smaller gray level intensity differences in pair elements. It has maximum value when all elements in the image are same.

$$\text{Homogeneity} = \frac{\sum_{i,j=0}^{n-1} c(i, j)}{(1 + (i - j)^2)} \tag{5}$$

Correlation: Returns a measure of how correlated a pixel is to its neighbor over the whole image. Range = [−1 1]. Correlation is 1 or −1 for a perfectly positively or negatively correlated image.

$$\text{Correlation} = \sum_{i,j=0}^{n-1} \frac{\{i \cdot j\} \times c(i, j) - \{\mu x - \mu y\}}{\sigma x \times \sigma y} \tag{6}$$

Entropy: Return the randomness of pixels as compared to its neighbor. It is also inversely proportional to energy.

$$\text{Entropy} = \sum_{i,j=0}^{n-1} - \log_2(c(i, j)) \cdot c(i, j) \tag{7}$$

Mean: Returns an estimate of the intensity of all pixels in the relationships that contributed to the GLCM.

$$\text{Mean} = \sum_{i,j=0}^{n-1} i \cdot c(i, j) \tag{8}$$

Standard deviation: The deviation from mean of the intensities of all reference pixels.

$$\text{Standard deviation} = \sqrt[2]{\sum_{i,j=0}^{n-1} c(i, j) \cdot (i - \mu)^2} \tag{9}$$

During training, every image present in a dataset is passed through a neural network model for classification. An artificial perceptron is created for several hidden layers. During each epochs, feature extraction is done and accuracy is measured.

Dataset is divided into training and testing dataset in some fixed ratio (Here, 80% and 20%, respectively).

5 Result Analysis

The dataset used to develop this system is consisting of images of cotton leaves of alternaria disease, anthracnose disease, bacterial blight disease, and cercospora disease. Images for healthy cotton leaves are also available in the dataset. Sample images are shown in Fig. 2.

Fig. 2 Sample images of dataset

Table 1 Classification results for different images

Test case no.	Observed disease	Image captured by drone	Image captured by drone from surface	Image captured by mobile	Classification
1	Alternaria	Alternaria	Alternaria	Alternaria	True
2	Healthy	Alternaria	Alternaria	Alternaria	False
3	Alternaria	Alternaria	Alternaria	Alternaria	True
4	Alternaria	Alternaria	Alternaria	Alternaria	True
5	Healthy	Healthy	Healthy	Alternaria	False
6	Bacterial blight	Bacterial blight	Bacterial blight	Bacterial blight	True
7	Cercospora	Cercospora	Cercospora	Cercospora	True
8	Cercospora	Healthy	Healthy	Cercospora	False
9	Bacterial blight	Bacterial blight	Bacterial blight	Bacterial blight	True
10	Cercospora	Cercospora	Cercospora	Cercospora	True

Table 2 Classification accuracy for each disease

No.	Disease	Classification accuracy (%)
1	Alternaria	85
2	Bacterial blight	79
3	Cercospora	70
4	Anthracnose	61

Table 1 shows the classification results for images captured with different methods, like, by drone, by keeping drone on surface, and by mobile. Table 2 shows the overall classification results for different diseases in cotton plant.

6 Conclusion

This system is developed for the farmers and researchers to early detection of disease in cotton plant by examining the leave. Through this system, we have tried to highlight the problems associated with the cultivation of cotton and causes of low yield loss in the developing countries like India. But, there is still a scope of improvement. This system detects only disease caused by fungi and bacteria. More disease can be included that occurred on the cotton ball. Also, a mobile application can be developed to facilitate this system to get connected with end user in way of uploading image through an app and getting notification of detected disease.

References

1. Mainkar P, Ghorpade S, Adawadkar M (2015) Plant leaf disease detection and classification using image processing technique. Int J Innov Emerg Res Eng 2
2. Premalatha V, Valarmathy S, Sumithra MG (2015) Disease identification in cotton plant using spatial FCM & PNN classifier. Int J Innov Res Comput Commun Eng 3
3. Bagdage A (2018) Crop disease detection using machine learning: Indian agriculture. Int Res J Eng Technol 5
4. Tijare R, Khade P, Jain R (2015) The survey of disease identification of cotton leaf. Int J Innov Res Comput Commun Eng 3
5. Khirade S, Patil A (2015) Plant disease detection using image processing. In: International conference on computing communication control and automation
6. Gedraite E, Hadad M (2011) Investigation on the effect of a Gaussian Blur in image filtering and segmentation. In: Proceedings ELMAR-2011, pp 393–396
7. Ingole A (2015) Detection and classification of leaf disease using artificial neural network. Int J Tech Res Appl 3:331–333
8. Oktavianto B, Purboyo TW (2018) A study of histogram equalization techniques for image enhancement. Int J Appl Eng Res 13:1165–1170

Sentiment Analysis of Regional Languages Written in Roman Script on Social Media

Nisha Khurana

Abstract Increasing popularity of smart phones, economic Internet packages and social media have changed the way of life the people are living. Social media plays an important role in today's generation. They use it for their own promotion, entertainment and even to an extent they share their sorrows and joys on social media. At the same time, the way of expressing themselves on social media has also changed a lot. People prefer to express themselves in their own regional language. As a result, the way of writing on social media has been changed. This has also resulted into the emergence of new languages which is blend of any regional language and English language. The opinions are written in regional language but in Roman script. With the availability of such massive opinionated data on social media, it is of no use if such data is not organized and used properly. This huge data may be used for the machines to learn and make them capable of taking the decisions like human beings. This paper discusses about the importance of opinion mining in blend of two languages and hence proposes a neural network architecture that can be trained on such opinions as negative or positive.

Keywords Opinion mining · Sentiment analysis · Neural networks · Machine learning

1 Introduction

1.1 Mind-Boggling Use of Social Media

Because of the mammoth use of social media, the data that is produced everyday online is mind-boggling. People use it for opinions whenever they want to buy a product. They look for the reviews online and decide. Feedback plays an important

N. Khurana (✉)
Computer Engineering Department, Gandhinagar Institute of Technology, Gandhinagar, India
e-mail: nisha.khurana@git.org.in

K. Kotecha et al. (eds.), *Data Science and Intelligent Applications*, Lecture Notes on Data Engineering and Communications Technologies 52, https://doi.org/10.1007/978-981-15-4474-3_13

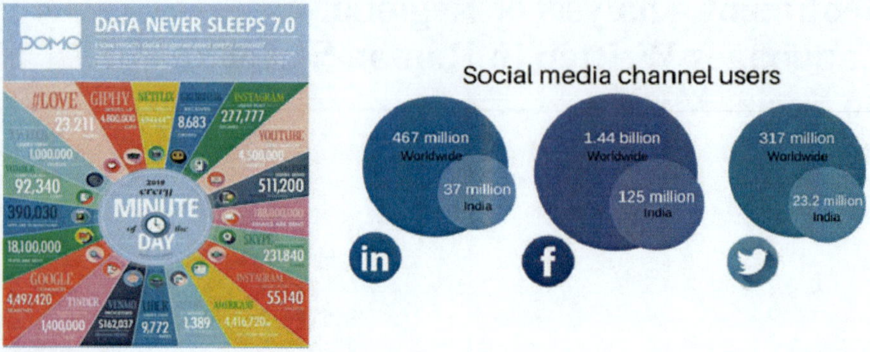

Fig. 1 Stats of use of social media [3, 19]

role in every sector whether it is public or private. Organizations use it to get the feedback from their customers and accordingly frame the business policies. Similar way law makers use it to frame and suggest new laws for the betterment of society, and none the less in the epoch of social media, it plays a very significant role [1, 2].

As per the "Data Never Sleeps 7.0" an article by doco.com [3], 2.5 quintillion bytes of data is created each day and is growing rather multiplying at a very fast pace. Alone in India, among the population of 1.36 billion, 70% that is 230 millions are the active social media users. As a result, masses of data are produced online every day. The data produced is so scattered that it is of no use if not used and organized properly. At the same time, data is of vital importance to any organization. Hence, data mining, data analytics has become important to these organizations [4] (Fig. 1).

1.2 Importance of SA

There are so many online media available which not only give the opinion of a particular product but also provides the comparison of products. It helps users in decision making effortlessly while sitting at home [2]. It gives the ability to clench the attitude of a consumer toward any product or any event and accordingly take the decision upon positive, negative, or neutral feedback [2].

Sentiment analysis is the computational study of sentiments expressed by the people [5]. It helps organizations in taking decisions, framing various policies, etc. [6]. A lot of research has been done on opinion mining, but the majority of research has been done on English sentiments as this is one of the most spoken languages globally [7, 8].

1.3 Evolving Way of Writing on Social Media

People use social media to express themselves, and they prefer to do it in their own native or regional language. As a result, the way of writing on social media has changed drastically. The use of hybrid languages has become the trend on social media [9]. People prefer to express sentiments in their language but in roman script. One of the main reason could be the widely available interface in any device is in English alphabets only. The other reason may be the typing is always easy in roman script as compared to any Indian regional language [10].

As a result, blend of two languages, also called hybrid languages, is becoming popular on social media as people feel comfortable to express themselves in these hybrid languages [11]. This blend is generally the regional language with English language. Many of the language literature persons have shown their concern on the rise of such hybrid languages. As per them, the languages are losing their identity. India has a cultural diversity and hence so in languages. Indian constitution gives official recognition to 22 regional languages along with Hindi. The most of the sentiment analysis related research has been done on English language only. This paper discusses about the sentiments written in these regional languages but in roman script on social media although the research has been done on one language that is Gujarati with the blend of English.

2 Proposed Architecture

A neural network has been proposed that is capable of classifying the messages or tweets in two categories as positive or negative [12]. The network has been developed in Python with Keras which has a powerful computational library. No work has been proposed at all for any regional language of India. The unique feature of this study is that mining is proposed to be done on the social media data which is specifically in hybrid languages [2, 13]. The study is proposed to be carried out on one language, for example Gujlish, Gujarati written in roman script, specifically. It classifies the messages in two categories, i.e., either positive or negative. The same study may be extended on any Regional language written in roman script. It also generates the graphs through which classification of positive and negative sentiments may be observed [14–16].

2.1 Generating the Corpus [2]

Generating a large amount of data for this project was really a challenging task. The specific data required for this study has been collected majorly from social networking microblogging websites like Twitter, Facebook and messenger WhatsApp [2]. Many

Web scraping tools are available in the market which really helps in gathering the data without writing a single line of code. These tools may be freely available or if paid may provide few days free trial for Web scraping [17].

Even social networking sites also provide their corresponding APIs through which data may be collected easily with a little programming from a particular ID. So, Twitter API and Facebook API have been used to collect data from these two social networking sites [18]. Apart from this import.io, Web scraping tool has been used to collect the required data from various sites. Following is the summary list of tools used in data collection [2]:

- Import.io (Web scraper tool)
- Facebook Graph API
- Twitter Streaming API
- Beautiful Soup (Python library)
- WhatsApp.

3 Experimental Results of Neural Network in Regional Language

Figure 2 is the snapshot at the time of fitting the model on regional data. It shows the sample size, number of epochs, and the value of each metrics at the end of each epoch. The model is trained for almost five thousand of Gujlish samples for a number of 60 epochs. Metrics like loss, accuracy, validation loss, and validation accuracy were recorded while training the model [2].

The number of epochs is set to avoid the chances of overfitting or underfitting of the model. During an epoch, an entire dataset is passed forward and backward

```
5026/5026 [======================] - 68s - loss: 0.2293 - acc: 0.9741 - val_loss: 0.2411 - val_acc: 0.8348
Epoch 41/60
5026/5026 [======================] - 64s - loss: 0.2190 - acc: 0.9778 - val_loss: 0.2300 - val_acc: 0.8676
Epoch 42/60
5026/5026 [======================] - 76s - loss: 0.2089 - acc: 0.9788 - val_loss: 0.2289 - val_acc: 0.8788
Epoch 43/60
5026/5026 [======================] - 80s - loss: 0.2076 - acc: 0.9792 - val_loss: 0.2258 - val_acc: 0.8898
Epoch 44/60
5026/5026 [======================] - 82s - loss: 0.1982 - acc: 0.9801 - val_loss: 0.2131 - val_acc: 0.9211
Epoch 45/60
5026/5026 [======================] - 91s - loss: 0.1979 - acc: 0.9809 - val_loss: 0.2110 - val_acc: 0.9319
Epoch 46/60
5026/5026 [======================] - 89s - loss: 0.1976 - acc: 0.9808 - val_loss: 0.2068 - val_acc: 0.9420
Epoch 47/60
5026/5026 [======================] - 92s - loss: 0.1872 - acc: 0.9819 - val_loss: 0.2023 - val_acc: 0.9541
Epoch 48/60
5026/5026 [======================] - 87s - loss: 0.1867 - acc: 0.9819 - val_loss: 0.1978 - val_acc: 0.9659
Epoch 49/60
5026/5026 [======================] - 78s - loss: 0.1862 - acc: 0.9910 - val_loss: 0.1921 - val_acc: 0.9760
Epoch 50/60
5026/5026 [======================] - 78s - loss: 0.1752 - acc: 0.9918 - val_loss: 0.1901 - val_acc: 0.9862
```

Fig. 2 A glimpse of number of epochs running

through the model only once. Since one epoch is too big to be given to model in one go, it is divided into batches of 32 (default size) each [2]. There is no thumb rule to decide the number of epochs. As the number of epochs increases, weights are changed that number of times and the gradient descent (learning rate) of the model increase. The curve (learning rate) goes from underfitting to optimal to overfitting curve. The four metric values loss, acc, val_loss, and val_acc for training set and validation set are recorded. After each epoch, loss decreases and accuracy increases. Following are the epochs during model generation (Gujlish) [2].

Below is the plot of epochs versus loss of training dataset and val_loss of validation set. If validation loss is too lower or too higher than the training set loss, the model is said to be either underfitted of overfitted. For an optimal model, the values of both the losses should be close by. The graph clearly shows the decline of loss of training as well as testing data with the passing number of epochs [2].

Below is the plot of epoch number versus accuracy of training dataset and val_acc of validation set. The acc refers to the learning of training dataset, and val_acc refers to the learning of validation set. Since training dataset is already known to the model and validation data is unseen by the model, acc commonly stays little higher than val_acc [2]. If validation accuracy is too lower or too higher than the training set accuracy, the model is said to be either underfitted of overfitted. The plot clearly shows the rise in the accuracy of testing and training data as the each epoch passes by which means that model is learning well during each epoch [2] (Figs. 3, 4, 5, 6, and 7).

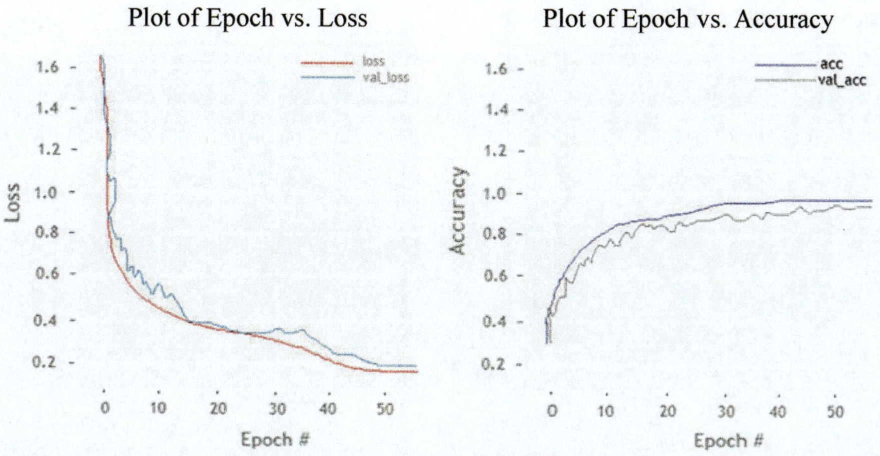

Fig. 3 Plot of epoch versus loss and epoch versus accuracy [2]

You entered the Statement: khoobj saras
And its sentiment is: positive Sentiment; 99.45799112319946% confidence

Fig. 4 Positive sentiments in Gujlish [2]

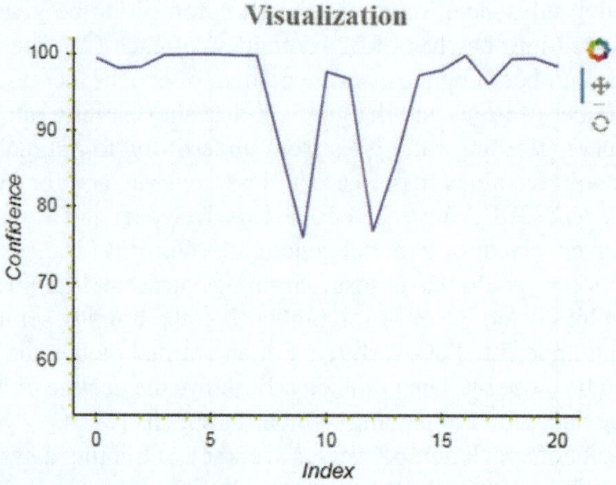

Fig. 5 Graph of positive sentiments with confidence [2]

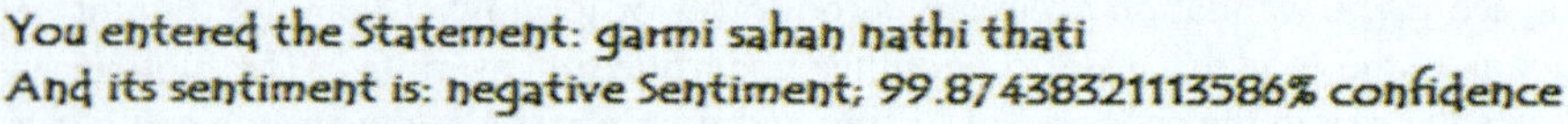

Fig. 6 Negative sentiments in Gujlish [2]

Fig. 7 Graph of negative sentiments with confidence [2]

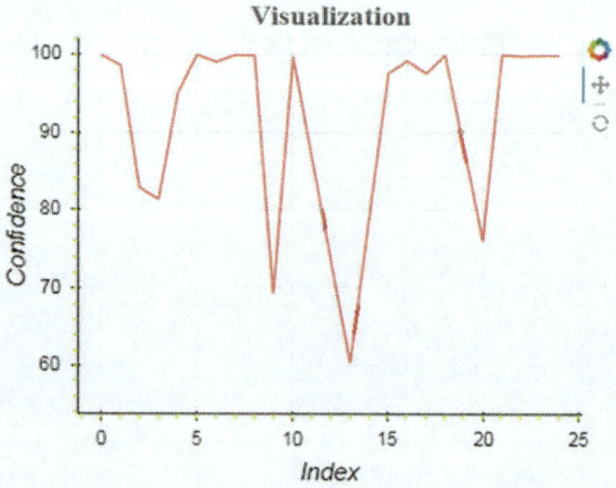

4 Conclusions

In this research, I have proposed a system that calculates the sentiments of a regional language blended with English. The trend of blend of languages is blooming, that too English, because of ease of typing, with the language in which user is comfortable in expressing their sentiments. Hence, this system has been performed, and no work has been proposed in this domain till date. The experiments were done on one regional language, Gujarati. The same work may be extended to other regional languages of India. No research work is complete without limitations. This analysis also has limitation of finding the sentiments containing sarcasm. Sometimes the sarcasm is even out of understanding of human being, and machines too are unable to find out the emotion of any said sentence in a particular situation.

References

1. Ranjan D, Arjariya T, Gangwar T (2017) Trend analysis through hashtags popularity level using hadoop with hive. Int J Innov Res Sci Eng Technol 6(5)
2. Nisha K. Sentiment analysis and opinion mining on social media. PhD thesis
3. https://www.domo.com/learn/data-never-sleeps-7
4. Appel O, Chiclana F, Carter J (2015) Main concepts, state of the art and future research questions in sentiment analysis centre for computational intelligence. Acta Polytech Hung 12(3):87–108
5. Cambria E, Schuller B, Liu B, Wang H, Havasi C (2013) New avenues in opinion mining and sentiment analysis. IEEE Intell Syst 28:15–21
6. Gamon M, Aue A, Corston-Oliver S, Ringger E (2015) Pulse: mining customer opinions from free text. In: Advances in intelligent data analysis VI. Springer, New York, pp 121–132
7. Liang P, Dai B (2013) Opinion mining on social media data. In: 14th international conference on mobile data management
8. Liu B, Zhang L (2012) A survey of opinion mining and sentiment analysis. In: Mining text data. Springer, Boston, MA
9. Mhaiskar R (2015) Romanagari: an alternative for modern media writings. Bull Deccan Coll Res Inst 75:195–202
10. Si A (2010) A diachronic investigation of Hindi–English code-switching, using Bollywood film scripts. Int J Biling 15(4):388–407. https://doi.org/10.1177/1367006910379300
11. Banerjee A: Romanagari can form system for language learning. Indian Express Newspaper. http://archive.indian express.com/news/-romanagari-can-form-system-for-language-learning/1105393/. Liu B (2013) Sentiment Analysis Tutorial, given at AAAI-2011
12. Pan B, Lee L, Vaithyanathan S (2002) Thumbs up? sentiment classification using machine learning techniques. In: Proceedings of the conference on empirical methods in natural language processing, pp 79–86
13. Farra N, Challita E, Assi RA, Hajj H (2010) Sentence-level and document-level sentiment mining for arabic texts. In: International conference on data mining workshops
14. Carenini G, Murray G, Ng R (2011) Methods for mining and summarizing text conversations. Morgan Claypool, San Rafael, CA
15. Carenini G, Ng R, Pauls A (2006) Multi-document summarization of evaluative text. In: Proceedings of EACL
16. Carenini G, Ng R, Pauls A (2006) Interactive multimedia summaries of evaluative text. In: Proceedings of IUI, pp 124–131
17. Denecke K (2008) Using SentiWordNet for multilingual sentiment analysis. In: 24th international conference on data engineering workshop. https://doi.org/10.1109/icdew.2008.4498370
18. Bosco C, Patti V, Bolioli A (2013) Developing corpora for sentiment analysis: the case of irony and senti-TUT. IEEE Intell Syst 28(2):55–63. https://doi.org/10.1109/mis.2013.28
19. http://www.mantran.in/blog/comparison-social-media-platforms-b2b-marketing/

Inductive Learning-Based SPARQL Query Optimization

Rohit Singh

Abstract In any query optimization, the goal is to find the execution plan which is expected to return the result set without actually executing the query or subparts with optimal cost. The users are not expected to write their queries in such a way so that they can be processed efficiently; rather it is expected from system to construct a query evaluation plan that minimizes the cost of query evaluation. SPARQL is used for querying the ontologies, and thus, we need to implement SPARQL query optimization algorithms in semantic query engines, to ensure that query results are delivered within reasonable time. In this paper, we proposed an approach in which the learning is triggered by user queries. Then, the system uses an inductive learning algorithm to generate semantic rules. This inductive learning algorithm can automatically select useful join paths and properties to construct rules from an ontology with many concepts. The learned semantic rules are effective for optimization of SPARQL query because they match query patterns and reflect data regularities.

Keywords Semantic Web · SPARQL query · Query optimization · RDF · OWL

1 Introduction

Query optimization is required to increase the performance of the query engine in terms of cost of evaluation of the query results. The cost of evaluation can be reduced by selecting the query evaluation plan having lower number of computations. The users are not supposed to write the query in a cost-effective way, so query engine needs to modify the query execution plan and evaluate the query results by executing plan which will require minimum number of computations [1–4]. The performance of any conventional information system depends on the performance of query processing and database management; in the same way, the performance of the semantic Web-based information systems depends on knowledge base and SPARQL

R. Singh (✉)
Gandhinagar Institute of Technology, Gandhinagar, Gujarat, India
e-mail: rohit.singh@git.org.in

K. Kotecha et al. (eds.), *Data Science and Intelligent Applications*,
Lecture Notes on Data Engineering and Communications Technologies 52,
https://doi.org/10.1007/978-981-15-4474-3_14

query processing. Enterprises rely on information systems for their operations. When an enterprise seeks to provide service-oriented businesses, their existing information systems need to be given additional flexibility and integration according to the service requirements. However, these legacy systems are often independently developed databases that support often isolated or stove-piped business applications. The reuse of these systems to support new business applications is a major challenge to all enterprises. Heterogeneity is the main difficulty for query optimization, and there are various kinds of heterogeneity explained in [1, 5–7]. Thus, we can see that the SPARQL query optimization is a potential research problem for semantic Web-based intelligent information systems.

2 Related Work

Foundation of semantic Web databases is explained in [5, 8–10], and the efficiency of answering queries to description logic (DL) and rule-based Web ontologies is explained in [4, 11–14]. Tableau algorithm for DL and DL programming is explained in [15–17]. The SPARQL queries can be optimized by selectivity-based static basic graph pattern (BGP) Optimization. The BGP is a set of triple patterns where a triple pattern is a structure of three components which may be concrete (bound) or variable (unbound). The three components forming triple patterns are subject, predicate and object. The problem is to determine the order in which query engine should execute the triple patterns. The solution of this problem is to start executing the last triple pattern and then move to the second last and so on and at the end execute the first triple pattern. Hence, a static optimizer should reverse the triple patterns. The join over the subject variable will be less expensive, and the optimization leads to a better query performance. The cost of selecting a triple pattern t, $c(t)$ can be calculated as [18]:

$$c(t) = c(s) \times c(p) \times c(o) \tag{1}$$

where $c(t)$ is cost of selecting a triple pattern t, $c(s)$, $c(p)$ and $c(o)$ are cost of selecting subject, predicate and object variable, respectively. The input SPARQL query is decomposed in smaller subqueries each having a triple pattern. The selection of ordering of subqueries for the execution is based on the cost estimation of the triple pattern associated with that particular subqueries. The query which is having minimum cost estimation will be selected first; after that the triple pattern having least among the rest of triple pattern will be selected and so on till all the triple patterns are selected. So, according to this order, subqueries are arranged and executed which is optimal for the given input SPARQL query.

3 Experiments for SPARQL Query Execution

SPARQL queries are executed using Seasame tool which supports complex query execution (other tools like protégé does not support complex queries) without any optimization. The ontology used for executing the SPARQL queries is as follows:

- agriOnt_ver4_alphatest.owl
- agriOnt_ver3_alphatest.owl
- AEZOnt_ver1.owl
- Open Linked Data

Several queries have been executed, and the results have been included for demonstration, Query-Variety returns the variety name of crop "cotton", Query-Variety-Zonewise returns the variety of cotton crop for a particular zone "North Gujarat" and Query-Pest-Details returns the pest name, pest category and scientific name and symptoms for which that particular pest is used for a particular crop "cotton". The executed queries with its results are given below:

Query-Variety (Fig. 1):

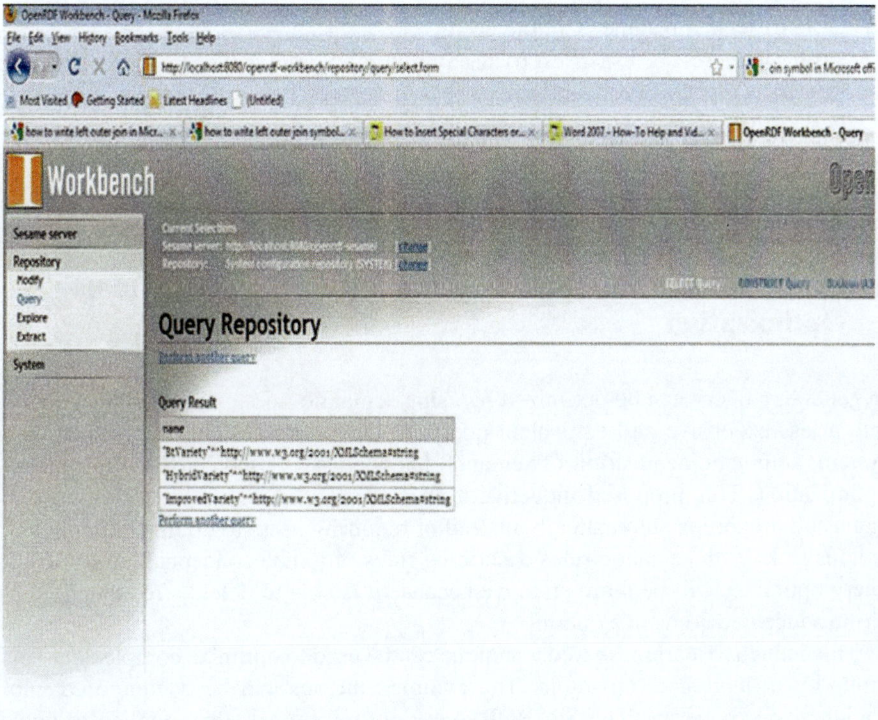

Fig. 1 Output for executed SPARQL query Query-Variety

```
PREFIX p1: <http://www.owlontologies.com/agriOnt.owl#> SELECT DISTINCT $name WHERE {
?Crop p1:isVarietyOf “Cotton”@en.
?Crop p1:VarietyType ?name
}
```

Query-Variety-Zonewise (Fig. 2):

```
PREFIX p1: <http://www.owlontologies.com/agriOnt.owl#> SELECT DISTINCT $Crop
WHERE {
$Crop p1:isVarietyOf “Cotton”@en.
$Crop p1:hasCropZone $n.
$n p1:CropZoneName “Central_Gujarat” @en.
}
```

Query-Pest-Details (Fig. 3):

```
PREFIX p1: <http://www.owlontologies.com/agriOnt.owl#>
SELECT DISTINCT ?Pest ?PestCategory ?ScientificName ?Symptom
WHERE {
?Crop p1:hasCropName “Cotton”@en.
?Crop p1:hasPest ?P est. ?Pest p1:hasSymptom ?Symptom.
?Pest p1:isTypeOf ?PestCategory.
?Pest p1:hasscientificname ?ScientificName.
}
```

4 Proposed Learning-Based Technique for SPARQL Query Optimization

In general, a query can be optimized by using semantic rule to reformulate a query into a less expensive and equivalent query. In our proposed approach, a learning system, automatic acquisition of semantic knowledge, is used for SPARQL query optimization. This proposed inductive learning algorithm can select relevant join paths and properties automatically instead of requiring users to do this difficult and tedious task. With semantic rules describing rules of joined concepts, the SPARQL query optimizer will be more effective because it is able to delete a redundant join or introduce new joins in a query.

This inductive learning-based technique can be used to optimize complex queries group-by or aggregate operators. The complex queries can be decomposed into conjunctive subqueries. The SPARQL query optimization is then applied to optimize each subquery and propagate constraints among them for global optimization.

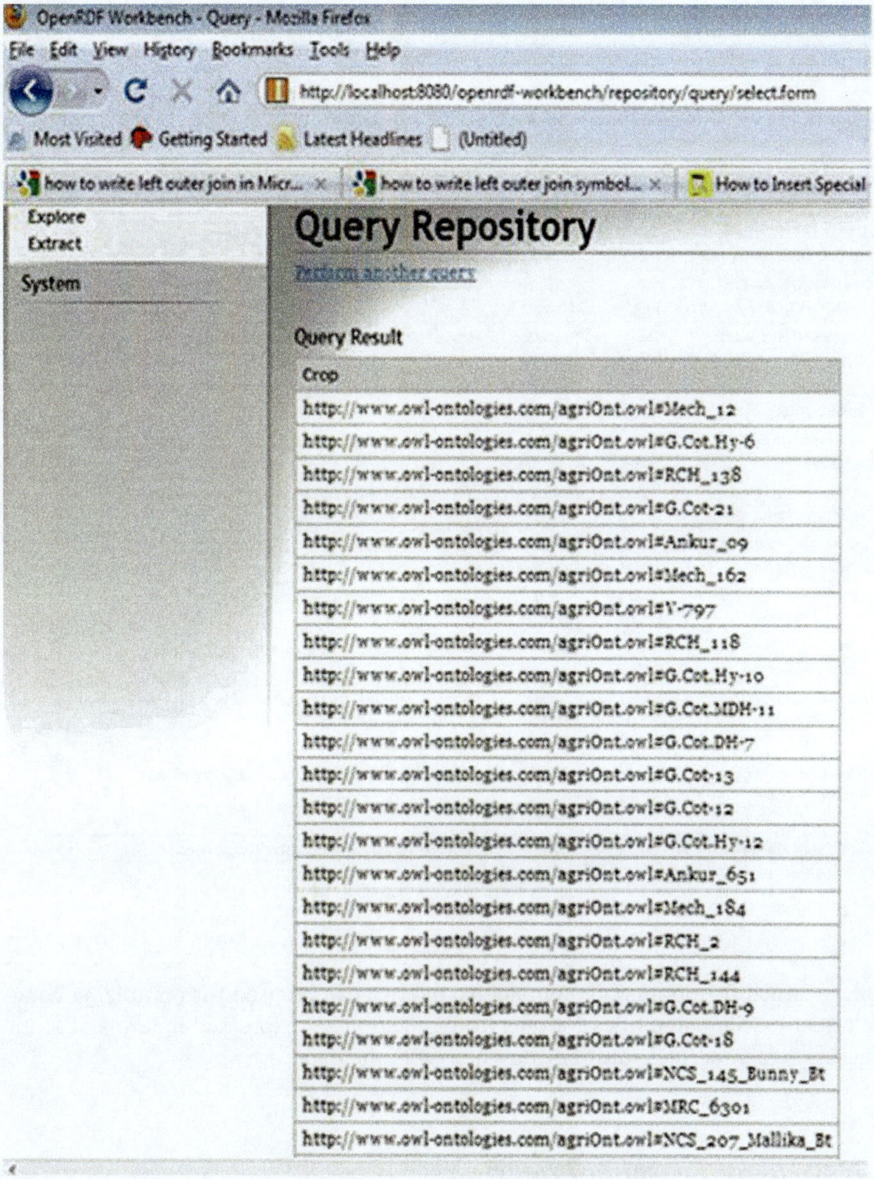

Fig. 2 Output for executed SPARQL Query-Variety-Zonewise

Query Repository

Perform another query

Query Result

Pest	PestCategory	ScientificName	Symptom
http://www.owl-ontologies.com /agriOnt.owl#SpodopteraLeafworm	"Leaf Feeder"@en	"Spodoptera litura"@en	http://www.owl-ontologies.com/agriOnt.owl#Defoliate_Leaves
http://www.owl-ontologies.com /agriOnt.owl#SpodopteraLeafworm	"Leaf Feeder"@en	"Spodoptera litura"@en	http://www.owl-ontologies.com/agriOnt.owl#Skeletinise_Leaves
http://www.owl-ontologies.com /agriOnt.owl#Spodoptera_Leaf_Worm	"Leaf Feeder"@en	"Spodoptera litura"@en	http://www.owl-ontologies.com/agriOnt.owl#Defoliate_Leaves
http://www.owl-ontologies.com /agriOnt.owl#Spodoptera_Leaf_Worm	"Leaf Feeder"@en	"Spodoptera litura"@en	http://www.owl-ontologies.com/agriOnt.owl#Skeletinise_Leaves
http://www.owl-ontologies.com /agriOnt.owl#Whiteflies	"Sucking Pest"@en	"Bemisia tabaci"@en	http://www.owl-ontologies.com/agriOnt.owl#curl_Upward
http://www.owl-ontologies.com /agriOnt.owl#Whiteflies	"Sucking Pest"@en	"Bemisia tabaci"@en	http://www.owl-ontologies.com/agriOnt.owl#HoneyDew_Deposit_On_Open_Bolls
http://www.owl-ontologies.com /agriOnt.owl#Whiteflies	"Sucking Pest"@en	"Bemisia tabaci"@en	http://www.owl-ontologies.com/agriOnt.owl#Reduced_PlantVigour
http://www.owl-ontologies.com /agriOnt.owl#Whiteflies	"Sucking Pest"@en	"Bemisia tabaci"@en	http://www.owl-ontologies.com/agriOnt.owl#Reduced_Plant_Vigour
http://www.owl-ontologies.com /agriOnt.owl#Whiteflies	"Sucking Pest"@en	"Bemisia tabaci"@en	http://www.owl-ontologies.com/agriOnt.owl#Honey_Dew_Deposit_On_OpenBolls
http://www.owl-ontologies.com /agriOnt.owl#Whiteflies	"Sucking Pest"@en	"Bemisia tabaci"@en	http://www.owl-ontologies.com/agriOnt.owl#Leaves_Cur_lUpward
http://www.owl-ontologies.com /agriOnt.owl#Jassids	"Sucking Pest"@en	"Amrasca biguttula biguttula"@en	http://www.owl-ontologies.com/agriOnt.owl#Leaf_Have_HopperBurn
http://www.owl-ontologies.com /agriOnt.owl#Jassids	"Sucking Pest"@en	"Amrasca biguttula biguttula"@en	http://www.owl-ontologies.com/agriOnt.owl#LeafBecome_Yellowish
http://www.owl-ontologies.com	"Sucking	"Amrasca	http://www.owl-ontologies.com/agriOnt.owl#Curl_Downwards

Fig. 3 Output for executed SPARQL Query-Pest-Details

Rules learned for optimizing conjunctive queries can be used for optimizing complex queries. Semantic rules for query optimization are expressed in terms of Horn clauses. Semantic rules must be consistent with the data.

4.1 General Learning Framework

Organization of ontology with a SPARQL query optimization and a learning system is illustrated in Fig. 4 [8]. The optimizer uses semantic rules in a rule bank to optimize input queries and then sends optimized queries to the knowledge base to retrieve data. When the knowledge base encounters a complex input query, it triggers the learning system to learn a set of rules from the data and then saves them in the rule bank. These rules will be used to optimize future queries. The system will gradually learn

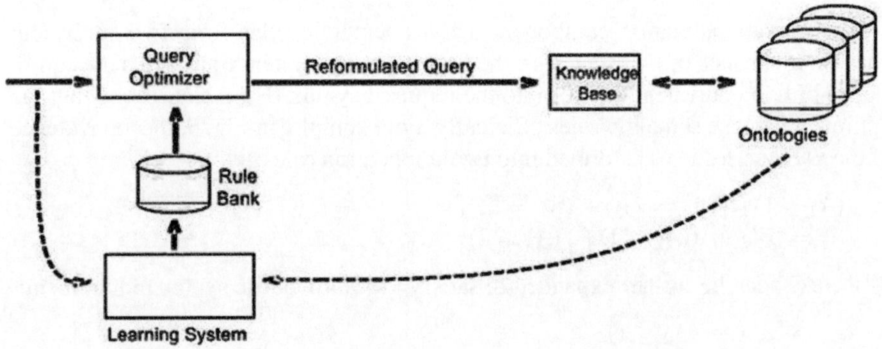

Fig. 4 Structure of the ontology system with SPARQL query optimizer and learner

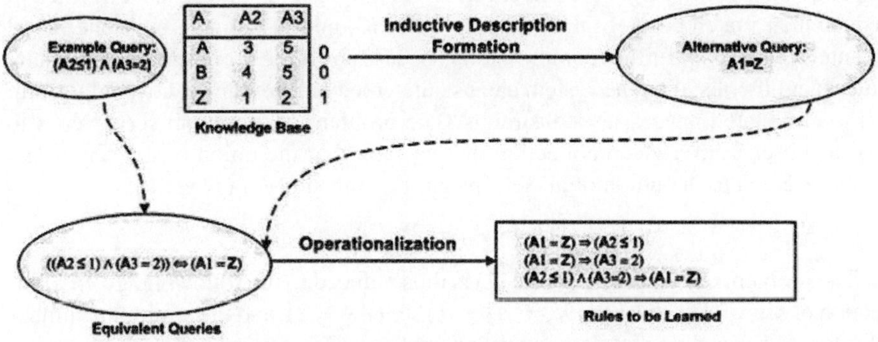

Fig. 5 A simplified learning scenario

a set of effective rules for optimization. Figure 5 illustrates a simplified scenario of our learning framework. This learning framework consists of two components, an inductive learning component and an operationalization component. The system applies an inductive learning algorithm to induce an alternative query equivalent to the input query with a lower cost. The operationalization component then takes the input query and the learned alternative query to derive a set of semantic rules. In Fig. 5, triples in the ontology are labeled as 1 if they satisfy the input query and 0 otherwise. The learned alternative query must cover all 1's triples but no 0 triples so that it retrieves the same data as the input query and is equivalent to the input query. Given a set of data triple classified as 1 or 0, the problem of inducing a description that covers all 1 data instance but no 0 is known as supervised inductive learning in machine learning. Since a query is a description of the data to be retrieved, inductive learning algorithms that learn descriptions expressed in the query language can be used in our framework.

The operationalization component derives semantic rules from two equivalent queries. It consists of two stages. In the first stage, the system transforms the equivalence of two equivalent queries into the required syntax (Horn clauses) so that the optimizer can use semantic rules efficiently. For example, in Fig. 5, the equivalence of the two queries is transformed into two implication rules:

1. $((A_2 \leq 1) \bigcap (A_3 = 2)) \rightarrow (A_1 = \text{'Z'})$
2. $(A_1 = \text{'Z'}) \rightarrow ((A_2 \leq 1) \bigcap (A_3 = 2))$

Rule (2) can be further expanded to satisfy the Horn clause syntax requirement:

3. $(A_1 = \text{'Z'}) \rightarrow (A_2 \leq 1)$
4. $(A_1 = \text{'Z'}) \rightarrow (A_3 = 1)$

After the transformation, the proposed rules (1), (3) and (4) satisfy syntax requirement. In the second stage, the system tries to compress the antecedents of rules to reduce their match costs. In this example, rules (3) and (4) contain only one literal as antecedent, so no further compression is necessary. If the proposed rule has many antecedent literals, then the system can use the greedy minimum set cover algorithm [7] to eliminate unnecessary constraints. The problem of minimum set cover is to find a subset from a given collection of sets such that the union of the sets in the subset is equal to the union of all sets. Negating both sides of (1) yields:

5. $\neg(A_1 = \text{'Z'}) \rightarrow \neg(A_2 \leq 1) \cup \neg(A_3 = 2)$

The problem of compressing rule (1) is thus reduced to the following: given a collection of sets of data that satisfy: $\neg(A_2 \leq 1) \cup \neg(A_2 = 2)$, and the minimum number of sets that cover the set of data satisfying $\neg(A_1 = \text{'Z'})$. Suppose the resulting minimum set that covers $\neg(A_1 = \text{'Z'})$ is $\neg(A_2 \leq 1)$, we can eliminate $\neg(A_3 = 2)$ from rule (5) and negate both sides again to form the rule:

$$(A_2 \leq 1) \rightarrow (A_1 = \text{'Z'}).$$

4.2 *Learning Alternative Queries*

Our inductive learning algorithm starts from an empty hypothesis of the concept description to be learned. The algorithm proceeds by constructing a set of candidate constraints that are consistent with all 1 valued triples and then using a gain/cost ratio as the heuristic function to select and add candidates to the hypothesis. This process of candidate construction and selection is repeated until no 0 valued triples satisfy the hypothesis.

The top-level algorithm of our inductive learning is shown in Fig. 6. The input of the algorithm is a user query Q and the ontology concepts. The primary concept of a query is the concept that must be accessed to answer the input query. For example, the primary concept of Q22 is geoloc because the output variable ?name of the query

```
INPUT Q = input query; OL = Ontology Concepts;
  BEGIN
    LET D = Primary concept of Q; LET AQ= Alternate query(Initially empty);
    LET T = Set of triple pattern candidate constraints (Initially empty);
    Construct triple pattern candidate constraints on D and add them to T;
    REPEAT
      Evaluate gain/cost of triple pattern candidate constraints in T;
      LET t = triple pattern candidate constraint with the highest gain/cost in T;
      IF gain(t)>0 THEN
        Merge t to AQ, and T=T-t;
        IF AQ⇔Q, THEN RETURN AQ;
        IF t is a join constraint on a new concept D' THEN
          Construct triple pattern candidate constraints on D' and add them to T;
        END IF;
      END IF;
    UNTIL gain(t)=0;
   RETURN fail, because no AQ is found to be equivalent to Q;
END
```

Fig. 6 Inductive algorithm for learning alternative queries

is bound to a property of geoloc. If the output variables are bound to properties from different concepts, then the primary concept is a concept derived by joining those concepts. Initially, the system determines the primary concept of an input query and labels the triples in the concepts as 1 or 0. A triple is 1 if it satisfies the input query; otherwise, it is 0.

4.3 Constructing and Evaluating Candidate Constraints

For each property of the primary concept, the system can construct an internal disjunction as a candidate constraint by generalizing property values of triples having value. The constructed constraint is consistent with 1 valued triples because it is satisfied by all triples which are having value 1. Similarly, the system considers a join constraint as a triple pattern candidate constraint if it is consistent with all 1 valued triples. This can be verified by checking whether all 1 valued triples satisfy the join constraint.

Table 1 Cost estimates of constraints in a query

Internal disjunctions, on non-indexed property	$\lvert D_1 \rvert$
Internal disjunctions, on indexed property	I
Join, over two non-indexed properties	$\lvert D_1 \rvert \cdot \lvert D_2 \rvert$
Join, over two indexed properties	$\lvert D_1 \rvert \cdot \lvert D_2 \rvert \max(I_1, I_2)$

After constructing a set of candidate internal disjunctive constraints and join constraints, the next step is to evaluate which one is the most promising and add it to the hypothesis. The evaluation function is the ratio of gain/cost where gain part of the heuristic is defined as the number of excluded 0 valued triples in the primary concept, and the cost of the function is defined as the estimated evaluation cost of the candidate constraint.

The evaluation cost of individual constraints can be estimated using standard query size estimation techniques [7]. A set of simple estimates is shown in Table 1. For an internal disjunction on a non-indexed property of a concept D, a query evaluator has to scan the entire concept to find all satisfying triple patterns. Thus, this evaluation cost is proportional to D, the size of D. If the internal disjunction is on an indexed property, then its cost is proportional to the number of triple patterns satisfying the constraint, denoted as I. For join constraints, let D_1 and D_2 denote the concepts that are joined, and I_1, I_2 denote the number of the distinct property values used for join. Then, the evaluation cost for the join over D_1 and D_2 is proportional to D_1 D_2 when the join is over properties that are not indexed, because the query evaluator must compute a cross product to locate pairs of satisfying triple patterns. If the join is over indexed properties, the evaluation cost is proportional to the number of instance pairs returned from the join, D_1 and D_2 $\max(I_1, I_2)$. This estimate assumes that distinct property values distribute uniformly in triple patterns of joined concepts.

5 Searching the Space of Candidate Constraints

When a join constraint is selected, a new concept and its properties are introduced into the search space of triple pattern candidate constraints. The system can consider adding constraints on properties of the newly introduced concept to the partially constructed hypothesis. We adopt a search method that favors triple pattern candidate constraints on properties of newly introduced concepts. That is, when a join constraint is selected, the system will estimate only those candidate constraints in the newly expanded level, until the system constructs a hypothesis that excludes all 0 valued triples (i.e., reaches the goal) or no more consistent constraints in the level with 1 valued triples gain are found. In the later case, the system will backtrack to search the remaining constraints on the previous levels. This search control bias takes advantage of underlying domain knowledge in the schema design of ontology. A join constraint

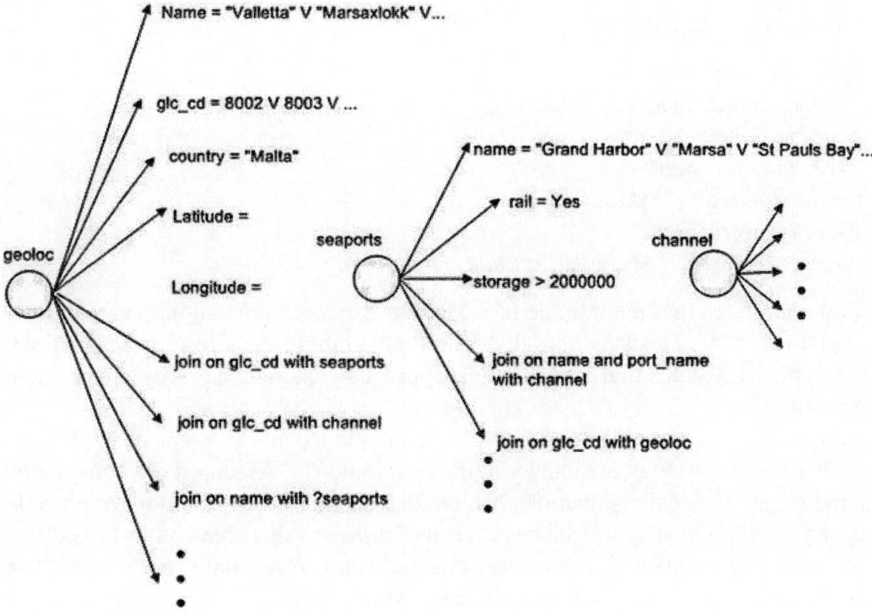

Fig. 7 Triple pattern candidate constraint to be selected

is unlikely to be selected on average, because an internal disjunction is usually much less expensive than a join. Once a join constraint (and thus a new concept) is selected, this is strong evidence that all useful internal disjunctions in the current level have been selected, and it is more likely that useful candidate constraints are on properties of newly joined concepts.
Example:
The schema of an example ontology with two concepts and their properties are shown in Fig. 7. In this ontology, the concept *geoloc* stores data about geographic locations, and the property *glc_cd* is a geographic location code.
Schema:

geoloc(name, glc_cd, country, latitude, longitude),
seaport(name, glc_cd, storage, silo, crane, rail).

Semantic Rules:

R1: geoloc(_, _, "Malta", ?latitude, _)?latitude $\Rightarrow$ *35.89.*
R2: geoloc(_, ?glc_cd, "Malta", _, _) $\Rightarrow$ *seaport(_, ?glc_cd, _, _, _, _).*
R3: seaport(_, ?glc_cd, ?storage, _, _, _) $\wedge$ *geoloc(_, ?glc_cd, "Malta", _, _)*
$\Rightarrow$*?storage > 2000000*
Q1: answer(?name):
geoloc(?name, ?glc_cd, "Malta",_),
seaport(_, ?glc_cd, ?storage, _, _, _), ?storage > 1500000.

The equivalent queries are:

Q21: answer(?name):
geoloc(?name, ?glc_cd, "Malta", _, _),
seaport(_, ?glc_cd, _, _, _, _).
Q22: answer(?name):
geoloc(?name, _, "Malta", _, _).
Q23: answer(?name):
geoloc(?name, _, "Malta", ?latitude, _), ?latitude $\leq$ *35.89.*

Rule R1 states that the latitude of a Maltese geographic location is greater than or equal to 35.89. 21 states that all Maltese geographic locations in the ontology are seaports. R3 states that all Maltese seaports have storage capacity greater than 2,000,000 ft^3.

Consider the geographic ontology schema mentioned above. Some example constraints for this ontology are shown in query C_3 and C_4. Among these constraints, C_0 and C_1 are internal disjunctions, which are constraints on the values of a single property. A triple of seaport satisfies C_0 if its *?storage* value is less than 150,000. A triple of geoloc satisfies C_1 if its ?city value is "Tunisia" or "Italy" or "Libya". The other form of constraint is a join constraint, which specifies a constraint on values of two or more properties from different concepts. A pair of triples of geoloc and seaport satisfy a join constraint C_2 if they share common values on the property *glc_cd* (geographic location code). The constraints used in example queries are as:

C_0: *seaport(?name,?storage, _, _, _), ?storage* $\leq$ *150000.*
C_1: *geoloc(?name1, _, ?cty, _, _), member(?cty, ["Tunisia", "Italy", "Libya"]).*
C_2: *geoloc(?name1, ?glc_cd, _, _, _), seaport(?name2, ?glc_cd, _, _, _, _).*

For property country, the system can generalize from the 1 valued triple, a triple pattern candidate constraint:

geoloc(?name, _, ?cty, _, _), ?cty = "Malta",

It is because the country value of all 1 valued triples is Malta.

Suppose the system is verifying whether join constraint C_2 is consistent with the 1 valued triple. Since for all 1 valued triples, there is a corresponding triple in seaport with a common glc_cd value; the join constraint C_2 is consistent and is considered as a triple pattern candidate constraint.

In this example, a new concept seaport is introduced to describe the 1 valued triples in geoloc. The search space is now expanded into two levels, as illustrated in Fig. 7. The expanded constraints include a set of internal disjunctions on properties of seaport, as well as join constraints from seaport to another concept. If a new join constraint has the maximum gain/cost ratio and is selected later, the search space will be expanded further. Figure 7 shows the situation that when a new concept, say channel, is selected, the search space will be expanded one level deeper. At this moment, triple pattern candidate constraints will include all unselected internal disjunctions on properties of geoloc, seaport and channel, as well as all possible joins with new concepts from geoloc, seaport and channel.

C_3: geoloc(?name, _, "Malta",,).
C_4: geoloc(?name, ?glc_cd, _, _, _), seaport(_, ?glc_cd, _, _, _, _).

Suppose *geoloc* is 30,000, and seaport is 800. The cardinality of *glc_cd* for *geoloc* is 30,000 again and for *seaport* is 800. Suppose both concepts have indices on *glc_cd*. Then, the evaluation cost of C_3 is 30,000, and C_4 is (30,000 × 800)/30,000 = 800. The gain of C_3 is 30,000 – 4 = 29,996, and the gain of C_4 is 30,000 – 800 = 29,200, because only 4 triples satisfy C_3 (refer Fig. 7) while 800 triples satisfy C_4. (There are 800 *seaports*, and all have a corresponding *geoloc* triple.) So, the *gain/cost* ratio of C_3 is 29,996/30,000 = 0.9998, and the *gain/cost* ratio of C_4 is 29,200/800 = 36.50. So, the system will select C_4 and add it to the hypothesis. Since C_4 was selected, the system will expand the search space by constructing consistent internal disjunctions and join constraints on seaport. Assuming that the system cannot find any candidate on *seaport* with 1 valued. It will backtrack to consider candidates on geoloc again and select the constraint on country (Fig. 7). Now, all 0 valued triples are excluded. The system thus learns the query:

Q3: answer(?name):
geoloc(?name, ?glc_cd, "Malta", _, _),
seaport(_, ?glc_cd, _, _, _, _).

The operationalization component will then take the equivalence of the input query Q22 and the learned query Q3 as input:

geoloc(?name, _, "Malta", _, _) ↔
oloc(?name, ?glc_cd, "Malta", _, _) ∩ aport(_, ?glc_cd, _, _, _, _).

and will deduce a new rule that can be used to reformulate Q22 to Q3:

geoloc(_, ?glc_cd, "Malta", _, _) → aport(_, ?glc_cd, _, _, _, _)

This is equivalent to rule R2. Since the size of *geoloc* is considerably larger than that of *seaport*, the next time when a query asks about geographic locations in Malta, the system can reformulate the query to access the seaport concept instead and speed up the query answering process.

5.1 Complexity of the Proposed Algorithm

When a new concept D is introduced as a primary concept or by selection of a join, the number of concept scans is bounded by $(1 + J(D)) + (A(D) + J(D))$, where $J(D)$ is the number of legal join paths to D and $A(D)$ is the number of properties of D. Constructing candidate constraints requires scanning the concept $1 + J(D)$ times, because constructing all internal disjunctions on D needs one scan over D and constructing join constraints needs an additional scan over each joined concept. Each iteration of *gain/cost* evaluation and selection needs to scan D once. In the worst case, if all candidate constraints are selected to construct the alternative query, it will require scanning the concepts $A(D) + J(D)$ times. Since usually a query involves a small number of concepts and expansions in learning are rare, the number of concept scans is linear with respect to the number of properties in most cases.

6 Conclusion

The knowledge required for SPARQL query optimization can be learned inductively under the guidance of input queries. In the described general learning framework, inductive learning is triggered by queries, and an inductive learning algorithm for learning from many concepts of the ontologies is proposed. The learned semantic knowledge for SPARQL query is producing substantial cost reduction for a real-world ontological system as shown using an example. It may be possible that the result of a SPARQL query is directly inferred from a rule or a chain of rules. In that case, there is no need to access the ontology, and saving will be much more. It may also possible that when a SPARQL query is sent to system, the system infers that it is unsatisfiable for a rule or chain of rules. In that case also, there is no need to access the ontology. This will also lead to avoidance of irrelevant access to ontology, and saving will be very high.

References

1. Silberschatz A, Korth F, Sudarshan S (2011) Introduction. Database system concepts, 6th edn. McGraw Hill, New York (chapter 1)
2. Jarke M, Koch J (1984) Query optimization in database systems. ACM Comput Surv 16(2):111–152
3. Silberschatz A, Korth F, Sudarshan S (2011) Query optimization. Database system concepts, 6th edn. McGraw Hill, New York (chapter 13)
4. Pérez J, Arenas M, Gutierrez C (2006) Semantics and complexity of SPARQL. In: Cruz I, Decker S, Allemang D, Preist C, Schwabe D, Mika P, Uschold M, Aroyo L (eds) Proceedings of the 5th international semantic web conference. Lecture Notes in computer science, vol 4273. Springer, Berlin, pp 30–43
5. Gutierrez C, Hurtado CA, Mendelzon A (2004) Foundations of semantic web databases. In: Proceedings of the ACM symposium on principles of database systems (PODS), pp 95–106
6. Adaptive server enterprise, performance and tuning guide: vol 2: query optimization, query tuning and abstract plans (2002). http://citeseerx.ist.psu.edu/viewdoc/download?doi=10.1.1.135.4127&rep=rep1type=pdf
7. Ullman D (1988) Principles of database and knowledge-base systems, vols I and II. Computer Science Press, Palo Alto, CA
8. Horrocks I, Patel-Schneider PF, van Harmelen F (2003) From SHIQ and RDF to OWL: the making of a web ontology language. J Web Semant 1:7–26
9. Zhang Z (2005) Ontology query languages for the semantic web: a performance evaluation. Department of Computer Science, University of Georgia, USA
10. Maganaraki A, Karvounarakis G, Christophides V, Plexousakis D, Anh T (2002) Ontology storage and querying. Technical report 308, Foundation for Research and Technology Hellas, Institute of Computer Science, Information Systems Laboratory

11. Ruckhaus E (2004) Efficiently answering queries to DL and rules web ontologies. W3C
12. Haarslev V, Wesse M (2004) Querying the semantic web with racer + nRQL. In: Proceedings of the KI, 2004 international workshop on ADL'04, Ulm, Germany
13. Möller R, Haarslev V (2001) In: Proceedings of the international workshop on description logics. DL2001, 2001, pp 132–141
14. Magkanaraki A, Karvounarakis G, Anh T, Christophides V, Plexousakis D (2002) Ontology storage and querying. Technical report 308, Foundation for Research and Technology Hellas, ICS, Information System Laboratory
15. Baader F, Sattle U (2001) An overview of tableau algorithms for description logics. Stud Logica 69(1):5–40
16. Grosof BN, Horrocks I, Volz R, Decker S (2003) Description logic programs: combining logic programs with description logic. In: Proceedings of the twelfth international world wide web conference, pp 48–57
17. Haussler D (1998) Quantifying inductive bias: AI learning algorithms and Valiant's learning framework. Artif Intell 36:177–221
18. Bernstein A, Kiefer C, Stocker M (2007) OptARQ: a SPARQL optimization approach based on triple pattern selectivity estimation. Technical report 2007.03, Department of Informatics, University of Zurich, Switzerland

Impact of Information Technology on Job-Related Factors

Virendra N. Chavda and Nehal A. Shah

Abstract During the last two decades of the twenty-first century, information technology played a vital role on various companies and their work environments, employees, individuals and society. With the advancement and growth of telecommunications, customized business software as well as digital computing, there have been changes in the workplace scenario and effects on individual job characteristics. This paper tries to identify the effect of information technology on various job factors like job satisfaction, work–life Balance, health and safety, performance and productivity. Data was collected with the help of a survey questionnaire based on Likert scale from 100 IT employees. Exploratory factor analysis and regression analysis are used to identify the effect of information technology on job factors like job satisfaction, work–life balance, health and safety, performance and productivity.

Keywords Information technology · Job satisfaction · Work–life balance · Health and safety · Performance and productivity

1 Introduction

Information technology is the application of computers, various technologies, devices and software to handle the mass quantity of data. Information technology is concerned with managing and processing of information with the help of computers, devices and various communication technologies. IT helps to adapt and change the dynamics of business by providing new learning aspects and helps to compete with modern business scenarios and economy. Last two decades of the twenty-first century witnessed maturation of digital computing with the various telecommunication technologies, and this directly have impact on the various work environments of

V. N. Chavda (✉) · N. A. Shah
Gandhingar Institute of Technology, Moti Bhoyan, Gandhinagar 382721, India
e-mail: chavdavirendra@gmail.com

N. A. Shah
e-mail: nehalshahmba@gmail.com

K. Kotecha et al. (eds.), *Data Science and Intelligent Applications*, Lecture Notes on Data Engineering and Communications Technologies 52,
https://doi.org/10.1007/978-981-15-4474-3_15

companies, employees and their relations, individuals and society as a whole. Peter Drucker has rightly identified that with the advancement of technologies, there will be direct effect on industry, companies and employees. Because of IT enhancement, digital data are generated at very fast rate which are easily accessible and interrelated. With the incorporation of robotics and artificial intelligence by information technology, digital business (E-business) as well as digital economy has started. These factors lead to identify the effect of information technology on modern business and employees. Information technology impacts at organization can be summarized in following terms:

- Internal processes of the companies
- Human resources of the company
- Structure of organization
- Change in relationship between companies and their various stake holders (customers, suppliers, investors).

1.1 Effect of Technology on Job-Related Factors

Job satisfaction deals with the expectations and rewards that job provides to the employee. Job satisfaction will be formed with the factors like wages, working conditions, relations between employees, employment supervision, grievances handling and fair treatment by employer. With the use of information technology, i.e., computer-aided manufacturing, virtual reality, expert systems companies enhanced their customer service and improved communications. Companies established high performance work systems with the use of technology-enabled human resource management practices. Companies have identified various categories of work, training, programs and reward system based on the new technology, and this leads to satisfaction and dissatisfaction among employees. Work–life balance is a state of equilibrium set by persons form their work conditions and their personal lifestyle. New technologies enable us to work more efficiently and more quickly than earlier. It allows employees to work at flexible timing, remote working even allowed to work from home. This leads to imbalance in work–life balance. According to report of World Health Organization, due to information technology lifestyle, diseases [a subset of non-communicable disease (NCDs)] like obesity, diabetes, anxiety, eye diseases, muscular and joint pains, chronic lung disease and cardiovascular diseases develop among employees. With the use of technology, organizations have improved individual as well as group performance. Companies use advanced technologies to identify and evaluate employee's performance which encourage the employees to learn new technology and gain knowledge. Compared to young technological savvy generation, old generation people find it annoying and reluctant to adapt the new changes and ultimately effect on their job performance. Changes in technology helps the companies to work harder when high demand with shifts and reduce their production when demand is low. With the rise of new technology in low period time, old technology becomes obsolete, and employees have to find the ways to cope up with the changes.

2 Literature Review

According to Morgan [1], the use of IT provides opportunity to be innovative at when we work, where we work and the way we work. Marler and Dulebhohn [2] developed a model based on employee self-service and usage of technologies. The outcome of model suggests that individual, technological and organizational factors are related to individual intentions to use technology. Lawless and Anderson [3] identify that technological advancement makes employees more effective and companies more valuable. Li and Deng [4], identify that technological advancement increase firm's performance and profitability. Howcroft and Taylor [5] identify that with the use of technology, labour utilization and scheduling of work affect on how employees work and lead to the melting of organizational boundaries. Turner [6] studied the effect of system implementation on employee's job performance and found that initial employees are positive towards the technological change and with the passing of time, they felt that they need training and management increased their job duties. He also found that there is increase in job responsibilities without any salary hike. Attar and Sweis [7] identify that information technology will raise the employee job satisfaction. Arnetz [8] in his study identifies that those who are working with telecommunications systems and video display technology like to suffer bodily, mental and psycho-physiological reaction. According to Alam [9], there is direct relationship between increase in technological complexity and productivity. With increase of technology, this will create techno stress and ultimately has negative impact on productivity. Imran et al. [10] identify that technological advancement has positive effect on employee performance if proper training has been provided. Mitchell [11] identifies that technology helps them for speed and ease of task completion which leads a balance between work life and personal life.

3 Research Objective

To study effect of technology on job factors like job satisfaction, work–life balance, health and safety, performance and productivity. To identify which factor is the most affected by adoption of new technology.

4 Research Methodology

- Research Approach and Nature of Data: For gathering primary data, survey approach was used
- Research Instrument: For this research, questionnaire was used. The questionnaire was designed using Likert scale
- Sample unit: Employees of IT companies
- Sample Area: Ahmedabad
- Sample Size: A total of 150 respondents
- Sample Procedure: Non-probability convenience sampling.

5 Data Analysis and Interpretation

The questionnaire has 25 questions divided in five sections. The first section has five questions comprising of job satisfaction, then five questions each for productivity, work–life balance, performance and health and safety. The responses of 150 respondents were analysed with the help of SPSS software.

5.1 Reliability Test Using Cronbach's Alpha

The Cronbach's Alpha score ranges from 0.711 to 0.91. All the values are above 0.70. In present study, all the values are above 0.8. It shows that the data collected is valid and reliable (Table 1).

Table 1 Reliability statistics

Variable	Cronbach alpha
Job satisfaction	0.886
Work–life balance	0.812
Health and safety	0.839
Performance	0.822
Productivity	0.801

Source Primary data

Table 2 KMO and Bartlett's test

Kaiser–Meyer–Olkin measure of sampling adequacy		0.801
Bartlett's test of sphericity	Approx. Chi-square	8206.225
	df	149
	Sig.	0.000

5.2 *Factor Analysis*

5.2.1 Suitability of Factor Analysis (for the Research)

Kaiser–Meyer–Olkin (KMO) measure indicates the suitability of factor analysis. Here, the value is 0.801 which shows that the factorization is possible.

High values of the KMO score 0.801 (above 0.5 and up to 1.0) and the Bartlett's test were significant (Chi 8206.225, df = 149; as per Table 2). This implies that the correlations between pairs of variables can be explained by other variables and that factor analysis was found suitable for this research. To determine the method of factor analysis, principal components analysis was used. The purpose was to obtain the minimum possible number of factors, referred as *principal components*, accounting for maximum variance in the data, for further multivariate analysis. Table 3 shows the exploratory factor analysis output of primary data. It shows final factor output with their respective factor loadings.

Total variance of each variable which constitute by initial Eigen values and rotation sums of squared loadings was calculated using varimax rotation. After rotation, the components are combined into one group of performance having Eigen value 4.867, and % of total variance is 47.036%. The components combined into group of job satisfaction have Eigen value 3.963, and % of total variance is 36.462%. The components combined into group of health and safety have Eigen value 2.715, and % of total variance is 29.529%. The components combined into group of productivity have Eigen value 2.38, and % of total variance is 18.665%. The components combined into group of work–life balance have Eigen value 1.251, and % of total variance is 11.949%.

5.3 *Regression Analysis*

Table 4 shows the output of various regression analyses between effects of technology on job-related factors. The result shows that the relationship between technology and performance is quite reasonable $R = 0.798$ and 63.6% variance (Adjusted R^2). 63.6% variance can be explained by performance. Similarly, technology have variations in job satisfaction is 48.9%, in health and safety is 33.8%, in productivity is 24.2% and in work–life balance is 23.1%.

Table 3 Exploratory factor analysis

Factors	Loading	Eigen value	% of variance
Factor 1: performance		4.867	47.036
My job scopes are improved	0.774		
My skill, knowledge and competencies are improved	0.765		
My work becomes more efficient and effective	0.679		
New technology helps to identify advanced and improved trainings	0.637		
Factor 2: job satisfaction		3.963	36.462
The technological changes have given better work opportunities	0.908		
New technology helps to handle number of responsibilities at one time	0.856		
New technological changes keep me committed towards work	0.799		
New technology helps my work to be appreciated	0.708		
Factor 3: health and safety		2.715	29.529
Working conditions becomes good and safe	0.788		
Technology reduces health problems	0.758		
Technology reduces stress level	0.728		
Factor 4: productivity		2.38	18.665
It takes less time to finish my work	0.745		
All work commitments are justified	0.732		
Flexibility and integration in production increased	0.699		
Work productivity has been increased	0.627		
Factor 5: work–life balance		1.251	11.949
I do not feel stressed while working	0.815		
Due to technological changes, working hours reduced	0.801		
Quality of work life has improved	0.784		
After work, I can also focus my personal life	0.702		

Source Primary data

Table 4 Regression analysis output

Dependent variable	R	R^2	F value	B	t	Sig.
Performance	0.798	0.636	222.482	0.423	6.201	0.000
Job satisfaction	0.699	0.489	232.499	0.354	7.508	0.000
Health and safety	0.582	0.338	124.426	0.325	4.350	0.000
Productivity	0.492	0.242	152.116	0.312	6.387	0.000
Work–life balance	0.481	0.231	149.212	0.299	4.271	0.000

Source Primary data

The above regression analysis output shows that the technology plays a major role in performance. Performance is highly affected by introduction of various technology, followed by job satisfaction, health and safety, productivity and lastly in work–life balance.

5.4 Findings and Recommendations

From the analysis, it is found that by the introduction of new technology, there is effect on various job-related factors. Performance is highly affected followed by job satisfaction, health and safety, productivity and work–life balance. Majority of the firms used latest technology to enhance their productive efficiencies through reduction of processing time and costs. Based on the present study, following recommendations are identified:

- Awareness of new technology is must before implementation.
- New technology must be introduced and implemented after through training and under expert guidance.
- Employees should be aware and informed well in advance.
- Employer must ready to justify the changes among employees and other stake holders
- Only trained and expert people must implement the new technology.

5.5 Limitations and Scope of Future Research

This study focuses on the IT employees of Ahmedabad. Majority of the respondents have prior duties and assigned work due to that they are not ready to waste their time in answering questions. Important informations are not provided by them regarding the various technologies that they are using and other work-related duties. It was not easy to make understand the different employees about importance and relevance of present research. Future research can be done by prior engaging the different company's managers and convincing them the importance of the study that will help employee to understand the study. This study focuses on all available technologies and its effect on job-related factors, and in future studies, there must be separate identification of technology and its effect on specific job-related factors.

6 Conclusion

IT employees have shown that there is high effect of technology on selected job factors in present study. Results shows that the effect of various job factors is affected to maximum extent on use of mail, availability of various data software, reducing time in data analysis and report making, presenting reports via various visualization software. Due to various technologies, efforts are less on recalling the information. Due to constant changes and upgradation of technology, expectations of employees as well as organizations are rising at very high level. This might lead in reduction of performance in future.

References

1. Morgan J (2014) The future of work: attract new talent, build better leaders, and create a competitive organization. Wiley, Hoboken, NJ
2. Marler JH, Dulebhohn JH (2005) A model of employee self-service technology acceptance. Res Pers Hum Resources Manag 24:137–180
3. Lawless MW, Anderson PC (1996) Generational technological change: effects of innovation and local rivalry on performance. Acad Manag J 39:1185–1217
4. Li Y, Deng S (1999) A methodology for competitive advantage analysis and strategy formulation: an example in a transitional economy. Eur J Oper Res 118:259–270
5. Howcroft D, Taylor P (2014) Researching and theorising 'new' new technologies. New Technol Work Employ 29(1)
6. Turner KM (2017) Impact of change management on employee behavior in a university administrative office. Walden Dissertations and Doctoral Studies Collection, Walden University
7. Attar GA, Sweis RJ (2010) The relationship between information technology adoption and job satisfaction in contracting companies in Jordan. J Inf Technol Constr 15:44–61
8. Arnetz B (1997) Technological stress: psycho physiological aspects of working with modern information technology. Scand J Work Environ Health 23:97–103
9. Alam MA (2016) Techno-stress and productivity: survey evidence from the aviation industry. J Air Transp Manag 50:62–70
10. Imran M, Maqbool N, Shafique H (2014) Impact of technological advancement on employee performance in banking sector. Int J Hum Resource Stud 4(1)
11. Mitchell P (2007) The effect of technology in the workplace: an investigation of the impact of technology on the personal and work life of mid-level managers of the fairfax county police department. University of Richmond UR Scholarship Repository

A Study on Preferences and Mind Mapping of Customers Toward Various Ice Cream Brands in Ahmedabad City

Nehal A. Shah and Virendra N. Chavda

Abstract Due to the geographical condition of India, the consumption of ice cream reached more than 2 L per annum. Due to heavy demand of ice cream by Indian consumers, helps the country to increase in packaged offerings as well as various parlors. With the advancement of technology and various frozen alternatives, customers are enjoying the ice cream in all the seasons. Most of the occasions, festivals, fun and enjoyment resulted in rocket sales of ice cream resulted in compound average growth rate of more than 15% in last decade. The present Indian ice cream market of organized and unorganized valued at 3000 Cr. The branded market comprises of home players as well as international players like Amul, Vadilal, Havmore, Kwality Walls and others. All these companies have their exclusive outlets as well as availability at every premier locations of India. Considering all this it is required to identify brand awareness, positioning and customer perceptions about various brands present in Indian market. The present study focuses on various factors affecting choice of ice cream brands in the city of Ahmedabad as well as their various promotion strategies used to lure the customers.

Keywords Brand awareness · Customer preferences · Ice cream · Perceptual/mind mapping

1 Introduction

An ice cream (derived from the earlier iced cream or cream ice) is combination of milk and cream with ingredients like fruits, nuts and other flavored ingredients. Sugar, colors as well as artificial ingredients are mixed to develop various flavors of ice cream. No natural materials are used to develop the ice cream. All these mixtures

N. A. Shah (✉) · V. N. Chavda
Gandhinagar Institute of Technology, Gandhinagar, Gujarat, India
e-mail: nehalshahmba@gmail.com

V. N. Chavda
e-mail: chavdavirendra@gmail.com

K. Kotecha et al. (eds.), *Data Science and Intelligent Applications*, Lecture Notes on Data Engineering and Communications Technologies 52,
https://doi.org/10.1007/978-981-15-4474-3_16

are slowly rotated by a machine to add air and to avoid any crystals in mixture; this will generate a mixture which is semi-solid. This semi-solid foam is used to make scoop.

The meaning of the phrase "ice cream" varies from one country to another. Phrases such as "frozen custard," "frozen yogurt," "sorbet," "gelato" and others are used to distinguish different varieties and styles. In some countries, such as the USA, the phrase "ice cream" smears only to a specific variety, and most governments regulate the commercial use of the various terms allowing to the relative quantities of the main ingredients. Products that do not come across the criteria to be called ice cream are labeled "frozen dairy dessert" instead. In other countries, such as Italy and Argentina, one word is used for all variants. Analogues made from dairy alternatives, such as goat's or sheep's milk, or milk substitutes, are existing for those who are lactose intolerant, allergic to dairy protein, or vegan [1].

1.1 Ice Cream Industry Scenario in India

In India, due to de-legalization in ice cream sector in 1997, the total size of ice cream market increased to more than 16 billion. Out of this, 30% market is in organized sectors which are valued at 5 billion Rs. HUL have intake of around 50% in this organized sector. The cooperative brand Amul has market share of 35%, while remaining market shared by Vadilal and Havmor with 9% of the market share. All the companies are trying hard to increase their market share by adding various flavors of ice cream in their product category.

2 Literature Review

2.1 History of Ice Cream

The ancient civilizations have used cold products with the use of ice for thousands of years. As per the BBC history reports, around 200 BC, China has able to create the frozen mixture of cold products with milk. King Tang of Shang with more than 90 men created frozen items with milk, flour and other ingredients. The Roman Emperor Nero (37–68 AD) had carried ice from the mountains and combined with fruit toppings. These are initial development toward the current ice cream. This item is known as "faloodeh" in Iran and basically made from wheat and milk [2].

2.2 Customer Perspective

Perceptions about how products perform on outstanding attributes are more vital to consumers' purchase behavior than actual product attribute performance. Attributes are those extents of a product that define a given consumption experience. They represent the building blocks that consumers use to make product decisions and form purchase decision [3]. They simply define "attributes" as those dimensions that form the criteria for evaluating a particular consumption experience, however, narrow or broad that experience may be. Products are known to represent a bundle of attributes, such as packaging, labeling, brand name, etc. [4].

3 Research Methodology

3.1 Objectives of the Study

- To study different factors which are considered at the time of purchasing the ice cream.
- To know the customer preferences regarding flavors for ice cream.
- To study mind/perceptual mapping of customer about ice cream.

3.2 Significance of the Study

- The study will help ice cream companies in improving their selling strategy and positioning strategies across all market.
- The study will help mfg. to focuses precisely on to understand the customer needs and individual perceptions toward different ice cream brands.

3.3 Research Plan

The study is descriptive in nature. The researcher adopted this research design to gather information from the respondents to assess the preferences of customers and factors affecting among the purchase of ice cream. The study was conducted with 306 samples from Ahmedabad city of Gujarat state. The sampling method adopted for the study was non-probability convenient sampling where the researcher can select the sample elements based on the ease of access. Structured questionnaire was prepared to survey the respondents. The primary data has been collected through a structured

Table 1 Instances of ice cream usage

Particulars	Percentage (%)
Party	21
Shopping	19
Festivals	25
Entertainment	31
Others	3
No response	1

questionnaire from the respondents. The secondary data has been collected from the books, journals, magazines and Web sites.

4 Data Analysis and Interpretations

4.1 Consumption Situation for Ice Cream

From Table 1, we can conclude that the entertainment is having highest percentage as consumption situation.

4.2 Preference of Flavors for Ice Cream

Table 2 reveals that mostly consumers like the ice cream which is more than one flavor, i.e.: 169.

Table 2 Flavors of ice cream

Particulars	Responses
Ice cream with plain flavor	147
Ice cream with premium flavor having a lot of nuts and cashews, etc.	122
Sundaes ice cream (dessert)	97
Ice cream with more than one flavor	169
No response	3

Table 3 Factors affecting selection of ice cream (in %)

	Extremely imp.	Imp.	Moderate	Least imp.	Not at all imp.	No response
Brand name	45	41	6	1	1	6
Flavor	60	21	9	2	1	7
Price	11	20	29	20	12	8
Packaging	7	17	32	21	14	9
Quality	85	3	1	0	3	8

Table 4 Favorite flavor of ice cream

Particulars	Percentage (%)
Chocolate	32
Vanilla	11
Peppermint	6
Strawberry	13
Fruit flavors	13
Most flavors with nuts	23
Other	1
No response	1

4.3 *Factors Affecting Selection of Ice Cream*

Table 3 shows that quality is the most important factor for selection of ice cream followed by flavor.

4.4 *Your Favorite Ice Cream Flavor*

Table 4 concludes that 32% customers like to consume chocolate ice cream followed by 23% as nuts flavor.

4.5 *Which Type of Promotion Would Attract You to Buy More Ice Cream?*

Table 5 shows that BOGO offer is the most preferred followed by discount coupons.

Table 5 Promotional preferences for ice cream

Particulars	Percentage (%)
Buy 1 get 1 free (BOGO)	47
Discount coupons	26
Occasional privileges (birthday discount)	14
Membership privileges	11
Any other	0
No response	2

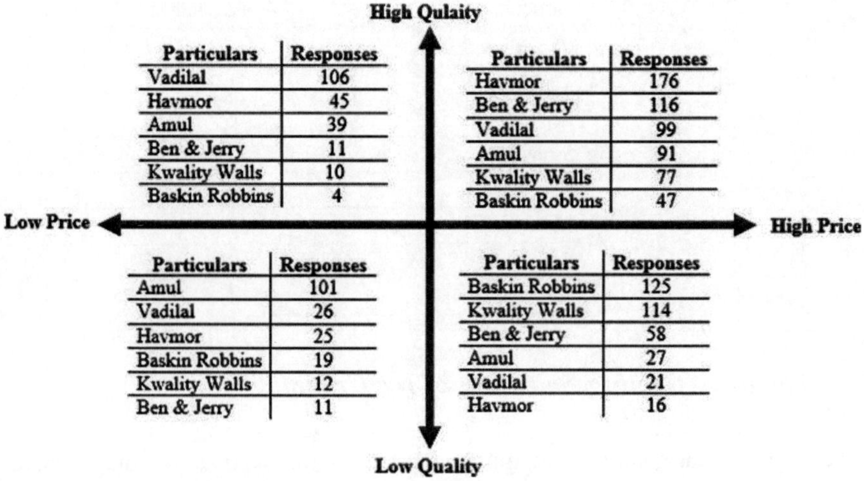

Fig. 1 Mind/perceptual map of various ice cream brands in the mind of respondents of Ahmedabad city

4.6 *Mind/Perceptual Mapping*

See Fig. 1.

5 Hypothesis Testing (Chi-Square)

H_0: Consumption of ice cream is independent form gender.
H_1: Consumption of ice cream is dependent form g**ender**.

From Table 6, it can be seen that the chi-square is 0.001 and on comparing it with Alpha 0.05, it can be seen that the chi-square is <Alpha. So reject the null hypothesis H_0. It means that consumption of ice cream is dependent form gender.

Table 6 Consumption of ice cream and gender

	Value	df	Asymp. sig. (2-sided)
Pearson chi-square	18.210	2	0.001
Likelihood ratio	11.177	2	0.059
Linear-by-linear association	0.465	1	0.500
No. of valid cases	206		

Table 7 Selection of ice cream and age

	Value	df	Asymp. sig. (2-sided)
Pearson chi-square	51.039	27	0.048
Likelihood ratio	51.091	27	0.048
No. of valid cases	306		

$\mathbf{H_0}$: Selection of ice cream flavor is independent from age.
$\mathbf{H_1}$: Selection of ice cream flavor is dependent from **age**.

From Table 7, it can be seen that the chi-square is 0.048 and on comparing it with Alpha 0.05, it can be seen that the chi-square is <Alpha. So reject the null hypothesis H_0. It means that the selection of ice cream flavor is dependent from age.

$\mathbf{H_0}$: Purchase of ice cream is independent from monthly income.
$\mathbf{H_1}$: Purchase of ice cream is dependent from monthly i**ncome**.

From Table 8, it can be seen that the chi-square is 0.000 and on comparing it with Alpha 0.05, it can be seen that the chi-square is <Alpha. So reject the null hypothesis H_0. It means that the purchase of ice cream is dependent from monthly income.

$\mathbf{H_0}$: Selection of flavor for ice cream is independent from gender.
$\mathbf{H_1}$: Selection of flavor for ice cream is dependent from ge**nder**.

Table 8 Purchase of ice cream and monthly income

	Value	df	Asymp. sig. (2-sided)
Pearson chi-square	119.333	27	0.000
Likelihood ratio	90.340	27	0.000
No. of valid cases	306		

Table 9 Flavors of ice cream and gender

Table.	Value	df	Asymp. sig. (2-sided)
Pearson chi-square	30.430	27	0.004
Likelihood ratio	27.360	27	0.013
No. of valid cases	306		

From Table 9, it can be seen that the chi-square is 0.004 and on comparing it with Alpha 0.05, it can be seen that the chi-square is <Alpha. So reject the null hypothesis H_0. It means that selection of flavor for ice cream is dependent from gender.

6 Findings

- As per survey 90% respondents were consuming an ice cream.
- As per the survey most of customers' means 31% customers were mostly consume ice cream for the entertainment purpose.
- Most of the customers were generally prefer ice cream with more than one flavor and majority customers were purchased from retail outlet.
- Research analyzed that most of the market is covered by Havmor, Vadilal followed by Amul brand.
- As per the survey most important factor for selection while purchasing an ice cream is quality followed by flavor and brand name.
- According to the customer's perspective, chocolate and most flavors with nuts flavors were highly favorite in market and they mostly preferred to have an ice cream in family pack.
- As per the survey 47% customers were attracted from buy 1 get 1 free offer (BOGO) promotion offer to buy an ice cream.
- As per the survey Havmor is providing high quality high price, Vadilal provide high quality low price, Amul provide low quality low price and Baskin Robbins provide low quality high price.
- Hypothesis testing revealed that demographic factors like age, gender, income plays most important role for ice cream consumption.
- Mind mapping depicted that Havmor provides "high quality high price," Vadilal provides "high quality low price," Amul provides "low quality low price" and Baskin Robbins provides "low quality high price."

7 Conclusion

Ice cream remains one of the most complex foodstuffs currently available. Not just from a structural, formulation and manufacturing level, but also in the context of consumer perception and eating behavior. From this research, it is evident that "Havmor" is most popular brand of ice cream in Ahmedabad city followed closely by Vadilal and Amul. While other brands like Kwality Walls and Ben and Jerry are not that much popular. Findings suggested that "quality" is the most important factor while choosing an ice cream followed by "flavor." Respondents reveled that "chocolate" is the most liked flavor followed by "flavors with nuts." As far as the promotional strategy is concerned, "BOGO" is the best promotional strategy to attract the customer for the ice cream. Really in the modern times, where ready-to-eat foods are in great demand; ice cream makes an excellent choice for being unique as a cold dessert.

References

1. Bikram M, Mandeep K (2013) Exploring branding strategies of FMCG, services and durables brands: evidence from India. J Prod Brand Manag 22(1):6–17
2. Acebron L, Mangin JL, Dopico DC (2000) A proposal of the buying model for fresh food products: the case of fresh mussels. J Int Food Agribus Mark 11(3):75–96
3. Puth G, et. al (2000) Consumer perceptions of mentioned product and brand attributes in magazine advertising. J Prod Brand Manag 8:38–49
4. Mueller S, Szolnoki G (2010) Food Qual Prefer 21:774–783

A Review on Big Data with Data Mining

Mayur Prajapati and Shreya Patel

Abstract Data mining is a procedure of driving hidden, unknown but potentially convenient information from massive data. While big data presents technologies for assembling, processing, analyzing and extracting convenient data from very huge volumes of structured and unstructured data processed by various sources at high speed, big data has great impacts on scientific discoveries and value creation. This paper defines the 5Vs of Big data along with the distinction of big data and big data analytics followed with the architecture of big data. Some tools representing Hadoop ecosystem are also presented.

Keywords Big data · Big data analytics · Data mining · Hadoop

1 Introduction

Data mining is an approach for discovering interesting patterns as well as descriptive and understandable models from huge scale data. DM can be used to find correlations or patterns among dozens of fields in large relational database [1]. Some of the data mining application in big data includes analyze bottlenecks in the system, fraud detection in the system, the large data processing system, architect highly scalable distributed systems, find unexpected relationships between different variables, real-time analysis of the system [2]. Big data analytics is complicated method of examining massive and varied datasets, to uncover information like hidden patterns, unknown correlations and market trends that facilitate organizations and make knowledgeable business decisions. The importance of big data analytics is directed by specialized analytics systems and software system, also as high-powered computing systems,

M. Prajapati
Shankersinh Vaghela Bapu Institute of Technology, Gandhinagar, India
e-mail: mayur.prajapati@bapugkv.ac.in

S. Patel (✉)
Gandhinagar Institute of Technology, Ahmedabad, India
e-mail: shreya.patel@git.org.in

K. Kotecha et al. (eds.), *Data Science and Intelligent Applications*, Lecture Notes on Data Engineering and Communications Technologies 52,
https://doi.org/10.1007/978-981-15-4474-3_17

big data analytics overture varied business advantages, which counts: New revenue opportunities, more practical marketing, higher client service, improved operational potency, competitive benefits over rivals.

1.1 What Is Big Data?

Big data in simple terms rectifies the datasets because of their abundant volume, for example, in the order of magnitude of petabytes, the definition is related to the fact that the dataset is too huge to be managed without using new algorithms or technologies. It indicates huge volume of data that can be structured, semi-structured and unstructured. It consists of 5 Vs [3, 4], i.e.,

- Volume: It invokes a quantity of knowledge which contains huge integer once it involves massive data.
- Variety: It invokes heterogeneous sources like social media, net server logs, etc.
- Velocity: It invokes too how briskly knowledge is expanding, knowledge is exponentially expanding and at an awfully quick ratio.
- Veracity: It invokes to the inconsistencies and uncertainty of knowledge like social media suggests that if the information is often trusty or not.
- Value: It invokes to the information that we have a tendency of storing and processing is worth and have a tendency of getting benefit from this huge quantity of knowledge.

Based on this, big data can be analyzed for understanding that results in higher choices and strategic business moves. It conjointly aims to an abundant amount of coarse and unstructured data from numerous authorities. Due to its immense amount and tremendous truthfulness essence, it usually needs a lot of estimating power to assemble and interpret. The content is decoded through varied digital mediums like mobile, Internet, social media, etc., and is then employed by employment to create appropriate choices, whereas data analytics provides operational vision into employment. It includes ancient information and so attracts out assumptions from them to seek out improved results to complicated employment issues. Data analysis is handled at a lower level, whereby data associated with the matter are apparently inspected over and defined with a selected aim in mind [2].

1.2 Distinction Between Big Data and Big Data Analytics

Think of big data as a library you visit when the information to answer your question is not readily available. Data analysis is like the book you are taking and browse to find the answers to your question [2]. This is the fundamental difference between them. Data analysis is more focused than big data because instead of gathering huge piles of unstructured data, data analysts have a specific goal in mind and sort out the relevant data to find ways to obtain data. While, big data is a compilation of huge amounts of data that requires a lot of filtering to get useful information. Another notable difference is big data uses complicated technology tools like parallel computing and other automation tools to handle big data. Data analysis uses predictive and statistical modeling with relatively simple tools.

2 Correlation Between Data Mining and Big Data

The relevance of big data defines what you get out of information and not what we have. We evaluate information to reduce value and time. Challenges are keeping such an oversized amount of information expeditiously. And how do we process and extort valuable information from this immense quantity of information at intervals of a given timeframe? And the result for the similar is Hadoop and Spark framework. Comparison of the big data and data mining is delineated as below:

- Data Mining chiefly focuses on several details of data, whereas big data focuses on several relationships between data.
- Data mining represents lots of detail, whereas big data represents lots of relationships.
- Data mining expresses the information about the data, whereas big data expresses why of the information.
- Data mining may be used for static data, whereas big data refers to an outsized quantity of knowledge sets.
- Data mining is a method for analyzing knowledge, whereas big data is a plan than an explicit term.
- Data types such as structured data, relational and dimensional database are used in data mining, whereas data types such as structured, semi-structured and unstructured data (in NoSQL) are used in big data.
- In data mining analysis chiefly applied for statistical analysis, concentrate on prediction and discovery of business factors on a little scale, whereas in big data analysis chiefly for applied data analysis, concentrate on prediction and discovery of business factors on a giant scale.
- In data mining, the results display mainly strategic decision making, whereas in big data, the results display the dashboards and prognostication measures.

Based on the difference, the conclusion states that big data solely refers to an oversized quantity and every one the large data solutions depend on the supply of information. It may be a combination of business intelligence and data mining. Briefing that big data is the quality and data mining is the manager of that is accustomed to offer helpful results.

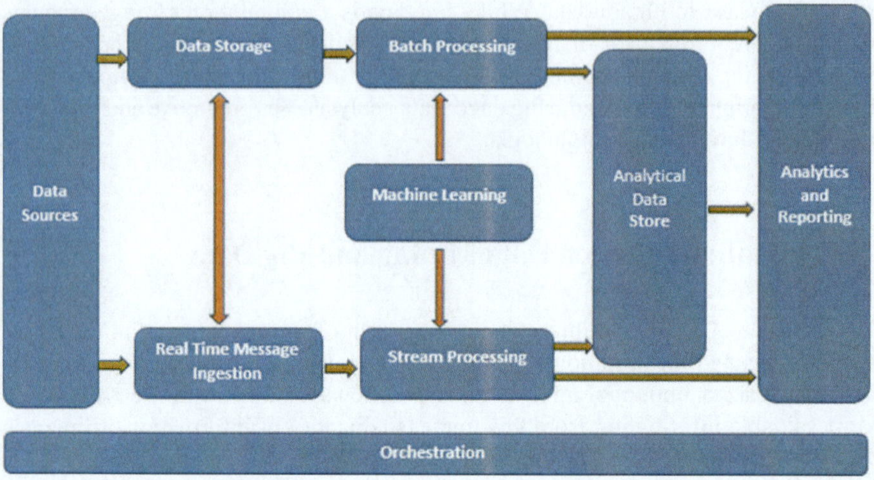

Fig. 1 Big data architecture [2]

3 Architecture of Big Data

Big data architecture focuses to absorb and process huge amounts of data so that it can be analyzed for business purposes. The architecture can be evaluated with the model for a big data solution based on the business needs of an organization. Big data architecture is drafted to handle the following types of tasks:

- Batch processing of big data sources.
- Real-time processing of big data.
- Predictive analytics and machine learning (Fig. 1).

4 Tools and Techniques

There are many software tools are used for processing and analysis of big data framework. Nowadays, the Hadoop software ecosystem is popular for big data [5]. It also implements the MapReduce Technology of Google which provides automatic parallel data processing on clusters of computers. The brief description of some components of the Hadoop ecosystem is given below:

MapReduce: MapReduce is a Google technology designed to process massive knowledge, implements (in Java) Google's distributed computing model for parallel computing with massive knowledge, many petabytes, in computer clusters. Operating mode splits into two parts [6]: Map and Reduce. While the primary half consists of distributing the computer files that are pre-processed to at least one computer known as the master node and then divides into many clusters mostly known as worker

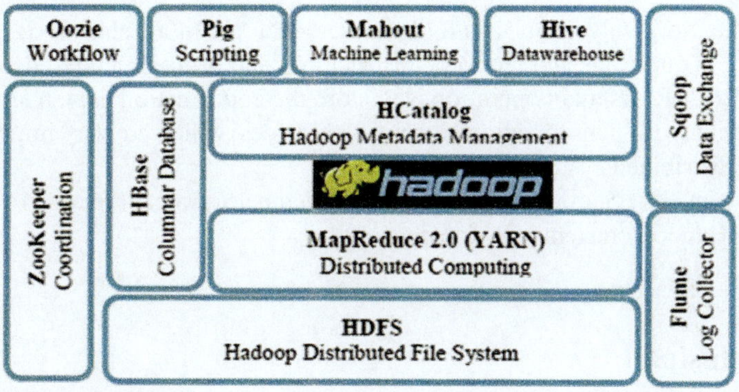

Fig. 2 Hadoop ecosystem [9]

nodes for parallel processing. The Reduce part consists of pre-processed knowledge by collecting all sub-results to supply one final feedback.

CASSANDRA: Cassandra could be a tool developed by Facebook; it is a column-oriented NoSQL database. It reinforces the MapReduce process and is especially renowned for its capability to promote data access for an oversized volume of records [7].

HBase: It is non-relational columnar distributed database designed to run on top of Hadoop Distributed Filing System which is the core of Hadoop [8]. It is a distributed file system for the storage and management of data/knowledge warehouses. HDFS stores massive files scaled to petabytes across multiple distributed nodes or clusters. To do this, it splits the computer files into blocks, and then, it allocates these blocks on servers in numerous locations. The TCP/IP level is employed for communication [9]. HDFS is fault-tolerant, and failure of any element does not have an effect on the system performance (Fig. 2).

MAHOUT: Mahout is a software developed over Hadoop using MapReduce paradigm, and it includes a group of helpful algorithms for filtering, classification, machine learning, key algorithms and clustering a distributed dataset [10, 11].

PIG: Also called Apache PIG is a platform that generates a sequence of MapReduce programs [12]. The programming language is termed 'PIG Latin.' It focuses to extend the performance of Hadoop and MapReduce by providing a programming language permitting quicker process. Provides support for acting SQL-like queries to distributed databases to Hadoop.

HCATALOG: HCatalog is a key element of Hive. It is a table management system enabling the storage of knowledge in any format regardless if it is structured or not [13]. It supports the property functioning of the Hadoop parts, like Pig, MapReduce, Streaming and Hive.

IN-MEMORY: In-memory process refers to treatments that are performed within the RAM of the instrumentation, instead of an arduous disk. The benefit of the in-memory process is that it permits urgent access to the correct data. However, these data are not kept within the future, which may cause archiving issues.

NOSQL: Not Only SQL is all databases that are not relative. It provides capabilities to questions and recovers unstructured and semi-structured data [8].

ZOOKEEPER: Its main operation is to store the coordination data, naming, providing distributed synchronization and cluster services, which are very important for a range of distributed systems.

Components like Sqoop and Flum included in the ecosystem are used to transmit data to the Hadoop clusters and vice versa [9].

5 Conclusion

Big data has recently emerged as an extremely promising paradigm for the analysis of the huge volumes of heterogeneous data. It serves as a good basis for several organizations and governments in numerous sectors that intend to automatically process and extract valuable insights that help in decision making. In this paper, we survey the state of the art of big data, its highlighted essential concepts, the distinction between data mining and big data, key tools and technologies as well as research directions which would focus on group of challenges that will be considered by big data tools. We hope our work will propose further research in this area.

References

1. Thakur B, Mann M (2014) Data mining for big data: a review. Int J Adv Res Comput Sci Softw Eng 4(5):469–473
2. Data Analytics. https://talentedge.com/blog/difference-between-big-data-and-data-analytics/
3. Baaziz A, Quoniam L (2013) How to use big data technologies to optimize operations in Upstream Petroleum Industry. Int J Innov 1(1):19–29
4. Thuan L (2018) A framework for five big V's of big data and organizational culture in firms 2018. In: IEEE international conference on big data, pp 5411–5413
5. Apache Hadoop. http://hadoop.apache.org
6. Dean J, Ghemawat S (2004) MapReduce: simplified data processing on large clusters. In: Proceedings of the 6th conference on symposium on operating systems design & implementation (OSDI'04), vol 6, pp 137–150
7. Apache Cassandra. http://cassandra.apache.org
8. Apache HBase. http://hbase.apache.org
9. Alguliyev R, Imamverdiyev Y (2014) Big data: big promises for information security. In: IEEE 8th international conference on application of information and communication technologies
10. Apache Mahout. http://mahout.apache.org
11. Wang L, Wang G (2015) Data mining applications in big data. Comput Eng Appl 4(3):143–152
12. Apache Pig. http://www.pig.apache.org/
13. Apache HCatalog. https://cwiki.apache.org/confluence/display/Hive/HCatalog

Big Data and Its Application in Healthcare and Medical Field

Yash Gandhi, Archana Singh, and Raxit Jani

Abstract The massive growth in data is increased day by day. This massive growth of data or high volume of data with high velocity and different variety is known as big data. Analyzing, processing and storing of this data, we require efficient techniques because the traditional techniques are not able to deal with massive growth of data. This is a research based on the big data in the health and medical field which can be used to transform the gathered data (unstructured data) into several diagnostics and predictions for personalized medicine and personalized healthcare options for identifying the accurate disease or symptoms. This will have an impact in the medical field to as it will be helpful to the doctors and authorities to take decision at the time of operation and diagnosing the patients based on the symptoms. In this paper, we develop a specialist framework through which the specialists and patients can be associated for all intents and purposes, alongside that searching for answer for storing this huge information in a progressively compelling and packed structure.

Keywords Big data · Apache Hadoop · HDFS · Map reduce · R

1 Introduction

1.1 An Introduction Toward Big Data

According to the analysis of different research papers related to big data, current research is on to analyze the different formats of dataset using big data. It provides two basic features few lines of coding and minimum time required for the execution, so it motivates the analyzer for using big data. It provides common data operation

Y. Gandhi (✉)
Master of Engineering in Computer Engineering, Gandhinagar Institute of Technology, Gandhinagar, Gujarat, India
e-mail: yashgandhi805@gmail.com

A. Singh · R. Jani
Assistant Professor (CE), Gandhinagar Institute of Technology, Gandhinagar, Gujarat, India

K. Kotecha et al. (eds.), *Data Science and Intelligent Applications*, Lecture Notes on Data Engineering and Communications Technologies 52,
https://doi.org/10.1007/978-981-15-4474-3_18

(filter, union, distinct, etc.) with minimum time [1]. Big data provides facility to process large amount of data with minimum time and high speed. Big data uses Hadoop Distributed File System (HDFS) to store high-volume data and Map reduce to process data. The issues of big data are research requires data in.csv format only, so if data is in another format, then we need to convert it in.csv format, this conversion takes time, and it is difficult when it is in high volume, storing of high-volume data requires more memory which may lead to high cost, difficult to provide high quality of service when volume of data is increased, extracting of meaningful knowledge from the dataset is difficult as data is in high volume, variety and velocity, maintaining the accuracy of data is major issue, and sharing of information, privacy, security is difficult to maintain [2].

2 Proposed Work

2.1 Why Image Processing

The basics of image processing, for example, image, picture investigation and comprehension, picture changes, pressure strategies, Optical Character Recognition and its applications [3], for example, video and 3D illustrations immovability, remote sensing, pattern appreciation, visual substance examination, biometrics, statistical picture handling, multimedia interfacing and virtual reality, face identification and medicinal picture preparing are talked about in this paper. This examination will assist the specialists with working on different fields, for example, picture preparing, issue identification in industrialized industries and therapeutic picture division [4].

The greatest impediment of everyone of these calculations is that the precision of these calculations is subjected to the goals nature of camera and viewpoint among camera and the objective article [5]. It is likewise seen that at certain points, the outcomes were not precise past a specific scope of camera. The eventual fate of advanced picture handling includes new smart, computerized mechanized robots made totally by explore researchers in various countries of the world [6]. It remembers progressions for different advanced picture expert applications. Because of advancements in picture preparing and other related technologies, there will be a huge number of robots on the planet in a couple of many years of time length, changing the manner in which the world is overseen [7]. In future, picture preparing innovation will be further developed, and the visual arrangement of man can be duplicated. The future patterns in remote detecting will point toward different demonstrated sensors that can record a similar scene in numerous ghastly channels [8]. Illustrations information is additionally getting massively significance in the field of advanced picture and sign preparing applications.

3 Image Processing in CT Scan (Brain)

Medicinal resonance imaging (MRI) and computerised Tomography (CT) have established most winning therapeutic imaging modalities for discovery of neurological abdominal muscle normalities and conclusion of drain, ischemic stroke and tumors with help of picture information investigation and perception. These qualities have made MRI to be a significant instrument in clinical and careful conditions diagnosing especially tumors [9]. Another methodology, CT filters are pre-predominantly utilized for separating drain brought about by seeping in the mind or ischemic stroke brought about by a blockage of blood stream in the cerebrum [10]. CT check encourages touchy data with faster examining time that makes it increasingly reasonable for detection of drain and stroke.

There are certain steps to perform the image processing as follows: Upload/Locate the images/folders, fetch the size of the uploaded image or folder, algorithm processing to reduce the image by using the reduction technique, output image is stored in the separate location with the reduced size, and an image is shown below which was uploaded to convert into reduced format [3].

4 Results and Discussions

A dataset of CT scan is taken with the reference from the Qure.ai. Here, using these images, the conversion part will be executed which will reduce the size of the images through which the collection of datasets can be reduced and will be easily used to train the machine, store the data in the form of datasets by saving the ample amount of storage which is going to reduce the cost of the existing system or storage [1]. Here, I had implemented the image processing using the framework of asp.net which is robust and easy to use. There are certain steps to perform the image processing as follows:

1. Upload/Locate the images/folders
2. Fetch the size of the uploaded image or folder
3. Algorithm processing to reduce the image by using the reduction technique
4. Output image is stored in the separate location with the reduced size.

An image is shown in Fig. 1 which was uploaded to convert into reduced format.

Now, this image will be processed to keep its original format and reduce the size in our proper format which is the goal to implement this work. Algorithm is used in the application by using the tinify package. Here, using this image, we had generated the output file of an image, histogram image and the pixel differentiation images (Fig. 2).

As you can see, there is a difference in the size and reduced image, but the quality is maintained as it was in the original images.

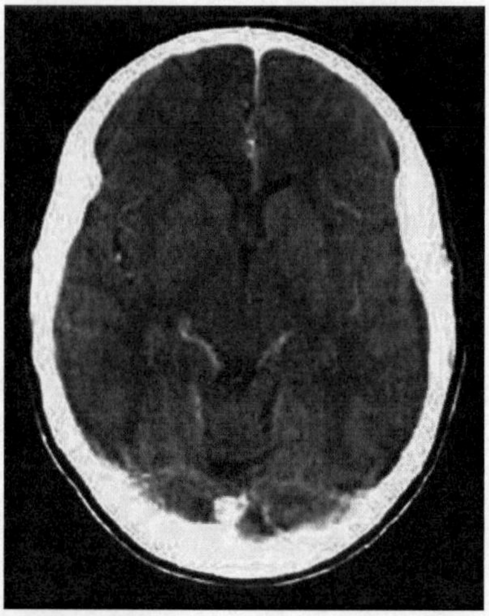

Fig. 1 Sample image of CT scan which is selected to reduce its size

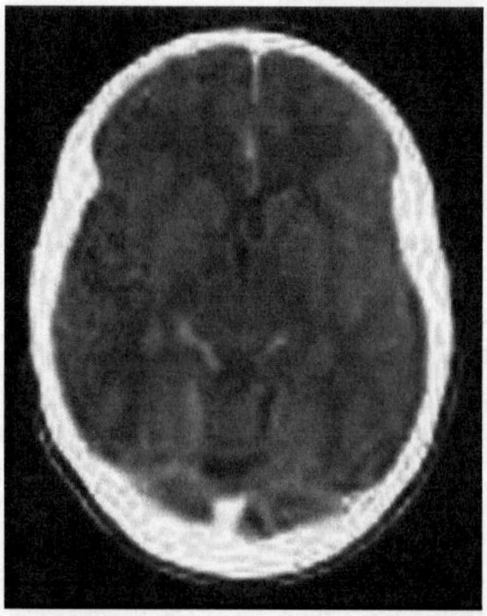

Fig. 2 Sample image generated (resized) using algorithm

Figure 3 denotes the difference between the two images after the images are processed. The pink highlighted part defines the outline of the image with the difference in the percentage of images. Figure 4 is the histogram of the images which gives the output max value as a 26 which is quite negligible. This gives the enhanced images after the image processing of single image. In this way, whole dataset can be configured to be processed and will be stored with minimal amount of storage. The image is reduced by maximum percentage having the originality of the image. If we have the collection of 5000 images and its total storage is around 1 GB so with the compression ratio up to 600 MB is quite efficient to use and store.

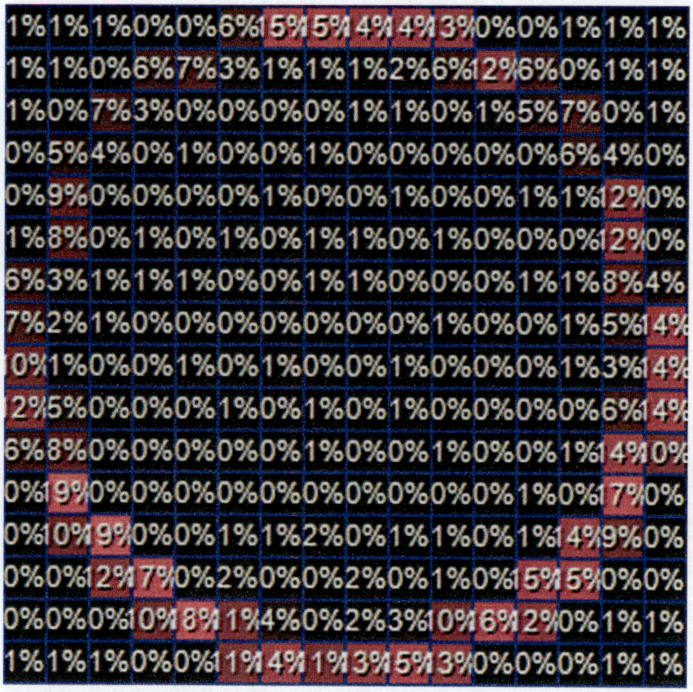

Fig. 3 Sample pixel image showing the percentage in difference to original and generated one compared to size

5 Conclusions

Based on the above-mentioned research, we can reduce the total size of the datasets using the image reduction technique through which the storage space problem is solved. Here, using the volume in big data we had compress the size of the datasets. The volume of data generated by modern IT and the healthcare system has been growing and is driven by the reduced costs of data storage and processing architectures and the need to extract valuable insights from data to improve business processes, efficiencies and services to consumers. This is the future option for the researchers about to use the volume characteristics in the big data over healthcare and medical field.

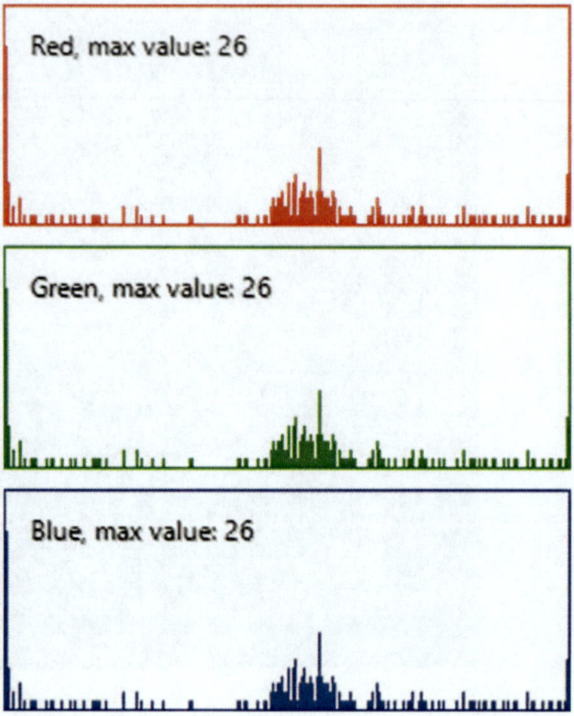

Fig. 4 Sample histogram image showing the percentage in difference to original and generated one compared to size

References

1. Sinha A, Hripcsak G, Markatou M (2009) Large datasets in biomedicine: a discussion of salient analytic issues. J Am Med Inform Assoc 16(6):759–767
2. Scruggs SB (2015) Harnessing the heart of big data. Circ Res 116(7):1115–1119
3. Rumsfeld JS, Joynt KE, Maddox TM (2016) Big data analytics to improve cardiovascular care: promise and challenges. Nat Rev Cardiol 13(6):350–359
4. Slobogean GP (2015) Bigger data, bigger problems. J Orthop Trauma 29:S43–S46
5. Lavecchia A (2015) Machine-learning approaches in drug discovery methods and applications. Drug Discov Today 20(3):318–331
6. Iavindrasana J (2009) Clinical data mining: a review. Yearb Med Inform 18(01):121–133
7. Al Hamid HA (2017) A security model for preserving the privacy of medical big data in a healthcare cloud using a fog computing facility with pairing-based cryptography. IEEE Access 5:22313–22328
8. Tanaka S, Tanaka S, Kawakami K (2015) Methodological issues in observational studies and non-randomized controlled trials in oncology in the era of big data. Jpn J Clin Oncol 45(4):323–327
9. Binder H, Blettner M (2015) Big data in medical science—a biostatistical view: part 21 of a series on evaluation of scientific publications. Dtsch Ärztebl Int 112(9):137
10. Kumar S, Singh M (2019) Big data analytics for healthcare industry: impact, applications, and tools. Big Data Min Anal 2(1):48–57
11. Shaikh TA, Ali R (2017) A dynamic approach to medical data for smart Indian healthcare. In: 2017 international conference on big data analytics and computational intelligence (ICBDAC). IEEE

WMDeepConvNets Windmill Detection Using Deep Learning from Satellite Images

A. Mridula and Shashank Sharma

Abstract In today's world, satellites are vital nodes in collecting and sharing information. Such information is in the form of images and is accurate. It is of use only if analyzed properly. However, the sheer volume of information in the form of images discourages analysis by manual scrutiny. Therefore, computers are taking over the job of image processing and analysis. First step in image processing is image detection and identification. Next step is tracking the image if need be. The algorithms developed so far have inherent disadvantage of dependency on white or plain background as edge tracking was dependent upon pixel delineation. So, these algorithms enjoyed limited success. 'Detection of windmills using satellite images through neural networks' is an attempt to detect windmills in satellite images. It is known that the satellite images set do change in every one to three months. This will be beneficial to get the data of any new windmill installed in any area and subsequently predict the energy production. This project can further be implemented to detect solar power panels and further predict their energy production as well.

Keywords ANN · CNN · Object detection · Satellite image · Deep learning

1 Introduction

Remote sensing or capturing real-time pictorial data of earth's terrain have started in this century in a big way. With the improved hardware on satellites like cameras with high resolution image sensors, zooming capabilities, etc., accuracy of the data obtained has improved a lot. With high resolution imagery, now the challenge is to interpret the big data timely and accurately so as to make sensible use of the data [1].

A. Mridula · S. Sharma (✉)
Department of IT, Manipal University Jaipur, Jaipur, India
e-mail: shashank.sharma@jaipur.manipal.edu

A. Mridula
e-mail: a.mridula12@gmail.com

K. Kotecha et al. (eds.), *Data Science and Intelligent Applications*, Lecture Notes on Data Engineering and Communications Technologies 52,
https://doi.org/10.1007/978-981-15-4474-3_19

In this paper, the objective is to identify windmills from satellite data automatically and localize the windmills in the image by defining enclosing boxes around them. Windmills have huge shadows, and this identifiable pattern is used as the signature for identifying the windmills through machine learning.

Windmills spread over large area, and identification by visual scrutiny is both cumbersome and unwieldy. Machine learning can help in a big way in this objective. Windmills are coming up in a big way. Identification of windmills, mapping them with reference to wind flow and other landmarks will greatly help in calculation of energy production on real-time basis. This methodology can also be used to train classifiers to detect solar panels, etc., from the satellite images [2]. Such intelligent mapping will be of immense use in the generation of renewable energy.

1.1 Background Overview

Neural Network—The concept of neural networks as the name suggests came from that of a human brain. A human brain consists of millions of neurons each having specific tasks assigned to them. These neurons together help a human to process various information on a regular basis. The concept of artificial neural networks is an aid to machine learning based on this theory. Most of the ANNs have some sort of learning method [3]. These methods can also be referred as learning rules. These rules are similar to the learning process of that of a human (Fig. 1).

Convolutional Neural Networks—These are made up of neurons, each of which is encoded to validate certain specific parameters. These parameters themselves are weights which CNN 'learns' with iterations [4]. CNN therefore a combination of the 'learning' of all these neurons which weigh and compare data, sometimes with built in nonlinearity [5].

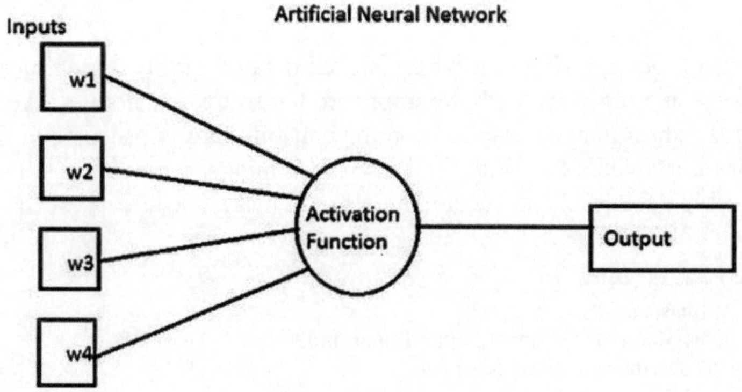

Fig. 1 Artificial neural network

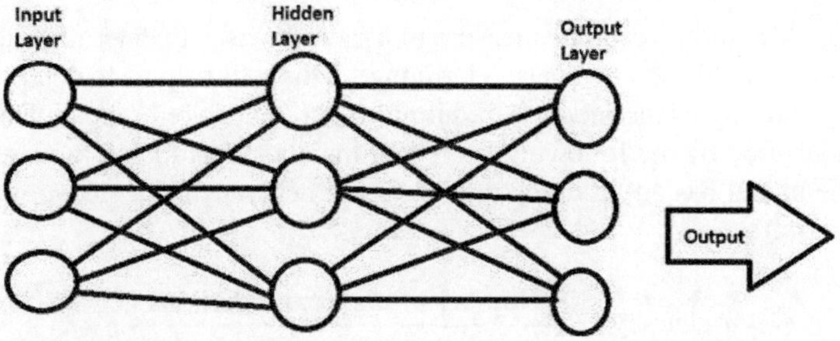

Fig. 2 Architecture overview

1.2 Architecture Overview

CNN comprises multiple layers of neurons. Each layer is holistically integrated with previous layer [6]. This enhances the process of machine learning and also improves accuracy (Fig. 2).

2 Methodology

Object detection: If the entire image is fed to the CNN at once, it will require tremendous computational power for analysis. Therefore, a sliding window method is employed, wherein a window of fixed size slides over the image, and one part is identified at a time from the sliding window. When the process is completed, an accurate interpretation of the image is available. Moreover, the process of machine learning is enhanced by such objective orientation of CNN [7]. This process therefore identifies one part at a time, and the net result is more accurate and reliable [8].

Activation Function: Whether a neuron should be activated or not is an objective decision output of activation function. This is achieved by a method of calculating weighted sums. Mathematically, it is explained as below.

$$z(1) = Y(1)X + b(1)$$
$$a(1) = z(1)$$

Here,

$z(1)$ is representative of vectorial output of layer 1
$Y(1)$ is representative of vectorial output of weights assigned to neurons of respective layers, i.e., y1, y2, y3 and y4
X is representative of vectorial input features, i.e., i1 and i2
b is representative of vectorial bias assigned to neurons in layers, i.e., b1 and b2
$a(1)$ is representative of vectorial form of any linear function.

Accuracy Measure: Accuracy measure of a model is used to determine how accurately it can detect objects in a particular image. This will help us to determine how much more training of the model is required to get the desired output. This can be further understood by the following [9]: From the above confusion matrix, one can say that the model has an accuracy rate of 99.6% accuracy. (The above matrix is a sample matrix.)

$$\text{Accuracy} = \frac{\text{Number of correct predictions}}{\text{Total number of predictions made}}$$

$$\text{Accuracy} = \frac{\text{True positive} + \text{False negative}}{\text{Total number of sample images}}$$

See Table 1.

Creation of Dataset: The neural network learns just as a human and hence the name neural network. To enable this process of learning of the neural network, a dataset that consists of images with good resolution of windmills and its shadow is necessary [10]. We concentrated our efforts in the dry arid region of Rajasthan for identifying the shadows. Our requirement was a minimum of 1500 images. Naturally, larger dataset than this would yield better results. We resorted to data scraping techniques to load Google satellite images. After a sample download, we extracted the coordinates from an atlas. Then, we batch downloaded the images according to these coordinates. Here, we have used images in.png format and not in.jpg format (to retain the image quality) (Fig. 3).

Table 1 Parameters

	Predicted positive	Predicted negative
Actual positive	2	2
Actual negative	0	996

Fig. 3 Sample training images

3 Proposed Approach

Mobilenet-ssd: The idea is to reduce the computational complexity by splitting the filtering process into depthwise convolution and pointwise convolution. These are performed in the order as above, and the features are grouped. This process is much faster. The full architecture of MobileNet V1 consists of a regular 3 × 3 convolution as the very first layer, followed by 13 times the above building block. There are no pooling layers in between these depthwise separable blocks (Fig. 4).

Faster R-CNN: Faster R-CNN uses two modules, i.e., the region proposal network (RPN) for proposing regions and the second module Fast R-CNN detector makes use of these proposed regions. Thus, it works as a single, unified network for object detection.

TensorFlow: The TensorFlow Object Detection API is an open source framework which makes it easy to use, design, learn and deploy object detection models. A config file to define various parameters of a network can be customized for this purpose. For our problem, we used Faster R-CNN algorithm applied to mobilenet-ssd.

TensorBox: This model has a smaller architecture and uses transfer learning and is quicker than Google API. With the use of TensorBox, time taken per step has reduced. After 110,000 iterations (approximately 8 h) of training, the results were accurate on test images. The accuracy of our model for the purpose was around of 99% on the validation set.

Fig. 4 Mobilenet ssd

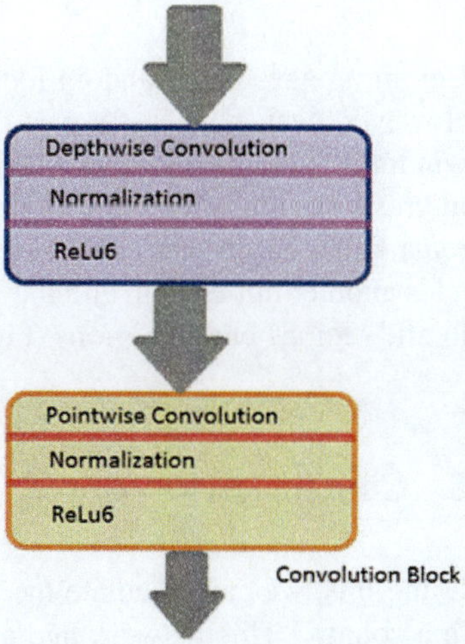

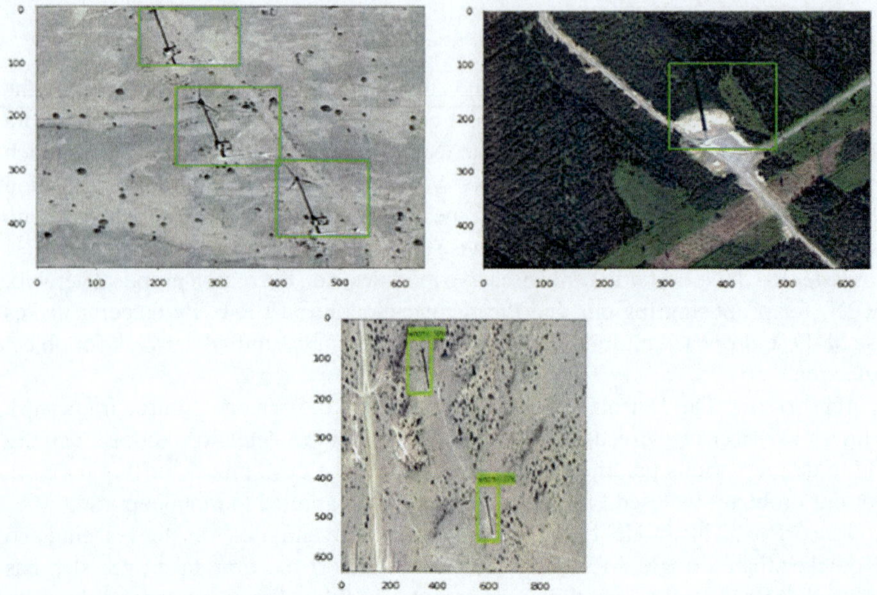

Fig. 5 Sample prediction on random images

4 Result

Certain images have random localizations which show the need for larger dataset. However, even with just a part of the shadow being visible in the image, still the windmill could be detected. Multiple windmills could also be easily detected. Even in grassy regions, it was interesting to note that it gave accurate results. This is the remarkable outcome as we trained our neural network on dry arid regions. Thus, it adds another dimension, and the mode is successful in detecting windmills not only in arid regions but all regions (Figs. 5 and 6).

5 Conclusion

Using this, we can calculate the approximate power generation through windmills in a country. This network and average wind speed can be used together to detect areas with high potency for power generation through windmills and further be used to install windmills in those areas to maximize the energy generation. What has been demonstrated in this paper has immense reach and has no bounds. In this era of computers solving every problem, this network will contribute toward providing solution to quite a few of them. It is also clear that as we increase the data set to include a greater variety of windmills, it will increase the predicting power for a windmill

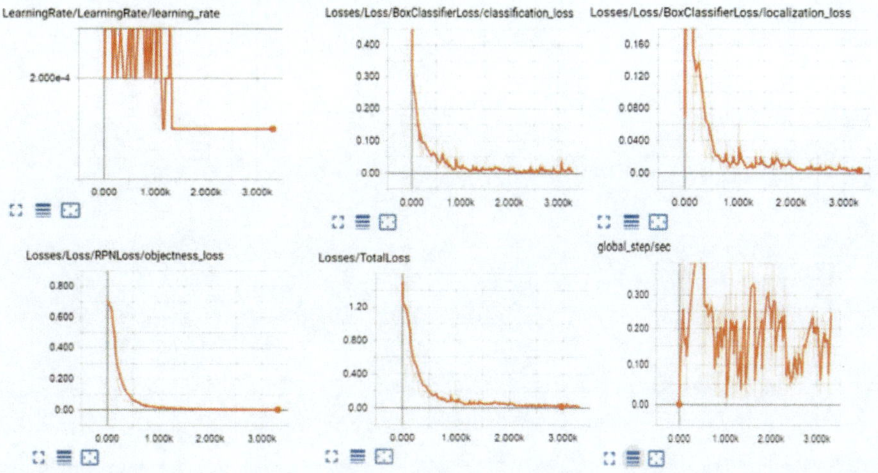

Fig. 6 Results

placed in different landscapes or backgrounds than what it has been particularly trained for. We can also extend this network to detect solar panels on roof tops, and with the help of data on sun's position, suggest better placement of the solar panels for increased efficiency.

References

1. Vakalopoulou M, Komodakis N, Karantzalos K, Paragios N (2015) Building detection in very high resolution multispectral data with deep learning features. In: Geoscience and remote sensing symposium (IGARSS). IEEE
2. Jurie F, Razakarivony S (2016) Vehicle detection in aerial imagery: a small target detection benchmark. J Vis Commun Image Represent 34:187–203
3. Krishnan A, Larsson J (2016) Vehicle detection and road scene segmentation using Deep Learning
4. Chen X, Liu CL, Xiang S, Pan CH (2014) Vehicle detection in satellite images by hybrid deep convolutional neural networks. IEEE Geosci Remote Sens Lett 11:1797–1801
5. Yuan J (2016) Automatic building extraction in aerial scenes using convolutional networks
6. Wani MA, Afzal S (2017) A new framework for fine tuning of deep networks. 16th IEEE international conference on machine learning and applications
7. Wani MA, Yesilbudak M (2013) Recognition of wind speed patterns using multi-scale subspace grids with decision trees. Int J Renew Energy Res 3:458–462
8. Joshi B, Hinai AA, Baluyan H, Woon WL (2014) Automatic rooftop detection using a two-stage classification. In: 16th international conference computer modeling and simulation, UKSim-AMSS
9. Mujtaba T, Wani MA (2018) Object detection from satellite imagery using deep learning. In: 5th IEEE international conference on computing for sustainable global development
10. Mnih V (2013) Machine learning for aerial image labeling. Doctoral dissertation, University of Toronto

Quality Grading Classification of Dry Chilies Using Neural Network

Nitin Padariya and Nimisha Patel

Abstract Due to lack of precise information for quality of crops produce, the overall growth of agriculture sector in India is limited. In the present work, the deep convolutional neural networks (CNNs) were used for image classification of dry chilies with distinguishable quality of farm produce. Our unique approach facilitates traders to visually evaluate the quality of chilies from the high-quality images received from producers in farms. The quality of chilies was determined from the color. The database was prepared following image capture, data collection and image analysis. The results showed that our model achieved precision between 65 and 75%, for separate class tests, on average 66%. The global objective of this project was to develop an innovative tool to investigate the quality of chilies from their color using image analysis to improve its applicability for commercial use in edible products.

Keywords Agriculture · Dry chilies · Convolutional neural network

1 Introduction

In India, agriculture is considered to be the most prominent and versatile occupation; however, it is highly vulnerable to variability and climate change with the highest impact on fresh produces, and therefore, the businesses which depend on agriculture growth have been affected. Unfortunately, the overall contribution of agriculture to Indian economy has reduced due to lack of knowledge, understanding and adaption of new and useful technologies and infrastructures [1]. In order to increase yield and quality of agriculture produces, artificial neural networks (ANN) to automate the farming of small properties need to be deployed in India. Because of incapability of linear systems for solving the complicated problems, ANN has proven ability

N. Padariya (✉) · N. Patel
Sankalchand Patel University, Visnagar, India
e-mail: nitinpadriya@gmail.com

N. Patel
e-mail: nimishapatelce@gmail.com

K. Kotecha et al. (eds.), *Data Science and Intelligent Applications*, Lecture Notes on Data Engineering and Communications Technologies 52,
https://doi.org/10.1007/978-981-15-4474-3_20

Fig. 1 Dry chilies images of **a** A1 grade, **b** A2 grade and **c** A3 grade collected from the field

and versatile tool to use feed-forward back propagation network to solve the same. The ANN models have been developed as an alternative and effective tool to more accurately predict yield of rice [2], maize [3] and few other crops worldwide. Most likely, ANN models have not been proposed to use for assurance of quality of farm produces using image analysis.

In the present study, we have investigated the deep convolutional neural networks (CNNs) for qualitative grading classification of dry chilies for quality assurance. CNNs are very crucial in handling the experimental data and that can further be used for image processing of data collected. In order to learn and recognize the patterns, the CNNs are exhibiting forward input signals stimulating network capability. CNNs are functioning on local understanding of images, shared weights and subsampling to ensure some degree of shift and distortion invariance [4]. Custom model in CNN was developed to classify the images for identification of quality chilies.

2 Materials and Methods

2.1 Data Collection and Processing

Different types of qualitative chilies images were collected from the various fields of the chilies. Those images were collected in a.jpg format. The training data matrix of graded images was by trained agronomist. Using backhand Python 3.7, CNN framework was deployed for qualitative grading classification from the image analysis. Figure 1 shows the three representative images which were collected from the various chili fields are graded by (a) A1, (b) A2 and (c) A3.

2.2 Neural Network Training

The graded images were used to generate data matrix and used throughout the study. With multilayers in CNN, the model was trained successfully for three different grades. Using RGB color bands in one array, all the images of chilies were investigated. Furthermore, the image values obtained were normalized in 0–1 to setting up the training model with CNN. The sigmoid and relu activation function was used for CNN to evaluate the images. The convolution neural network model is implemented in Python 3.7 environment. The developed model was further used to read different images, and their outcome was depicted herein. The model was compiled based on given different epoch size that is according to the data obtained and the accuracy of model to predict. The matrix of training data is tabulated in Table 1.

3 Result and Discussion

The CNN model was generated from the training and testing results for classification of quality graded dry chilies. The predicated values rely on the structure of CNN models and the input data. Our results show that the probability of A1, A2 and A3 grades is 0.381, 0.435 and 0.426, respectively, in tested images. Figure 2 shows the tested image of dry chilies. The highest accuracy of the model tested remained around 66.70% for the dry chili images that were evaluated.

In the present study, the neural network was trained using training datasheet, whereas the performance of the network was evaluated using testing dataset. Later on, the optimum activation function was figured out from the individual test cases which were ranked according to the magnitude of RMSE and CC. The minimum of RMSE and maximum of CC were considered as optimum for the current CNN model. The classification based on texture and physical abnormalities of dry chilies provides higher accuracy within stipulated time. Due to higher accuracy and time efficient approach, our neural network model can easily be scaled up for chili traders in India and worldwide.

4 Conclusions

In conclusion, this is the first time to our knowledge that the CNN network has been used with different type of activation functions to predict the quality of dry chilies from image analysis which is a unique approach allowing to reduce efforts, cost of transportation for an expert to do quality check and time. With the accuracy of 66.70%, the quality of dry chilies using the image analysis was performed which facilitates the traders to choose the quality of farm produce cost effectively and time efficiently. These results show that the accuracy of the quality assurance varies with the input data and CNN model structure. Further efforts will be invested to improve the efficiency and accuracy of model for better quality assurance in our future work.

Table 1 Data matrix of training data

Id	Genre	A1	A2	A3	Id	Genre	A1	A2	A3	Id	Genre	A1	A2	A3
1	[‘A2’]	0	1	0	7	[“A1”]	1	0	0	13	[‘A3’]	0	0	1
2	[‘A2’]	0	1	0	8	[‘A3’]	0	0	1	14	[‘A2’]	0	1	0
3	[‘A2’]	0	1	0	9	[‘A3’]	0	0	1	15	[“A1”]	1	0	0
4	[‘A3’]	0	0	1	10	[‘A2’]	0	1	0	16	[‘A3’]	0	0	1
5	[‘A3’]	0	0	1	11	[‘A3’]	0	0	1	17	[‘A2’]	0	1	0
6	[‘A2’]	0	1	0	12	[‘A2’]	0	1	0					

Fig. 2 Tested image of dry chilies

References

1. Dahikar SS, Rode SV, Deshmukh P (2015) An artificial neural network approach for agricultural crop yield prediction based on various parameters. Int J Adv Res Electron Commun Eng 4(1):94–98
2. Ji B, Sun Y, Yang S, Wan J (2007) Artificial neural networks for rice yield prediction in mountainous regions. J Agric Sci 145(3):249–261
3. Adisa OM, Botai JO, Adeola AM, Hassen A, Botai CM, Darkey D, Tesfamariam E (2019) Application of artificial neural network for predicting maize production in south africa. Sustainability 11(1145):1–17
4. Kamilaris A, Prenafeta-Boldu FX (2018) A review of the use of convolutional neural networks in agriculture. J Agric Sci 156(3):312–322

Email Classification Techniques—A Review

Namrata Shroff and Amisha Sinhgala

Abstract Email has become a significant correspondence medium. Official, personal, social and promotional and other messages hit our mail box every day. From the research, it has discovered that the normal office specialist gets 121 messages for every day. Now and then because of flooding of messages in inbox, a portion of the some mails stay unattended, so on the off chance that messages are characterized into top need folders, at that point, the issue of unattended or unanswered mail will be tackled. In this paper, we identified the key features of email classification are temporal, behavioral, single email multinomial valued, content and local and global features. Also datasets, techniques and tools in various email classification like spam, phishing, multifolder and machine generated email classification were studied. Different email classifiers provide different mechanisms for classification. Challenges in email classification are discussed. From the study, it is found that J48 classification algorithm works the best for spam and ham email classification. In comparison with various email service provider, Microsoft Outlook filters the mail based on many criteria.

Keywords Email classification · Feature in email classification · Spam and phishing · Email classification

1 Introduction

Email has been a precise and prevalent correspondence system as the number of Internet users increases day by day. Along these lines, email organization is a significant and growing issue for people and organizations. Emails can be of type official, personal, social, promotional, machine generated and so on. Application areas of

N. Shroff (✉)
Gujarat Technological University, Chandkheda, Gujarat, India
e-mail: namrata@gecg28.ac.in

A. Sinhgala
S.V.I.T Vasad, Vasad, India

K. Kotecha et al. (eds.), *Data Science and Intelligent Applications*,
Lecture Notes on Data Engineering and Communications Technologies 52,
https://doi.org/10.1007/978-981-15-4474-3_21

email classification are spam email classification, phishing email classification, multifolder classification and machine generated mail classification. The unsolicited email messages, known as spam, are a case of abuse. Spam is sending spontaneous mass email, i.e., sending unrequested mails to the multiple recipients. Phishing is an act of trying to get the personal, financial information of the user using deceptive emails. Multifolder classification is one in which user can organize his/her emails into multiple folders according to sender, by subject, by occasion, by dates and many more criteria. Having emails organized makes the searching easy and saves time of an individual. Machine generated email is usually made by scripts and embedded in an image, sends by commercial group or organizations. A machine generated mail contributes to 90% of nonspam Web mail traffic.

Email classification uses concepts of text mining, natural language processing and semantic analysis to classify mails. The primary objective behind organizing an email is in busy schedule sometimes important email goes unattended, so if mails are organized properly, then the highlighted and unread email can be easily distinguished. Also, emails categorization, spam recognition, common subject grouping and identification of machine generated mails can be done. To achieve all the mentioned task, feature selection should be properly done.

This paper is organized as Sect. 2 contains related work in the area of email classification, Sect. 3 key features contributing the email classification and work done in spam, phishing, multifolder and machine generated email classification, Sect. 4 contains classification done by different email service providers, and Sect. 5 contains challenges in email classification.

2 Related Work

In [1], author discussed the email classification application areas, datasets used in those application area, various features sets used in those applications, classification techniques and work done in that techniques and performance metrics. The email is classified into five main application areas, namely spam email classification, phishing email classification, spam and phishing, multifolder categorization and other related application areas. Datasets used in these application areas were PU, Phishing Corpus, Spam Assassin, Enron datasets. The most widely used features were email header part, email body part, URL, JavaScript, behavioral and term-based features. In this review, author mentioned five different classification techniques, they were supervised machine learning, semisupervised machine learning, unsupervised machine learning, content-based and statistical learning. According to author, supervised machine learning is the most widely used technique, and in that, SVM was the most frequently used technique and showed the best performance, followed by decision tree and Naïve Bayes technique. The frequently used performance measures were precision, recall, accuracy, f-measure, false positive rate, false negative rate and error rate to determine the performance of email classifiers.

From [2], a personal large dataset of emails is collected and assembled. Several approaches are evaluated to cluster the emails based on their contents. Manual or supervised classification can be much more reliable and effective. However, in many cases, there is a need to perform such process automatically. Several clustering methods were evaluated based on samples' selection or based on utilizing the complete emails' dataset.

In this paper [3], four classifiers including neural network, SVM, Naïve Bayesian and J48 were tested to filter spams from the dataset of emails. All the emails were classified as spam (1) or not (0). That was the characteristic of the dataset of email for spam filtering. J48 is very simple classifier to make a decision tree, but it gave the efficient result in the experiment. Naïve Bayesian classifier also showed good result, but neural network and SVM did not show good result compared with J48 or Naïve Bayesian classifier. Neural network and SVM were not appropriate for the dataset to make a binary decision. From this experiment, author found that a simple J48 classifier can provide better classification result for spam mail filtering.

In [4], author introduces two email data representation approaches in preprocessing phase, and then, they identified five email mining tasks namely contact analysis, spam detection, email categorization, email visualization and email network property analysis. For each of the five tasks, they discussed the commonly used techniques.

3 Key Features in Email Classification

The majority of the volume of email traffic was analyzed for one month, and it is noted that more than 60% of the overall traffic originated from mass senders in multiple domains (manually classified via editors) such as social (Facebook, linked in, etc.), commerce (eBay, Amazon, etc.), finance, shopping, travel, etc., as illustrated [5]

Features are the characteristics that describe the measurement of email behavior. It is very important to extract and select the useful features in email classification to develop accurate and efficient classifier. Features can be divided taking into account the structure of email like email header, email body, email URL, email JavaScript, behavioral, term-based, rule-based features, etc. Features used in email classification areas are as shown in Fig. 1 (Table 1).

Spam emails are classified on the basis of single email multinomial valued features and content-based features. Phishing emails are classified on the basis of behavioral, single email multinomial valued and content features. Multifolder email is classified on the basis of content and local and global features, and machine generated mails are classified on the basis of content features. Table 2 shows the dataset, techniques, tools used in phishing email classification, spam email classification, multifolder classification and machine generated mails classification.

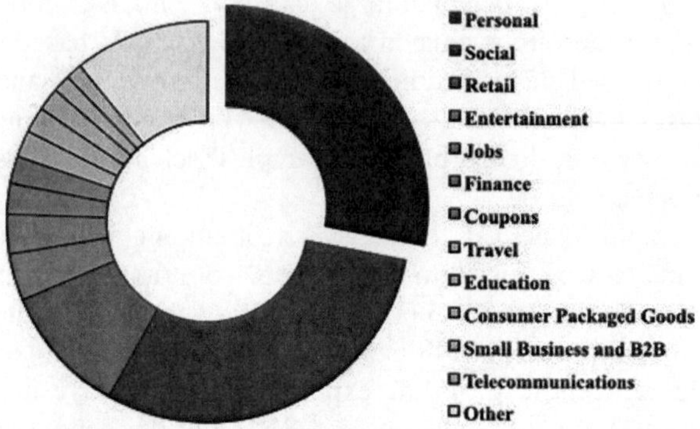

Fig. 1 Categories of email traffic

Table 1 Types of email classification features

Temporal features	Behavioral features	Single email multinomial valued features	Content features	Local and global features
Simple temporal features such as the day of the week and the time of the day the message was received from the message timestamps Temporal sequential patterns: The events in the sequence are ordered by their timestamps: $t_1 < t_2 < \cdots < t_n$	Behavioral features include words extracted from the sender's and recipient's action	If email contains script tags, embedded image, hyperlinks, file attachments, etc., then they are single email multinomial valued features	Gathering the relevant features from subject and body. Taking into account character length, no. of URLs and email address subdomain	Local features take the behavioral pattern of the specific user over similar past messages. Global features refer to entire dataset activity tracing

4 Different Email Provider Provides Different Mechanism for Classification

Setting up email filters, or arranging discussions into organizers, is a proficient method to deal with your email inbox. For example, one might need to have a job filter to isolate every one of the notifications from his/her regular messages. One could likewise do this for bulletin memberships, messages from relatives, or messages identified with work. It is likewise an incredible apparatus to naturally waste or stamp certain messages as spam. As should be obvious, there is a great deal of favorable circumstances for separating your messages. Here is how Gmail, Yahoo and Outlook provide the classification.

Table 2 Literature survey

Citation	Paper title	Dataset	Techniques	Tools	Conclusion
[6]	Detection of phishing emails using data mining algorithms	Phishing email 4559—from Nazario, and 4559 were selected from spam assassin	The J48 algorithm was used, which is a powerful knowledge representation and reasoning algorithm based on the C4.5 algorithm	mbx2eml, which is a free tool used to split emails grouped in mbox file format and stores each email in a separate file	Random Forest, J48 and PART with accuracy rates of 98.87%, 98.11% and 98.10%, respectively. The lowest false positive rate was registered for the PART algorithm at 0.26%
[7]	Email phishing detection and prevention by using data mining techniques	.eml extensions file for emails. These files contain the information of mail in header, body and attachment	MLAPT, SVM	WEKA	The system that decides if emails received have any element of phishing behavior is MLAPT decision making system. SVM is successfully applied in phishing detection filtering, black listing mechanism and pattern recognition methods
[8]	Proposed efficient algorithm to filter spam using machine learning techniques	Personal emails	WEKA	Techniques used are C4.5 decision tree classifier, multilayer perceptron and Naïve Bayes classifier	Multilayer perceptron showed higher efficiency than Naïve Bayes classifier algorithm and J48 with a low rate of false positive
[9]	Spam filtering email classification (SFECM) using gain and graph mining algorithm	Test email samples	Java	LingerIC spam filter and content-based email classification model	Using Linger and content-based classification model, proposed solution provides 100% accuracy at filtering spam from a set of mixed emails
[10]	Automatic categorization of email into folders: Benchmark experiments on Enron and SRI Corpora	Enron and SRI datasets	Maximum Entropy (MaxEnt), Naïve Bayes (NB), Support Vector Machine (SVM), Winnow	Preparing the text data Creating word dictionary Feature extraction process Training the classifier	Fast and simple-to-implement Winnow classifier performs insignificantly better than the more popular and more complex-to-implement SVM and MaxEnt methods

(continued)

Table 2 (continued)

Citation	Paper title	Dataset	Techniques	Tools	Conclusion
[11]	Editing training data for multi-label classification with the k-nearest neighbor rule	Multimedia classification (Emotions), bioinformatics (Yeast) and text categorization (Medical, Enron and Webpage)	EMLkNN and Rank-SVM methods	MATLAB	Edited nearest neighbor for multi-labeled data is an efficient editing method
[12]	Automated extractions for machine generated mail	Yahoo mail traffic (travel domain)	Structure-based clustering		Structural clustering method for messages and devised a modular automated method for creation of extraction rules at the cluster level
[13]	Learning effective embeddings for machine generated emails with applications to email category prediction	Gmail consists of data from 273,000 anonymized users and 35 million B2C emails spanning 90 days	LSTM with one hidden layer, MLP		With the pre-trained embedding, the predictive performances of LSTM and MLP for the two email category prediction tasks can be significantly improved and that the pre-trained embedding can be used to address cold-start problems

In Gmail, after opening Gmail and log in, one has to click on filters and blocked addresses from setting menu. Click create a new filter. For filtering in Gmail, the criteria are as follows:

(1) Search: from where to filter inbox, send mails or all mails
(2) From: Senders email address
(3) To: Receivers email address
(4) Subject
(5) Has the word, does not have, has attachment, do not include chats, size of mail, date within.

Are the various parameters for filtering the email into specific folder/label (Figs. 2 and 3).

In Yahoo, as compared to Gmail, it has few filters, they are From, To/CC, Subject and Body field to customize. For applying filters, one has to open Yahoo email inbox, go to setting by clicking the gear icon on the top right corner and selecting the filter tab on the left and then add a new one.

In Outlook, one can apply filter by opening your inbox, selecting the message to apply filter and then clicking on three dot menu on the top right side and choosing the

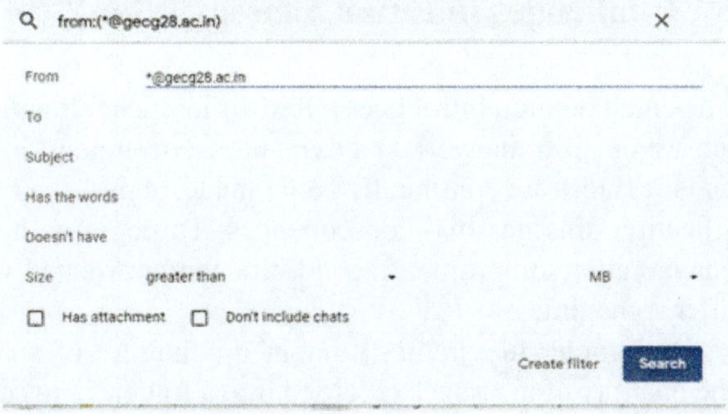

Fig. 2 Applying filters in Gmail

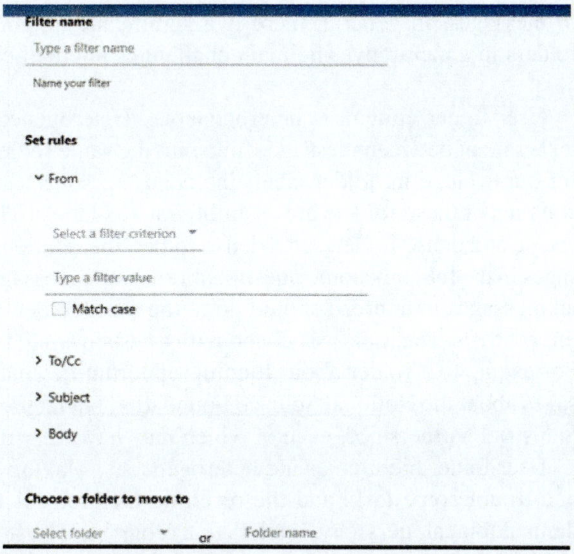

Fig. 3 Applying filters in Yahoo

create rule option. Automatically, based on the selected mail, the filter criteria will be filled if one wishes to change he/she can change it by clicking X icon. Outlook provides more filtering criteria than Yahoo. One can select the word in subject, subject or body or in sender's address and so on. After setting the condition, one has to select the actions. One can move/copy the message to a folder or delete it.

Looking at the comparison of email filters provided by three email provider Gmail, Yahoo and Outlook, it can be observed that Gmail and Outlook offer more filtering criteria than Yahoo. If filters are applied properly and relevant actions are assigned to them, then much of the task is reduced, unwanted mails can be handled, and important emails cannot go unattended.

5 Challenges in Email Classification

The email domain introduces a few difficulties for an intuitive documenting specialist. An email framework is a dynamic environment. In an email framework, folders and messages are continually being made, demolished and redesigned. For text classification, this has two consequences. To begin with, the quantity of classes may change after some time. Second, the content related with class labels may change after some time.

User folder hierarchies contain a countless of sparse folder sparcity refers to the quantity of messages recorded into a folder. Lotus discovered 35% of the normal client's folder contain less than three messages. On the other hand, it is not uncommon for folder hierarchies to contain one or more dense folders, which contain hundreds of messages. In general, there is a significant amount of variance in the sizes of folders in a hierarchy. The main challenges in email classification are as discussed [14]:

User folder content is heterogeneous. Heterogeneity refers to the similarity of text content between messages filed into the same folder. In the mail stores collected for our evaluation, folder labels included 'Personal' and 'Fun.' We can assume the contents of these folders are as ambiguous as labels. This can be a problem for both sparse and dense folders. In a dense folder, the relationships between different messages in the folder become questionable as more messages accumulate. Occasionally, all messages in a thread should go to the same folder, but other times the topic in a thread drifts. The topic associated with a certain email folder can also shift over time. For example, a folder about funding opportunities may at first contains only messages about the National Science Foundation but later only gets new messages about industrial partnerships, each of which may have very different word distributions.

User folder hierarchies are heterogeneous. Making a lot of documenting folders is a troublesome task, and the resulting hierarchy is unlikely to resemble a well-defined topical hierarchy. Folders in a progressive system may cover in content, and a subfolder may not speak to a subtopic of the parent organizer content.

6 Conclusion

From the above study, it is found that lot of work is been carried out in classifying the spam and phishing emails. Spam emails are classified on the basis of single email multinomial valued features and content-based features. Phishing emails are classified on the basis of behavioral, single email multinomial valued and content features. Multifolder email are classified on the basis of content and local and global features, and machine generated mails are classified on the basis of content features. The J48 classification algorithm is the best suited for spam classification. Challenges in email multifolder classification are, users inbox is dynamic, user's inbox contains heterogeneous stuff and the content changes frequently.

References

1. Mujtaba G, Shuib L, Raj RG, Majeed N, Al-Garadi MA (2017) Email classification research trends: review and open Issues. IEEE Access 5:9044–9064
2. Alsmadi I, Alhami I (2015) Clustering and classification of email contents. J King Saud Univ Comput Inf Sci 27(1):46–57
3. Youn S, McLeod D (2007) A comparative study for email classification. Advances and innovations in systems, computing sciences and software engineering. Springer, Dordrecht, pp. 387–391
4. Tang G, Pei J, Luk WS (2014) Email mining: tasks, common techniques, and tools. Knowl Inf Syst 41(1):1–31
5. Ailon N, Karnin ZS, Liberty E, Maarek Y (2013) Threading machine generated email. In: Proceedings of 6th ACM international conference on web search data mining, pp 405–414
6. Smadi S, Aslam N, Zhang L, Alasem R, Hossain MA (2016) Detection of phishing emails using data mining algorithms. In: 9th international conference on software, knowledge, information management and applications
7. Şentürk Ş, Yerli E, Soğukpnar İ (2017) Email phishing detection and prevention by using data mining techniques. In: 2nd international conference on computer science and engineering (UBMK), pp 707–712
8. Aski AS, Sourati NK (2016) Proposed efficient algorithm to filter spam using machine learning techniques. Pacific Sci Rev A Nat Sci Eng 18(2):145–149
9. Chae MK, Alsadoon A, Prasad PWC, Sreedharan S (2017) Spam filtering email classification (SFECM) using gain and graph mining algorithm. In: 2nd international conference on anti-cyber crimes, pp 217–222
10. Bekkerman R, McCallum A, Huang G (2004) Automatic categorization of email into folders: benchmark experiments on Enron and SRI corpora. Science 80(418):1–23
11. Kanja S. Editing training data for multi-label classification with the k-nearest neighbor rule. https://www.hds.utc.fr/~tdenoeux/dokuwiki/_media/en/publi/paaa2015.pdf
12. Di Castro D (2018) Automated extractions for machine generated mail. In: WWW '18 companion: the 2018 web conference companion, vol 2, pp 655–662
13. Sun Y, Garcia-Pueyo L, Wendt JB, Najork M, Broder A (2019) Learning effective embeddings for machine generated emails with applications to email category prediction. In: Proceedings—2018 IEEE international conference on big data (Big Data), vol ii, pp 1846–1855
14. Brutlag JD, Meek C (2000) Challenges of the email domain for text classification. In: Proceedings of the seventeenth international conference on machine learning

A Novel Approach for Credit Card Fraud Detection Through Deep Learning

Jasmin Parmar, Achyut Patel, and Mayur Savsani

Abstract In the current scenario of the world, digitalization looks expanding ubiquity as a result of consistent, helpful and advantageous utilization of e-commerce. It has demonstrated to be a slam gasp and simple method of installment passage. Purchasers pick online installment and e-shopping, due to time comfort. Because of this, an enormous measure of online business use, there is an immense addition in credit card frauds moreover. Fraudsters attempt to abuse the card and straightforwardness of online installments. To beat the fraudster's action become fundamental, the primary point is to verify credit card transactions; in this way, individuals can utilize e-exchanges securely and effectively. To identify the charge card misrepresentation, there are different methods which depend on deep learning, logistic regression, naïve Bayesian, support vector machine (SVM), neural network, artificial immune system, nearest neighbor, data mining, decision tree, fuzzy logic-based system, genetic algorithm and so on.

Keywords Fraud · Credit card · Deep learning

1 Introduction

Ongoing investigations show that the number of bank exchanges by means of charge cards raised radically and with it the number of cheats and card burglary. The charge card is the mainstream and assumes a significant job in e-commerce and online cash exchange which is developing. Because of the developing use of charge card, fraudsters attempt to discover more chances to submit cheats that can make colossal misfortunes cardholders and banks. Credit card fraud takes many forms, namely

J. Parmar (✉) · A. Patel
Saurashtra University, Rajkot, Gujarat, India
e-mail: jasminparmar91@gmail.com

M. Savsani
Symbiosis Statistical Institute, Symbiosis International (Deemed) University, Pune, Maharashtra, India

K. Kotecha et al. (eds.), *Data Science and Intelligent Applications*, Lecture Notes on Data Engineering and Communications Technologies 52,
https://doi.org/10.1007/978-981-15-4474-3_22

[1]: offline credit card fraud: Happens when the plastic card is stolen by fraudsters, using it in stores as the actual owner. Online credit card fraud: A popular and very dangerous fraud, credit cards' information is stolen by fraudsters to be used later in online transactions by Internet or phone.

Fraud means obtaining services/goods and/or money by unethical mean and is a growing problem all over the world nowadays. Fraud deals with cases involving criminal purposes that, mostly, are difficult to identify. Credit card fraud detection is one of the most explored domains of fraud detection. Numerous authorization techniques are used to prevent credit card frauds, such as signatures, credit card number, identification number, cardholder's address and expiry date. However, these techniques are not enough to hinder credit card fraud. Therefore, there is a need to use fraud detection approaches which analyze data that can detect and eliminate credit card fraud.

Today, with the development of web-based business, it is on the web that half of all credit card fraud is led. Deceitful exercises cannot be adequately identified with basic example coordinating calculations. Productive fraud discovery is expected to assert the exactness and small size bogus identification. Since credit card fraud has enormously expanded, it is essential to comprehend the strategy for recognizing and finding credit card fraud. In actuality, the information or tasks with inconsistencies are insufficient to effectively distinguish frauds. The main practical arrangement is robotization by PC. The PCs can be utilized to evaluate credit card exchanges as "suspicious" and this can be performed by basic factual techniques. Along these lines, customary strategies do not do the trick, and more current methods like machine learning and AI are required.

Another attribute of this space is that the dataset is lopsided, that is, the "not fraud" class is substantially more successful than the "fraud" class. Significant contemplations incorporate how quick the cheats can be distinguished, what number of styles/kinds of frauds are identified, regardless of whether the location was in on the web/constant (occasion driven) or group mode (time driven).

In the present fraud discovery frameworks, the "suspicious" exchange is stamped and afterward physically examined by human administrators for the last decision to disambiguate questionable cases.

There are specific challenges with credit card fraud detection because it is challenging. However, there are several constraints associated with the problem. In what follows, we mention a few important points [2]:

- Unavailability of datasets
- Dynamic fraudulent behavior
- Highly skewed dataset
- Right evaluation parameters.

2 Types of Fraud

Bankruptcy fraud: Bankruptcy fraud educates the utilization with respect to a credit report from acknowledging agency as a wellspring of data seeing the candidates' open records just as a potential implementation of a bankruptcy model. It is the most troublesome kinds of fraud for anticipating liquidation misrepresentation which implies utilizing a credit card while being indebted.

Theft fraud/counterfeit fraud: Theft fraud implies utilizing a card that is not yours. The culprit will take the card of another person and use it, and however, many occasions as would be prudent before the card is blocked. The sooner the proprietor will respond and contact the bank, the quicker the bank will take measures to stop the criminal. At a certain point, one will duplicate your card number and codes and use it through specific sites, where no signature or physical cards are required.

Application fraud: In many banks, to be qualified for a credit card, candidates need to finish an application structure. Every one of those qualities might be utilized while scanning for copies. Instead of utilizing factual procedures, another method simple to actualize is cross-coordinating. Conversely, personality wrongdoing, as it is named, is executed by genuine crooks filling application information incorrectly intentionally.

Behavioral fraud: It happens when subtleties of genuine cards have been gotten deceitfully and deals are made on a "cardholder present" premise. These businesses incorporate phone deals and Internet business exchanges, where just the card subtleties are required.

3 Theoretical Background of Deep Learning

Deep learning is an amazing asset to make forecast a significant outcome. It is computer programming that imitates the system of neurons in a mind. It is a subset of AI and utilizes deep neural systems. A deep neural system gives the best in class precision in numerous assignments, from object discovery to discourse acknowledgment. They can adapt naturally, without predefined information expressly coded by the developers.

The process of deep learning like: Understand Problem > Identify Data > Select Algorithm > Trail the Model > Test Model

There are two types of classification under deep neural network, namely shallow neural network which has only one hidden layer between the input and output. Another is deep neural network which has more than one layer. Deep learning methods are broadly classified as (Fig. 1).

Existing techniques: Artificial neural networks (ANNs) are multilayer completely associated neural nets that resemble the figure beneath. They comprise an input layer, multiple hidden layers and an output layer. Each node in one layer is associated with

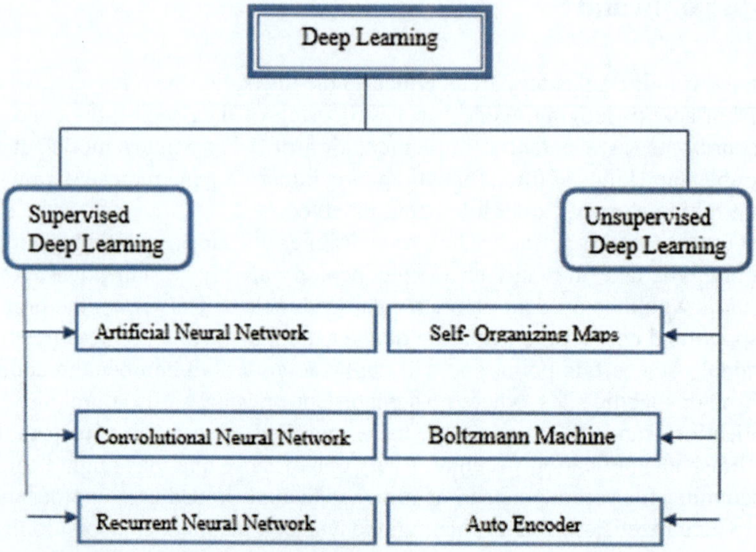

Fig. 1 Classification of deep learning methods

each other node in the following layer. We make the system more profound by expanding the quantity of hidden layers (Fig. 2).

The building block of a neural network is the neurons. An artificial neuron works much the same way the biological one does. Artificial neural networks (ANNs) are computing systems that are designed a similar way to human brain analysis and process the information. Single artificial neurons called a perceptron.

Deep learning auto-encoder: Auto-encoder neural system is an unsupervised learning calculation that applies back-engendering, settings the objective qualities to be equivalent to the sources of info, for example, it utilizes $Y(i) = X(i)$. Stacking layers of auto-encoders produce more deep engineering known as stacked or deep

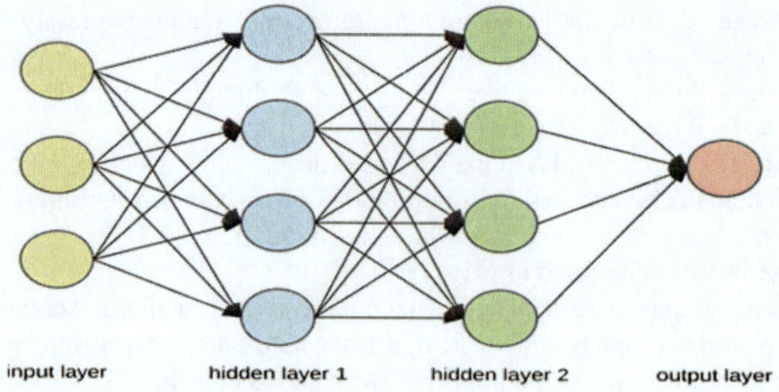

Fig. 2 Basic flow of ANN with two hidden layer

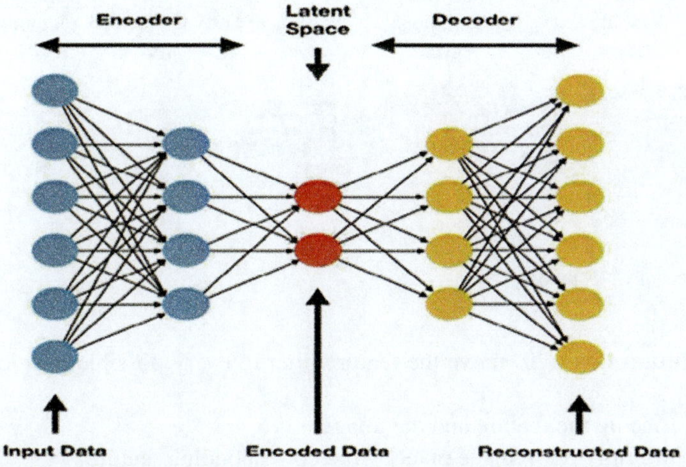

Fig. 3 Basic flow of deep auto-encoder

auto-encoders. Auto-encoders are neural systems with equivalent info and output. It is made out of two sections: 1. *Encoding network*: This part of the network compresses the input into a latent space representation. The responsibility of the encoder layer is to take input and reduce it in some form. 2. *Decoding network*: This layer decodes the encoded input back to the original dimension. The decoded output is some form of original input, and it is reconstructed from the latent space representation (Fig. 3).

Feed-forward neural network: Neural system rose up out of an extremely well-known AI calculation named perceptron. Deep feed-forward systems, likewise regularly called feed-forward neural networks or multilayer perceptrons (MLPs), are the quintessential deep learning models. Joining numerous layers of perceptrons are known as multilayer perceptrons or feed-forward neural systems. With this sort of engineering, data streams in just a single course, forward. That is to say, the data's streams start at the information layer, go to the "covered up" layers and end at the yield layer. The system does not have a circle. Data stops at the yield layers.

There are two different classes of network architectures: ***Single-layer feed-forward (neurons are sorted out)***: Neurons with this sort of activation work are additionally called fake neurons or straight edge units. It very well may be prepared by a straightforward learning calculation that is generally called the delta rule. ***Multilayer feed-forward (in non-cyclic layers)***: The units of these systems apply for a sigmoid work as an actuation work. It very well may be prepared by a learning calculation that is normally called back-propagation.

Convolutional neural network: In case of convolutional neural network (CNN or ConvNet), the neuron in a layer will only be connected to a small region of the layer before it, instead of all of the neurons in a fully connected manner. It is a special type of feed-forward artificial neural network which is inspired by the visual cortex. CNN has the following layers:

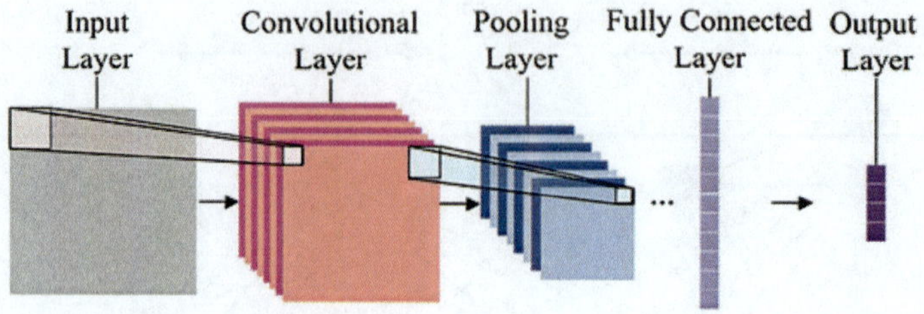

Fig. 4 Layers of CNN model

Convolution layer: To move the feature/filter to every possible position on the image.

Step-1: Line up the feature and the image

Step-2: Multiply each image pixel by the corresponding feature pixel.

ReLU layer: To remove every negative value from the filtered images and replaced it with zero's. This is done to avoid the values from summing up to zero. ReLU transform function only activates a node if the input is above a certain quantity, while the input layer is below, the output is zero, but when the input rises above a certain threshold, it has a linear relationship with the dependent variable. Function of ReLU like:

$$F(x) = 0 \text{ if } x < 0$$
$$\{x \; if \; x \geq 0$$

Pooling layer: To shrink the image stack into smaller size

Fully Connected layer: Final layer, where the actual classification happens. Here, we take our filtered and shrunken images and put them into a single list (Fig. 4).

Recurrent neural network: Recurrent neural networks (RNNs) are a type of neural network where the output from the previous step is fed as input to the current step. It can handle sequential data, considers the current input and also the previously received inputs, can memorize previous inputs due to its internal memory. Important parameters that affect the performance of RNNs are activation function, dropout rate and loss function [3] which is illustrated in Fig. 2 for a diagram of an RNN (Fig. 5).

Long short-term memory systems: Standard RNNs experience the ill-effects of evaporating or detonating inclination issues. To address these issues, the long short-term memory architecture was proposed. LSTMs contain a memory cell which keeps up its state after some time. Also, gating units are utilized to direct the data stream into and out of the memory cell. All the more explicitly, an input entryway can enable the info sign to change the cell state or counteract it (i.e., sets the input gates to zero). An output gate can permit the cell state to influence neurons in the shrouded layers or square it. An overlook door empowers the cell to recollect or overlook its past state. Initially, the relative importance of each component is not clear [4]. Important LSTM parameters that affect the quality of the output are the number of neurons in

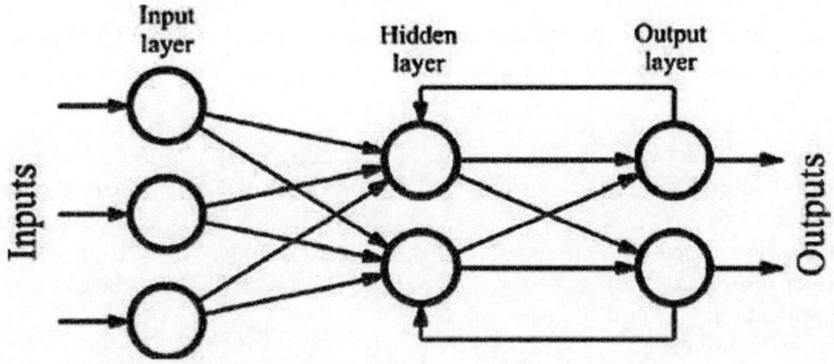

Fig. 5 RNN with one hidden layer

the hidden layers, activation function and inner activation function and dropout rate [5].

Gated recurrent unit: A gated recurrent unit makes each intermittent unit adaptively catch conditions of various time scales. Albeit like an LSTM, the GRU has gating units that tweak data stream into the unit, nonetheless, GRUs do not have separate memory cells. Unlike LSTMs, GRUs expose their whole state each time [6]. The same parameters that affect LSTMs apply to GRUs too [7].

4 Comparative Analysis of Recent Study

After the literature review, the deep learning methods were studied for credit card fraud detection. Following Table 1 reflects relative analysis of various deep learning methods for credit card fraud detection. Studies of parameter evaluation from different deep learning methods were analyzed. This parametric evaluation gives some new ideas to propose a novel approach for deep learning method.

5 Proposed Approach

After the study of comparative analysis of deep learning methods for credit card fraud detection, different parameters were evaluated using different deep learning methods like DNN with auto-encoder, feed-forward neural network, CNN, ANN, RNN, LSTM and GRU which were studied. The literature study helps us to find that some more parameters can be evaluated using supervised deep learning methods. Following are some of the objectives to find out after the detailed review from different deep learning methods:

- To achieve high nonlinearity of dataset
- To work on a large amount of data

Table 1 Comparative analysis of deep learning methods for credit card fraud detection

S. No.	Example	Learning paradigm	Techniques	Challenges
1	Title: An Effective Real-Time Model for Credit Card Detection Based on Deep Learning [8] Publication: ACM-Conference, 2019	Unsupervised	Deep neural network (DNN) with auto-encoder	Cannot effectively handle confusion matrix parameter Highly unbalanced dataset
2	Title: Ensemble Learning for Credit Card Fraud Detection [2] Publication: ACM-Conference, 2018	Unsupervised	Feed-forward neural networks	Limited to dataset having numeric value
3	Title: Credit Card Fraud Detection Using Convolutional Neural Networks [9] Publication: Springer-Conference, 2016	Supervised	Convolutional neural network	Data imbalance is too high
4	Title: Deep Learning Detection Fraud in Credit Card Transactions [10] Publication: IEEE Journal, 2018	Supervised	Artificial neural network (ANN), recurrent neural network (RNN), long short-term memory (LSTM), gated recurrent units (GRU)	Hyper-parameter like momentum, batch size, number of the epoch, dropout rate not included

- To analyze lots of features of dataset
- To work on mostly unlabeled data in the dataset
- To analyze various other parameters of deep learning methods.

Detailed steps for deep learning involving pre-processing of dataset, feature extraction, training the dataset, reduction in errors, testing the dataset and generating the output are shown in the following proposed system (Fig. 6).

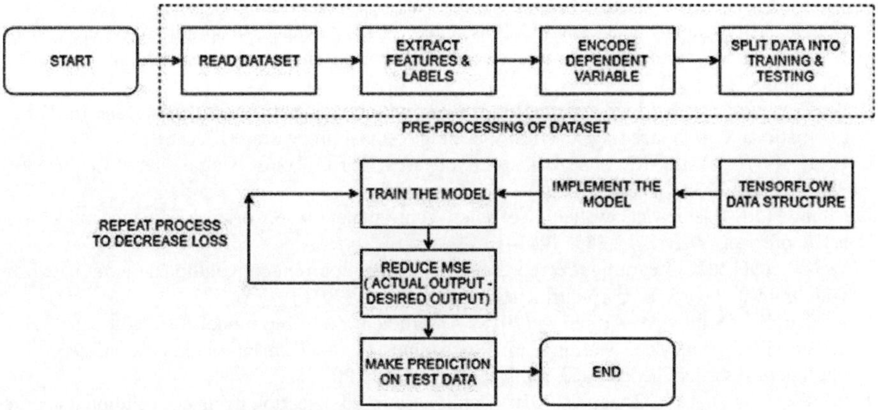

Fig. 6 Proposed system for credit card fraud detection with deep learning method

6 Conclusion

Credit card fraud is a demonstration of criminal deceptive nature. This article has audited late discoveries in the charge card field. This paper has distinguished the various sorts of frauds, for example, bankruptcy fraud, counterfeit fraud, theft fraud, application fraud and behavioral fraud and talked about measures to identify them. Such measures have included pair-wise coordinating, decision trees, clustering systems, neural networks and genetic algorithms. From a moral viewpoint, it very well may be contended that banks and credit card organizations should endeavor to recognize every deceitful case. However, the amateurish fraudster is probably not going to work on the size of the expert fraudster, and thus, the expenses to the bank of their identification might be uneconomic. The bank would then be looked at with a moral situation. As the subsequent stage in this examination program, the center will be upon the execution of a "suspicious" scorecard on a genuine informational collection and its assessment. After the investigation of a similar examination of deep learning strategies for credit card fraud recognition, there is a tremendous extension for various parameter assessments. A portion of the hyper-parameters like confusion matrix, momentum, batch size and number of epoch can be taken care of adequately.

References

1. Abdallah A, Maarof MA, Zainal A (2016) Fraud detection system: a survey. J Netw Comput Appl 68:90–113
2. Sohony I, Pratap R, Nambiar U (2018) Ensemble learning for credit card fraud detection. In: Proceedings of the ACM India joint international conference on data science and management of data. ACM, pp 289–294

3. Bengio Y, Boulanger-Lewandowski N, Pascanu R (2013) Advances in optimizing recurrent networks. In: IEEE International conference on acoustics, speech and signal processing (ICASSP). IEEE, pp 8624–8628
4. Wu Z, King S (2016) Investigating gated recurrent networks for speech synthesis. In: IEEE international conference on acoustics, speech and signal processing (ICASSP)
5. Greff K (2016) LSTM: a search space odyssey. IEEE Trans Neural Netw Learn Syst 28(10):2222–2232
6. Chung J (2014) Empirical evaluation of gated recurrent neural networks on sequence modeling. arXiv preprint arXiv:1412.3555 (2014)
7. Wen Y et al (2016) Learning text representation using recurrent convolutional neural network with highway layers. arXiv preprint arXiv:1606.06905 (2016)
8. Abakarim Y, Lahby M, Attioui A (2018) An efficient real-time model for credit card fraud detection based on deep learning. In: Proceedings of the 12th international conference on intelligent systems: theories and applications. ACM, p 30
9. Fu K, Cheng D, Tu Y, Zhang L (2016) Credit card fraud detection using convolutional neural networks. In: International conference on neural information processing. Springer, Cham, pp 483– 490
10. Roy A, Sun J, Mahoney R, Alonzi L, Adams S, Beling P (2018) Deep learning detecting fraud in credit card transactions. In: 2018 Systems and information engineering design symposium (SIEDS). IEEE, pp 129–134

The Art of Character Recognition Using Artificial Intelligence

Thacker Shradha, Hitanshi P. Prajapati, and Yatharth B. Antani

Abstract This research paper investigates the uses of AI in the problem that occurred during defining the machine-printed characters. It uses backpropagation net which uses 84 attribute font styles and examined over two different printing font styles. These results which are retrieved from the test are compared with the results got on the identical data by a more used and suitable path.

Keywords Text recognition · OCR neural networks · AI

1 Introduction

The need for a machine recognizes some form of autonomous and semiautonomous *optical character recognition* for decennary. Nowadays, we have numbers of algorithms that perform the task with its vitality and flaws. The variations in more than one different OCR algorithms are: The first is an article eradication method by traditional AI techniques for allotment and the second, a neural network approach with virtually no proceedings explored by this paper [1]. The base idea of this research was first to run the same data on two distant most suitable algorithms and record the variations of the algorithms, running side by side [1]. The test executed by the authors at the first time using neural net method was used in this article with the similar attribute set and following an identical off-process by this method [2]. Thus, the comparison

T. Shradha (✉) · H. P. Prajapati
Department of Computer Science and Engineering, ITM Universe, Vadodara, Gujarat, India
e-mail: thackershradha95@gmail.com

H. P. Prajapati
e-mail: hitanshi1695@gmail.com

Y. B. Antani
Cyber Security and Incident Response, Institute of Forensic Science, Gujarat Forensic Sciences University, Gandhinagar, Gujarat, India
e-mail: yatharth.antani.01@gmail.com

K. Kotecha et al. (eds.), *Data Science and Intelligent Applications*,
Lecture Notes on Data Engineering and Communications Technologies 52,
https://doi.org/10.1007/978-981-15-4474-3_23

of the two different methods was not instigated by arbitrary characteristic in dataset or off-balance gimmick that would be favored by a method over the other method [2].

There were three eighty-five character fonts used by the dataset. The attribute verdict used for all the fonts is basically 8 * 8. The first as well as second case alphabets (upper case and lower case both) with all numbers and a handful motley asterisk marks were used in the dataset [3]. The sets are used by an old CBM C128 computer as factual display characters, so they represent the actuality of data. In scrupulous, any real examples of allowable surrogate forms of symbols included by the sets, and no real attempt was done by the algorithm to statistically deal with this all facts. Thus, you get the result lesser than the originals, and so, a fair analysis would be prorated with these stand-in type as they would apparently be implemented as special cases in a marketable OCR program [1].

AI is the future of the world and the best field in research work. The most preferable use of AI is in security sector. We can use speech identification and face identification system. It is also used in image processing system.

2 AI—Background

Data Mining: Finding the knowledge or analyzing the real patterns from lots of data generated every day.

Expert Systems: Computer program which is having expert thought process similar as the thought process of a human intelligence for showdown.

Neural Networks: It is a tool which is based on the mantel process analogy.

Fuzzy Logic: It is based on comparative reasoning theory.

Artificial Life: Revolutionary computing techniques which is similar to human thinking and swarm intelligence.

Artificial Immune System: The logic which is developed by computer with the help biological privileged system.

AI also used in system of pattern identification, handwriting identification, speech and face recognition, artificial imagination, computer perception, virtual reality as well as image processing, natural language processing, theory of gaming and strategic planning, gaming AI and predicate logic, translation and chatter bots [3].

3 Preprocessing

The reason for which this research paper designed is not to hide the actual article point wrenching algorithm in deep. The summarized content of its architecture and working was listed for amenity [1]. The point in the base of the article point wrenching procedure is to find the attributes based on the characteristics that are identical to the articles used to recognize the characters by humans. The exposition is that when the procedure does not classify a character (as done by all programs), it should choose an attribute that would approve to be a legitimate estimation by a human, because it is easy to solve mistakes for humans that were generally done by the humans [2]. To use this documentation, the program would examine the whole attribute matrix of 8 × 8 and evaluate all occupied pixel. The prompt precinct of the pixel would be analyzed, and pixels that liked to be eligible for notification were highlighted as shown in Fig. 1 "Character Display" [3]. The primordial C128 ROM attribute set was handled in this demeanor, and using them as a higher standard, attributes could be recognized by correlating with the number of entries in the given "*dictionary*" [4]. Comparing the sum of shortest spaces between two featuring vertices of the attribute distinguishing the prominent part of vocabulary. The guess was that the vocabulary of attributes with the tiniest addition along a pre-defined verge [4]. Neglecting the efforts to allow various prominent made by this algorithm, it also chastised on ignoring the absent or excess attributes but not tremendously. Revamping was prevented of the upmost probable outcome; it is just a try to test the preformed idea of the procedure. The neural network connecting different sources was not designed for the system; the whole differentiation methods of two OCR are balanced. The outcome of the prominent parts featuring algorithm is highlighted in Fig. 2 [5] (Fig. 3).

The neural network exploits three isolated poses. The first pose includes translating the binary attribute in a friendly, and the second pose will take the outcome of the first pose and train the reverse proliferation network on outcome, displaying all the outcome weights and general network information [4]. Taking the outcome of second step the 3rd step made a network. Then, it executed a complete attribute data with the help of the network and displayed knowledge to the attributes which are included in the set [4]. As all three programs were practical, it makes sense to implement neural net OCR. The code which is ready for preprocessing of the article

Fig. 1 AI tree

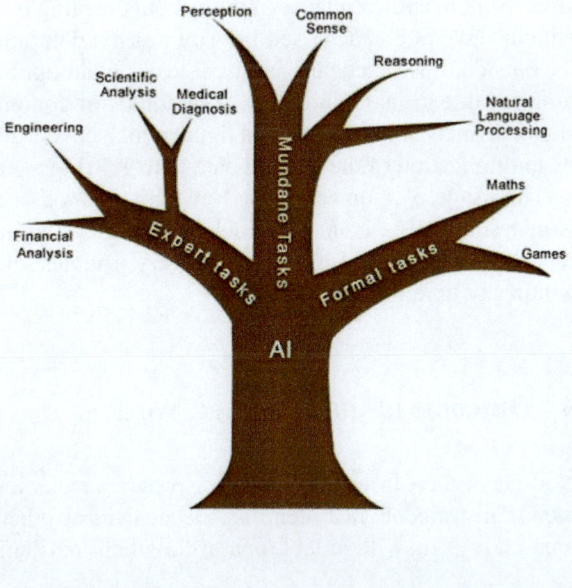

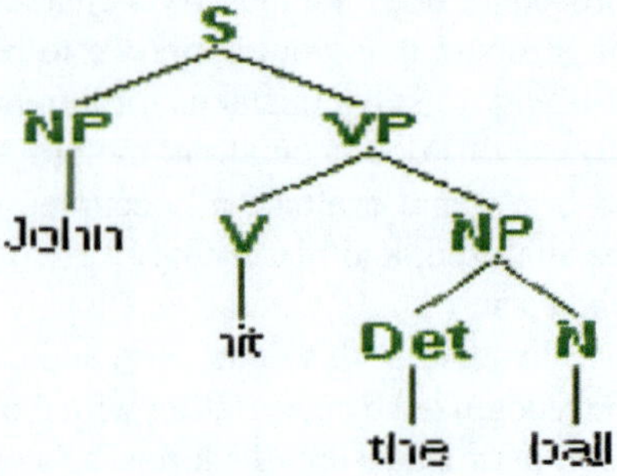

Fig. 2 Natural language processing

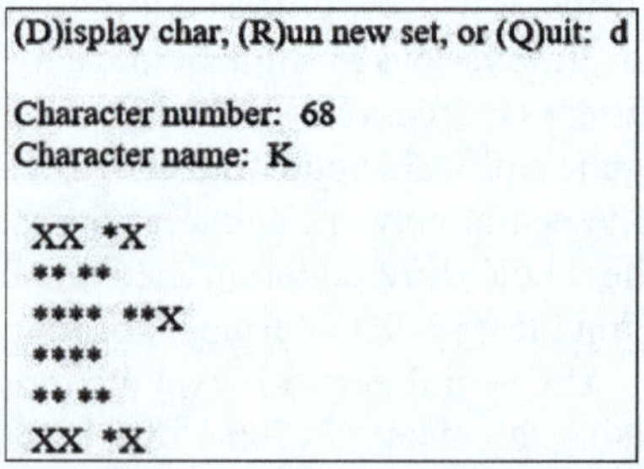
(D)isplay char, (R)un new set, or (Q)uit: d

Character number: 68
Character name: K

XX *X
** **
**** **X
****
** **
XX *X

Fig. 3 Character display

outlining OCR algorithm can be obtained from separating the first step and removing the area of comparison between these two algorithms. The second step was removed because of slow processing. Therefore, several machines can be devoted to learning, while others are used to analyze the outcomes [4].

There were many difficulties that came across to train the neural net. Major diversity in behavior of overall learning can be obtained by minor changes. More step sizes result in entire collapse of the system enabling it to mingle in meaningful spans with the help of character set. Priority was not determined for a good step size from the outside. It is practically impossible to obtain optimum results from the learning program due to its requirement of hundreds of computing hours on a Sun SPARC station. Ninety six is not an ideal figure which covers all points; the figure is obtained by multiplication of the entered data with 1.5 (Figs. 4 and 5).

The usage of momentum or Newton's methods could practically decrease the issue training. The issue of training sticks out as an unwanted asset that the article removing the shortfall in steps; it is very troublesome to design a neural network within any time limit.

4 Outcome of the Proposed Work

Nowadays, there is not any prevailing program either on private instruments or on the mobile instruments that identifies Devanagari or other handwritten attributes. This high-end system with suitable neural classifiers and dictionary classifiers can improve

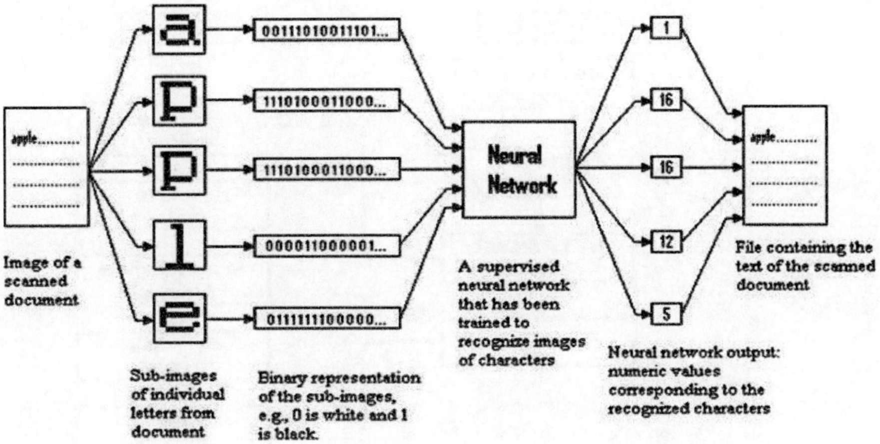

Fig. 4 Neural networks ontology

Fig. 5 Character recognition

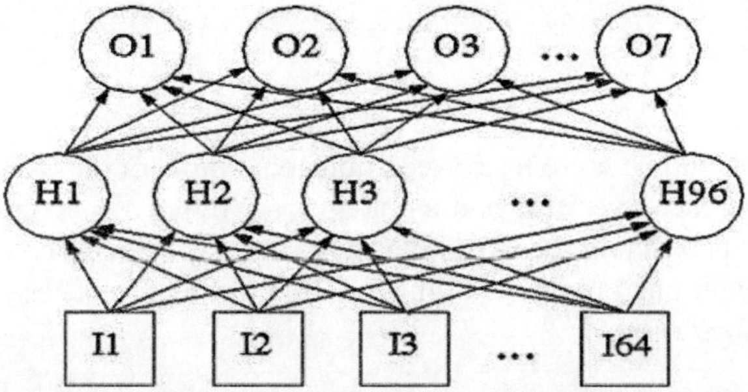

the recognition rate to 90% and above which can be stated as a good beginning stage for enforcing it to particular instruments with more improvisation in identifying words (Fig. 6).

Recognition Devanagari or other Character Script: The image being input should contain only and only Devanagari script. The cause for this constraint is that the algorithm is based on distributional properties and topological as well as geometrical properties of the various Hindi character sets. Character set or Matras should not be overlapping: The characters should not come in touch with each other. Though partition of overlapping characters has been one of the hard jobs in text recognition, this is the only hardest area of research. Given image should be devoid of handwritten scripts (ardhaksr): Handwritten script (Ardhakshers) were ignored from the identifying domain due to the following two main causes, the tremendous difference in the handwritten scripts like I and WE causes a huge difference. For example: Example shows how neural networks and AI are used in different applications similar to artificial creativity, pattern recognition, facial animation, autonomous walker (AW) and swimming eel, computer visioning, VR and processing of images. Image

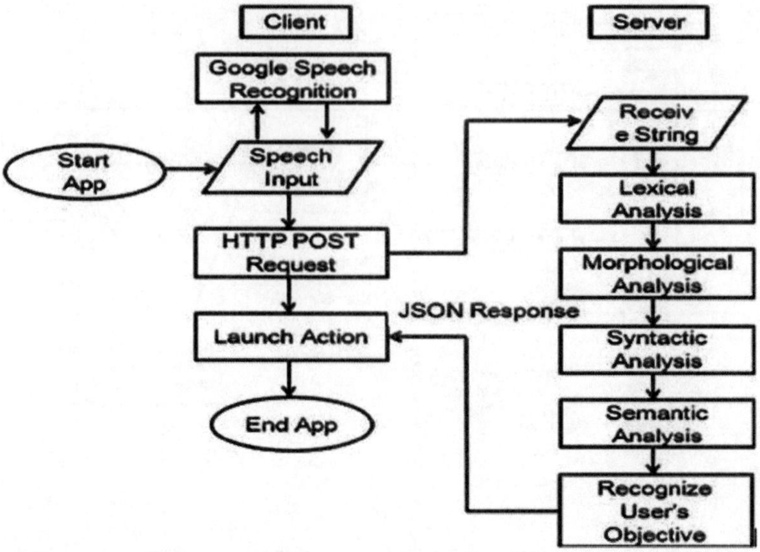

Fig. 6 Speech recognition DFD

should be noise free: Prediction about input of the Hindi text image should be interference free. This prediction does not remove the complexity of issue as this is just a part of preprocessing phase. An interfered image can be made interference free by inputting pre-defined functions and techniques. These were abandoned due to time limit (Fig. 7).

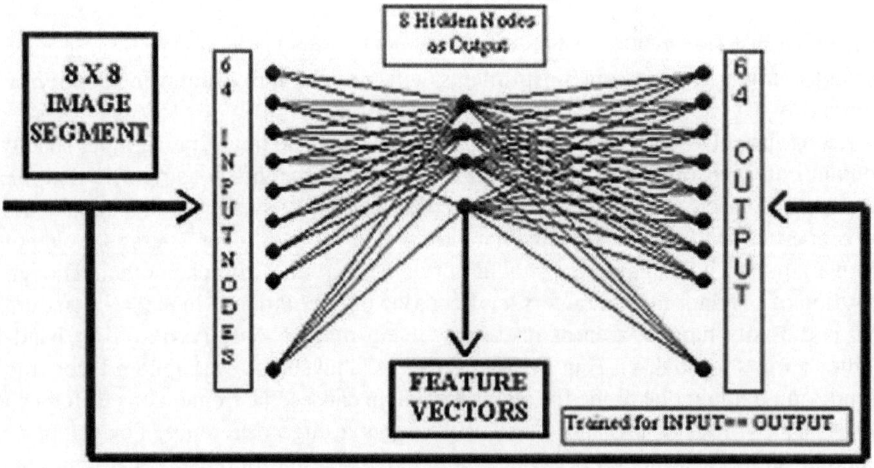

Fig. 7 Hindi handwritten character recognition

5 Conclusion

The research paper defines the uses of neural networks and artificial intelligence to create an intelligent behavior. How AI and NN are mixture of physiology and philosophy in science of computer. An AI creates machinery expert in engaging with human behavior.

References

1. Madhavnath S, Vijaysenan D, Kadiresan TM (2007) LipiTk: a generic toolkit for online handwriting recognition. In: International conference on computer graphics and interactive techniques, no 13
2. Joshi N, Sita G, Ramakrishnan AG, Madhvanath S (2015) Machine recognition of online handwritten devanagari characters. In: Proceedings eighth international conference on document analysis and recognition, vol 2, pp 1156–1160
3. Santosh KC, Nattee C (2006) Structural approach on writer independent nepalese natural handwriting recognition. IEEE
4. Xu AR, Yeung D, Shu W, Liu J (2002) A hybrid post-processing system for handwritten chinese character recognition. Int J Pattern Recogn Artif Intell 16(6):657–679
5. Kompalli S, Setlur S, Govindaraju V (2009) Devanagari OCR using a recognition driven segmentation framework and stochastic language models. Int J Doc Anal Recogn 12(2):123–138
6. Tappert CC, Suen CY, Wkahara T (1990) The state of the art in on-line handwriting recognition. IEEE Trans Pattern Anal Mach Intell 12(8):787–808

Banana Leaves Diseases and Techniques: A Survey

Ankita Patel and Shardul Agravat

Abstract Agricultural productiveness is an important sector within the Indian economy. Early detection of disease in plants is very important to improve the quality and production of crops. As stated, precaution is better than cure. Manual detection of diseases in plant leaves (within a farm) is time consuming and requires excessive effort; instead, a computerized method can save time and efforts and also can improve overall production. Banana is one of the most common fruits in the world and their growths are largely affected by various diseases. This causes decrease in production, as the diseases are not identified at an early stage. This paper provides a brief review of main diseases (illnesses) in banana plant leaves, and it also explains the image processing techniques that are used in the process of disease identification and classification using machine learning techniques in banana plant leaves. This research will be helpful to farmers to take necessary actions against the diseases and reduce the loss of crop yield and thus increase agricultural production.

Keywords Agriculture · Image processing · Plant diseases

1 Introduction

The agricultural sector performs an important role in economic development by providing rural employment. Crop production is vital factor that is directly associated with the economy and people. Crop diseases have become a risk to food security. Diseases in plant leaf cause severe problems in terms of financial losses and less crop production. Identifying diseases from the images of the plant is one of the most demanding analyses in the agriculture field. As the population is increasing, to fulfill

A. Patel (✉) · S. Agravat
Shantilal Shah Engineering College, Sidsar Campus, Vartej, Bhavnagar 364060, India
e-mail: patelankita294@gmail.com

S. Agravat
e-mail: shardulagravat@gmail.com

K. Kotecha et al. (eds.), *Data Science and Intelligent Applications*, Lecture Notes on Data Engineering and Communications Technologies 52,
https://doi.org/10.1007/978-981-15-4474-3_24

the requirement of food, crop production should also be increased. If proper care is not taken for the diseases, then it may result in the reduction of quality and quantity of the crop.

Banana is one of the important crops in the world. Some of the banana plant diseases, whose symptoms are seen on leaves, are Sigatoka disease, Panama disease, Moko disease, Speckle disease, Bugtok disease, Rust disease, Spot disease, Streak virus, Mosaic virus, etc. [1]. The proper detection and recognition of the disease are very important in applying the required fertilizer. Diseases in plant leaf can be detected from their leaves' images by using image preprocessing, segmentation, feature extraction and classification and many machine learning techniques. There are several techniques to detect the leaf that are: CIELAB color space, grayscale matching, color-based gradient matching, histogram analysis, etc. Some techniques to detect or classify the diseases are support vector machine (SVM), k-nearest neighbor (KNN), artificial neural network (ANN), convolution neural network (CNN), etc.

Image processing and machine/computer vision technology have become more potential and more important to many areas in agricultural technology [2]. Measuring the extent of plant unwellness (disease) is the essential thing within the analysis of disease management, yield loss estimate, sickness resistance and breeding.

2 Banana Plant Leaf Diseases

Diseases cause characteristic damage on the surface of crop elements, e.g., leaf, root, blossom, fruit, flower, stem, etc. Banana plants are affected by various fungal, bacterial and viral diseases [3]. Some diseases as follows:

Panama Disease [1, 4]: Panama disease is the type of fungal disease. Leaves begin to yellow, beginning with the oldest leaves and acquiring toward the middle of the banana. This sickness is deadly. It is transmitted through water, wind, moving soil and farm instrumentation (Fig. 1).

Moko Disease [1, 4]: Moko disease is the type of bacterial disease. This illness is the chief illness of banana. It is transmitted via insects, machetes and alternative farm tools, plant detritus, soil and root contact with sick plants. The sole positive defense is to plant resistant cultivars. Dominant infected bananas are long, expensive and resistant (Fig. 2).

Sigatoka Disease (Yellow Sigatoka/Black Sigatoka/Black leaf streak) [1, 4]: Sigatoka disease is the type of fungal disease. Yellow Sigatoka is one of the most serious diseases affecting the crop of banana. Initial symptoms appear on the leaves in the form of light yellowish spots. Leaf color also transforms to dark brown. Later on, the middle of the spot dies which turns in lightweight gray spots enclosed by a brown ring, and in severe cases, the leaves may die. Rainfall, cold and temperature determine the spread of the disease on the plant leaf. Condition responsible for mass infection area is the season with temperature higher than 21 °C (Fig. 3).

Fig. 1 Panama disease [5]

Fig. 2 Moko disease [6]

Fig. 3 Sigatoka disease [7]

Anthracnose [4]: Anthracnose disease is the type of fungal disease. This sickness attacks on banana plants at all stages of growth. This disease first attacks the flowers, skin and then ends of banana heads. This disease turns fruit black (Fig. 4).

Mosaic Disease [4]: Mosaic disease is the type of viral disease. It can be characterized by typical mosaic symptoms on the plant leaves. Plants with mosaic disease are simply recognized by their short growth and deformed leaves with spots. The

Fig. 4 Anthracnose disease [8]

Fig. 5 Mosaic disease [9]

early symptoms can be first seen on young leaves as light green or yellowish streaks or bands giving a spotted look (Fig. 5).

Streak Disease [1, 4, 10]: Streak disease is the type of viral disease. A distinguished symptom in banana streak virus (BSV) is yellow streaking leaves, which increases continuously and give black-streaked look in older leaves. The virus of this disease is transmitted principally through infected planting materials (Fig. 6).

3 Methodology

There are mainly five steps [3, 12] involved in detection of diseases in plant leaf, which include,

1. Capturing an image (image acquisition)
2. Performing preprocessing steps on acquired image (image preprocessing)
3. Performing image segmentation by appropriate technique (image segmentation)

Fig. 6 Streak disease [11]

4. Extracting required features (feature extraction)
5. Classifying image according to extracted features (classification)

Image Acquisition: First, the images are acquired using an image capturing device with required resolution for better quality. The image quality and type itself are responsible for the efficiency of the classifier, which decides the strength of the algorithm.

Image Preprocessing: After acquiring image, the next step is image preprocessing to improve the image information, hide undesired distortions and enhance the required important features. Image preprocessing includes color conversion, resizing an image, image enhancement, etc.

Image Segmentation: In the next step, appropriate segmentation technique is applied on the image. Various techniques including edge-based, threshold-based, model-based, feature-based, cluster-based [3] are available for segmentation.

Feature Extraction: In the next step, portion affected by disease is extracted. Then, required features are extracted, and based on them, verification is done with the means of a given sample. Some of the commonly considered features include color, shape, size, texture, etc. Nowadays, most of the research is targeting plant leaf texture because it is one of the most important features in classifying plants. Using texture features, diseases in the plants can be classified into different types.

Various classification techniques can be used to classify the results, and some of the texture-based classification techniques are included in Table 1.

4 Conclusion

Crop disease detection is a major challenge in the agriculture domain. The aim of this work is to help farmers in improving the agricultural productivity by detecting the diseases, those which are infecting the crop at the early stage. The image processing

Table 1 Texture classification techniques comparison [3]

Techniques	Advantages	Disadvantages
Support vector machine(SVM) [1–3, 13]	Simple geometric interpretation Sparse solution It may be robust, even with biased training sample	Training phase is slow Difficult to recognize the structure of the algorithm Large no. support vectors are needed from a training set to perform the classification tasks
Back-propagation neural network (BPNN) [1, 3]	Implementation is easy Can be used in wide range of problems	Slow learning rate Difficult to identify the number of layers and neurons required
K-nearest neighbor (KNN) [3]	Simple classifier, no training phase is required Can be used with smaller datasets	It is expensive to test every instance Irrelevant inputs may greatly affect the results
Radial basis function (RBF) [3, 12]	The training phase is faster Hidden layer is easier to interpret	Slower in execution when speed is a factor
Probabilistic neural network (PNN) [3]	Can resist noisy inputs Instances can be classified by more than one output Can work with changing data.	Training takes more time Network structure is more complex Requires lot of memory for training data
Convolutional neural network (CNN) [14]	Very fast Highest accuracy of image classification	Large no. of training data is needed
Artificial neural network (ANN) [3, 15]	Easy to implement Applicable to a wide range of problems	Slow learning High processing time

techniques/algorithms can be used along with the classification systems to detect the plant diseases accurately which can increase the crop production.

References

1. Surya Prabha D, Satheesh Kumar J (2014) Study on banana leaf disease identification using image processing methods. IJRCSIT
2. Tzotsos A, Argialas D (2008) Support vector machine classification for object-Based. In: Image analysis. Springer, pp 663–677
3. Gavhale KR, Gawande U (2014) An overview of the research on plant leaves disease detection using image processing techniques. J Comput Eng 16:10–16
4. http://agropedia.iitk.ac.in/content/banana-diseases-their-control. Accessed on 26 Nov 2019
5. https://www.plantmanagementnetwork.org/pub/php/management/bananapanama/. Accessed on 2 Nov 2019

6. Dataset: https://github.com/godliver/source-code-BBW-BBS. Accessed on 3 Dec 2019
7. https://appliedecology.cals.ncsu.edu/absci/2016/05/sigatoka-on-banana/. Accessed on 13 Dec 2019
8. http://oxfarm.co.ke/tag/anthracnose-disease-in-bananas/. Accessed on 29 Nov 2019
9. https://www.ctahr.hawaii.edu/oc/freepubs/pdf/PD-101.pdf. Accessed on 29 Nov 2019
10. Karthik G, Praburam N (2016) Detection and prevention of banana leaf diseases from banana plant using embeeded linux board. IEEE
11. https://betterbananas.com.au/2018/01/17/streaks-on-leaves/. Accessed on 29 Nov 2019
12. Devaraj A, Rathan K, Jaahnavi S, Indira K (2019) Identification of plant disease using image processing technique. IEEE
13. Islam M, Dinh A, Wahid K, Bhowmik P (2017) Detection of potato diseases using image segmentation and multiclass support vector machine. IEEE
14. Fang T, Chen P, Zhang J, Wang B (2019) Identification of apple leaf diseases based on convolutional neural network. Springer
15. Shah N, Jain S (2019) Detection of disease in cotton leaf using artificial neural network. IEEE

Human Activity Recognition Using Deep Learning: A Survey

Binjal Suthar and Bijal Gadhia

Abstract Human activity recognition refers to predict what a person is doing from series of the observation of person's action and surrounding conditions using different techniques. It is an active research area providing personalized support for various applications and its association with a wide range of fields of study like medicinal services, dependable automation developing, and smart surveillance system. This paper provides an overview which gives idea about some existing research methods on human activity recognition. It describes a general view on the state of the art for human activity recognition and shows comparative studies between existing research which consist of various methods, evaluation criteria, and features. It also comprises benefits and limitations of various methods to provide researchers to propose new approaches.

Keywords Human action · Deep learning · CNN · RNN · Vision · Machine learning

1 Introduction

Action recognition intends to recognize the activities and objectives from a series of observations and the natural conditions of at least one specialist. When trying to recognize human activities, one must decide the kinetic states of an individual, with the goal that the PC can proficiently perceive this movement. Handcrafting features from the time series data based on preparing AI models and fixed-size windows has

---

B. Suthar (✉)
Research Scholar, CE Department, Government Engineering College, Sector-28, Gandhinagar, India
e-mail: sutharbinjal@gmail.com

B. Gadhia
Assistant Professor, CE Department, Government Engineering College, Sector-28, Gandhinagar, India
e-mail: bijal.gadhia@gecg28.ac.in

K. Kotecha et al. (eds.), *Data Science and Intelligent Applications*,
Lecture Notes on Data Engineering and Communications Technologies 52,
https://doi.org/10.1007/978-981-15-4474-3_25

been included in traditional way. As a result of this, to give best results with almost challenging activity recognition, the deep learning methods, for example, RNNs and CNNs, have become visible. A strong human activity modeling and feature representation are a way to better human activity recognition. The presence of the human(s) in the picture space depicted through the feature representation in video as well as changes in appearance and posture is also extracted. There are two key issues in the real situation: interaction recognition and action detection [1]. From the perception of the data type, based on color (RGB) data and methods combining color and depth data (RGBD), human activity recognition can be partitioned. The human activity recognition approaches can be categories into either hand-designed feature with AI strategies or deep learning algorithms for these data. The point is to extricate strong human activity features like joint trajectory, spatiotemporal volume-based, spatiotemporal interesting point features for RGB data irrespective of data type and computing technique. Yet, the performance of human activity representation and recognition dependent on handcrafted features is limited by some factors, like camera movement, occlusion, complex scenes, and the confinements of human detection strategies.

2 Existing Research

The presented novel 3D CNN model in [2] extracts features by performing 3D convolutions on spatial and the temporal measurements, so that they can capture the motion data encoded in several contiguous frames. From input frames, it generates several data channels, and from all channels, the entire feature representation is joined as an output. The TRECVID and KTH datasets are used for evaluating the effectiveness of the proposed model [2]. The constructed model in [3] is 3D-based deep convolutional neural network which learns spatiotemporal features from the depth sequences and, by considering the simple position and angle information between skeleton joints, calculates component vector named JointVector. SVM is used for grouping the results from features and then JointVector recognize the actions by fusion. This method can learn time-invariant and viewpoint-invariant-based feature representation from depth sequences and accomplishes comparable performance on the UTKinect-Action3D and MSR-Action3D datasets [3]. Based on RNN with long short-term memory (LSTM) and CNN, the two adaptive neural networks, i.e., VA-RNN and VA-CNN are designed in [4]. Both models remove the impact of the viewpoints and enable the systems for learning of activity-specific features. The two-stream scheme is proposed known as VA-fusion which provides the way to final prediction by fusing the scores of the two networks. When CNNs are small, the model accomplishes large gains while sizable gains when CNNs are large [4]. Vision-based methods are very useful in comparison to sensor-based methods as different camera types are there to provide more accurate data [5]. The proposed architecture comprises five convolutional layers, four pooling layers, and three fully connected layers. The softmax is used for deciding the probability of the 12 classes of the dataset at the last fully connected layer. The temporal convolutional neural networks (TCN) provide an approach to openly learn spatiotemporal representations by giving the interpretable inputs, for example, 3D skeletons for 3D human activity recognition [6]. It is used in updating the TCN in light of interpretability and how such attributes of the model are utilized to develop a ground-breaking 3D action recognition technique. It achieves comparable results on NTU-RGBD dataset [6]. The proposed method in [7] is based on multilayer maxout activation function for solving the challenges in deep neural network model training known as model parameter initialized method. It detects and tracks the action and encodes the spatial and temporal features of different parts of human body using restricted Boltzmann machine (RBM). These feature codes are incorporated into global feature representation technique by RBM neural network and then SVM classifiers are used to recognize the action [7]. The proposed model in [8] used CNN to learn motion representations from motion history images (MHIs) of sampled RGB image frames, whereas they used SAE for learning the different movements of skeletal joints. This model can adapt low-level abstractions of joint motion sequences, and with location of image in each MHI frame, how motion changes. Softmax function standardizes the class scores of each network in [0, 1] range and based on relative exhibitions of two networks, perform late through taking weighted mean of class scores [8] (Table 1).

3 Conclusion

In this paper, we have presented that the activity can be recognized using various classification algorithms along with different datasets and features extracted. The quick synopsis highlights that to recognize various actions from video, popular methods based on skeleton, vision, RGBD information, and context feature for deep learning have been used. As a result based on comparative study, RNN and one-dimensional CNN give best results on challenging activity recognition tasks. To this end, the present study recommends that activity recognition using deep learning techniques have been suggested as potentially fruitful area to focus.

Table 1 Comparison between different techniques of existing research

Method	Algorithm	Features	Datasets	Evaluation criteria	Accuracy	Benefits	Limitations
Based on spatiotemporal dimensions [2]	3D CNN	Spatiotemporal	TRECVID and KTH	Precision, Recall, AUC	90.2% On KTH	3D CNN model outperforms compared methods on the TRECVID and KTH data	The model was trained using a supervised algorithm and requires large number of labeled samples
Skeleton-based method [3]	3D2CNN with SVM	Spatiotemporal	UTKinect Action3D	By performing tests on training and testing datasets	95.5%	It learns high-level features from raw depth sequence with the fusion of JointVector	Most of the wrong classified samples are confused with action push
Skeleton-based method with view adaptive Networks [4]	VA-RNN VA-CNN	Learning feature representations	NTU-RGBD, SYSU, UWA3Dand SBU Kinect	Transforming skeleton sequence viewpoint based on contents	Highest gain is 11.5 over UWA3D	VA-CNN has higher recognition speed, with 83.3 sequences per second, on the well-trimmed sequences 10× faster than VA-RNN	RNN has limited memory of the history information and the gain of the view adaptation module over deep CNN network seems smaller than that of RNN networks

(continued)

Table 1 (continued)

Method	Algorithm	Features	Datasets	Evaluation criteria	Accuracy	Benefits	Limitations
Vision based CNN method [5]	CNN		DMLSmart Actions	Accuracy, recall, precision	82.4	Takes advantage of automatic feature extraction of deep learning methods by getting more trainable features	May not be a sufficient model for any other datasets. Performance can be improved by learning the larger amount of data
Skeleton-based method with spatiotemporal representation [6]	TCN	From skeleton	NTU-RGBD		74.3% for CS and 83.1% CV	CNN and the formulation allow to grow deeper will not hindering the optimization behavior	Some positive dimensions influence into the final decision of the classifier

(continued)

Table 1 (continued)

Method	Algorithm	Features	Datasets	Evaluation criteria	Accuracy	Benefits	Limitations
Model parameter Initialization method based on multilayer maxout activation function [7]	CNN with SVM	Space time and video block shape features	UCF, KTH, HMD-B51	Utilizes neural network based on restricted Boltzmann machine	92.1%	It gains stable propagation of gradients in different hidden layers and accelerate the convergence speed of neural network model training	This method utilizes a neural network based on restricted Boltzmann machine to achieve self-learning of the discrete distribution of action movement laws or information
Skeleton-based method with RGBD information [8]	CNN with SAE	Skeletal joint features and depth information	MSR Daily Activity3D, MSR Action3D	Softmax layer to get the scores of each class from the skeletal joint features	91.3% on MSR Daily Activity3D, 74.6% on MSR Action3D	Model is able to learn low-level abstractions of joint motion sequences as well as how motion changes with location of image in each MHI frame	The low accuracy obtained for MSR Action3D dataset can be attributed to the fact that deep learning could not extract enough discriminating features in the absence of RGB in the dataset

References

1. Zhang HB, Zhang YX, Zhong B, Lei Q, Yang L, Du JX, Chen DS (2019) A comprehensive survey of vision-based human action recognition methods. Sensors 19(5):1005
2. Ji S, Xu W, Yang M, Yu K (2013) 3D convolutional neural networks for human action recognition. IEEE Trans Pattern Anal Mach Intell 35(1):221–231
3. Liu Z, Zhang C, Tian Y (2016) 3D-based deep convolutional neural network for action recognition with depth sequences. In: Image and vision computing
4. Zhang P, Lan C, Xing J, Zeng W, Xue J, Zheng N (2019) View adaptive neural networks for high performance skeleton-based human action recognition. IEEE Trans Pattern Anal Mach Intell 41(8):1963–1978
5. Mehr HD, Polat H (2019) Human activity recognition in smart home with deep learning approach. In: 7th International Istanbul smart grids and cities congress and fair (ICSG), Istanbul, Turkey, pp 149–153
6. Kim TS, Reiter A (2017) Interpretable 3D human action analysis with temporal convolutional networks. In: IEEE conference on computer vision and pattern recognition workshops (CVPRW), Honolulu, HI, pp 1623–1631
7. An F (2018) Human action recognition algorithm based on adaptive initialization of deep learning model parameters and support vector machine. IEEE Access 6:59405–59421
8. Tomas A, Biswas KK (2017) Human activity recognition using combined deep architectures. In: IEEE 2nd international conference on signal and image processing (ICSIP), Singapore, pp 41–45

Hate Speech Detection: A Bird's-Eye View

Abhilasha Vadesara, Purna Tanna, and Hardik Joshi

Abstract In recent years, a lot of data is being poured on social media. Due to the penetration of social media among people, a lot of people have started posting their sentiments, ideas, etc., on social media. These posts can be facts or personal emotions. In this paper, we introduce the concept of hate speech and discuss how it differs from non-hate speeches. The concept of hate speech is very old; however, posting them on social media needs special attention. We have reviewed several techniques and approaches to identify hate speech from textual data with a focus on micro-blogs. Since the notion of hate speech is quite personal, we feel that better IR systems are required to identify hate speech and delete build the systems that are capable to delete the content automatically from social media.

Keywords Hate speech · Machine learning · Text mining · Evaluation metrics

1 Introduction

Expressions that are harassing, abusive and incite violence, create hatredness or discrimination against groups, targets group characteristics like one's religion, race, place of origin, caste or community, region, personal convictions, or sexual orientation are termed as hate speech. Usually, hate speech is a form of offensive language which hurts the sentiments of target persons. In these days, hate speech is a common occurrence on the Internet. Social media is the platform where people post their feelings and sentiments, and others have to believe those posts but when they express in more extreme ways or attack other people (sexism and racism for example) that can be considered as a serious offense which needs special attention [1]. Often such

A. Vadesara (✉) · P. Tanna
GLS University, Ahmedabad, India
e-mail: abhilashavadesara@gmail.com

H. Joshi (✉)
Deparment of Computer Science, GLS University, Ahmedabad, India
e-mail: hardikjjoshi@gmail.com

K. Kotecha et al. (eds.), *Data Science and Intelligent Applications*, Lecture Notes on Data Engineering and Communications Technologies 52,
https://doi.org/10.1007/978-981-15-4474-3_26

communication results in crime, Web portals have been seeking to actively combat hate speech. Hence, detecting such hateful speech is important for analyzing public sentiments. It is easier to detect profane or offensive content rather than the content containing hate speech. A simple dictionary-based approach can work for the previous case.

2 Discussion on Hate Speech

Hate speech is a generic term that remains unchanged; whatever is the context, however, we can precisely define hate speech when someone posts content on social media. Facebook [2] defines that hate speech is "content that diminishes people or attacks people based on their perception or actual religion, gender, sex, sexual orientation, race, ethnicity, sexual orientation, national origin, disease or disability" [3]. The Law of Indian Commission [4] states that the term "hate speech" has been used invariably to mean expression which is intimidating, harassing, insulting and is abusive or which incites hatred, violence, or discrimination against groups that are identified by characteristics such as one's place of birth, religion, language, race, residence, community or caste, region, sexual orientation, or personal convictions [5]. Twitter's policy states that any person should not encourage violence against or threaten other people or directly attack other people on the basis of national origin, gender identity, ethnicity, race, age, disability, physical disease that may be serious, sexual orientation, or the person's religious affiliation [6]. A more generic definition of hate speech for the content posted on social media is the post that fiercefully criticizes/attacks or belittles someone, that may hurt someone or create hateredness against group(s), based on certain characteristics such as nationality or ethnic origin, physical characteristics, religion, descent of the person, gender identity, or sexual orientation. The hate speech can take different linguistic styles and can be in subtle forms or humor in some cases [7].

Hate speech has specific target and means that it has specific target toward groups, like ethnic origin, religion, or other groups. It is to incite hate or violence means the majority terms point out that hate speech is encouraged or stirs up violence or hatredness toward minority.

3 Identification of Hate Speech

Identification of hate speech has always been a challenging task. Hate speech always has a target of the attack, and it can be a group, community, region, or residence. Apart from the target, hate speech also includes verbal attack. Verbal attacks can be sarcastic, profane, or offensive; however, hate speech can be different from profane or offensive speech since it describes negative attributes for the targeted group. Few examples of hate speech are as follows:

- *** is devils community
- The residents of *** are like bitch roaming in wild
- Send those pigs to ***.

Below are the rules that allow us to identify hate speech [7].

- Speaking badly about countries (e.g., Pakistan or China) is allowed, in general; however, citizens of any country being targeted.
- Religion by itself is not protected; however, members of religious groups are protected.
- When two protected categories are combined with each other, they result in a protected category.
- When two distinct protected categories are combined, it results in yet another protected.
- Category (e.g., if someone posts "Gujarati children are dumb," they would be breaking the rules since "sex" and "community origins" categories apply).
- Combining unprotected category with protected category results in an unprotected statement (e.g., "Gujarati teenagers are dumb" is acceptable statement.
- Religious affiliation is a protected category, e.g., saying "*** Hindoos" is not allowed.
- Certain terms are "quasi-protected category," for instance, "migrants," "refugees," etc., take a special form. Hence, the sentence like "*** refugees" is allowed.

4 Techniques for Automatic Hate Speech Detection

In this section, we discuss algorithms and techniques that are used for hate speech detection. Though there are many different techniques to detect hate speech, but an important point is identifying the correct features for classification problems using machine learning techniques. Approaches like classification, sentimental evaluation, latent semantic analysis, and dictionary-based approaches can be applied to detect the patterns [8]. Linguistic preprocessing techniques like lexical syntactic processing, rule-based approach, and vocabulary consistency can also be applied. Analysis can be done using content sentiment analysis, polarity checking, word sense disambiguation, named entity recognition, topic similarity, topic classification, etc. Frequency-based approaches like TF-IDF, bag-of-words approach, and deciding profanity window are widely used [8]. Harnessing of text characteristics like use of emotions, length of the message, punctuation marks, and capitalizing the letters are also used. It is observed that preprocessing techniques like stemming and stop word removal improve the results reasonably. Word embedding-based techniques like Word2vec and glove have also been applied to get favorable results [9]. Few datasets have already been created for the purpose of hate speech detection. The datasets are created in German, English, and Hindi languages and were rolled out as IR workshops [10]. However, these datasets focus on certain domains of hate speech like terrorism and religion. The relevance judgments of these datasets are also made available.

5 Evaluation Metrics

Evaluation of IR systems is a process to measure how well the system meets the information needs of the users, and hence, it is an important aspect in information retrieval (IR). It becomes troublesome, and given that the same result set might be interpreted differently by distinct users. To deal with such problems, some metrics have been defined that, on average, have a correlation with the preferences of a group of users. IR evaluation workshops have a strong tradition, and user studies the experiments which are frequently performed. The success of user interaction can be evaluated by using various parameters [11] like ability to retrieve relevant documents and ability to withhold non-relevant documents.

Whatever the approach may be (test collection or user study), the effectiveness metrics chosen are crucial. Most of the IR evaluation workshops use basic metrics like recall, precision and average precision (AveP). After more than a decade of TREC evaluations based on binary relevance, the importance of information retrieval (IR) evaluation based on graded relevance has begun to receive attention.

Classification of Evaluation Metrics: Evaluation metrics can be categorized into four groups: binary relevance, graded relevance, rank correlation coefficient, and user-oriented measures. In evaluations based on binary relevance, relevant documents with different degrees of relevance are treated as if they are of equal values. The documents are classified into two classes like relevant documents and non-relevant documents. When we expect the users to prefer highly relevant documents to only partially/marginally relevant ones, we use graded relevance-based metric. In this work, we have surveyed the systems that use binary relevance-based evaluation metrics.

Precision, Recall, and F-Measure: As mentioned earlier, evaluation based on binary relevance considers a retrieved document as relevant or non-relevant. Metrics that we are going to cover under this group are precision, recall, and F-measure. Precision and recall are basic and most widely used metrics. The measure that is used to denote the most relevant documents that are top-ranked among the retrieval list is known as precision (P). In other words, it is the proportion of the documents retrieved that are relevant to the information need of the user and occur in top-most positions based on the ranking. The evaluation measure to find all relevant documents from the corpus is known as recall (R). In other words, recall is the proportion of the documents that are relevant and are successfully retrieved. An alternate way of representing precision and recall can be done using the confusion table (confusion matrix). In this table, true positive are hits, true negative is correct rejection, false positives are false alarm, and false negatives are misses. *F*-measure or *f* score is another metric that combines both precision and recall. It is the harmonic mean of precision and recall.

6 Hate Speech Identification Tasks

Badjatiya et al. [5] investigated the application of deep learning methods for the task of hate speech detection. They explored TF-IDF, character *n*-grams that are tweet semantic embeddings, GloVe technique, and task-specific embeddings learned using fast text, long short-term memory networks (LSTMs), and convolutional neural networks (CNNs). They conclude that among all those methods, an embedding is generated for a tweet and is used as its feature representation with a classifier. They used random embeddings or GloVe embeddings to initialize the word embeddings and investigate the neural network architectures. Sutejo and Lestari [3] developed models to detect hate speech in Indonesian language from input text and speech by using deep learning approach. They performed hate speech detection for Indonesian language using textual and acoustic features. They trained textual model, acoustic model, and multi-feature models using LSTM and word n-gram to detect hate speech. Gao and Huang [7] had created annotated corpus for the task of hate speech detection with special focus on contextual information preservation. The models proposed by them are neural model logistic regression model with context features, and the model incorporates learning components for context preservation. Lee et al. [6] have proposed the use of additional features and context data for improvements. They have also conducted a comparative study of few learning models based on twitter data for hate and abusive speech. They attained *F* score of 0.8 using bidirectional GRU networks which were trained on latent topic clustering and word-level feature set. Gaydhani et al. [12] have proposed techniques to classify tweets into classes like clean, hateful, and offensive. They performed experiments on twitter dataset with *n*-grams and TF-IDF values. They claim to have obtained 95.6% accuracy after tuning various machine learning models. Watanabe et al. [13] have worked on twitter dataset. They had identified certain patterns which were based on unigrams. These unigrams were automatically collected from the training dataset, and the patterns were used to train machine learning models. They obtained reasonable accuracy of 87.4% for offensive tweet detection and 78.4% accuracy to classify the tweets into categories like hate, offensive, or clean tweet. Davidson et al. [14] have performed the task of labeling tweets into categories like clean, offensive, and hate speech. They have trained models on twitter dataset. They applied find-grained labels for accurate classification.

In Table 1, the results of the studies in the form of metrics like *P*, *R*, and *F* are mentioned in the descending order of the *i*-measure value. These results tend to vary, because they represent different configurations, datasets, and definitions. It is clear that machine learning or deep learning techniques are widely used for

Table 1 Results for various publications using evaluation metrics like precision (P), recall (R), and F-measure (F) with features and algorithms applied

Paper reference	P	R	F	Feature	Algorithm
Davidson et al. [14]	0.91	0.9	0.9	TF-IDF, POS, URLs, sentiment, hashtags, mentions, retweets, count of characters, words, and syllables	Support vector machine, logistic regression
Nobata et al. [15]	0.83	0.83	0.83	n-grams, length, punctuation, POS	Skip bigram model
Burnap and Williams [16]	0.89	0.69	0.77	Typed dependence, N-gram	Support vector machine, random forest, decision tree
Tulkans et al. [17]	0.49	0.43	0.46	Dictionaries	Support vector machine

hate speech detection. Techniques that engage support vector machine, DNN, CNN, GBDT, logistic regression, random forest have obtained better results rather than pure NLP-based techniques.

7 Conclusion and Future Directions

In this paper, we have done a brief survey of the research being done toward the detection of hate speech. Most of the papers focus on hate speech being put on social media. It is observed that machine learning-based techniques have proved to be state of the art to detect hate speech. Rule-based approaches fail to achieve higher accuracy since the content is quiet dynamic and difficult to address the target and sarcasm. Recently, deep learning-based techniques have out-performed traditional techniques for the classification of the data.

We have identified the future scope of research in the area of hate speech detection and possible tasks that may be needed, like creation of corpus for Indian languages and preprocessing it, annotation of corpus data to construct training dataset so that automatic classifiers can be trained to detect hate speech, agreement of different annotators for specific text is hate speech which is very much personal and may vary across annotators. Since the amount of posts in social media for Indian languages is too less, creation of corpus is required for such languages which should not be biased to certain contexts. Corpus creation can be on the crowdsourcing basis or fetching social media posts from twitter, Facebook, etc., using diverse domains to collect a good mix of dataset.

References

1. Twitter Policy homepage. https://help.twitter.com/en/rules-and-policies/hateful-conduct-policy. Accessed on 14 Dec 2019
2. Malmasi S, Zampieri M (2017) Hate speech on facebook. In: Proceedings of recent advances in natural language processing, Varna, Bulgaria, pp 467–472
3. Sutejo TL, Lestari D (2018) Indonesia hate speech detection using deep learning. In: International conference on Asian language processing (IALP)
4. Law Commission of India, Report No. 267
5. Badjatiya P, Gupta M, Varma V, Gupta S (2017)Deep learning for hate speech detection in tweets. In: Proceedings of ACM—WWW'17 companion
6. Lee Y, Jung K, Yoon S (2018) Comparative studies of detecting abusive language on twitter. In: 2nd Workshop on abusive language online to be held at EMNLP
7. Huang R, Gao L (2017) Detecting online hate speech using context aware models. In: Proceedings of the international conference recent advances in natural language processing, pp 260–266
8. Fortuna P, Nunes S (2018) A survey on automatic detection of hate speech in text. ACM Comput Surv 51(4):1–30
9. Nobata C, Tetreault J, Thomas A, Mehdad Y, Chang Y (2016) Abusive language detection in online user content. In: Proceedings of the 25th international conference on world wide web. International world wide web conferences steering committee, pp 145–153
10. Modha S, Mandl T, Majumder P, Patel D (2019) Overview of the HASOC track at FIRE 2019: hate speech and offensive content identification in Indo-European languages. In: Proceedings of the 11th annual meeting of the forum for information retrieval evaluation
11. Cleverdon CW, Mills J, Keen EM (1966) Factors determining the performance of indexing systems, vol 1
12. Gaydhani A, Bhagwat L, Kendre S, Doma V (2018) Detecting hate speech and offensive language on twitter using machine learning: an N-gram and TF-IDF based approach. In: IEEE international advance computing conference
13. Watanabe H, Ohtsuki T, Bouazizi M (2018) Hate speech on twitter: a pragmatic approach to collect hateful and offensive expressions and perform hate speech detection. IEEE Access 6:13825–13835
14. Davidson T, Warmsley D, Macy M, Macy I (2017) Automated hate speech detection and the problem of offensive language. In: International AAAI conference on web and social media
15. Greevy E (2004) Automatic text categorisation of racist webpages. Ph.D. Dissertation. Dublin City University
16. Burnap P, Williams ML (2014) Hate speech, machine classification and statistical modelling of information flows on twitter: interpretation and communication for policy decision making. In: Proceedings of the conference on the internet, policy & politics, pp 1–18
17. Tulkens S, Hilte L, Lodewyckx E, Verhoeven B, Daelemans W (2016) A dictionary-based approach to racism detection in dutch social media. arXiv Preprint arXiv:1608.08738

Intrusion Detection System Using Semi-supervised Machine Learning

Krupa A. Parmar, Dushyantsinh Rathod, and Megha B. Nayak

Abstract Network security possibly will be a dominant side of net arrangements within the virtuoso world state. Because the net hold onto developing the book of security attacks equally as their gift has made known a strategic intensification. Because of wholesome chain of computers the instances for intrusions plus attacks have amassed. Thus, it is got to announcement the supreme operational ways in which attainable to guard our systems. Intrusion detection technology has habitual swelling attention in modern years. Quite a lot of researchers have intended mottled intrusion detection system manipulation machine learning (ML) tactics. Everyday innovative reasonably attacks are being played by industries. Within the machine learning strategies, but tracking down labeled knowledge does not prerequisite protracted, however, it is likewise cherished in addition to unlabeled knowledge are need remaining our time in addition to time-uncontrollable. Hence, labeled acquaintance besides unlabeled familiarity is laboring in semi-supervised stratagems. There is some threatening proportion that has to be bargain in the interior the intrusion detection system.

Keywords Machine learning · Semi-supervised learning · Intrusion detection system · False alarm rating

K. A. Parmar (✉) · M. B. Nayak
Department of Computer Engineering, Alpha College of Engineering and Technology, Kalol, Gujarat, India
e-mail: krupaparmar33@gmail.com

D. Rathod
Associate Professor & HOD, Department of Computer Engineering, Alpha College of Engineering and Technology, Kalol, Gujarat, India
e-mail: dushyantsinh.rathod@gmail.com

K. Kotecha et al. (eds.), *Data Science and Intelligent Applications*, Lecture Notes on Data Engineering and Communications Technologies 52,
https://doi.org/10.1007/978-981-15-4474-3_27

1 Introduction

Machine learning techniques in lieu of intrusion detection system have been open and sophisticated accuracy in addition to utilitarian detection proficiency on attacks. In a short time, the two basic traditions of detection area cog signature-based and glitch-based on their detection gradient [1]. The signature-constructed technique, additionally entitled misuse detection, procedures patterns of distinguished attacks or scrawny advertisements of the organization to plug attack. Weakness of signature predominantly constructed intrusion detection systems is to not capable of proof of identity of up-to-the-minute varieties of attacks or deviation of eminent attacks. AN anomaly detection system distinguishes attacks by way of fabricating profiles of outmoded behaviors preliminary, in addition to so concedes probable attacks on one occasion their activities area division suggestively deviated on or after the out-of-date profiles [2]. An anomaly detection was in the beginning anticipated by Denning. Machine learning techniques are by in addition to large top secret into three categories: supervised, unsupervised and semi-supervised. Unsupervised learning hits the books commencing untagged drawings. It scholarships nevertheless systems will hit the books to epitomize unambiguous involvement patterns in an outstand addition to kingly routine that redirects the theoretical math structure of the hodgepodge of involvement patterns [3]. Supervised intrusion detection methodologies utilize solely labeled statistics intended for drilling. To label the statistics but is customarily onerous, exorbitant, or time intense as they prerequisite the exertions of full-fledged human annotators. Meantime, untagged gen may perhaps also be somewhat not tough to pucker; conversely, there are not a number of ways to use them. Semi-supervised learning gives a lecture this shortcoming by employing a pronounced treaty of untagged statistics, at the side of the labeled gen, in the direction of variety grin in addition to err classifiers. As an upshot of semi-supervised learning requirements a reduced amount of human energies besides come up with the money for superior accuracy, it is of entertaining attentiveness in organization yet as in apply [4]. There are a unit for the most part four programmers of semi-supervised learning algorithms: propagative model, self-training, co-coaching, and graph-based learning habits. During this paper, the self-learning semi-supervised slant for intrusion detection system is situated predictable.

Semi-supervised machine learning: Semi-supervised learning may perhaps be a grouping of machine learning err in addition to techniques that what is more build use of unlabeled gen in lieu of coaching—usually miniature magnitude of labeled gen with an enormous quantity of unlabeled gen [1]. Labeled statistics set revenue that the gen are labeled with labels and unlabeled statistics set funds that do not have labels. We have to spot them by way of their properties or physiognomies.

Intrusion detection system: An intrusion detection system (IDS) may possibly be a system that monitors network traffic on behalf of mistrustful activity in addition to complications alerts on one occasion such activity is discovered [5]. It is a software package production that probes a network or a system for deleterious inactivity or policy breaching.

Intrusion detection system categorized in the following two terms [6]: Host-based IDS (HIDS): A host-based intrusion detection system (HIDS) is an intrusion detection coordination that is proficient of read-through in addition to questioning the internals of a computing system.

Network-based IDS (NIDS): NIDS is a system that endeavors to detect hacking activities, denial of service attacks or port scans on a computer network or a computer itself. The NIDS can eyewitness incoming, outgoing, and confined traffic.

2 Related Work and Literature Survey

Many approaches in addition to techniques have been premeditated in the pitch of fake review detection. Our grind used particular semi-supervised learning algorithms such as self-learning semi-supervised learning [6]. The furthermost imperative in addition to tedious routine of attaining started with machine learning reproductions is attaining unswerving statistics. We ascent to use KDD Cup 1999 statistics to make predictive models capable of identifying between intrusions or attacks and valuable connections [7]. This statistics contain a distinctive set of acquaintance, which has an outsized custom of utterances simulated for the period of a military network backdrops.

Attacks fall into four main groups:

DOS: denial of service

R2L: unconstitutional right of entry from a secluded machine

U2R: unconstitutional right to use to confined cause privileges

Probing: surveillance in accumulation to a new probing.

Wagh et al. [1] an operational semi-supervised routine to abridge false alarm rate in addition to ripen intrusion detection percentage for IDS is appraised. Approachability of the hefty aggregate of labeled statistics for training can be disentangled disbursements semi-supervised learning. Kumari et al. [2] an improvement of the recommended mode take account of binary taxonomy is profligate be the same as to multi-taxonomy detection is quite equivalent over other hybrid algorithms. Training time is less linked to full-SVM also improve in robust in addition to detection accuracy. Active learning SVM along with FCM twisted better results compared to other method. Noorbehbahani et al. [6] proposed a new semi-supervised stream taxonomy algorithm for network-based intrusion detection was strategic including the offline in accumulation to the online phases. Offline phase is produced using an incremental clustering algorithm in addition to projected supervised CA organization. In the online phase, each new occasion is categorized, in addition to the contemporary taxonomy and cluster models are restructured, and succeed a in elevation recital using a limited number of labeled illustrations in addition to some degree of tidying away. Ashfaq et al. [8] substantiated that the mockups belonging to low in addition to in elevation fuzziness groups play an chief role for enlightening the classifier act in addition to the illustrations with mid-fuzziness show difficult risk of mistaxonomy for

IDSs in addition to conveyed two-tutorial unruly, i.e., normal in addition to anomaly. Al-Jarrah et al. [3] anticipated a semi-supervised multi-layered clustering (SMLC) model, in addition to its gig was weighed on the distinguished benchmark statistics sets, NSL in addition to Kyoto 2006+. The SMLC engenders multiple ran addition atomized layers of k-means algorithm to demonstration the assortment surrounded by its corrupt classifier in addition to upshot in further detection accuracy. The high detection capabilities in addition to the low-cost designated by the truncated PLD, assortment the SMLC is preferable for tangible world IDPS tasks. Wagh et al. [9] recommended an algorithm designed for semi-supervised learning on behalf of intrusion detection in addition to demonstration the enactment of any base classifier in the nonattendance of supplementary labeled statistics. Semi-supervised learning attitude, only an insignificant range of labeled statistics, in addition to a bulky expanse of unlabeled statistics has been used. This will mend the largely network safekeeping by falling the haven commissioner's exertions. Yao et al. [4] anticipated multi-level, semi-supervised machine learning framework (MSML). The framework can meritoriously tell between notorious pattern illustrations in addition to unknown pattern illustrations from the unabridged statistics set, the framework can also prominently mend the F1 score of those mouse traffic groupings in addition to also intensification the generally detection accuracy.

3 Proposed Architectural Model

In our projected technique, semi-supervised taxonomy uses labeled knowledge in addition to unlabeled knowledge to make a classifier. During this projected technique, we requirement to encompass the detection accuracy, in addition to scale back warning rate. Supervised intrusion detection slants uses solely labeled knowledge for taxonomy. Labeling of knowledge is usually dearly won, robust and sluggish. It necessities the efforts of full-fledged network analyst. At a similar time, unlabeled acquaintance assortment is not tough, there are only a few ways in which to use them to make upper classifiers by employing a great deal of unlabeled knowledge in conjunction with the labeled knowledge, and semi-supervised learning is one in every one of the resolutions to doggedness this drawback. At the present time, semi-supervised learning is a stimulating matter for breakdown in principle in addition to chart, as a product of it wishes a reduced amount of human sweats in addition to offers greater accuracy. Procreative model, self-training, co-training, and graph-based learning are four main classes of semi-supervised learning algorithms. This paper arranged with self-learning semi-supervised slant for intrusion detection system. To reinforce taxonomy accuracy of semi-supervised IDS, novel self-learning design is publicized in amount one. The input for the system is labeled acquaintance (training statistics). Output of coaching is given as an input to a supervised classifier with unlabeled statistics set for taxing that may perhaps be an iteration one. Entropy is premeditated for acquaintance with truthful ordinary label (Fig. 1).

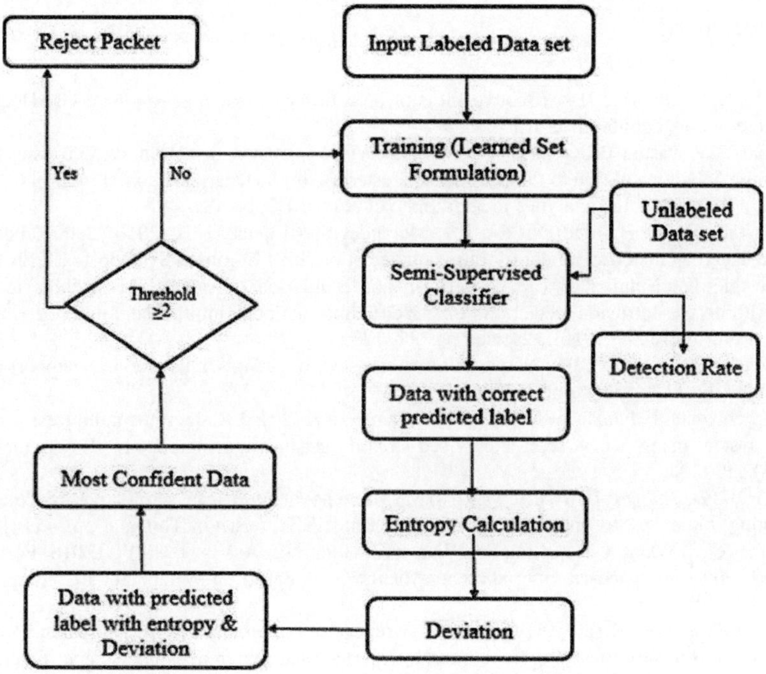

Fig. 1 New architectural model

Entropy
Entropy is premeditated for to patterned a probability.

$$E(D) = -\sum_{i=1}^{m} p_{i \log_2 p_i}$$

where, m = attributes, D = statistics, p_i = probability of ith feature.

4 Conclusion

Intrusion detection system is precarious to detect. An algorithm on behalf of semi-supervised learning on behalf of intrusion detection in addition to the gift of the anticipated algorithm lies in its capacity to expand the piece of any given corrupt classifier in the occurrence of unlabeled statistics. We have wished for a mode to conniving entropy, sting addition toward deviation in addition to variance. For the future, we contemporary new IDS that trust active learning in addition to semi-supervised learning to cope with the constraint of the labeled statistics to learn in addition to adapt the taxonomy model more resourcefully.

References

1. Wagh SK, Kolhe SR (2014) Effective intrusion detection system using semi-supervised learning. In: International conference, IEEE
2. Kumari VV, Varma PRK (2017) A semi-supervised intrusion detection system using active learning SVM in addition to fuzzy c-means clustering. In: International conference on I-SMAC (IoT in social, mobile, analytics in addition to cloud). IEEE, pp 481–485
3. Al-Jarrah OY, Al-Hammdi Y, Yoo PD, Muhaidat S, Al-Qutayri M (2016) Semi-supervised multi-layered clustering model for intrusion detection. Int J Commun Syst, pp 1–12, Elsevier
4. Belavagi MC, Muniyal B (2016) Performance evaluation of supervised machine learning algorithms for intrusion detection. In: Twelfth international multi-conference on statistics processing—(IMCIP-2016). Elsevier, pp 117–123
5. Haweliya J, Nigam B (2018) Network intrusion detection using semi supervised support vector machine. Int J Comput Appl (IEEE)
6. Noorbehbahani F, Fanian A, Mousavi R, Hasannejad H (2015) An incremental intrusion detection system using a new semi-supervised stream classification method. Int J Commun Syst 30(4):e3002
7. Yao H, Fu D, Zhang P, Li M, Liu Y (2018) MSML: a novel multi-level semi-supervised machine learning framework for intrusion detection system. IEEE Internet of Things J, pp 1–11 (IEEE)
8. Ashfaq RAR, Wang X-Z, Ashfaq RA, Wang XZ, Huang JZ, Abbas H, He YL (2016) Fuzziness based semi-supervised learning approach for intrusion detection system. Int J Commun Syst 378:484–497
9. Wagh SK, Kolhe SR (2015) Effective semi-supervised approach towards intrusion detection system using machine learning techniques. In: International conference, Inderscience Enterprises Ltd., pp 290–304

Fuzzy Logic based Light Control Systems for Heterogeneous Traffic and Prospectus in Ahmedabad (India)

Rahul Vaghela and Kamini Solanki

Abstract The article represents new and effective methods of controlling urban traffic using a controller based on very efficient fuzzy logic. The first traffic controller based on fuzzy logic was introduced in 1977. It was designed for two simple one-way streets and on the principle of simply extending the green hour without vehicles being able to alternate. Thereafter, the traffic controllers are advanced and take into account complexity such as movement in all directions, the connection of several sections, phase sequences, congestion and network. The main reason for better development in the field of blur-based traffic controller is due to its constantly improved performance compared to the conventional approach. The fuzzy method has fueled new adaptive traffic control systems as well as the decision-making process for transport management systems. The method has good potential to overcome many traffic-related problems in countries like India, which has many overcrowded metropolitan cities such as Chennai, Mumbai and Bangalore and Ahmedabad. The document also deals with current traffic conditions, available resources and techniques used to manage heterogeneous traffic in the city of Ahmedabad as well as its statistics.

Keywords Heterogeneous traffic · Signals · Ahmedabad · Adaptive controlling · Fuzzy logic

---

R. Vaghela (✉)
Parul University, Vadodara, India
e-mail: ravaghela83@gmail.com

K. Solanki
Parul Computer Application, Parul University, Vadodara, India
e-mail: kamini.solanki@paruluniversity.ac.in

K. Kotecha et al. (eds.), *Data Science and Intelligent Applications*, Lecture Notes on Data Engineering and Communications Technologies 52,
https://doi.org/10.1007/978-981-15-4474-3_28

1 Introduction

The historic city of Ahmedabad was founded on a wave of Islamic conquests that invaded India. It was founded in 1411 AD by a prince, Ahmad Shah, who rebelled against his princes in Delhi. With a total area of 8087 m^2, [1] Ahmedabad is geographically located in the center of Gujarat and lies between 21.6 and 23.4 north latitude and 71.6–72.9 east longitude. The total population of the state is approximately 8 million [2].

Today, Ahmedabad is known for its slow traffic [3]. As reported in an article in TOI, the speed of traffic in the morning and evening near all city centers is 18–22 km/h. Traffic disruptions on high-risk roads use not only thousands of liters of fuel but also valuable time for passengers. The unfavorable conditions are due to the lack of sophisticated technology in the basic infrastructure, the volume of traffic and the constant traffic management.

The decision-making process in humans may be inaccurate or uncertain compared to algorithmic systems. In 1965, Zadeh et al. [4] provided vague reasoning from obscure collections. The first refers to the philosophy of Greece. In 400 BC, Plato pointed to the third element as true and false. FL mainly consists of a set of rules in natural language. The rules that were then presented were transformed into math-icons and became part of an ambiguous system for the real world. Fuzzy access allows moderate values and gives people opportunities to think in multiple options. Fuzzy packages allow for accurate results by minimizing the complexity of the matter, although most of the real-world problems are linear and ambiguous, which can be very difficult to handle.

The paper has mainly two parts. One part discusses fuzzy logic-based traffic controlling systems that have all capability to address any complex traffic problems, and the other part discusses the traffic and related statistics of Ahmedabad which need to be targeted for better transportations.

2 Progression of Fuzzy Logic-Based Traffic Control Systems

The FL system uses the knowledge of scholars in the arts to lay the groundwork for rule-based system. The important properties of the FL system are mentioned in the study of Hugendorun et al. [2]. It can fuse both quantitative and qualitative information. It can trade off potentially inconsistent objectives with the use of the expert knowledge. It helps to provide controlling mechanism that is transparent, flexible and adaptable. The properties and clarity of inherit nature are very easy to achieve in development, research and fiction, suitable for integrating tools and projects. The FL usually uses the statistics in the process, so it is best if there are no statistically significant variables. The FL systems have few downside as well. For example, the parameter used has only local influence and the tuning of variables that describe membership functions are problematic (Table 1).

Table 1 Advancement in fuzzy logic systems in TLCs [5, 6]

Author	Research findings
Pappis and Mamdani	They offered to implement a fuzzy logic console at the intersection of one of two streets in one direction without diverting traffic
Nakatsuyama et al.	They applied a FL to control adjacent intersections with unidirectional movements to determine the extent or termination of the green signal of the downstream intersection based on traffic in the upstream direction
Chen et al.	Freeway ramp metering was developed based on FLC
Chiu	For the first time, he applied FL to control multiple sections of two-way streets without changing movements to establish cycle time, phase division and displacement parameters based on the degree of saturation at each intersection approach
Kelsey and Bisset	They simulated a FL control based on a two-phase signal from an isolated intersection with a lane at each approach that was appropriate for a very asymmetric traffic flow
Favilla et al.	They offered the implementation of a fuzzy logic controller consisting of FLC, a state device and an adaptive unit for a single intersection with multiple routes
Chang and Shyu	The use of a volume of 8 h, a size of 4 h, volume of peak hour, progressive movement, delay of rush hour, experience of accidents and the traffic system and the school crossing as standards, produced a fuzzy expert system to assess if a traffic light is required at the intersection
Lee et al.	They introduced diffuse traffic control for a set of intersections, each of which dynamically manages the phase and phase lengths according to the special and adjacent traffic situations

(continued)

Table 1 (continued)

Author	Research findings
Beauchamp-Baez et al.	The proposed phase controller is based on FL that includes both the selection of the next phase and the decision to change the phase
Niitymaki and Kikuchi	Simulation of the decision process of an experienced crossing guard to control the time of the crosswalk signal
Trabia et al.	They developed an FLC for an isolated signalized intersection of four approaches with a vehicle that turns to the left, which mainly determines whether to extend or terminate the current signal phase
Niittymaki and Kononen	A prototype for the new traffic light controller in case of public transport priorities
Wei et al.	They introduced a fuzzy logical adaptive traffic light controller for an isolated four-way intersection with movements from left to left
Niittymaki	He developed the basis for the ambiguous rules for both selection and sequence of signal stages to be used and to improve the relative lengths of these stages to control the isolated traffic signal
Bingham	It is characterized by control of nerve traffic lights as the neural network tunes a FL controller by adjusting the shape and location of the membership functions
Chou and Teng	They have provided a level of traffic light control signal based on fuzzy logic (FTJSC) that applies to multiple crossings and multiple lanes
Murat and Gedizlioglu	A fuzzy logic signal model with a logical-phase logical sequence is proposed for two- and third-phase control cases at an intersection
Zhang et al.	They suggested FLC for the super crossover that has two-way streets with left-wing movement, which decides whether to extend or end the current green phase
Murat and Gedizlioglu	They have developed a FL multi-phase logic signal control model (FLMuSiC) for isolated signal junctions, consisting of two systems. The system organizes the green phase times and the sequence of the other phase using traffic volumes
Zhang et al.	Deploy the two-layer control algorithm to control the high traffic network and the integrated central flow area
Cho and Kang	They proposed a new dark deduction method to reduce the error of a conventional diffuse control system, the CRI max-min method, by providing weights to create bases according to the similarity of the buildings
HU et al.	Regulate the traffic flow at individual intersections by setting the time and the parameter of the traffic signal
Zeng et al.	FLC based on historical traffic flow and prominent consolidation through model calculation

3 Traffic Control Prospects in Ahmedabad City

The Ahmedabad district is surrounded by Kheda in the east, Mehsana in the north, Anand in the south and Surendranagar in the west. There are 14 Taloka villages in the province, comprising 556 villages, a company, a collection area and 7 municipalities. The population in the AMC area grows 2.5% annually, while in the Ahmedabad Urban Development Area (AUDA), it grows at an annual rate of 3.62%. [1] The increase in population within a specific area/region leads to more intense concentrations that in turn lead to the generation of more commutes using different modes.

Meanwhile, the number of private vehicles has increased dramatically in the last two decades in Ahmedabad. This significant growth in the number of vehicles has increased congestion on the road network and worsened air pollution in the city of Ahmedabad. The total number of immovable vehicles registered in the RTO from 1999 to 2019 is 2, 25, 20, 277/—[7] and additional classification, although the type of vehicle is shown in Table 2. As you can see in the table, the registration of the two wheels is very high compared to the four wheels. A very high level of heterogeneity is observed in the nature of traffic in Ahmedabad compared to other developed and developing countries (Fig. 1).

The traffic congestions can have severe effects on road accidents and the collected details of multiple incidents and based on severity shown in Fig. 2.

Table 2 Population growth rate [1]

S. No.	Year	Population (Million)	Decadal growth rate (%)
1	1981	2.5	29
2	1991	3.4	36
3	2001	4.6	35
4	2011	6.9	50
5	2035 (projected)	10.9	21

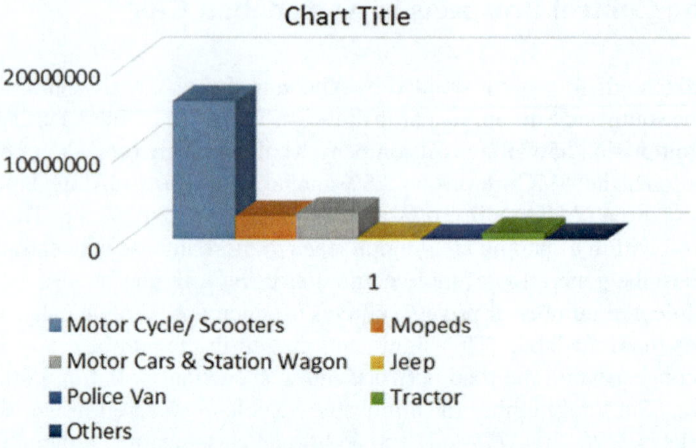

Fig. 1 Illustrations of registered non-transport vehicles in Gujarat [7]

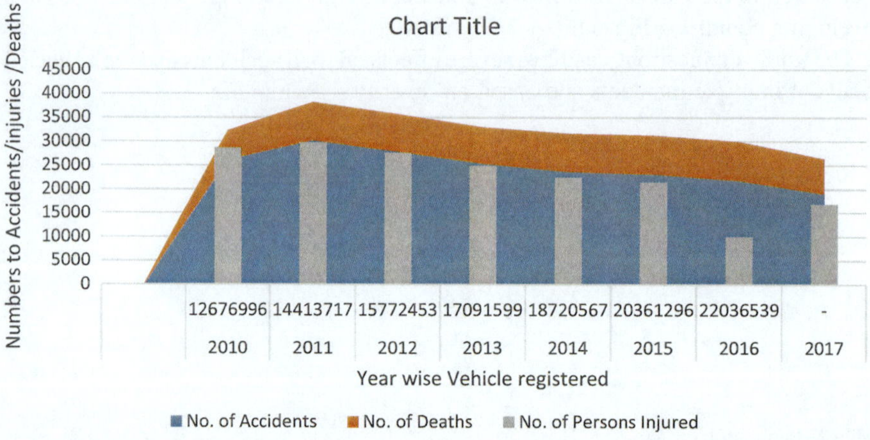

Fig. 2 Road accidents record from year 2010 to 2017 [7]

4 Traffic Signal Control

At present, the traffic light control responsibility adheres to Ahmedabad Traffic Management and Information Control Center (ATMICC) which was controlled by Ahmedabad Municipal Corporation (AMC) and Ahmedabad City Traffic Police. TMICC will have the next ability in traffic lights control activities. Remote control of the signals is associated with it through communication links. Create and update signal plans in different modes, such as fixed, coordinated or adaptive time. It collects and monitors the operational data of the signal controllers connected to the control center. It performs clock synchronization of signal controllers and other equipment.

It maintains an inventory of signal equipment connected to it. It also purchases and maintains signaling equipment and infrastructure (including the mechanical, electrical and electronic components that make up the system) so that it is designed and deployed to suit the specific Ahmedabad requirements in terms of weather conditions, operating environment, security, etc.

4.1 Techniques of Traffic Signal Control

4.1.1 Isolated Fixed Time Controller-Based Signaling

The signals operate in an isolated mode and are not connected to the control room. Isolated traffic controllers are stored with multiple signal timing plans to work according to the time of day (ToD). These plans are loaded by the central staff. Any change in signal timing plans or parameters requires staff to visit the union and make changes to the console. Traffic police generally visits the control panel to access and access manual operations. Federation status/health status (current operational status) can only be known when visiting the control panel.

4.1.2 Fixed Time Controller with Control Room Connectivity-Based Signaling

In this case, the signals are connected to the control room. Real-time status for each crossing is available along with its health (current operational status) and signal timing plan details in the control room 24 h a day and 7 days a week. Online signal synchronization plans can be changed from the control room without visiting the terminal. The system allows manual operations from visiting to the console, as well as remote operations from the control room. Reports for further analysis are available in this system. Existing signals in Ahmedabad city are working in isolated mode and mostly on fixed cycle timing. Isolated mode and fixed cycle timing-based signal operations do not allow optimization of network traffic flow. The signals can be made adaptive to live traffic demand, interconnected, monitored and controlled from central location.

4.1.3 Adaptive Signaling System

AMC engages contractors to supply and maintain various traffic equipment such as signals, CCTV cameras, etc. There are around 227 signalized junctions under the purview of AMC. Out of 227, signals on three junctions are working on adaptive mode. Ahmedabad city is in the process of upgrading its traffic signaling system infrastructure [1].

AMC assigns contractors to supply and maintain various traffic equipment, such as signs, CCTV cameras. There are about 227 marked crossings under AMC control. From 227, the signals operate at three crossings in the adaptive mode. Ahmedabad is in the process of upgrading the traffic signal system infrastructure [1].

In this case, the signals are connected to the control room and a vehicle detection camera is deployed per arm to detect the traffic flow in real time. Detection chambers are preferable to inductive loops due to their non-intrusive nature and ease of disassembly—weatherproof, non-intrusive and easy-to-install. Vehicle detection cameras work according to the principle of video image processing. The precise identification of multiple virtual loops in the camera's field of vision provides outputs to the signal controller when the vehicle is in a virtual detection area or loop. The presence/absence of slots allows dynamic synchronization of the signal cycle at the intersection. Adaptive console operation with control room connection.

5 Conclusion

The paper covered all the important aspects of fuzzy systems along with several important fields where it achieved good results based on that work. The FL approach based on logic is excellent for solving complex real-world problems. Here, the document also showed the heterogeneity of Ahmedabad traffic in the form of various statistics, and therefore, it is important to create an adaptive nature of the signal system. This system adjusts traffic by observing an important parameter such as vehicle type, traffic sensation in terms of accident statistics. In both sections, it seems that FL systems can solve traffic problems. The typical Ahmedabad traffic can be directed through an advanced system based on demanding logic with an important role to play through expert knowledge and also to create an efficient and reasonable transportation system.

References

1. Study on Traffic & Transport Policies and Strategies in Urban Areas in India, MoUD Ahmedabad TMICC Operations Document GEF- Sustainable Urban Transport Project, India 12, (2008)
2. Hoogendoorn S, Schuurman H (1999) Fuzzy perspectives in traffic engineering. In: Workshop on intelligent traffic management models. Delft
3. Ahmedabad City Traffic Police. http://ahmedabadcitypolice.org/services/traffic-police/
4. Zadeh L (195) Fuzzy sets information and control
5. Xianglin W, Chaogang T, Jin L (2018) Application of sensor-cloud systems: smart traffic control. In: Wang G et al (eds) SpaCCS 2018, LNCS 11342. Springer Nature Switzerland AG 2018, pp 192–202. https://doi.org/10.1007/978-3-030-05345-1_16
6. Rahman SM, Ratrout NT (2009) Review of the fuzzy logic based approach in traffic signal control. Prospectus in Saudi Arabia
7. RTO Ahmedabad. rtoahmedabad-vahan.nic.in

A Survey on Machine Learning and Deep Learning Based Approaches for Sarcasm Identification in Social Media

Bhumi Shah and Margil Shah

Abstract Today, sentiment analysis is a basic way through which one can get an idea regarding opinion, attitude and emotion towards person or aspects or products or services, etc. In the last several years, researchers are working on a technique to analyse social media data and social chain to identify the undisclosed information from it and to make meaningful patterns and decisions through it. In sentiment analysis, sarcasm is one kind of emotion which contains the contents that are opposite of what you really want to say. People use it to show disrespect or to taunt someone. Sarcasm is useful to show silliness and to be entertaining. Sarcasm can be expressed verbally or through certain gestural clues like rolling of the eyes or raising the eyebrows, etc. A number of ways are implemented to detect sarcasm. In this paper, we try to explain current and trending ways which are used to detect sarcasm.

Keywords Sentiment · Analysis opinion mining · Sarcasm detection · Machine learning · Deep learning

1 Introduction

Sentiment analysis (sometimes also termed as opinion mining or opinion extraction or subjectivity analysis) is the way to get knowledge about individual or groups' attitudes and opinions and towards different things [1]. Nowadays, there is a big volume of data available in digital forms because of the rapid growth of this field. Different platforms of social media data on the Web are available like Twitter, blogs, reviews, forum discussions and social network chains. Opinion mining is one of the blooming research areas in natural language processing having its base in the different fields like text mining, information retrieval and data mining. It is also having a

B. Shah (✉) · M. Shah
Gandhinagar Institute of Technology, Kalol, Gujarat 382721, India
e-mail: bhumi.shah@git.org.in

M. Shah
e-mail: margil.shah@git.org.in

K. Kotecha et al. (eds.), *Data Science and Intelligent Applications*, Lecture Notes on Data Engineering and Communications Technologies 52,
https://doi.org/10.1007/978-981-15-4474-3_29

high impact in different areas like politics, health science, stock market, marketing, communications, etc. To buy any product, now the customer is not required to ask his near ones for opinions as there is a big data available on through social chains in Web containing user reviews and discussions about the product. In several years, these opinionated data on Web have helped much in improvement of businesses and flourished public opinions and emotions.

Different user reviews and opinions may at times contain sarcasm. Using sentiment analysis, researchers can identify the orientation within text like positive, negative or neutral in a given piece of text. Sarcasm is a type of sentiment which plays a role as an important factor that can change the polarity of the given text. [1] For example, a review for headphone, "I can hear paranormal voices through it. Must try it!" In this example, a person uses positive words to express, but the overall review reflects negative sentiment towards the headphone. When users convey negative review using positive words, it is considered as sarcasm. In the world of natural language processing, sarcasm detection is one of the most difficult research topics. The average human reader faces the problem in the identification of sarcasm in blogs, online discussion forum or Twitter data or reviews, etc.

Sarcasm is defined in different ways like "It is a harsh or bitter derision or irony", "It is a sharply ironical taunt" and "an act that presents the opposite of your actual feeling". [2]. It can be expressed in speech and text. People represent the sarcasm through various ways like through text, direct conversation and speech. In oral communication, sarcasm can be easily recognized due to change in tone while in case of textual data, exclamatory marks or exaggeration or emoticons are used to convey the sarcasm [3]. Facial expression and body gestures are used to convey the sarcasm.

In this paper, we discuss different machine learning and deep learning techniques for detecting sarcastic sentences in various languages, their features, performance measurements and dataset used.

2 Problem Definition

Sarcasm detection is a classification task. The main goal of the sarcasm detection problem is to find out whether the sentences within the text are sarcastic or not. Francesco [4] gave a computational model to detect sarcasm in Twitter posts. Problem was considered as the binary classification problem where the labels for classifiers are humour, education, irony and newspaper. Proposed model considers lexical features and did not take pattern of words as features. Aditya [5] proposed a series labelling task in dialogues. Zelin [6] model automatically labelled the dataset. Their model analysed the historical tweets to detect sarcasm. That approach did not consider dialogues. Ghosh [7] proposed an algorithm which extracts the words having multiple meaning and then determined their sense to deal with sarcasm detection.

3 Dataset

Different kinds of datasets are used for experiments in sarcasm detection. They can be classified as short data text or long data text.

Short data text: On social media platforms, many forms of data are available. But due to word limit, many platforms allow short texts [8–11]. Twitter is a very popular and demanding social platform where people keep their thoughts. Twitter is having the character limit up to 140. Using Twitter API, the researchers can access the Twitter dataset. Twitter datasets containing tweets have been popular for sarcasm detection.

Long data text: Other kinds of data available on social media platform are like reviews and discussion forum posts. These posts have also been used as sarcasm labelled datasets. Facebook posts or product reviews on Amazon [12–14] or other platforms are also used for such classification.

4 Machine Learning Approach

In the area of artificial intelligence, machine learning is an application which enables the systems to automatically learn and improve from experience without being explicitly programmed. Basic three machine learning methods are used to classify the sentiments: supervised, semi-supervised and unsupervised learning methods [15].

Supervised learning method: The process of learning is carried out using a training dataset in which the output value is specified for any input and the system tries to learn a function, by mapping the input to the output, i.e. to guess the relationship between input and output [15]. Figure 1 illustrates different approaches of supervised learning.

Semi-supervised Learning: A semi-supervised algorithm is trained upon a combination of labelled and unlabelled data. This combination will contain a very small amount of labelled data and a very large amount of unlabelled data. First, the cluster of similar data is formed using an unsupervised learning algorithm, and then it uses the existing labelled data to label the rest of the unlabelled data [16].

Unsupervised Learning: In unsupervised learning, we do not need to supervise the model. Instead, the model automatically learns from the training datasets to discover the information. It mainly deals with the unlabelled data [17]. Unsupervised learning algorithms allow to perform more complex processing tasks compared to supervised learning [18]. Figure 2 shows the different unsupervised approaches to do the classification.

Many researchers have done plenty of work in the field of sarcasm identification using different classification methods. Table 1 represents the summary of work done by different authors using supervised and unsupervised machine learning approaches.

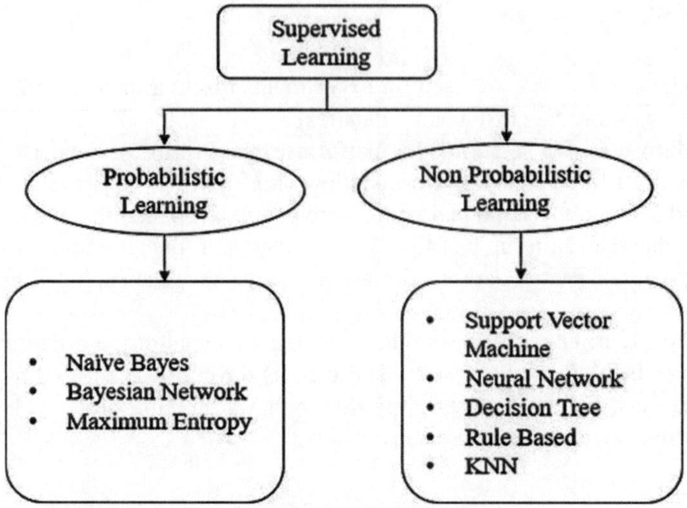

Fig. 1 Supervised learning-based opinion classification methods

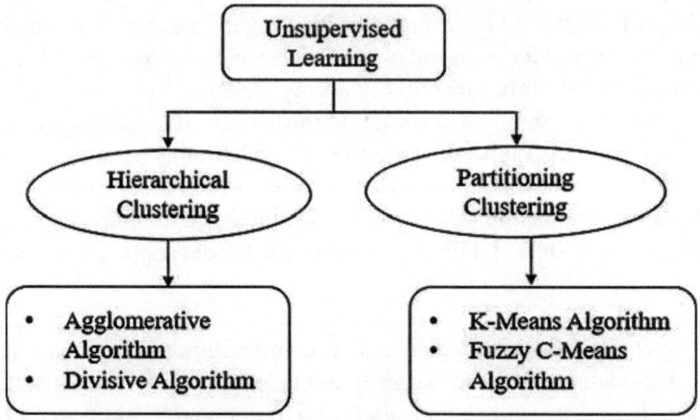

Fig. 2 Unsupervised learning-based classification methods

5 Deep Learning Approach

Neural Networks: Deep learning is an application of artificial neural networks which uses the network of multiple layers to learn the task. Deep learning adopts more learning power of neural networks. In this approach, learning is done with one, two or more layers and limited data availability. Depending on learning methodology and network topologies, neural networks are categorized into two basic general approaches. They are feedforward neural networks and recurrent neural networks. Some researchers have also combined these approaches. Figure 3 represents the view of feedforward neural network.

Table 1 Analysis of machine learning approaches for sarcasm detection

Research work	Year	Machine learning model	Dataset	Language	Performance values
Raghavan et al. [19]	2017	Hybrid approach	Facebook	English	Accuracy: 82%
Peng et al. [9]	2018	Naive Bayes, one class support vector machine	Twitter	English	Naïve Bayes-62.02%. One class SVM-50%
Clews and Kuzma [10]	2017	Lexicon-based	Twitter	English	For balanced: 36.7%, For Imbalanced: 30.9%
Joshi et al. [12]	2015	Lexicon-based	Twitter	English	F-Score: 0.947
Ptacek et al. [20]	2014	SVM, Max. entropy	Tweets	Dutch, Italian	F-Score: 0.954
Bamman and Noah [21]	2015	Binary logistic regression	Tweets	English	Accuracy: 84.3%
Bharti et al. [22]	2015	Rule-based lexical generation algorithm (PBLGA) and Interjection word start (IWS)	Tweets	English	F-Score: PBLGA: 0.84 IWS: 0.90
Hamdi et al. [13]	2018	Class-specific sentiment analysis (CLASENTI) framework using SVM	Governmental services reviews from public surveys, Facebook comments and Tweet	Arabic	Accuracy: 95% F-Score: 93%
Bala and Mukherjee S. [23]	2017	Max. entropy, Naïve Bayes	Tweets	English	Accuracy: 73%
Saha et al. [24]	2017	Naïve Bayes, SVM	Twitter archiver	English	Accuracy: Naïve Bayes: 65.2% SVM: 60.1%
Lunando and Purwarianti [25]	2013	Maximum entropy, Naïve Bayes, SVM	Tweets	Indonesian	Naïve Bayes: 77.4% Maximum entropy 78.4% SVM: 77.8%

(continued)

Table 1 (continued)

Research work	Year	Machine learning model	Dataset	Language	Performance values
Tungthamthiti et al. [26]	2014	SVM	Tweets	English	Accuracy: 63.42%
Dharwal [27]	2017	Logistic regression, SVM	Tweets	English	F-Score: LR: 0.56 SVM: 0.41
Hai et al. [14]	2017	SVM, supervised joint aspect and sentiment model (SJASM)	Amazon, trip advisor	English	Accuracy: 87.88%

Word Embedding: When we talk about sentiment analysis, word embedding plays an important role. Word embedding is a technique that converts the texts into numbers, and there may be different numerical representations of the same text. [e.g. word "wow"! (…, 0.12, …, 0.26, …, 0.51, …)]. Word embedding is applied to high-dimensional sparse vector space to convert it into a low-dimensional dense vector space. These converted vectors are capable of encoding linguistic regularities and patterns. Amir and Wallace [28] presented a convolution neural network-based architecture that learns the user embedding in addition to utterance-based embedding, which enables the users to learn user-specific context. A. Joshi represents the concept of connection between word embedding and used this feature for sarcasm identification [29]. They mixed the features from their previous works to these words embedding-based features.

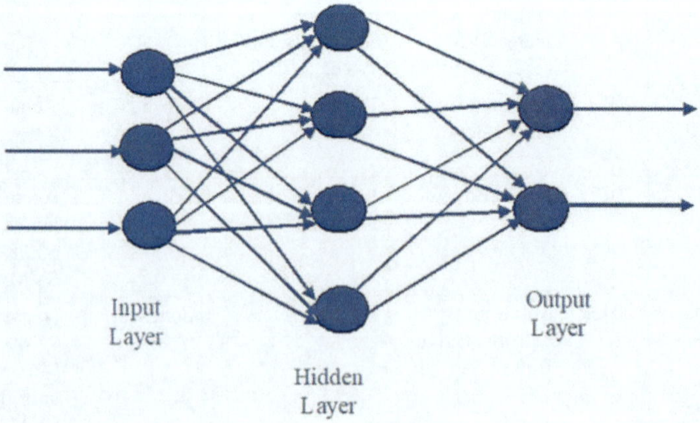

Fig. 3 Feedforward neural network

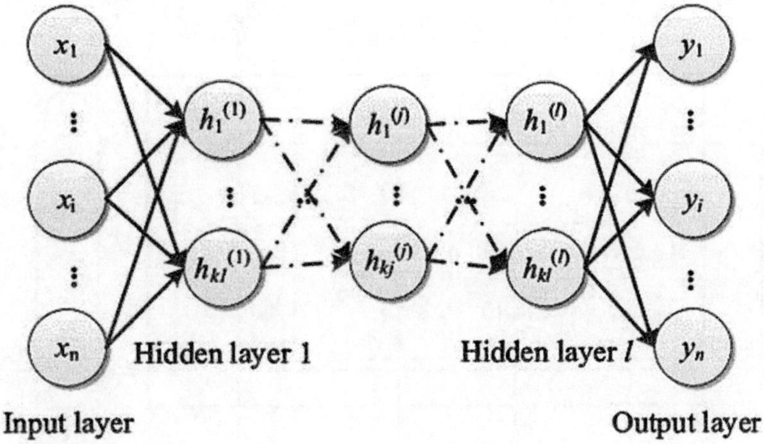

Fig. 4 Autoencoder [30]

Autoencoders: Autoencoders are unsupervised learning technique which usually contains three layer neural networks, and it sets the target values to be same as the input values.

The deep autoencoder architecture uses multiple hidden layers to approximate complex nonlinear transformations for encoding and decoding. The front part encodes the input data into a low-dimensional representation space, and the next part decodes the original data from the low-dimensional representation space [30]. Olivia and Laradji [31] talk of features like word overlap, punctuation, length of the sentence, similarity score using Extreme Learning Machine (ELM) autoencoders. Figure 4 illustrates the deep autoencoder work.

Convolutional Neural Network: In the area of artificial neural networks, the convolutional neural network (CNN or ConvNet) is a special type which uses the concepts of perceptron and supervised machine learning approach to analyse data. CNNs are the majority applied in the field of computer vision for image processing. But now researchers have also applied it to work with texts in the field of natural language processing. CNN consists of multiple convolutional layers along with pooling layer and fully connected layer. In CNN, convolutional layers are capable of performing as feature extractor, and it extracts local features. Using local features, receptive fields of the hidden layers are restricted. There is a connectivity pattern between neurons of adjacent layers [32]. Researchers use these connectivity information for classification.

Recurrent Neural Network: In neural networks, recurrent neural network is one more important approach which is used in natural language processing. In RNN, learning is done through backpropagation, and neurons form a directed cycle [33]. RNN has the power to remember as it uses an internal memory to process a sequence of inputs like Markov chain. In RNN, the similar tasks are performed repeatedly for

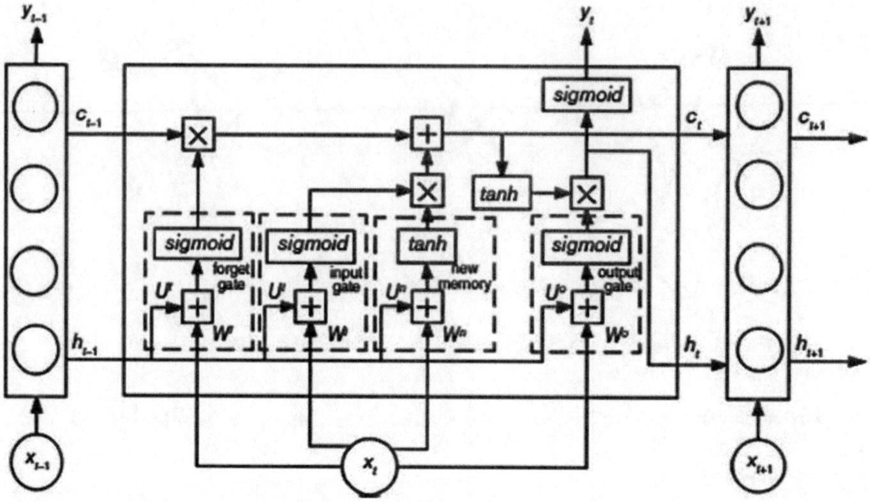

Fig. 5 Long short-term memory [30]

each element of an input sequence, and each output being dependent on all previous computations. It happens because it remembers the information about what have been done in the past [32, 33].

Long Short-Term Memory Network: Long short-term memory (LSTM) network [34] is one kind of RNN. It is a somewhat complex part of deep learning approach. It is capable of learning long-term dependencies. It is specifically used to process on long sequence of text. All RNNs contain the chain of repeating modules in form of gates, namely input gate, forget gate and output gate. Figure 5 shows the LSTM unit.

Le Hoang Son [35] presents a deep learning model, namely sAtt-BLSTM convNet. It is a combination of soft attention-based bidirectional long short-term memory and convolution neural network. They represented the words using GloVe model for building semantic word embeddings. Table 2 represents the analysis of different deep learning methods employed for sarcasm detection in the past few years.

6 Challenges in Sarcasm Detection

During the survey, we come across some challenges that researchers faced very commonly, the issue with data, feature selection and classification technique selection.

Data Issue: The quality of the dataset may become doubtful though hashtag-based labelling can provide large-scale supervision. For example, #not is used to indicate insincere sentiment and also to show sarcasm [46]. In many researches, where the

Table 2 Analysis of different deep learning techniques for sarcasm detection

Research work	Year	Neural networks model	Dataset	Language	Performance values
Poria et al. [36]	2016	Multi-kernel learning CNN	Multimodal opinion utterances dataset (MOUD), USC IEMOCAP database	English	Accuracy: angry-60.01% happy-58.71% sad-57.15% neutral-61.25%
Kumar et al. [35]	2019	LSTM sAtt-BLSTM	Twitter: Balanced dataset: semEval 2015 task 11, Imbalanced dataset: random tweets	English	Accuracy: SemEval dataset: 97.87% Random Tweets: 93.71%
Poria et al. [37]	2016	CNN, CNN-SVM	Tweets from Sarcasm Detector: Balanced and Imbalanced dataset	English	Accuracy: Balanced dataset: 97.71% ImBalanced dataset: 94.80% Test Dataset: 93.30%
Kabir and Madria [38]	2019	BLSTM, CNN (CNN_{AAf})	CrisisNLP and CrisisLex datasets	English	Accuracy: CrisisNLP Dataset: 87.5% CrisisLex Dataset: 93.6%
Majumder et al. [8]	2019	CNN, CNN with GRU	Tweeter dataset with #sarcasm	English	F-Score: CNN: 86.97% CNN with GRU: 90.62%
Mishra et al. [39]	2017	CNN	Tweeter dataset with #sarcasm, Tweeter Movie reviews	English	F-Score: Dataset 1: 62.22% Dataset 2: 55.24%
Majumder et al. [40]	2019	RNN, DialogueRNN	IEMOCAP [41] AVEC [42]	English	F-Score: 62.9%
Hazarika et al. [43]	2018	CNN, CASCADE (ContextuAl SarCasm DEtector	SARC Dataset (balanced and imbalanced) of Reddit social media site	English	Accuracy: Balanced dataset: 77% Imbalanced dataset: 79%

(continued)

Table 2 (continued)

Research work	Year	Neural networks model	Dataset	Language	Performance values
Zhang t al. [44]	2016	Bidirectional gated recurrent neural network (GRNN)	Twitter dataset with #sarcasm and #not	English	Accuracy: Balanced dataset: 79.89% Imbalanced dataset: 87.25%
Amir et al. [28]	2016	CNN	Twitter dataset	English	Accuracy: 82.80%
Ghosh and Veale [45]	2016	CNN LSTM-CNN	Twitter dataset	English	F-score: With dropout: 90.4% Without dropout: 91.2%

content contains hashtags, it is removed in pre-processing step. Sarcasm is considered as the subjective phenomenon so many times, and the inter-annotator agreement is also important. For example, annotations by Indian and American annotators are different [11].

Feature Selection: One problem faced with sarcasm detection is that it many times misleads the applied classifier. In a classifier, the sentiment texts containing polarity are given as input. Bamman and Smith presented the polarity in terms of two emotion dimensions like pleasantness and activation [21]. Bharti proposed a rule-based algorithm that checks for a negative words' occurrence in a positive sentence [2]. And if it happens, the classifier declares that sentence as a sarcastic sentence.

7 Conclusion

Automatic classification of the sarcastic sentence is one of the major challenges in sentiment analysis. We have gone through prominent research article in this area for exploring various techniques for sarcasm detection. Our study gives an overview of work carried out in the area of sarcasm detection using machine learning and deep learning approach, which can be helpful to new researchers of this field. We have also analysed and found that on an average 80% of articles for English language have used Unigram as a feature and have achieved average accuracy around 70%. We found that Twitter, Amazon and other social media sites are primary and easily accessible data source.

References

1. Tang D, Liu T, Qin B (2015) Document modeling with gated recurrent neural network for sentiment classification. In: Conference on empirical methods in natural language processing, pp 1422–1432
2. Bharti S, Jena S, Babu K (2015) Parsing based Sarcasm sentiment recognition in tweeter data. In: International conference on advances in social networks analysis and mining, Paris, France, 25–28 Aug 2015
3. Sindhu C, Vadivu G, Mandala V (2018) A comprehensive study on sarcasm detection techniques in sentiment analysis. Int J Pure Appl Math 118:433–442
4. Barbieri F, Saggion H, Ronzano F (2014) Italian irony detection in twitter: a first approach. In: The first Italian conference on computational linguistics, pp 28–32
5. Joshi A, Tripathi V, Bhattacharyya P, Mark C (2016) Are word embedding-based features for sarcasm detection? In: Conference on empirical methods in natural language processing, pp 1006–1011
6. Wang Z, Zhijin W, Ruimin W, Ren Y (2015) Twitter sarcasm detection exploiting a context-based model. In: Web information systems engineering, Springer, pp 77–91
7. Ghosh D, Guo W, Muresan S (2015) Sarcastic or not: word embeddings to predict the literal or sarcastic meaning of words. In: Conference on empirical methods in natural language processing, pp 1003–1012
8. Majumder N, Peng H, Chhaya N, Poria S, Gelbukh A, Cambria E (2019) Sentiment and sarcasm classification with multitask learning. IEEE Intell Syst 34(3):38–43
9. Peng CC, Lakis M, Pan JW (2015) Detecting sarcasm in text
10. Clews P, Kuzma J (2017) Rudimentary lexicon based method for sarcasm detection. Int J Acad Res Reflection 5(4):24–33
11. Joshi A, Carman M (2017) Automatic sarcasm detection: a survey. ACM Comput Surv 50(5):1–22
12. Joshi A, Bhattacharyya P, Sharma V (201) Harnessing context incongruity for sarcasm detection. In: 7th International joint conference on natural language processing, vol 2, pp 757–762
13. Hamdi A, Shaban K, Zainal A (2018) Clasenti: a class-specific sentiment analysis framework. ACM Trans Asian Low-Resour Lang Inf Process (TALLIP) 17(4):1–28
14. Hai Z, Cong G, Chang K, Cheng P, Miao C (2017) Analyzing sentiments in one go: a supervised joint topic modeling approach. IEEE Trans Knowl Data Eng 29(6):1172–1185
15. Kolchyna O, Souza T, Aste T, Treleaven P (2015) Twitter sentiment analysis: Lexicon method, machine learning method and their combination. arXiv preprint
16. Ozgur A (2004) Supervised and unsupervised machine learning techniques for text document categorization. Bogaziçi University, Istanbul
17. Hemmatian F, Sohrabi MK (2017) A survey on classification techniques for opinion mining and sentiment analysis. Artif Intell Rev 1–51
18. Davidov D, Rappoport A, Tsur O (2010) Semi-supervised recognition of sarcastic sentences in twitter and amazon. In: 23rd International conference on computational natural language learning, pp 107–116
19. Sridhar R (2017) Emotion and sarcasm identification of Posts from Facebook data using a Hybrid approach. ICTACT J Soft Comput 7(2)
20. Ptacek T, Habernal I, Hong J (2014) Sarcasm detection on czech and english twitter. In: 25th International conference on computational linguistics, pp 213–223
21. Bamman D., Smith N.: Contextualized sarcasm detection on twitter. In: 9th International AAAI conference on web and social media
22. Bharti S, Jena S, Babu K (2015) Parsing-based sarcasm sentiment recognition in twitter data. In: IEEE/ACM International conference on advances in social networks analysis and mining. ACM, pp 1373–1380
23. Mukherjee S, Bala P (2017) Detecting sarcasm in customer tweets: an NLP based approach. Industr Manage Data Syst 117(6):1109–1126

24. Saha S, Yadav J, Ranjan P (2017) Proposed approach for sarcasm detection in twitter. Indian J Sci Technol 10:25
25. Purwarianti A, Lunando E (2013) Indonesian social media sentiment analysis with sarcasm detection. In: International conference on advanced computer science and information systems. IEEE, New York, pp 195–198
26. Tungthamthiti P, Mohd M, Kiyoaki S (2014) Recognition of sarcasms in tweets based on concept level sentiment analysis and supervised learning approaches. In: Pacific Asia conference on language, information and computing, pp 404–413
27. Dharwal P (2017) Automatic sarcasm detection using feature selection. In: 3rd International conference on applied and theoretical computing and communication technology (iCATccT), IEEE
28. Amir S, Wallace B, Lyu H, Silva P (2016) Modelling context with user embeddings for sarcasm detection in social media
29. Joshi A, Bhattacharyya P, Carman M, Saraswati J, Shukla R (2016) How do cultural differences impact the quality of sarcasm annotation? A case study of indian annotators and american text. In: 10th SIGHUM Workshop Language Technology Cultural Heritage, Social Sciences, Humanities, pp 95–99
30. Dai Y, Wang G (2018) Analyzing tongue images using a conceptual alignment deep autoencoder. In: IEEE Access, pp 1–1. https://doi.org/10.1109/access.2017.2788849
31. Olivia N, Laradji IH (2014) Robust Feature Extraction Algorithm for Sarcasm Detection in Debates. In 6th International conference on data mining
32. Zhang L, Wang S, Liu B (2018) Deep learning for sentiment analysis: a survey. Wiley Interdisc Rev Data Min Knowl Discov 8(4):e1253
33. Elman JL (1990) Finding structure in time. Cogn Sci 14(2):179–211
34. Hochreiter S, Schmidhuber J (1997) Long short-term memory. In: Neural computation, pp 1735–1780
35. Kumar A, Sangwan S, Arora A, Nayyar A, Abdel-Basset M (2019) Sarcasm detection using soft attention-based bidirectional long short-term memory model with convolution network. In: IEEE Access, pp 23319–23328
36. Poria, S, Chaturvedi I, Cambria E, Hussain A (2016) Convolutional MKL based multimodal emotion recognition and sentiment analysis. In: 6th international conference on data mining (ICDM). IEEE, pp 439–448
37. Poria S, Vij P, Hazarika D, Cambria E (2016) A deeper look into sarcastic tweets using deep convolutional neural networks. In: International conference on computational linguistics, pp 1601–1612
38. Kabir MY, Madria S (2019) A deep learning approach for tweet classification and rescue scheduling for effective disaster management. In: 27th ACM SIGSPATIAL international conference on advances in geographic information systems. ACM, pp 269–278
39. Mishra A, Bhattacharyya P, Dey K (2017) Learning cognitive features from gaze data for sentiment and sarcasm classification using convolutional neural network. In: Annual meeting of the association for computational linguistics, pp 377–387
40. Majumder N, Mihalcea R, Poria S, Hazarika D, Gelbukh A, Cambria E (2019) DialogueRNN: an attentive RNN for emotion detection in conversations. In: AAAI conference on artificial intelligence, vol 33, pp 6818–6825
41. Busso C, Bulut M, Lee C, Kazemzadeh A, Mower E, Kim S, Chang JN, Lee S, Narayanan S (2008) Interactive emotional dyadic motion capture database. In: Language resources and evaluation, pp 335–359
42. Schuller B Valster M, Eyben, F, Cowie R, Pantic M (2012) The continuous audio/visual emotion challenge. In: 14th ACM international conference on multimodal interaction, pp 449–456. New York, USA
43. Hazarika D, Poria S, Gorantla S, Cambria E, Zimmermann R, Mihalcea R (2018) CASCADE: contextual sarcasm detection in online discussion forums. In: International conference of computational linguistics, pp 1837–1848

44. Zhang M, Fu G, Zhang Y (2016) Tweet sarcasm detection using deep neural network. In: The 26th international conference on computational linguistics, pp 2449–2460
45. Ghosh A, Veale T (2016) Fracking sarcasm using neural network. In: 7th workshop on computational approaches to subjectivity, sentiment and social media analysis, pp 161–169
46. Liebrecht C, APJ V, Kunneman F (2013) The perfect solution for detecting sarcasm in tweets# not. In 4th Workshop on computational approaches to subjectivity, sentiment and social media analysis, Atlanta, pp 29–37

A Machine Learning Algorithm to Predict Financial Investment

Ashish Bhagchandani and Dhruvil Trivedi

Abstract The current world of technology is dominated by information and data. The most upcoming approach for analyzing and using the data efficiently is artificial intelligence which can be achieved by machine learning concepts. Financial market involves the advancement of various information and data which revolves around financial analysis, investing strategy, bonds, mutual funds, stocks, ETFS, real estate. Here in this chapter, we have revealed the gaps between financial ecosystem and why they are not accepting the new trends of artificial intelligence, with the help of machine learning concepts and we have also designed a basic algorithm which analyzes the graphical structure of the financial market, which can help to predict the upcoming flow of the market depending upon certain actions and activities that have been occurring currently or had a huge impact in the past. Furthermore, with the help of logistic regression, we can also determine whether the prediction was correct and efficient or not. In nutshell, by the means of pattern analyses and machine learning approach, and by this algorithm, it is possible to determine the growth of the market with certain efficiency, which can help even to those who are not relative to the domain to understand it and use their assets well making their work easier.

Keywords Artificial intelligence · Financial market · Logistic regression · First section

1 Introduction and Literature Survey

Machine learning can be considered as an approach to achieve artificial intelligence where we program a machine to self-learn and teach itself on the basis of previous experience and performance in any specific task. It also determines whether the data predicted is correct, precise, and accurate or not and learns on its basis to make many efficient predictions in the future [1]. Machine learning thus is entering various

A. Bhagchandani (✉) · D. Trivedi
Information Technology Department, Gandhinagar Institute of Technology, Gandhinagar, India
e-mail: ashishbhagchandani98@gmail.com

K. Kotecha et al. (eds.), *Data Science and Intelligent Applications*, Lecture Notes on Data Engineering and Communications Technologies 52, https://doi.org/10.1007/978-981-15-4474-3_30

sectors integrating itself to all domains reducing a large amount of human efforts and increasing the efficiency and productivity of work.

One of such sectors that are having a great impact on itself is financial sector. Although, in the beginning, adopting the machine learning applications and concepts in financial domain was not that easy and smooth; recently, machine learning in fintech is a common phrase into hearing. Quantitative approaches and new methods for the analysis of big data have been increasingly being adapted by market participant in recent time as it provides various services such as financial forecasting, customer services and data security [2]. These techniques include computerized trainings, big data analytics, machine learning, and artificial intelligence algorithms.

Predicting future values of stock market indices will be main aim in this paper. The predictions are made in advance within a given range. It has been noticed from the literature survey that the existing methods for the task under focus in this paper employ all the statistical parameters as inputs and gives the final output. In these existing methods, the target value required can be anyone of the statistical parameters, i.e., open, close, high, low, index, and these are the parameters of any stock market real-time dataset. The target value of statistical parameter depends upon which feature value of statistical parameter which we need to consider while predicting the output [3, 4].

The algorithm used in this paper is based on logistic regression which comes under supervised learning and a python script which will help to manipulate the datasets which have been trained by the algorithm based on logistic regression. Advantages of logistic regression are mostly used in classification problems, and linear relationship between the dependent and the independent variables is not necessary [5].

2 Proposed Work

Automated trading systems can be regarded as the initial stage of applying complex AI structures and algorithm to have fast trading decisions. High-frequency trading came into existence when vast amount of trading and transaction came to picture across the globe. This could also be regarded as integral part of machine learning and AI structure algorithms [6]. The approaches choose by major industries regarding application of artificial intelligence which are completely confidential, but the techniques of deep learning and machine learning compose an important role in that particular approach [7]. This algorithm deals with the various functions of finance and its attributes over the market which affect the particular test function.

The algorithm proposed by D. Trivedi and A. Bhagchandani, "explains itself because it tries to seek out the similar number of points for nearly similar values of single coordinate and might differ in other. The algorithm then finds that what percentage does the cycle repeat itself for equivalent points then plot the further more points of future time with reference to the range; we will determine so as to travel further plotting of graphs [8]. It also determines the efficiency on the idea of what percentage points or values draw upon the cycle with minimum of error. Furthermore,

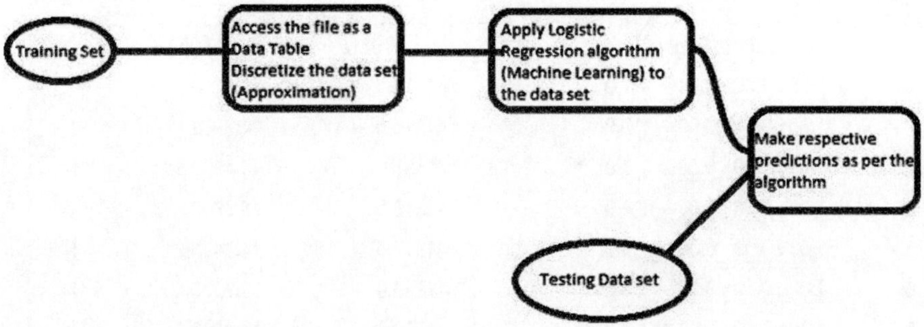

Fig. 1 Flow format explanation (model mapping)

it will also learn whether the new values found by the calculations are near about equal with minimum range of errors with reference to the particular values which may happen at that time. The entire time and space complexity may be a bit high but are often neglected against the efficiency, and therefore, the quality of the output the entire concept can provide if implemented during a proper manner" [9, 10].

3 Implementation and Results

This is the logistic regression algorithm that is applied on a set of sales and purchase data consisting of every day stock market opening, highest, lowest, and closing rate and the volume of the sale or purchase for the stocks on each specific data [11]. The training dataset consists of 10,000 tuples which trains the logistic algorithm to predict a range within which the possible future values would fall into in upcoming times. Currently, the highest possible value of a day would be predicted as regards with other features (Figs. 1, 2, and 3).

Here, the image describes a range, i.e., less than 57.855 (indicated in blue color), and the actual highest value on that specific date was falling under the respective range. The other endpoint for the ranges is 97.1725, and the logistic regression predicted the range correctly for that specific date. Hence, the logistic regression algorithm is trained most efficiently to provide the range in the most effective manner [12, 13] (Fig. 4).

4 Conclusion

The recoil of the businesses in the financial market is on the basis or volumes or trades, out of which, about 73% of all transactions are done by automated systems and decision-making algorithms according to recent statistics. Previously, the stock

	Logistic Regres	index	date	open	
1	1.00 : 0.00 : 0.00 →	1.0	41313.0	15.0700	15
2	1.00 : 0.00 : 0.00 →	2.0	41316.0	14.8900	15
3	1.00 : 0.00 : 0.00 →	3.0	41317.0	14.4500	14
4	1.00 : 0.00 : 0.00 →	4.0	41318.0	14.3000	14
5	1.00 : 0.00 : 0.00 →	5.0	41319.0	14.9400	14
6	1.00 : 0.00 : 0.00 →	6.0	41320.0	13.9300	14
7	1.00 : 0.00 : 0.00 →	7.0	41324.0	14.3300	14
8	1.00 : 0.00 : 0.00 →	8.0	41325.0	14.1700	14
9	1.00 : 0.00 : 0.00 →	9.0	41326.0	13.6200	13
10	1.00 : 0.00 : 0.00 →	10.0	41327.0	13.5700	13
11	1.00 : 0.00 : 0.00 →	11.0	41330.0	13.6000	13
12	1.00 : 0.00 : 0.00 →	12.0	41331.0	13.1400	13
13	1.00 : 0.00 : 0.00 →	13.0	41332.0	13.2800	13
14	1.00 : 0.00 : 0.00 →	14.0	41333.0	13.4900	13
15	1.00 : 0.00 : 0.00 →	15.0	41334.0	13.3700	13
16	1.00 : 0.00 : 0.00 →	16.0	41337.0	13.5000	14
17	1.00 : 0.00 : 0.00 →	17.0	41338.0	14.0100	14
18	1.00 : 0.00 : 0.00 →	18.0	41339.0	14.5200	14
19	1.00 : 0.00 : 0.00 →	19.0	41340.0	14.7000	14

Fig. 2 Trained dataset

market investment was expected to be a long-term investment for many years; however, the scenario currently is completely different as the average hold time period for any stocks is 22 s on Wall Street. The traditional trading mindset of investors is the key skills that help them to figure out the company's true worth. However, the algorithms only look at one thing, i.e., the prices. They analyze millions of possible deals to determine which would be profitable seeing it statistically. This in turn gives them the advantage of speed which humans could never cope with. To compete with the next algorithm over which has same technology and probably got as many set stations on set, it can be solved by quote stuffing, think of a machine as having a set amount of computational power it can handle X amount of data per microsecond in order to gain a competitive advantage over that machine we have to slow down that algorithm by simply stuffing data into the market and sending random quotes, bid or ask prices but because my algorithm originates this data I can simply ignore it and I gain a computational advantage over them. So future scope of these algorithms totally depends on its computational speed—the faster the algorithm will surely solve more trades in the given time limit. However, the output could be defined more efficiently

	Logistic Regression	index	date	open	high	low	close
10065	0.00 : 0.02 : 0.98 → ≥ 95.1...	10065.0	43129.0	200.1100	200.8600	197.8700	198.2300
10066	0.00 : 0.01 : 0.98 → ≥ 95.1...	10066.0	43130.0	197.2500	197.7300	194.8900	196.9000
10067	0.00 : 0.01 : 0.98 → ≥ 95.1...	10067.0	43131.0	197.1300	200.9600	196.7500	199.7600
10068	0.00 : 0.01 : 0.98 → ≥ 95.1...	10068.0	43132.0	199.1200	201.7500	198.0845	199.3800
10069	0.00 : 0.01 : 0.98 → ≥ 95.1...	10069.0	43133.0	197.3300	199.4000	195.4400	195.6400
10070	0.00 : 0.01 : 0.98 → ≥ 95.1...	10070.0	43136.0	194.0600	198.4600	188.0000	190.2700
10071	0.00 : 0.01 : 0.98 → ≥ 95.1...	10071.0	43137.0	186.5900	194.8200	182.1000	194.4700
10072	0.00 : 0.01 : 0.98 → ≥ 95.1...	10072.0	43138.0	193.8700	196.2400	192.0200	192.3400
10073	0.99 : 0.01 : 0.00 → < 57.855	10073.0	41313.0	44.7200	45.9000	44.4500	45.7000
10074	0.99 : 0.01 : 0.00 → < 57.855	10074.0	41316.0	45.9900	46.1400	45.7700	46.0800
10075	0.99 : 0.01 : 0.00 → < 57.855	10075.0	41317.0	46.1500	46.3500	45.9600	46.2700
10076	0.99 : 0.01 : 0.00 → < 57.855	10076.0	41318.0	46.2900	46.6300	46.0600	46.2600
10077	0.99 : 0.01 : 0.00 → < 57.855	10077.0	41319.0	46.0200	46.7000	45.9800	46.5400
10078	0.99 : 0.01 : 0.00 → < 57.855	10078.0	41320.0	46.3900	46.6250	45.8600	46.1750
10079	0.99 : 0.01 : 0.00 → < 57.855	10079.0	41324.0	46.4000	47.0400	46.2000	47.0100
10080	0.99 : 0.01 : 0.00 → < 57.855	10080.0	41325.0	46.9000	47.2700	45.7600	45.7900
10081	0.99 : 0.01 : 0.00 → < 57.855	10081.0	41326.0	45.7000	45.7700	44.8350	45.1200
10082	0.99 : 0.01 : 0.00 → < 57.855	10082.0	41327.0	45.1905	45.8900	45.1900	45.5200
10083	0.99 : 0.01 : 0.00 → < 57.855	10083.0	41330.0	46.1100	46.2700	44.7500	44.7700
10084	0.99 : 0.01 : 0.00 → < 57.855	10084.0	41331.0	45.0800	45.3100	44.6500	45.0000
10085	0.99 : 0.01 : 0.00 → < 57.855	10085.0	41332.0	44.5800	45.6200	44.3100	45.3800
10086	0.99 : 0.01 : 0.00 → < 57.855	10086.0	41333.0	45.2800	45.8200	43.6800	45.2200
10087	0.99 : 0.01 : 0.00 → < 57.855	10087.0	41334.0	45.0000	45.4400	44.4100	45.2300
10088	0.99 : 0.01 : 0.00 → < 57.855	10088.0	41337.0	45.2900	45.3450	44.8900	45.3000
10089	0.99 : 0.01 : 0.00 → < 57.855	10089.0	41338.0	45.5000	45.9300	45.5000	45.7700
10090	0.99 : 0.01 : 0.00 → < 57.855	10090.0	41339.0	45.8300	46.1900	45.8000	46.1200
10091	0.99 : 0.01 : 0.00 → < 57.855	10091.0	41340.0	46.2500	46.2575	45.5900	45.7100
10092	0.99 : 0.01 : 0.00 → < 57.855	10092.0	41341.0	45.9600	46.2700	45.5300	46.0500

Fig. 3 Predicted values of dataset

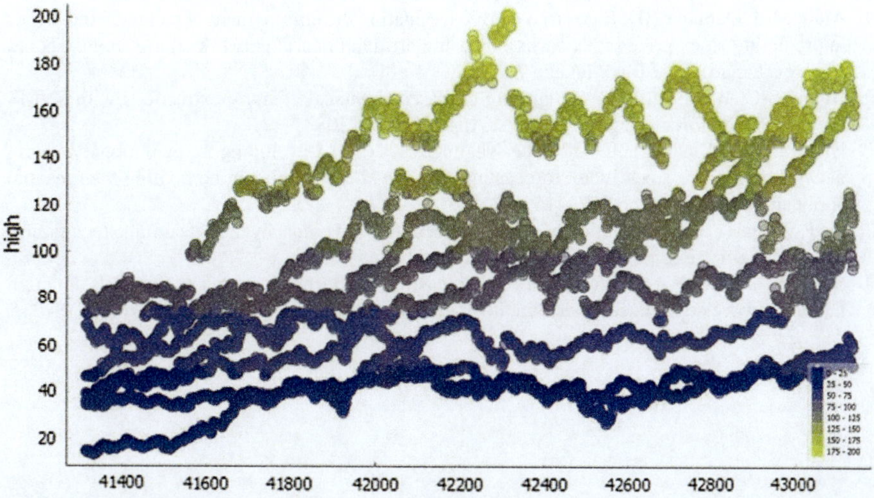

Fig. 4 Graph generated from predicted value of dataset

when other useful approaches are combined at initial stage. Furthermore, we are planning to create a multi-domain concept of machine learning and big data analysis so that vast amount of data can be provided for training set; this will further overcome the needs of expert advice from any side.

References

1. Lee Y-S, Tong L-I (2011) Forecasting time series using a methodology based on autoregressive integrated moving average and genetic programming. Knowl-Based Syst 24:66–72
2. Hadavandi E, Shavandi H, Ghanbari A (2010) Integration of genetic fuzzy systems and artificial neural networks for stock price forecasting. Knowl Based Syst 23:800–808
3. Asadi S, Hadavandi E, Mehmanpazir F, Nakhostin MM (2012) Hybridization of evolutionary Levenberg–Marquardt neural networks and data pre-processing for stock market prediction. Knowl-Based Syst 35:245–258
4. Cheng C, Xu W, Wang J (2012) A comparison of ensemble methods in financial market prediction. In 2012 Fifth international joint conference on computational sciences and optimization (CSO). IEEE, pp 755–759
5. Pai P-F, Lin K-P, Lin C-S, Chang P-T (2010) Time series forecasting by a seasonal support vector regression model. Expert Syst Appl 37:4261–4265
6. Kazem A, Sharifi E, Hussain FK, Saberi M, Hussain OK (2013) Support vector regression with chaos-based firefly algorithm for stock market price forecasting. Appl Soft Comput 13:947–958
7. Goldberg DE, Holland JH (1988) Mach Learn. https://doi.org/10.1023/A:1022602019183
8. Trivedi D, Bhagchandani A, Ganatra R, Mehta M (2018) Machine learning in finance. In: 2018 IEEE Punecon, Pune, India, pp 1–4 https://doi.org/10.1109/punecon.2018.8745424
9. Aldin MM, Dehnavr HD, Entezari S (2012) Evaluating the employment of technical indicators in predicting stock price index variations using artificial neural networks (case study: Tehran stock exchange). Int J Bus Manage 7
10. Huang S-C, Wu T-K (2008) Integrating ga-based time-scale feature extractions with SVMS for stock index forecasting. Expert Syst Appl 35:2080–2088
11. Hadavandi E, Ghanbari A, Abbasian-Naghneh S (2010) Developing an evolutionary neural network model for stock index forecasting. In: Advanced intelligent computing theories and applications. Springer, pp 407–415
12. Ou P, Wang H (2009) Prediction of stock market index movement by ten data mining techniques. Mod Appl Sci 3:P28
13. Shen W, Guo X, Wu C, Wu D (2011) Forecasting stock indices using radial basis function neural networks optimized by artificial fish swarm algorithm. Knowl-Based Syst 24:378–385

Artificial Intelligence: Prospect in Mechanical Engineering Field—A Review

Amit R. Patel, Kashyap K. Ramaiya, Chandrakant V. Bhatia, Hetalkumar N. Shah, and Sanket N. Bhavsar

Abstract With the continuous progress of science and technology, the mechanical field is also constantly upgrading from traditional mechanical engineering to the mechatronics engineering and artificial intelligence (AI) is one of them. AI deals with a computer program that possesses own decision-making capability to solve a problem of interest with imitates the intelligent behavior of expertise which finally turns into higher productivity with better quality output. From the inception, various developments have been done on AI system which nowadays widely implemented in the mechanical and/or manufacturing industries with broaden area of application such as pattern recognition, automation, computer vision, virtual reality, diagnosis, image processing, nonlinear control, robotics, automated reasoning, data mining and process control systems. In this study, review attempt has been made for AI technologies used in various mechanical fields such as thermal, manufacturing, design, quality control and various connected fields of mechanical engineering. The study shows the blend mixed of AI technologies like deep convolutional neural network (DCNN), convolutional neural network (CNN), artificial neural network (ANN), fuzzy logic and many more to control the process parameters, process planning, machining, quality control and optimization in the mechanical era for smooth development of

A. R. Patel (✉)
Chandubhai S Patel Institute of Technology, Charotar University of Science and Technology, Changa 388421, India
e-mail: amit.er1984@gmail.com

K. K. Ramaiya · C. V. Bhatia
Mechanical Engineering Department, Gandhinagar Institute of Technology, Gandhinagar 382721, India

H. N. Shah
Gandhinagar Institute of Technology, Gandhinagar 382721, India

S. N. Bhavsar
Mechatronics Engineering Department, G H Patel College of Engineering and Technology, Vallabh Vidyanagar 388120, India

K. Kotecha et al. (eds.), *Data Science and Intelligent Applications*, Lecture Notes on Data Engineering and Communications Technologies 52,
https://doi.org/10.1007/978-981-15-4474-3_31

product or system. With the implementation of AI in mechanical engineering applications, the error, rejection of components can be minimized or eliminated and system optimization can be achieved effectively turn in economical better quality products.

Keywords Artificial · Intelligence · Neural network · Machine learning · Automation

1 Introduction

For analyzing and extending human intelligence, in all the field of science and technology such as psychology, information and system science, cognitive science, space science, engineering, an artificial intelligence (AI) is an emerging technology and science available in a recent era [1, 2]. Continuous improvement in the established conventional technologies and introducing with complex information technology network groups analyzed and turn into new higher degree technology implemented in the mechanical and/or manufacturing field recently to minimize rejections, errors in the final products with the help of self-organized unit controlling all inputs parameters. In this regard, the future scope of AI can be characterized as an advanced intelligence at a higher degree of innovative thinking for the overall spectrum of the task for the engineering industry in such a high competitiveness in real-time market.

Nowadays combination of AI technology and mechanical/mechatronics engineering become a growing technology to upgrade the level of automation and intelligent manufacturing for continuous improvement [3]. Based on a set of optimized data and algorithms, machine learning is used to predict intelligent outcomes. In a normal situation, it rarely happens that information flows back from production to planning but lots of information from the production stage affect the planning tasks very effectively and reduced the product defects, unplanned downtimes and improve flexibility.

The purpose of this paper is to report the composition and development of the AI in the mechanical field such as thermodynamics, stress analysis, mechanics, fluid mechanics, dynamic analysis and control, parameter optimization, quality control, production engineering, process planning, process monitoring and diagnosis and allied fields like self-driven smart cars, drones, and automatized missiles were the results in the industry. With the help of drawing of components, assemblies, systems or layouts, computer-aided designing (CAD) and fabrication process [4]. Here, in this review, mainly focused on manufacturing, thermal and design stream of mechanical engineering.

2 Artificial Intelligence in Manufacturing

Chen et al. present three studies of intelligent manufacturing as defined by the Chinese Academy of Engineering, and the concept, characteristics and systemic structure of the intelligent machine tool (IMT) are presented in this paper. Three stages of machine tool evolution—from the manually operated machine tool (MOMT) to the IMT—are discussed, including the numerical control machine tool (NCMT), the smart machine tool (SMT) and the IMT. The new generation of intelligent manufacturing, which is motivated by deep integration of new-generation AI technology with advanced manufacturing technology, is becoming the core driving force for the new industrial revolution as shown in Fig. 1. This work examines the enabling principles of autonomous sensing and connection, autonomous learning and modeling, autonomous optimization and decision making and autonomous control and execution using big-data-based AI technology. It reveals that the essential characteristic of the IMT is that it can automatically generate, accumulate and utilize knowledge so as to achieve the goals for effective utilization and low consumption in the production process.

A control code also known as the intelligent code (I-code), which is embedded with the results of the optimum decision, is created for machining optimization, as shown in Fig. 1. Self-governing optimization and decision making are the process of forecasting the response of the machine tool, making decisions and eventually generating the corresponding and appropriate I-code [3]. Feng et al. propose an integrated method for intelligent green scheduling of the sustainable flexible workshop with edge computing considering an uncertain machine state. Putting forward a reasonable and effective green scheduling method for the sustainable flexible workshop which is suitable for the actual green manufacturing environment can contribute

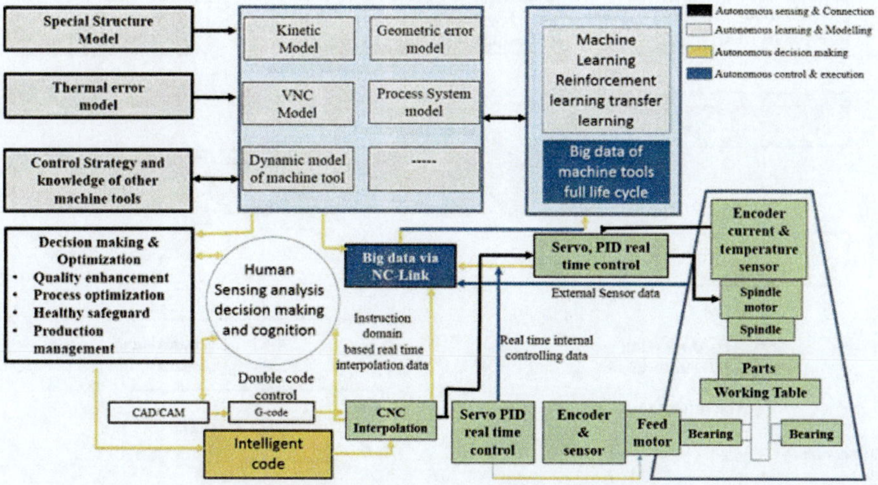

Fig. 1 Control principle of the IMT. VNC: virtual numerical controller [3]

greatly to the improvement of machine utilization rate, the shortening of the product makespan, the reduction of processing cost and energy consumption and effective utilization of resources [5]. Zhou et al. represent a cutting tool selection approach powered by deep learning, which contributes to effective and efficient improvement of the intelligence of the process of cutting tool selection. With the developed techniques, engineers could take an engineering drawing as input for cutting tool selection, instead of using a complex and elusive descriptor to express the information of special-shaped machining features. Finally, the feasibility of the proposed approach is demonstrated through the special-shaped machining features of a vortex shell work piece.

Figure 2 elaborates a cutting tool selection approach for special-shaped machining topographies of complex products by means of a deep learning approach, which increases the cleverness of the process of cutting tool selection. On another side, tool selection takes into account the machining feature expression, recognition and tool selection results [6]. Weigelt et al. show machine learning (ML)-based key technology in smart manufacturing. In comparison with common physical simulations, ML algorithms offer insight into complex processes without requiring in-depth domain knowledge. It is quite difficult to generate a model and control the innovative contacting process such as ultrasonic crimping, which is used in electric drives production.

Thus, the author transfers the potential of ML to the ultrasonic crimping manufacturing process and presents a conceptual design of an intelligent ultrasonic crimping process. To validate the proposed architecture, relevant ML algorithms for the prediction of the joint quality using visual features are selected. Figure 3 elaborates the idea of classifying the crimp quality according to their visual appearance. This comparison marks the learning technique as more robust regarding the appearance

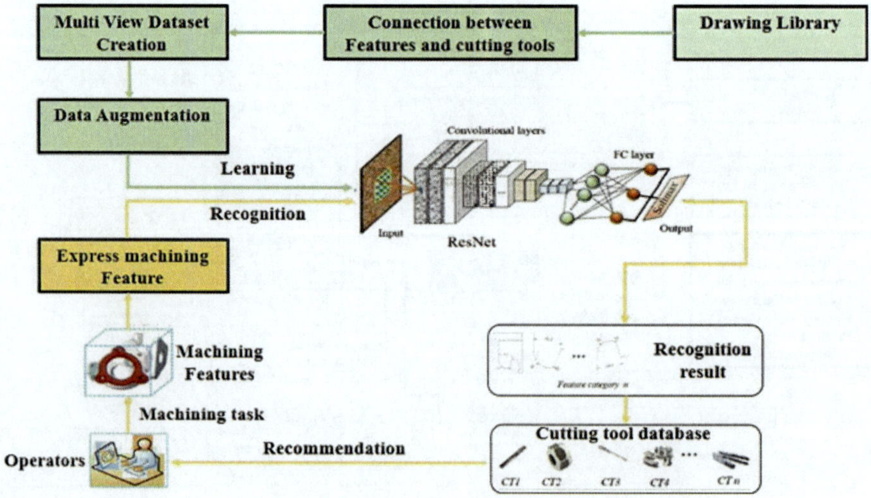

Fig. 2 Cutting tool selection methodology based on deep learning [6]

	Average Color Area of possible melting resides	Average Color Area of sonotrode influence	Length of possible melting residues	Visual classification using a CNN
Image	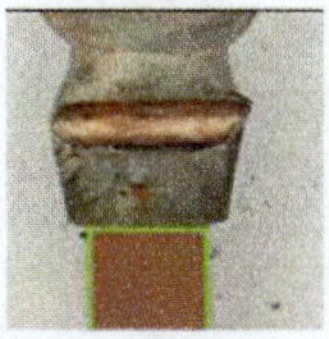	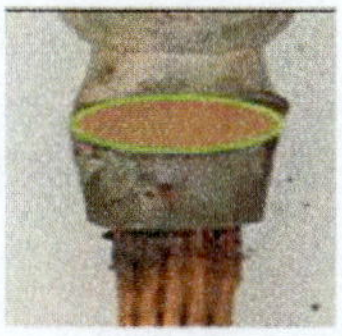		
Prediction quality	21% (69%)	38%	56%	Up to 91 %

Fig. 3 Prediction quality for different visual indicators [7]

of melting residues and easier to automate, as the relevant sections of the image are identified by the algorithm, which saves costly preprocessing of the image data. Regarding the accuracy, the convolutional neural network (CNN) performs much better than the presented deterministic approaches [7]. Wenkler et al. have carried work toward intelligent characteristic value determination for the cutting process. Information on milling processes usually flows only from planning to production. In mass production, the effort is put into optimizing a process, and in the production of individual parts, the information does rarely flow back from production to planning. As a result, a lot of process information is not used during the planning stage. The motivation comes to develop a method to generate characteristic values from the key performance indicators of production process. A machine data protocol monitors milling processes and recalculates specific cutting force. Influence parameters and the characteristic value are collected in a database. Based on the database, an ANN is trained, which predicts the specific cutting force.

The ANN system described above is adapted to the data basis. In industrial applications, the data basis is increasing endlessly, and hence, it is dynamic. For this cause, a dynamic ANN must be produced. The said target can be realized by developing a control loop that repeatedly adapts the ANN to the data basis as shown in Fig. 4 [8]. Wenbin et al. proposed a novel manufacturing resources organization method based virtual manufacturing cell to simplify the production planning process and improve planning quality. An optimization mathematical model is projected for (cell formation) CF problem of the cellular manufacturing system, and the objective role in this model is to decrease the total manufacturing system production cost. A novel artificial intelligence approach based on the endocrine regulatory mechanism is proposed to resolve the model. A case study is used to validate the usefulness and efficiency of the projected hormone-based approach. Generally, the proposed artificial intelligence approach for the cellular manufacturing system has numerous advantages as follows: (1) reduce the manufacturing cost and keep the balance between machines; (2) simplify the complications arising in the production plan and scheduling; and (3) improve the agility of the manufacturing system [9].

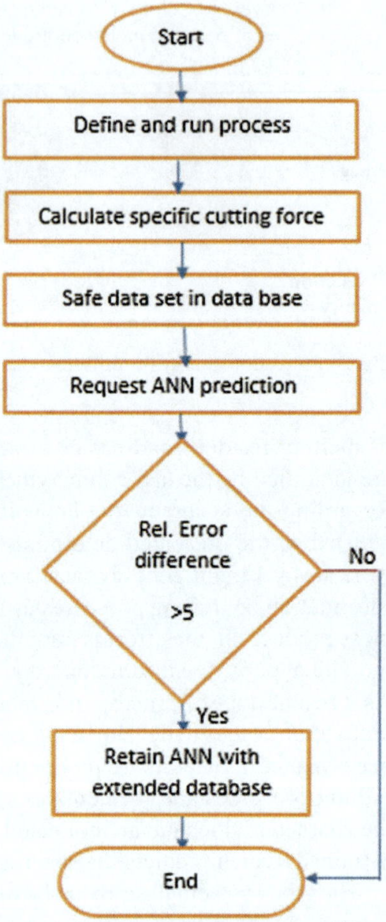

Fig. 4 Flowchart for continuous adaption of the ANN [8]

Yuyong et al. adopted an artificial neural network to model AWJ cutting stainless steel process based on experimental information. In order to accomplish clear-cut machining, it is necessary to accurately predict the cutting surface superiority in Abrasive water jet cutting (AWJ). A trained network which constitutes reasonable AWJ cutting parameters was produced. The ANN model for abrasive water jet cutting stainless steel process has adequate accuracy. Subsequently, precise machining can be attained by integrating this model into computer numerical controller. Based on the ANN model built for AWJ cutting stainless steel, simulation of AWJ cutting and optimization of cutting speed, cutting process parameters, etc. could be carried out. For actual applications, a reasonable model is designated automatically by integrated software of numerical controlled system based on the material to be machined and process requirements to meet the desires of customer [10]. Ivan reviewed the existing AI technology with a special view on potential applications in manufacturing engineering. The following are some of the areas of the AI research relevant to computer-based manufacturing systems: automatic planning which can be applied, for example, to robot programming; qualitative modeling and qualitative simulation of systems and processes using symbolic computation techniques; machine synthesis of new knowledge; AI languages and programming systems, Lisp, Prolong, symbolic computation rule-based programming object-oriented programming [11]. Concept of recursive Allen temporal algebra is fully described, along with its relationship to the smart scheduler program and the concept of smart scheduling in general was covered in this paper. Flexible manufacturing systems (FMS) are described, and their relationship with smart scheduling is explored. A case study that shows the utility of SSP in an FMS environment is presented in detail. Conclusions about the concept of smart scheduling are presented in closing. RATA has been formally defined, and some mathematical relationships relevant to its application have also been specified. The program is written in Pascal and is PC-based. The case study demonstrates how smart scheduling could be applied to a real-world FMS environment. This unique and powerful automatic scheduling tool has more flexible temporal relations and better fuzzy knowledge rule-representation capabilities than any other known scheduling system [12].

Lee et al. discussed the result of increased use of sensors and networked machines in manufacturing operations, and artificial intelligence techniques play a key role to derive meaningful value from big data infrastructure. These techniques can inform decision making and can enable the implementation of more sustainable practices in the manufacturing industry. In machining processes, a considerable amount of waste (scrap) is generated as a result of failure to monitor a tool condition. Therefore, an intelligent tool condition monitoring system is developed. To identify sustainability-related manufacturing trade-offs and a set of optimal machining conditions, an intelligent tool condition monitoring system was developed. Author proposed an intelligent tool condition monitoring system for a machine tool and also to identify an optimal set of machining conditions as a function of tool wear by optimizing trade-offs between different objectives—profit, quality and productivity. Since a tool's performance changes over the machining time, tool condition information is incorporated in the multi-objective optimization technique to identify trade-offs. The proposed monitoring system is expected to recommend a proper degree of tool utilization by maximizing a manufacturer's needs. The recommend values enable better decision making, which can also help to reduce the amount of the scrap by controlling product quality [13]. Wang et al. present concepts of an Internet-assisted manufacturing system for agile manufacturing practice. It consists of a CAD/CAM/CAPP/CAA (CNS) server which links to local FMS, or CNC or NC machines by means of cable connections. After a local user inputs the product information, the CNS can generate complete CAD/CAM/CAPP/CAA files and control the remote FMS or CNC machines to accomplish the production process [14].

3 Artificial Intelligence in Thermal Engineering

Cheng et al. give information pertaining to the development of artificial intelligence (AI) technology for improving the performance of heating, ventilation and air-conditioning (HVAC) systems. Functions, including weather forecasting, optimization and predictive controls, have become mainstream. The estimated average energy savings percentage and the maximum saving rations of AI-assisted HVAC control are 14.4% and 44.04%, respectively. In this study, the lower accuracy of prediction tools and the resulting poor energy savings of HVAC systems are hypothesized.

In Fig. 5, the typical energy savings percentage when using AI-assisted HVAC control is 14.02%. Grounded on the Normalize Harris Index (NHI), the predictable average energy savings percentage, deviations in energy savings and the maximum energy savings of AI-assisted HVAC control are 14.4%, 22.32% and 44.04%, correspondingly. Linking these outcomes with the investigational data of 14.02%, 24.52% and 41.0% as shown in the above figure, the errors are 3%, 9% and 7%, respectively [15]. Nasiri et al. propose a novel and accurate radiator condition monitoring and intelligent fault detection method based on thermal images and deep convolutional neural network.

The suggested CNN model directly uses infrared thermal images as shown in Fig. 6 as input to classify six conditions of the radiator, respectively normal, tubes blockage, coolant leakage, cap failure, loose connections between fins and tubes and fins blockage. The modified CNN model presented by the author showed accurate performance for the cooling radiator fault diagnosis under various combinations of

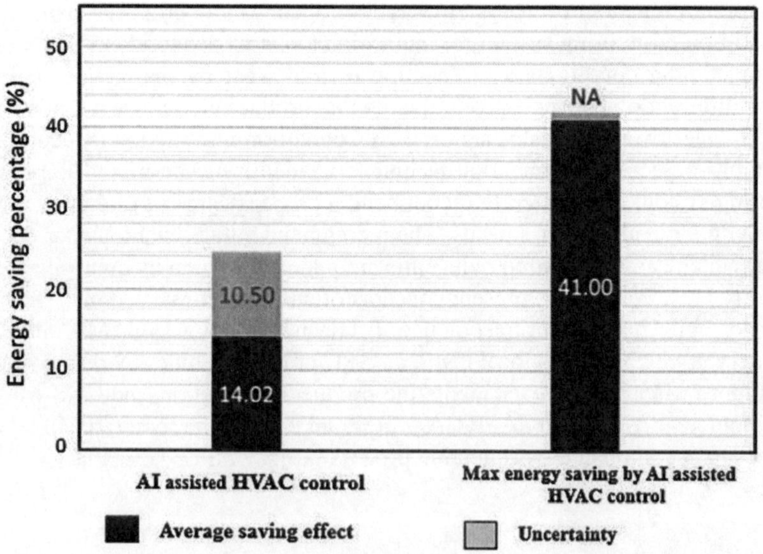

Fig. 5 The average energy savings of the 24 cases and the maximum energy savings attained by AI-assisted HVAC control [15]

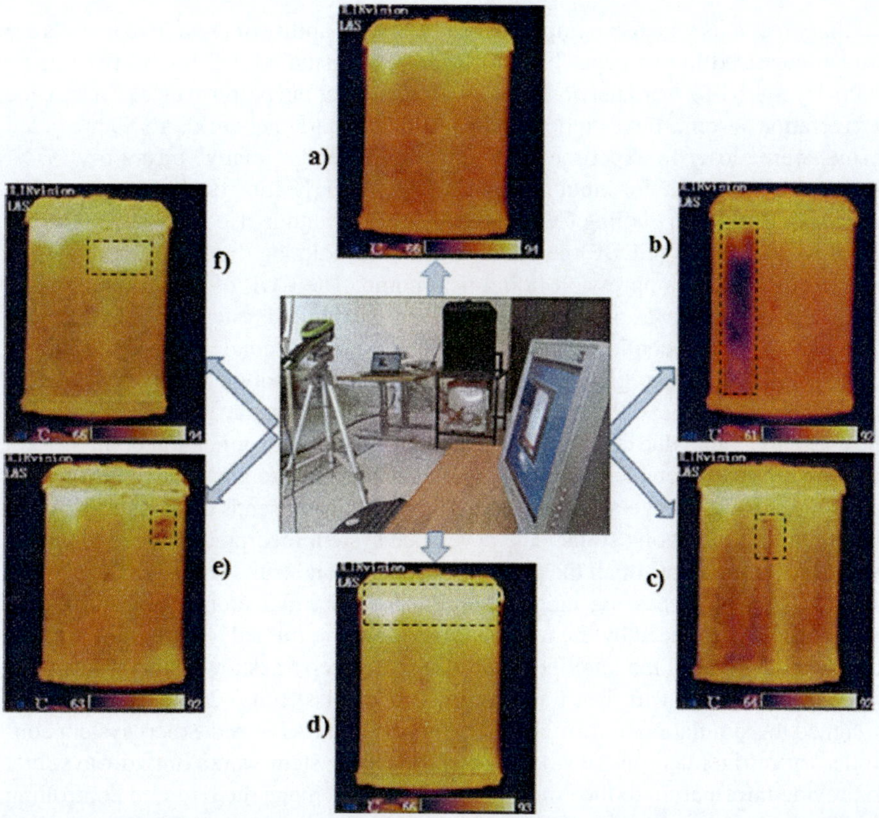

Fig. 6 Examples of the attained IR thermal images of the six conditions of the radiator. **a** Normal radiator, **b** radiator tubes blockage, **c** coolant leakage, **d** radiator cap failure, **e** loose connections between fins and tubes and **f** radiator fin blockage [16]

coolant temperature, flow rate and suction air velocity. From the results, it is evidenced that the deep CNN based model is working very effectively [16]. Han et al. describe transfer learning as an encouraging tool for improving the efficiency of focused problems by manipulating information from the preceding tasks that are dissimilar but comparable. The author presents an innovative transfer learning structure on the basis of a pre-trained convolutional neural network (CNN), where the feature transferability at different stages of the deep structure is debated and compared in two case studies. The outcomes show that the projected framework can transfer the features of the pre-trained CNN to the marked domain with diverse working conditions or fault types, and a high accuracy is attained for both cases [17]. Liu et al. review various techniques in hybrid intelligent wind energy forecasting models, eight types of mainstream shallow and deep learning-based intelligent predictors are classified. These intelligent predictors have their own merits and limitations and are suitable for different forecasting tasks, namely learning and metaheuristic optimization. Two auxiliary methods which can improve the forecasting ability of predictive models are also summarized in this paper [18]. Mohanraj et al. studied the suitability of using multi-layer feed-forward network (MLFFN) to predict the performance of a cascade refrigeration system. It was reported that artificial neural network (ANN) predicted results were closer to experimental results with average relative errors of 1.37%, 4.44%, 2.05%, 1.95% for input power, heating power, heating coefficient of performance (COP) and for cooling COP, respectively. The author also studied generalized radial basis function (GRBF) neural network for predicting the steady-state performance of a vapor–compression liquid heat pump. The COP of a heat pump using R22, LPG and R290 was predicted with reference to chilled water outlet temperature from the evaporator, cooling water inlet temperature to the condenser and evaporator capacity [19]. Shweta originates an AI-based hardware prototype which can sense the age of individual vegetables placed in the domestic refrigerator. The trigger is sent to the user's mobile in due course with the help of an android application or an SMS in case of unfound vegetables for 30 days. The received message contains information about the age of the vegetables inside. It also periodically sends data regarding what you have not been eating so far. The developed system incorporates three segments where the training data of all the possible vegetables and fruits are fed to the system. Images for image processing, aging algorithm and voice indicators to indicate the age of the vegetables placed by the user and what have we missed to eat or left unused. The method is one of the intelligent approaches to keep track on balanced nutritious value in our bodies with fresh and healthy vegetables/fruits [20]. Marchant et al. designed the simulation to form an integrant part of an AI-based expert system controller for refrigerated potato stores. The proposed system works in two ways, first to provide information to the controller to allow it to make the required controlling decisions, and second to provide data which cannot be avail from sensor inputs, either because of the physical difficulty of measurement or due to the affordability issues. The system has been developed in C++ and illustrates some of the advantages that can be achieved by using an object-orientated approach. Modeling the heat and mass transfer functions of both the crop and the store has also been carried out and is based on dividing the crop into capricious layers with plug airflow between them. The simulation has been run under different initial conditions and control strategies and the results compared with data logged at commercial stores. The results are satisfactory and considered acceptable for the application intended [21]. Teeter et al. describe a functional link neural network approach to performing the heating ventilation and air-conditioning (HVAC) thermal dynamic system identification. The author also presents the application of conventional and reduced-ordered function link neural networks for HVAC thermal dynamic system identification. A single-zone thermal system model was chosen for analysis. The system represents a simplification of an overall building climate control problem, but retains the distinctive features of an HVAC system. The practice and usage of neural networks for identification and control provide a means of adapting a controller online in an effort to minimize a given cost index [22]. Ogaji et al. trained an artificial neural network system to identify, segregate and evaluate faults in some of the modules of a single spool gas turbine. The tiered diagnostic approach adopted comprises a number of decentralized networks trained to handle specific tasks. All sets of networks were tried with data not used for the training practice. The results show that substantial benefits can be derived from the actual application of this technique. The designated methodology has been tested with data not used for training and generalization is found to be suitable for the actual application of this technique. Also, the level of accurateness achieved by this decentralized application of ANN shows a clear improvement over techniques that need just a solo network to accomplish fault detection, isolation and assessment [23].

4 Applications of Artificial Intelligence in Engineering Design

Akbani et al. describe a case study for defining natural frequencies of a cantilever beam with a crack using ANN. The case study points out that the artificial neural network (ANN) technique has the competency to reduce efforts and time when applied to complex engineering problems.

Table 1 based on x/L and a/h parameters (where L = Length of beam, a = crack depth, h = distance of crack from fixed end of beam) shows that the values of natural frequencies obtained by the developed ANN are in good agreement with those obtained using ANSYS. Integration of ANN in the case problem eradicates the need to remodel the beam over and again for diverse crack parameters and rerun the finite element (FE) analysis. The features of ANN can be utilized for the advantage of mechanical engineers in the application areas such as quality control, production planning, job shop scheduling, supply/demand forecasting, mechanism design and analysis, design optimization and several other areas [24].

Chen et al. concluded that to improve prognostics and diagnostics vast data collection should be analyzed and it is possible by the intelligent processing method based on AI and various sensing technologies. It will exploit the data from vibration and electrostatic sensors, and from this failure, time may be determined and find the nature of the failure [25]. Wang integrated the AI technology for the design of the bucket elevator design. Using this technology complex structure is possible to design easily with lesser time. The author also discussed the case representation method and used the fuzzy mathematics for the new product design. It will help in the rapid response to the need of market, reusing the available knowledge of design

Table 1 Case study table [24]

Case-1	$x/L = 0.2, a/h = 0.2$		
Frequency	ANN	ANSYS	Variation (%)
1	179.7553	180	−0.13613
2	1154.227	1133.2	1.821723
3	3210.578	3061.5	4.643325

and reduced the time of the design process [26]. Yildirim et al. analyzed the vibration, emission characteristics and noise level for the four-stroke four-cylinder diesel engine. Fuel used in this engine was sunflower, canola and corn biodiesel blends and H_2 is injected through the inlet valve using two different methods. One is support vector machines and the second is ANN. After analyzing both the method for the output results and error, the author concluded ANN is a good option instead of SVM for such analysis [27]. Sivasankari et al. analyzed the damages in the axle failure and joints of heavy load vehicles and find the detection and prevention method for the same. Heavy load vehicle needs regular proper checking of all the parts and fixed the error. Damages in the nuts, bolts and axle can be checked regularly using the sensors and the AI coding. Using this technology, damages can be eradicated with 100% accuracy [28]. Pratt et al. concluded that a numerical optimization is an important tool in many aspects of jet engine design, development and test. Early optimization work centered on structural optimization projects, while more recent applications are multidisciplinary in nature. A compressor/turbine performance optimization effort has produced an automated technique for the preliminary design of advanced gas turbine engines. By varying towpath design parameters, component efficiencies were maximized, subject to both structural and aerodynamic constraints. In a manufacturing problem, a technique was developed to inspect drill hole patterns and tolerances automatically and to compare them with corresponding template geometries. An interesting hybrid solution process combines the penalty function and constrained optimization techniques. An acoustic data reduction project has provided an automated procedure to match analytical models to engine data. Using parametric design techniques, prototypes have been rapidly generated for trade-off studies [29].

5 General Applications of Artificial Intelligence in Mechanical Engineering

Dhingra reported the prospects and applications of artificial intelligence in design and manufacturing processes such as component selection, design, reasoning, learning, perception, sensing, recognition, intuitions, creativity, analysis, abstraction, planning and prediction. This is characterized as the eventual fate of upkeep in which an intelligent framework can formulate the machines and frameworks to achieve the most remarkable execution and almost zero breakdowns with self-supportabilities. The author also concluded that AI has changed a designer's methodology toward taking care of complex building issues regardless of any field [30]. Huang Q studied the AI technology, including its development process, composition and also presented the concept of mechatronics and AI technology. Using AI technic, the author analyzed the fault diagnosis of hot forging press. The author also discussed limitations and remedies in mechanical and electronic engineering, such as the unsteady system, the reason for the problem is the deficient factor of the electronic information system. The author concludes that because of the increasingly terrible competition in the machinery industry, the hybrid intelligent design, monitoring, control, diagnosis system based on fuzzy logic, neural network, expert system, will be a new research hotspot in order to improve the level of its intelligent control [1]. Haidong et al. focus on the intelligent diagnosis, a new development in the machinery fault detection technology, which can efficiently analyze prodigious, collected data and automatically generate reliable diagnosis results. Among different intelligent diagnosis methods, artificial neural network (ANN) and support vector machine have been the most extensively applied in the past years. The author also proposes a novel method called deep wavelet auto-encoder with extreme learning machine for rolling bearing intelligent fault diagnosis. Results show that the proposed method can get rid of the reliance on manual feature extraction, which is more effective than the traditional methods and standard deep learning methods [31]. Shaonak et al. concluded a number of areas of work in which AI techniques and developments are being used to improve the design. The approaches to design and design systems are concealed, along with some techniques that are used. The incorporation of AI systems into a design environment is directly related to the industry who want to make better use of existing resources and enhance their capabilities rather than make them redundant. Main objective of AI is useful to speed up information flow and increase information and expertise availability so that the quality of decision making can be improved. The first step to put efforts to develop the theories, improvement of computing techniques to handle large amounts of imprecise data. Secondly, the integration of existing and future systems into a flexible and global environment for product lifecycle management is required. Thirdly, human–computer interfaces must continue to be developed so that they meet the requirements of the users and show the information in the most easily assimilated form [32]. Zajacko et al. have proposed a method to fully automate the process of quality control of produced tires by the manufacturer. At present, this task is performed by an operator who has an inspection stand consisting of a rotary mandrel, pneumatic drives, PLC controlling individual drives and lighting. The inspection stand serves only as a device assisting the operator in handling the checked tire. The entire assessment process is in the full proficiency of the operator and was carried out solely on the basis of a quality assessment corresponding to the range of knowledge of the operator of the product. The author tried to ensure complete automation of the control process. The deep convolutional neural network (DCNN) application in an automated quality control process will solve the most complex set of tasks by employing an automated error detection system on inspected tires. The inspection stand was drawn-out by a proposed camera system that will produce a huge amount of input data, DCNN was used to effectively extract and identify image elements and to automatically detect the existence of errors and abnormalities in the controlled product [33]. Nicola et al. extensively used bibliographic database queries to develop and present a complete summary of the applications of computational intelligence (CI) to mechanical engineering (ME). Tables and pie charts have been provided to investigate the actual research trends and, particularly, to identify those subjects of ME where the tools and the methods of CI have been relatively overlooked with no apparent reason. Among these fields, classical subjects such as machine design, mechanisms kinematic and dynamic and transmissions appear quite sympathetic to CI, which is believed to have a great prospective for these ME applications. Hence, the results presented in this paper encourage researchers to develop CI applications in these areas of mechanical engineering [34].

6 Conclusion

In this study, the field of AI is incorporated in mechanical applications combining process parameters and quality control of manufacturing processes. AI has the potential to be very effective tool for solving complex dynamical system at different loading conditions. From the preceding discussions and case studies, it is often seen that AI will facilitate save time and efforts for advanced issues, wherever analytical techniques are terribly tough and tedious to use. By studying the concept and technologies of AI in this work, we predict the very high demands on part of the practice to give the solution by applying the AI system in the mechanical field right from drawing to the development phase of the product. In general, applications of AI systems are widespread and it can be applied to any system that needs the replacement of human expertise to provide a useful solution. The identification of future research scope can be carried out in various applications of mechanical field combined with robotics and automation engineering by applying artificial intelligence.

References

1. Huang Q (2017) Application of artificial intelligence in mechanical engineering. In: 2nd International conference on computer engineering, information science & application technology (ICCIA 2017), vol 74, pp 855–860. https://doi.org/10.2991/iccia-17.2017.154
2. Zajacko I, Gal T, Sagova Z, Mateichyk V, Wiecek D (2012) Application of artificial intelligence principles in mechanical engineering. In: MATEC web of conference, vol 244, pp 1–7. https://doi.org/10.1051/matecconf/201824401027
3. Chen J, Hu P, Zhou H, Yang J, Xie J, Jiang Y, Zhang C (2019) Toward intelligent machine tool. Engineering 5(4):679–690. https://doi.org/10.1016/j.eng.2019.07.018
4. Carter IM (2018) Applications and prospects for AI in mechanical engineering design. Knowl Eng Rev 5(3):167–179. https://doi.org/10.1017/S0269888900005397
5. Feng Y, Hong Z, Li Z, Zheng H, Tan J (2019) Integrated intelligent green scheduling of sustainable flexible workshop with edge computing considering uncertain machine state. J Clean Prod. https://doi.org/10.1016/j.jclepro.2019.119070
6. Zhou G, Yang X, Zhang C, Li Z, Xiao Z (2019) Deep learning enabled cutting tool selection for special-shaped machining features of complex products. Adv Eng Softw 133(28):1–11. https://doi.org/10.1016/j.advengsoft.2019.04.007
7. Weigelt M, Mayr A, Seefried J, Heisler P, Franke J (2018) Conceptual design of an intelligent ultrasonic crimping process using machine learning algorithms. Procedia Manuf 17:78–85. https://doi.org/10.1016/j.promfg.2018.10.015
8. Wenkler E, Arnold F, Hanel A, Nestler A, Brosius A (2019) Intelligent characteristic value determination for cutting processes based on machine learning. Procedia CIRP 79:9–14. https://doi.org/10.1016/j.procir.2019.02.003

9. Wenbin G, Wang Y (2018) An artificial intelligence application for cellular manufacturing system inspired by the endocrine mechanism. IEEE, Chengdu, China. https://doi.org/10.1109/itnec.2017.8285049
10. Yuyong L, Puhua T, Daijun J, Kefu L (2010) Artificial neural network model of abrasive water jet cutting stainless steel process. In: IEEE international conference on mechanic automation and control engineering. Wuhan, China. https://doi.org/10.1109/mace.2010.5536724
11. Ivan B (1988) AI tools and techniques for manufacturing systems. Robot Comput Integr Manuf 4(1–2):27–31. https://doi.org/10.1016/0736-5845(88)90056-7
12. Scheduling in flexible manufacturing systems. In: Handbook on scheduling. International handbook on information systems. Springer, Berlin, Heidelberg (2007). https://doi.org/10.1007/978-3-540-32220-7_14
13. Lee WJ, Mendis GP, Sutherland J (2019) Development of an intelligent tool condition monitoring system to identify manufacturing tradeoffs and optimal machining conditions. Procedia Manuf 33(019):256–263. https://doi.org/10.1016/j.promfg.2019.04.031
14. Wang Z, Rajurkar KP, Kapoor A (1996) Architecture for agile manufacturing and its interface with computer integrated manufacturing. J Mater Process Technol 61(1–2):99–103. https://doi.org/10.1016/0924-0136(96)02472-7
15. Cheng CC, Lee D (2019) Artificial intelligence-assisted heating ventilation and air conditioning control and the unmet demand for sensors: Part 1. Problem formulation and the hypothesis. Sensors 19, 1131. https://doi.org/10.3390/s19051131
16. Nasiri A, Taheri-Garavand A, Omid M, Carlomagno G (2019) Intelligent fault diagnosis of cooling radiator based on deep learning analysis of infrared thermal images. Appl Therm Eng 163. DOI: https://doi.org/10.1016/j.applthermaleng.2019.114410
17. Han T, Liu C, Yang W, Jiang D (2019) Learning transferable features in deep convolutional neural networks for diagnosing unseen machine conditions. ISA Trans 93:341–353. https://doi.org/10.1016/j.isatra.2019.03.017
18. Liu H, Chen C, Lv X, Wu X, Liu M (2019) Deterministic wind energy forecasting: A review of intelligent predictors and auxiliary methods. Energy Convers Manag 195(May):328–345. https://doi.org/10.1016/j.enconman.2019.05.020
19. Mohanraj M, Jayaraj S, Muraleedharan C (2012) Applications of artificial neural networks for refrigeration, air-conditioning and heat pump systems—a review. Renew Sustain Energy Rev 16(2):1340–1358. https://doi.org/10.1016/j.rser.2011.10.015
20. Shweta AS (2017) Intelligent refrigerator using artificial intelligence. In: 11th International conference on intelligent systems and control (ISCO). IEEE, Coimbatore, India, pp 5–6. https://doi.org/10.1109/isco.2017.7856036
21. Marchant AN, Lidstone PH, Davies TW (1994) Artificial intelligence techniques for the control of refrigerated potato stores. Part 2: heat and mass transfer simulation. J Agric Eng Res 8(1):27–36. https://doi.org/10.1006/jaer.1994.1032
22. Teeter J, Chow MY (1998) Application of functional link neural network to HVAC thermal dynamic system identification. IEEE Trans Industr Electron 45(1):170–176. https://doi.org/10.1109/41.661318
23. Ogaji SOT, Singh R (2003) Advanced engine diagnostics using artificial neural networks. In: Proceedings of the IEEE international conference on artificial intelligence systems (ICAIS'02), Applied soft computing, vol 3, no 3, pp 259–271. https://doi.org/10.1016/s1568-4946(03)00038-3
24. Akbani I, Baghele A, Arya S (2012) Artificial intelligence in mechanical engineering : a case study on vibration analysis of cracked cantilever beam. In: IJCA Proceedings on national conference on innovative paradigms in engineering and technology (NCIPET), vol 8, pp 31–34
25. Chen SL, Craig M, Callan R, Powrie H, Robert W (2008) Use of artificial intelligence methods for advanced bearing health diagnostics and prognostics. In: 2008 IEEE aerospace conference, 1095–323X, Big Sky, MT, USA. https://doi.org/10.1109/aero.2008.4526604
26. Wang J, Huixue S (2007) Studies on CAD systems with artificial intelligence. In: Eighth ACIS international conference on software engineering, artificial intelligence, networking, and parallel/distributed computing, IEEE. https://doi.org/10.1109/snpd.2007.310

27. Yıldırım S, Tosun E, Calık A, Uluocak I, Avsar E (2018) Artificial intelligence techniques for the vibration, noise, and emission characteristics of a hydrogen enriched diesel engine. Energy Sources, Part A: Recover, Util, Environ Eff 41(18):2194–2206. https://doi.org/10.1080/15567036.2018.1550540
28. Sivasankari B, Akashkumar V, Elavenil M (2019) Auto detection of joints and axle failure in heavy load vehicles using artificial intelligence. In: 5th International conference on advanced computing & communication systems (ICACCS), IEEE, Coimbatore, India. https://doi.org/10.1109/icaccs.2019.8728469
29. Pratt TK, Seitelman LH, Zampano RR, Murphy CE, Landis F (1993) Optimization applications for aircraft engine design and manufacture. Adv Eng Softw 16(2):111–117. https://doi.org/10.1016/0965-9978(93)90056-Y
30. Dhingra M (2018) Prospects of artificial intelligence in mechanical. Int J Eng Technol Res Manage 2(4):36–38
31. Haidong S, Hongkai J, Xingqiu L, Shuaipeng W (2017) Intelligent fault diagnosis of rolling bearing using deep wavelet auto-encoder with extreme learning machine. Knowl-Based Syst 140:1–14. https://doi.org/10.1016/j.knosys.2017.10.024
32. Shaonak K, Mishra L, Saraswat U (2017) Impact of aritificial intelligence in the mechanical engineering. Int J Mech Prod Eng 5(7)
33. Zajacko I, Gal T, Sagova Z, Mateichyk V, Wiecek D (2012) Application of artificial intelligence principles in mechanical engineering. In: MATEC web of conferences, vol 244, pp 1–7. https://doi.org/10.1051/matecconf/201824401027
34. Nicola PB (2014) Applications of computational intelligence to mechanical engineering. In: IEEE 15th international symposium on computational intelligence and informatics (CINTI), Budapest, Hungary. https://doi.org/10.1109/cinti.2014.7028702

Genetic Algorithm Based Task Scheduling for Load Balancing in Cloud

Tulsidas Nakrani, Dilendra Hiran, Chetankumar Sindhi, and MahammadIdrish Sandhi

Abstract Due to on-demand services for online resources like processing power, storage, software, infrastructure, etc., provided by cloud computing, it becomes incredibly popular today. So, intensity of Web data is increasing day by day. To balance the load of different nodes is the biggest challenge in this era. Load balancing method makes sure that no any node is over utilized or underutilized. It is considered to be an optimization problem. This paper proposes a genetic algorithm-based task scheduling for load balancing. The proposed strategy is simulated using cloud analyst. The results demonstrate that proposed method is better than existing algorithm like Round Robin (RR), Equally Spread Current Execution Load algorithm (ESCELEA) and Throttled algorithm (TA).

Keywords Genetic algorithm · Load balancing · Cloud

1 Introduction

Cloud is a term that reflects the Internet that means something that is available on the Internet and every people can use it. So, cloud provides in demand computing services just like big IT companies, i.e. Amazon, Microsoft, Google provide these

T. Nakrani (✉) · D. Hiran
Pacific University, Udaipur, India
e-mail: nakranitulsidas@gmail.com

D. Hiran
e-mail: sigmapawan72@gmail.com

C. Sindhi
Nividous Software Solutions Pvt Ltd, Ahmedabad, India
e-mail: drcksindhi@gmail.com

M. Sandhi
Sankalchand Patel College of Engineering, Visnagar, India
e-mail: misandhi.mca@spcevng.ac.in

K. Kotecha et al. (eds.), *Data Science and Intelligent Applications*, Lecture Notes on Data Engineering and Communications Technologies 52, https://doi.org/10.1007/978-981-15-4474-3_32

types of services. This type of IT companies provides software as service, infrastructure as a service and platform as service. These types of services are taken by individuals as well as different organization or company. By using these types of services, total cost can be reduced. Cost of different hardware and software can be saved by individuals and organization [1]. Using this technology, industry can grow exponentially and so that it is accepted by most of the industry. As the size of this type of services is increasing, then it also increases the cost for management of those types of services. The demand of individual and different organization is not static, it can be dynamically changed, so this technology accommodates provisioning and de-provisioning on-demand. So that organization can avoid capital cost of different hardware and software. So it is the biggest challenge to maintain the load of nodes for maintaining the performance of nodes of service provider [2]. This type of problem is called load balancing problem. This is very important task to maintain when we use component parallels or on distributed systems.

In this work, we have tried to solve the problem of load balancing of distributed system of different hosts by analysing the response time of algorithm like Round Robin and similar others. We had also tried to find the alternative of these algorithms to minimize the response time of algorithm. Load balancing tries to balance the workload to different nodes of the distributed system. It checks that no node is overwhelmed or under loaded anytime. So that performance of the system can be increased. The proposed algorithm shows an improved response time compared to the some other algorithms. We arrange the rest work below. In Sect. 2, we mentioned the some of the work which are already done in this field which describe the methods of load balancing. In Sect. 3, it includes the methodology of proposed GA algorithm. In Sect. 4, we discussed the analysis of simulation result with the information about the cloud analyst simulation tool. Finally, Sect. 5 deals with conclusions and future scope.

2 Related Work

Much research has been conducted on load balancing procedures under distributed computing conditions [3]. The use of some existing scheduling technique such as Round Robin (RR), Equally Spread Current Execution Load (ESCEL) and Throttled (TA) for load balancing also exists in the literature. Sun et al. [3] have proposed a new load balancing policy, and they go in detail. They proposed a new model to regulate the circulation of information to advance the execution of distributed processing in serious applications, such as the mining of dispersed information. In this article, the flexible processing approach called genetic algorithm has been proposed. The cloud analyst visual simulator is used to analyse an algorithm. Several algorithms such as Round Robin (RR), Equally Spread Current Execution Load (ESCEL) and Throttled algorithm are compared with proposed genetic algorithm [4], Benlalia et al. have found that Equally Spread Current Execution Load algorithm will get better the performance of processing time and response time of data centre when implemented [5]. When different work is assigned to data centre using broker policy closest to data centre and Equally Spread technique is implemented on it, then there will be great improvement if comparison is made with the Round Robin scheduling technique in the data centres performance. Response time (RT) is measured to check the performance of different algorithms [6]. Parida et al. proposed Throttled Load Balancing algorithm provides the best response time, data centre processing time with small processing cost as compared to Round Robin and Equally Spread Current Execution algorithm [7]. Among different service broker policies, closest data centre is the best as it forwards the request to the closest data centre and thus results in smaller response time. In cloud computing, at worst, they need an exponential time that is not good enough. The other methods cannot assure sufficient answers for all items. Instead, these methods find a satisfying task to resolution the effort with high prospect. Aliya et al. say that in Equally Spread Current Execution algorithm, [8] a communication exists between the load balancer and the data centre controller for updating the index table foremost to an overhead. Further, this overhead causes delay in providing rejoinder to the arrived requests. Round Robin algorithm selects the load on random basis, [9], and therefore, condition arises as some nodes have to face heavy workload, and some nodes become free, i.e. with low workload.

3 Methodology

Although cloud computing is vigorous in view, this problem is formulated by assigning N task to M processor at a given time. There are some codes which will be used to implement the proposed algorithm for better understanding. We calculate processing unit vector (V_{pu}) for all processing unit. Each vector has the value which is the processing capacity. I_{ps} means no of instructions executed by processor per second, and it is measured in millions. If cloud provider is unable to provide service in time, then they need to pay some penalty to customer, and it is known as C_d, cost of late execution. N_{ij} is used to show number of instruction are in job. Symbol t is used for PaaS, SaaS and IaaS, which are the type of cloud service required to fulfil the request. C_{ei} is used to write instruction execution cost. T_{ja} is the symbol for job arrival time. V_{ju} is the unit vector for job, and T_{wc} is the completion time for worst case situation. w_1 and w_2 are used to show predefined weight.

$$V_{pu} = f(I_{ps}, C_{ei}, C_d) \tag{1}$$

$$\text{Likewise } V_{ju} = f\left(I_{ps}, T_{ja}, T_{wc}, t\right) \tag{2}$$

The following function (3) is an objective function which needs to minimize, and for that, CSP needs to distribute N jobs to all M processors.

$$Z = w_1 * C_{ei}, \left(N_{ij} * I_{ps}\right) + w_2 * T_{wc} \tag{3}$$

Here, value of w_1 and w_2 is difficult to decide, but the criterion could be that weight is as more as the factor is more general. Logic is the user's preference or significance for one factor especially over the other. Here, the optimization was performed on sets of data weights. Value of two different weights w_1 and w_2 is taken as 0.7 and 0.3, respectively. These weights are taken in such a way that its total becomes 1. Because of the more complexity of load balancing problem, it is very difficult to solve it by linear programming techniques. So, we have tried it to solve this problem with the help of soft computing techniques, called genetic algorithm. Using this technique, it becomes easy to solve such hard complex problem.

Genetic algorithm is AI technique which is based on natural selection and stochastic algorithm for the purpose of search and optimization problem. For large research space and heavy complexity, problems have been solved with the help of genetic algorithms. These are very useful, and also, it is stable to find optimal global solutions for such a type of problems. So, it is very useful to find the optimum solution of a problem of balancing a load of node in cloud environment because it is a problem of large research space. For load balancing problem, scheduling the tasks of workflow is important. So, the tasks are mapped with processing element in such a way that no node goes in overloaded or under loaded. So, when task is entered by cloud user, that task is performed on processing element based on assignment of processing element by load balancing algorithm like genetic or any other load balancing algorithm. Proposed genetic algorithm is explained in the following section.

3.1 Proposed Genetic Algorithm

The three main operations, such as selection, crossover and mutation, are implemented in simple GA. This algorithm gives the best results for complicated objective function as well as it gives the best result in local search problems. It can also handle the large research space and so that it can handle complex problem easily and gives the best results for these types of problems. Working of this algorithm is show below.

Selection: To represent a chromosome, each individual is converted into binary string. For that, all the possible solution in solution space is verified and converted into binary string. From these converted binary string, ten chromosomes are selected randomly for initial population.

Crossover: In this operation, new people or solution is created which is different than the people or solution is selected in previous phase. Using some fitness function, fitness value is calculated for each chromosome for selection of it for next procedure. Here, Eq. 3 is the fitness function which is used to find the fitness value of each chromosome. In crossover, different techniques are used like single point selection, two point selection and uniform selection. Here, single point selection is used for crossover. In this technique, chromosome is divided into two parts. First part is replaced by the second part of the second chromosome so that we can get the new chromosome and that is the different chromosome than firstly selected in previous methods.

Mutation: By this technique, some genes (bit) value is updated from 0 to 1 or from 1 to 0. So, the chromosome is updated by this process. It also used some probability of mutation. Here, very small value 0.1 is taken as a mutation probability. Mutation creates new pairs of chromosomes. The above-mentioned processor like selection, crossover and mutation process is repeated until optimal solution found or termination condition exists.

The proposed algorithm steps are as follows:
Step 1: Encode the population of processing unit into binary string.
Step 2: Generate the initial population randomly.
Step 3: Find the fitness value of each population using fitness function.
Step 4: Do the following sub-steps until final condition found or get the optimum solution 4(a): Select the chromosome with the lowest fitness value twice and skip the chromosome with the highest value. 4(b): Perform the single point crossover to get the new offspring. 4(c): Perform mutation operation to update the offspring with desired probability. 4(d): Put this new offspring for new population for next round. 4(e): Check the exit condition.
Step 5: Exit

4 Results and Analysis of Simulation

The proposed algorithm is implemented by cloud simulator named "Cloud Analyst".

4.1 Simulator—Cloud Analyst

Simulation is a model used to find the effect of large cloud environment. In cloud computing paradigm, it supports requirements like infrastructure, applications, programming environment, etc. Various simulators are available in world like Green Cloud, iCanCloud, Groudsim, DCsim, Cloud Analysts and Cloudsim [5, 10]. For this work, we have used cloud analyst because it is easy to understand and manage because it provides graphical interface. It is shown in the following Fig. 1a, b.

It is a GUI-based simulator, and we had made different experiment on it. To study and analyse the algorithm, we can configure simulation environment by setting some parameters [11]. Based on the parameter value, simulation result is generated and also shows in graphical form. We can also save the configuration in cloud analyst. In this simulation, it considers that total world is divided into total six regions; and it is the main six continents of the world. Total six "Userbases" are used from where user can send a request from main six region of the world. There is consideration of particular instance region for all the userbases, and there are number of users are registered online during peak hours. And also, we can write the numbers of user registered for off-peak hours. The following Table 1 demonstrates the information of the userbases which used for the testing. Each data centre which is used from

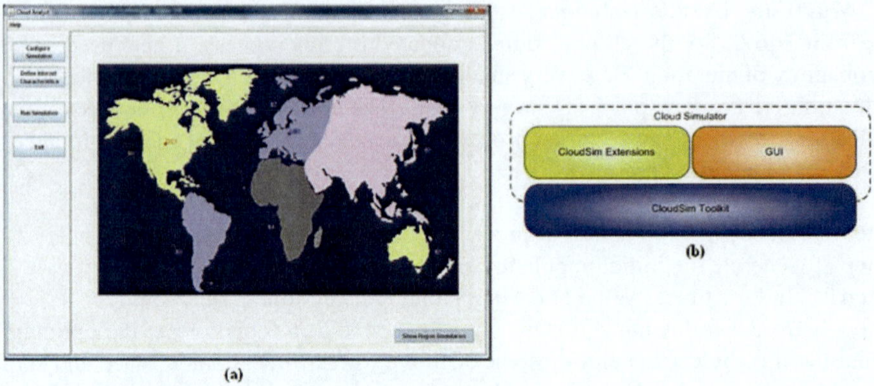

Fig. 1 **a** Cloud analyst with graphical user interface **b** cloud analyst architecture

Table 1 Userbase detail

S. No.	Userbase	Area/region	Peak users	Off-peak users
1.	1	0-North America	4100	850
2.	2	1-South America	5100	1250
3.	3	2-Europe	3400	750
4.	4	3-Asia	7800	1350
5.	5	4-Africa	1350	220
6.	6	5-Oceania	1450	455

simulation has fixed numbers of virtual machine. 4096 MB RAM and 100000 MB storage space has been configured for each virtual machine. Also, the CPU with execution speed of 10,000 MIPS is configured, and such a 4 CPU is configured in each machine.

4.2 Simulation Configuration

There can be different scenarios for testing through one centralized data centre in the cloud. So, every request received by any place in the world is processed by one data centre (DC) and that data centre has 25, 50, 75 virtual cloud configuration (CC) machines that is assigned to the request. This simulation setting is demonstrated in details in Table 2 which also shows the average response time (RT) in ms for different algorithm like ESCELA, RR, TLBA and GA. Performance analysis is shows in Fig. 1. In Table 3, it shows the combination of 25, 50 and 75 virtual machines for two data centres. Similarly, other performance analysis is mentioned by taking the different

Table 2 Simulation setting and overall average response time (in ms) for one data centre

Cloud configuration	1	2	3
	CC1	CC2	CC3
No of VMs	25	50	75
RT using GA	292.06	291.99	285.07
RT using ESCELA	292.06	292.05	292.09
RT using RR	292.09	292.07	292.09
RT using TLBA	292.07	292.13	292.07

Table 3 Simulation setting and overall average response time (in ms) for two data centres

Cloud configuration	1	2	3	4	5	6
	CC1	CC2	CC3	CC4	CC5	CC6
No of VMs	25	50	75	25, 50	25, 75	50, 75
RT using GA	284.12	284.98	285.05	285.02	284.99	285
RT using ESCELA	285.34	285.38	285.37	285.31	285.32	285.34
RT using RR	285.34	285.41	285.42	285.33	285.35	285.36
RT using TLBA	285.33	285.36	285.35	285.32	285.34	285.36

combination of cloud configuration with different number of used virtual machine for 3, 4, 5 and 6 data centres (Figs. 2, 3, 4, 5, 6 and 7; Tables 4, 5, 6 and 7).

Fig. 2 Performance analysis based on one data centre

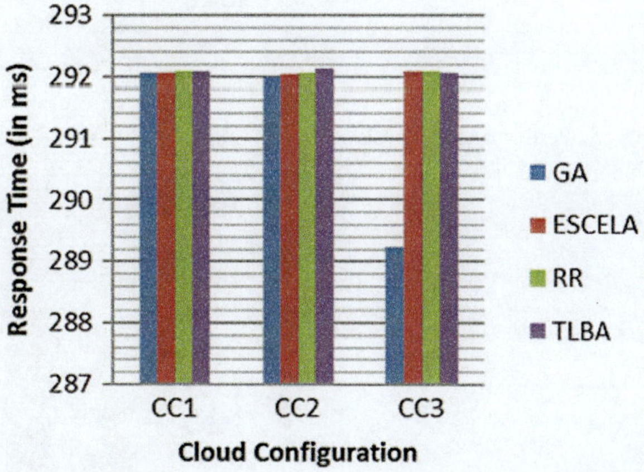

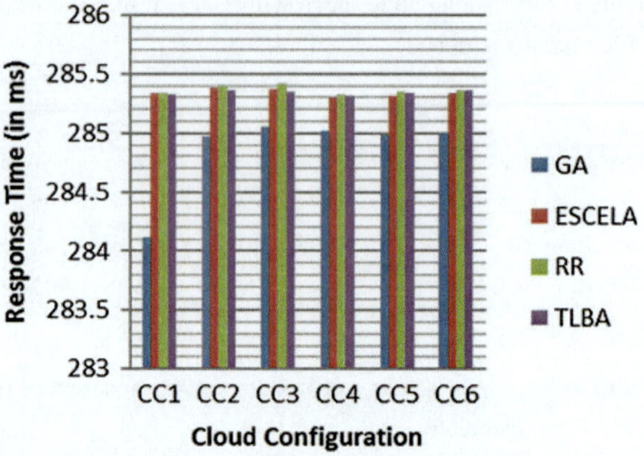

Fig. 3 Performance analysis based on two data centres

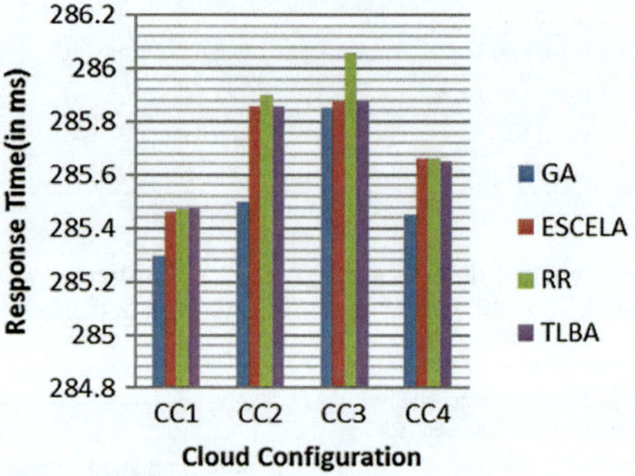

Fig. 4 Performance analysis based on three data centres

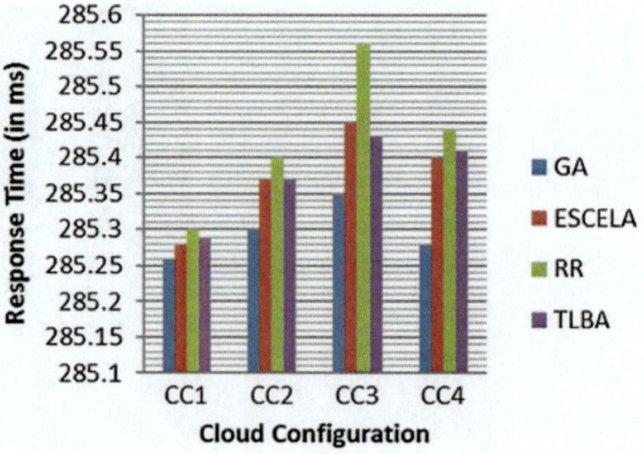

Fig. 5 Performance analysis based on four data centres

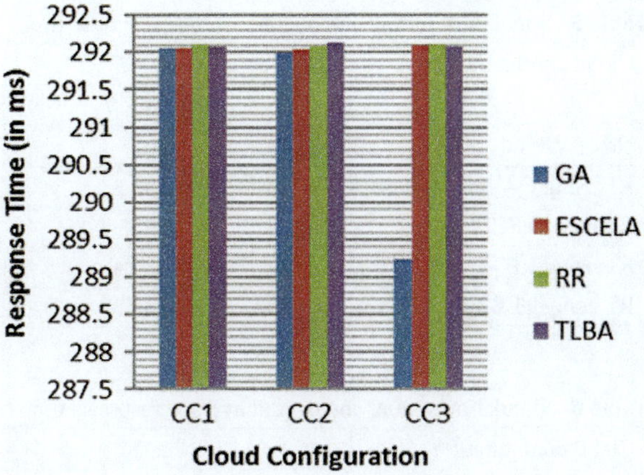

Fig. 6 Performance analysis based on five data centres

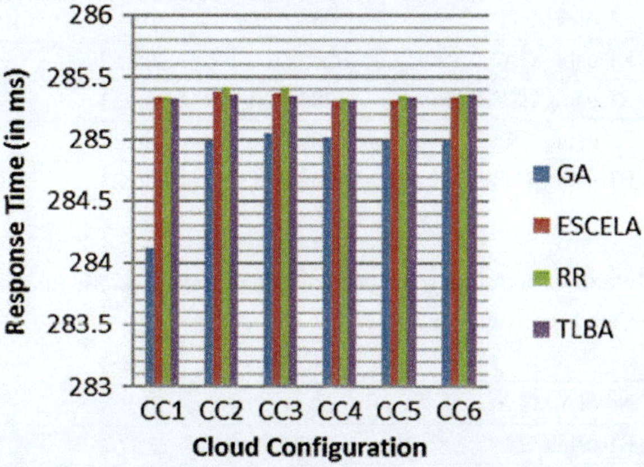

Fig. 7 Performance analysis based on six data centres

Table 4 Simulation setting and overall average response time (in ms) for three data centres

Cloud configuration	1	2	3	4
	CC1	CC2	CC3	CC4
No of VMs	25	50	75	25, 50, 75
RT using GA	285.3	285.5	285.85	285.45
RT using ESCELA	285.46	285.86	285.88	285.66
RT using RR	285.47	285.9	286.06	285.66
RT using TLBA	285.48	285.86	285.88	285.65

Table 5 Simulation setting and overall average response time (in ms) for four data centres

Cloud configuration	1	2	3	4
	CC1	CC2	CC3	CC4
No of VMs	25	50	75	25, 50, 75
RT using GA	285.26	285.3	285.35	285.28
RT using ESCELA	285.28	285.37	285.45	285.4
RT using RR	285.3	285.4	285.56	285.44
RT using TLBA	285.29	285.37	285.43	285.41

Table 6 Simulation setting and overall average response time (in ms) for five data centres

Cloud configuration	1	2	3	4
	CC1	CC2	CC3	CC4
No of VMs	25	50	75	25, 50, 75
RT using GA	285.56	285.59	285.6	285.59
RT using ESCELA	285.84	286.23	286.81	286.32
RT using RR	285.88	286.22	286.81	286.33
RT using TLBA	285.86	286.23	286.81	286.34

Table 7 Simulation setting and overall average response time (in ms) for six data centres

Cloud configuration	1	2	3	4
	CC1	CC2	CC3	CC4
No of VMs	25	50	75	25, 50, 75
RT using GA	285.51	285.67	285.58	285.54
RT using ESCELA	285.87	286.35	286.78	286.46
RT using RR	285.86	286.35	286.85	286.45
RT using TLBA	285.86	286.36	286.84	286.46

5 Conclusion

We proposed one genetic algorithm-based technique which is used for task scheduling. Task scheduling is done based on fitness function, and finally, optimization is achieved. So here, task scheduling is done in such a way that no node in cloud is overwhelmed or under loaded. So, the performance can be increased. Here, we compare our proposed genetic-based algorithm with different algorithms. To check the performance of algorithm, we only check the response time of each algorithm. To check it, we have configured the cloud analyst's simulation tool. By executing simulation, we can get the different results based on different configuration like numerous number of virtual machine like 25, 50, 75, etc. Finally, by observing this result, we conclude that proposed algorithm performs good results compared to some other algorithms. It also gives the assurance of quality to the customer for assigning job. Here, we perform this operation by assuming that all the jobs have the same priority, but it is not in the real situation. This may be accommodated in the job unit vector, and later, care should be taken for fitness function. Here, single crossover technique is used to find the different crossover, but we can use multiple crossover or some different techniques to improve the performance of this algorithm for future work.

References

1. Jadeja Y, Modi K (2012) Cloud computing-concepts, architecture and challenges. In: International conference on computing electronics and electrical technologies (ICCEET), pp 877–880
2. Dorigo M, Birattari M (2010) Ant colony optimization. Springer, US, pp 36–39
3. Sun J, Feng B, Xu W (2004) Particle swarm optimization with particles having quantum behaviour. In: Proceedings of the 2004 congress on evolutionary computation. vol 1, pp 325–331
4. Benlalia Z, Beni-hssane A, Abouelmehdi K, Ezati A (2019) A new service broker algorithm optimizing the cost and response time for cloud computing. Procedia Comput Sci 992–997
5. Tyagi N, Rana A, Kansal V (2019) Creating elasticity with enhanced weighted optimization load balancing algorithm in cloud computing. In: Amity international conference on artificial intelligence, pp 600–604
6. Swarnakar S, Raza Z, Bhattacharya S, Banerjee C (2018) A novel improved hybrid model for load balancing in cloud environment. In: 2018 Fourth international conference on research in computational intelligence and communication networks (ICRCICN), pp 18–22. IEEE
7. Parida S, Panchal B (2018) Review paper on throttled load balancing algorithm in cloud computing environment
8. Aliyu AN, Souley B (2019) Performance analysis of a hybrid approach to enhance load balancing in a heterogeneous cloud environment
9. Hirsch P (2019) Task scheduling using improved weighted round robin techniques
10. Alshammari D, Singer J, Storer T (2018) Performance evaluation of cloud computing simulation tools. In: 2018 IEEE 3rd International conference on cloud computing and big data analysis, pp 522–526
11. Rathore J, Keswani B, Rathore V (2019) Analysis of load balancing algorithms using cloud analyst. In: Emerging trends in expert applications and security, pp 291–298

Proactive Approach of Effective Placement of VM in Cloud Computing

Ashish Mehta, Swapnil Panchal, and Samrat V. O. Khanna

Abstract Virtual machine is one of the major areas of infrastructure as a service in cloud computing. The VM provision for service provider and user should be inexpensive. In last decades, number of scientists proposed various schemes. There is less opportunity for the service provider to use from the resource pooling. Cloud computing is providing hosting service over Internet, and to service provider, number of request of resource can be served using Internet. Recently, resource management is required to maintain autoscaling of resource and improving the efficiency of resource in cloud computing. There are various approaches available for workload predication that is based on single model approach. It is very critical to find the result on the basis of traditional model. Different methods and techniques were analyzed by us in order to identify the virtual machine allocation. We have defined a new dynamic resource allocation and policy-based improvement of the effective management of the resources. Our proposed implementation shows a better performance and improves the VM allocation with accuracy and less time consuming.

Keywords M allocation · VM placement · VM placement policy · CloudSim · OpenNebula

---

A. Mehta
Department of Computer Engineering, Indus University, Ahmedabad, Gujarat, India
e-mail: ashish.svit@gmail.com

S. Panchal (✉)
Gandhinagar Institute of Technology, Khatraj Chokdi, Gandhinagar, Ahmedabad, India
e-mail: swapnil.panchal@git.org.in

S. V. O. Khanna
Indus University, Ahmedabad, Gujarat, India

K. Kotecha et al. (eds.), *Data Science and Intelligent Applications*,
Lecture Notes on Data Engineering and Communications Technologies 52,
https://doi.org/10.1007/978-981-15-4474-3_33

1 Introduction

Cloud computing is the best model to pay as use. There are various amounts of resource availability back end that leads to the cloud computing recently. Data center is really important to manage all the business of various organizations. Data center provides platform to deploy user's application and provides the environment to manage the deployed application. Data center is managing key resource, and organization is a major issue to utilize data center resource. The virtual machine is hosted on single physical machine in cloud computing. Resource is provided to the user as virtual machines, and it is called virtual machine instances in infrastructure as a service.

The main key concept of cloud computing is virtualization. Large data center virtualizes their resource to improve resource utilization, performance, profit level and so on. Virtual machine is considered the individual entity to run users' various application as shown in Fig. 1.

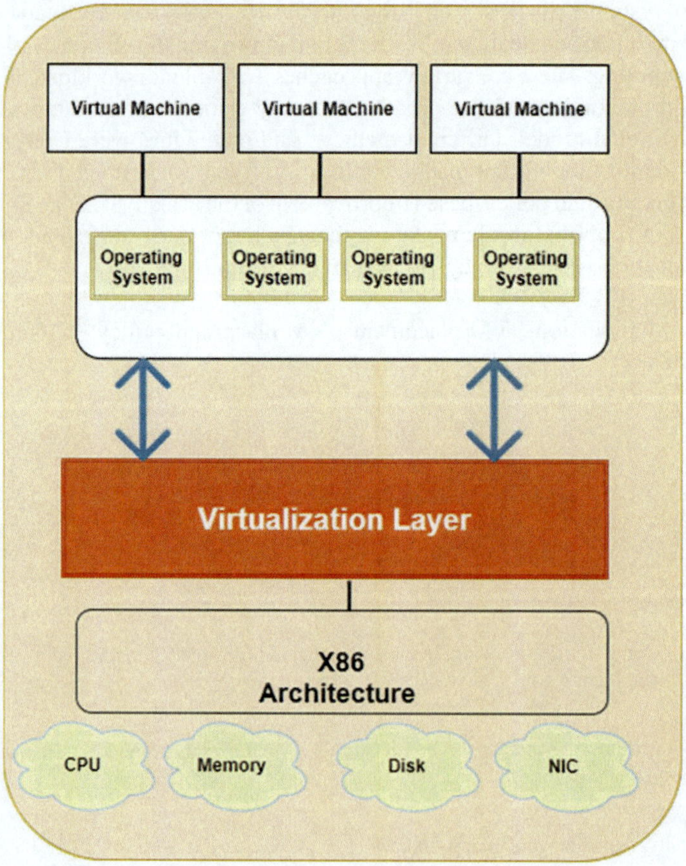

Fig. 1 Virtualization

The users can send request to the cloud service providers and based on the needs at initial stage create the virtual machine (VM). The dynamic resource is assigned in the server in which the virtual machine has been deployed. In which its create VM host server due to much overloaded, and then data center has to be installed more server in cloud. Hence, the current scenario generates more overhead for the users and cloud service providers including the performance and profit.

There are different types of services provided by cloud computing, but we are using various deployments of application using software as a service and suppose required infrastructure setup and virtualization along with virtual machine through infrastructure as a service [1]. There is a main advantage to utilize the services of scalability to divide the traffic horizontal and vertical and improve performance. There are different model and predication strategies utilized. The different classical models including various models like neural network (NN), SVM, Markov models and Bayesian models. [2–4].

There are various application operations to change the use of workload. For example, workload is associated with different online administration currently few ephemeral glib terns, for example periodic and busting [5, 6]. VM provisioning manages and deploys the user's application in infrastructure as a service.

1. VM is assigned according to the user's application requirement using the VM provisioning policy.
2. VM resource provisioning requirement of user is linked to the available resource, and using the scheduling available virtual machine is associated with physical machine.

The requirement of users and assigning the virtual machine to the needy user process is known as virtual machine provisioning. The virtual machine is dynamically available to the cloud provider and improves the performance in cloud computing. The given examples depend on the native user's qualities and essentially rely on the workload force. The basic of workload and the self comparisons have been seen into normal workload [7, 8]. The service is fetching the VM_i. During execution, VM_i will occupy some portion of physical memory at SS_i. In IAAS, the virtual machine is running individually on the physical machine, and it represents independent entity Table 1.

Table 1 Amazon EC2 defines VM instance type

VM instance type	VCPU	Memory (virtual)	Network type	Storage type
Medium	1	4	20	EBS
Large	2	8	20	EBS
X-Large	4	16	20	EBS
XX-Large	8	32	20	EBS
XXXX-Large	16	64	20	EBS

2 VM Allocation and VM Provisioning Based on Auction Model

The scheduler maintains the request for the VM instance [9]. The user is not receiving any extra incentive for the bid amount [10, 11]. Using truthful greedy and optimal mechanisms used for dynamic VM provisioning and VM Allocation for the resource, Amazon EC2 is considering multiple types of resources. The true valuation as requested bundle appreciates the users and is able to receive the incentives. The proposed model and analysis find out an optimal solution for VM allocation and VM provisioning dynamically to meet h market needs on time and create high values of cloud providers.

Service level objective will cause violation due to VM provisioning, which are connected with key area of punishment. It is required to maintain strategic distance with two main issues: (1) The predication of application resource requirement must be assumed fast pace of time so to change the resource allocation against requirement. (2) Selecting the amount of asset required to distribute is non-minor as long as application resource required recent changes with time [12, 13].

The best proposed model for dynamic VM resource allocation and VM provisioning fulfills the market needs on time and gives the high profit to the cloud providers. The proposed model is getting optimal solutions and also analyzed the use payment [9, 14].

The combinational auction-based mechanism was basically used for dynamically VM allocation and VM provisioning. Cloud service providers as well as users utilized dynamic resource allocation methods. On the basis of the bid request, the VM is getting granted [15]. An augmented shuffled frog leaping algorithm (ASFLA) is introduced and implemented for resource scheduling and provisioning in the IAAS cloud. The ASFLA performance is compared with shuffled frog leaping algorithm (SFLA) as well as PSO algorithm. The Java-based custom simulator has been used for possessing the efficiency of ASFLA as well as for different scientific work flow of assorted different size using Java-based custom simulator. The simulation effects can show improved performance as well as schedule time along with decrease execution cost. The practical analysis of SFLA (shuffled frog leaping algorithm), ASLA and PSO display the out performs and the other algorithm to reduced the overall execution values of assumed workflows [16].

There is actualization of cloud workload consideration module which is based on software as a service. Autoregressive integrated moving average (ARIMA) model is used to generate the accuracy for the workload predication model. The resource utilization and quality of service are major parameters to define the comparison impact of finished accuracy in cloud computing. The result of simulation shows average accuracy up to 91% [17].

In the combination of neural network and linear regression, they developed strategy to predication-based resource allocation and resource measurement. In this paper, the studies presented approaching for managing the resource request [18].

The improvement quality of service is the main required efforts. In cloud platform, the number of total efforts shows improving the quality of service [19].

Linear regression model is presented here based on workload predication approach and cloud architecture in cloud environment. The proposed mechanism of autoscaling utilized for scaling the virtual resource at various levels in cloud service as well as to combine both real-time scaling and prescaling [20]. The workload proposed approach is implemented in a way to divide the different workload in front of workload features and based on different predication model; the programming model is used for workload classification [21]. The ancient workload approach is used in a cluster based on new jobs. The predictive elastic resource scaling (PRESS) scheme approach is utilized for resource allocation in cloud services [22]. The PRESS scheme retrieves the dynamic similar patterns in software requirement and assigns automatic resource allocation [23]. The prediction system is monitoring real-time maximum usage of resource and autofeeds the values into various buffers according to the categories like resource type and tenure size [24].

Panneer et al. [25] Markov and Bayesian Modeling with seven hour Google cluster data proposed approach is mainly key objective to analyze the data for cloud various workload and the criteria of better performance mainly used for predication schemes. The new proposed approaches used for dynamic resource allocation have been proposed here along with different parameters. In traditional approach, there are some static parameters under the considerations such as CPU utilization, threshold value, workload for overloading conditions and a dynamic resource allocation schemes that overcome all over the issues [26].

CloudSim is implemented, and the proposed system shows that the proposed model can enhance the resource utilization and time. The cloud computing has mainly the strength for readability and scalability as well elasticity compared to the distributed computing system. It is dynamically allocated and manages the resource in the existing g environment. The algorithm is used to propose and to improve the performance of resource allocation and provision dynamically compared to the existing algorithm and generates better result.

The proposed scooter approach has been used for self management of cloud resources in cloud environment. The key resource management framework called scooter has been proposed and used to manage the self recovery for any fault. Scooter has few main properties like fault discovery, minimum execution cost, self optimizing the problem, self management, self protective and automatic detection with less human involvement [27].

There are various existing algorithms available for resource allocation and provisioning but the best, automatically, management of various resources in cloud has been served by scooter. The main key features of SCOOTER have reduced the involvement of human, and the core strengths are self proactive, utilize maximum resource usage and improve the efficiency(self optimize), to find maximum faults and try to overcome(self recovery), less execution time and cost, as well as service level agreement(SLA) [27].

There are number of real-time data of scooter that derive the practical approach of scooter that are better than any other autonomic resource management framework for quality of services like fault detection, intrusion detection rate, cost of execution, consumption of energy, throughput and amount of waiting time. Scooter schedules the resource automatically for better execution of heterogeneous workloads and maintains the user satisfaction which is achieved through service level agreement.

3 Proposed Algorithm for Resource Allocation

The proposed approach for VM allocation and provisioning is based on evaluating and analyzing workload along with the heterogeneous scalable data over the cloud computing. The architecture depicts workload production management and automatically schedules the resource according to request for solving the problem and provides the optimal solution.

The resource allocation architecture proposed approach is depicted in Fig. 2. Various components consist master slaves, clusters, cloud service providers and different load balancer.

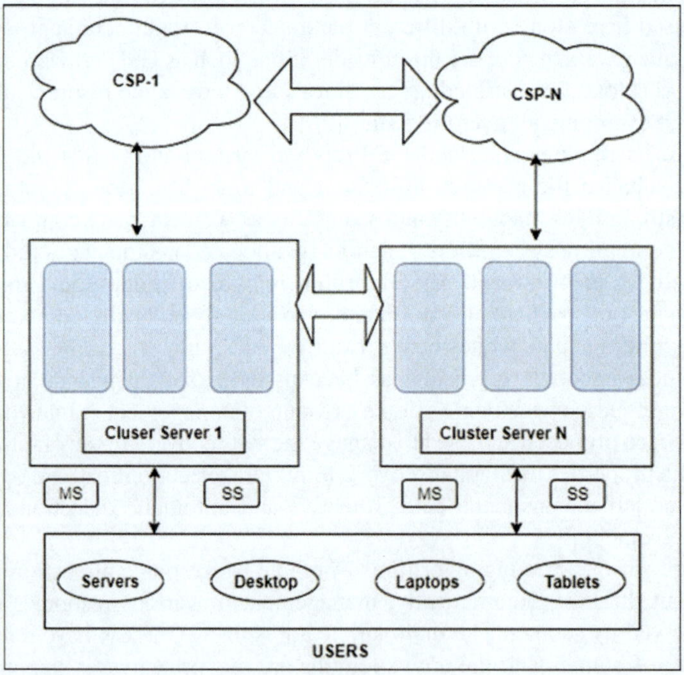

Fig. 2 Architecture of proactive resource provisioning

3.1 Proposed Architecture of Different Components

Users: In cloud computing, the main strength to compose the server and the physical dynamically available different resources are connected to each other. The user is assigned to define the high-performance machine to the service provider in server cluster. In the architecture Fig. 2, define the user is already exist in the cloud computing environment then request is to going to the most nearest cluster server and where sending request for virtual machine allocate to the slave server. The virtual machine is defined by the cloud service provider according to the user's needs in existing algorithm proposed load balancing of resource using master server. There are various parameters to measure the resource consumption. We will define the value of available load balancer in which allocated new virtual machine must be execute but there will be define the values of threshold and then decide to execute the allocated new virtual machine on which slave server.,

Master Server: The number of slave server is defined by master server. The master server is consisting with number of different component, but key part of load balance is checking the total consumption of resource, checking the threshold value and based on that load balancer will be finalizing on which server VM is executed on slaves server. In some other cases, if any server will be interrupt and there is not any free resource, then master will send the request to the most nearest server for the resource and will send request to cloud service provider.

Slave server: There are two main parts of cluster server which includes master server and number of different slave server in cloud computing. According to the user's need, cloud service provider dynamically allocates the *i*th resource to the server. Slave server and cloud service provider can assign the virtual machine randomly at run time, and virtual machine includes RAM, CPU, hard disk and other storage devices, which under the slave server created using cloud service provider.

Heterogeneous cloud service provider: The VM provisioning and any kind of resource to the users are provided by the cloud service provider. According to the users needs, cloud service provider creates the VM and also provides reference of unique ID to the master server and will decide which slave serve will execute the VM. Maintaining the resource is the key aspects of heterogeneous cloud service providers and is responsible for receiving request resource analysis and resource provisioning. In case when request is sent, most nearest service provider will allocate dynamically the resource. There are various roles performed by cloud service provider like opportunity to provide the physical resources, provide unique ID for VM, creating virtual machine, etc. Master server is controlled in all the virtual machine of the slave server. The heterogeneous cloud service provider also assigns the demanded resource if it is not available then find out from the nearest cloud service providers.

4 Proposed Algorithm

The existing algorithm and proposed proactive algorithm for allocation of resources in heterogeneous environment in cloud computing are mentioned below.

The main approach for proactive algorithm is utilized allocation of resource dynamically based on the user's needs and gives instant response for better improvement of performance to the user as well as direct allocation of resource. There are number of different existing algorithm compared to the proposed algorithm, and below listed parameters give better performance and better output. The listed below parameters in Table 2 are the following.

The proactive approach is representing to produce the maximum utilization of resource. Algorithm 1 represents input list of SS_i and MS_i and allocation of virtual machine on SS_i via cloud service provider as represented as output. The objective of Algorithm 1 is virtual machine allocation on SS_i using cloud service provider. The key objective is to provide resource utilization through SS_i and will send the request by MS_i to create the CSP for VM allocation. The CSP will create the VMi, generate the reference ID and send to the MS_i. If SS_i is already null, then MS_i will cross-check in the given list for free resource Ri(Z) on SS_i. If SS_i has required one-fourth part

Table 2 List of parameters

Shortened form	Detailed description
CSP_i	Cloud service provider which dynamically allocates virtual machine
CS_i	Both include number of cluster server that involve master server (MS_i) and slave server (SS_i)
MS_i	Both include number of master server that manage the list for slave server (SS_i)
SSi	Number of slave server involve for virtual machine (VM_i)
VM_i	Number of virtual machine for resource allocation
R_i	Number of resources like CPU, primary storage and other storage (R1, R2, R3, ... Rn)
Max_Cap	Represent maximum capacity
Ref	Represent the reference
Avl	Represent for resource available
Req	Represent demand for resource
Utl	Represent consumption of resource
Free R_i(C)	Represent virtual machine for free resource
Exe	Represent execution
Free R_i(Z)	Represent slave server free resource
N_Req_i	Represent new request for site
SVM_i	Represent sub virtual machine

of resource, then MS_i will allocate the resource Ref (VM_i) to SS_i and send to the required user though Ref (VM_i) to SS_i.

Algorithm 1: Allocation of VM_i on slave servers (SS_i) using cloud service provider CSP_i

Input: List for MS_i and SS_i
Output: increase efficiency of VM_i allocation
Step 1. For (i = 1, 2, 3, 4, 5.....)
Step 2. For each slave server ($SS_{i)}$ in cluster server (CS_i) do
Step 3. For each virtual machine(VM_i) in slave server (SS_i) do
Step 4. For each one slave server (SS_i) list on MS_i do
Step 5. Suppose If all slave server (SS_i) = zero or resources (R_i) at $N\_Req_i > VM_i$ then
i. MS_i request sends N_Req to CSP_i
ii. If for R_i(N_Req) on CSP_i for i = 0, 1, 2, 3, 4, 5.......n [End if statement]
Step 6. CSP_i generate the VM_i and request send back Ref(VM_i) to MS_i
a. Suppose If all slave server ($SS_{i)}$ = zero or empty then
i. Master server(MSi) dynamically allocate the Ref(VM_i) to SS_i
ii. Update the list of VM on SS
iii. Then first Forward the Ref(VM_i) to user else
b. MS check Free R (Z) on SS
c. Suppose If Free 4(R_i) (Ref(VM_i)) < R_i(Z)(SS_i) then
i. MS_i dynamically allocate the Ref (VM_i) to SS_i
ii. Update list of VM_i on SS_i
iii. First Forward the Ref (VM_i) to user.
Step 7. Return.

If the process of VM allocation is done through cloud service provider, then it is taking too much time. Here the proposed Algorithm 1 is representing the allocation of VM_i via MS_i, making a SVM_i on SS_i. If there are number of requesting for resource N_Req_i, then it will check the availability of resource using MS_i on free resource $R_i(C)$, and it must define N_Req_i <=Free $R_i(C)$. If the condition is satisfied, then SVM_i will be generated by MS_I and actually created reference is sent to cloud service provider Ref(SVM_i) to $CSP_{i,}$ otherwise the free resource Free $R_i(C)$ will check by MS_i and Free R_i request for reaming resource VM_{i+1}. If this request for free resource Free $R_i(C)$ is at ($N\_Req_i <= VM_i + VM_{i+1}$), then SVM_i will be created by MS_I after aggregating the free resource for virtual machine Free $R_i(C)$ at VM_i and VM_{i+1}. Hopefully, they need to check the free resource for VM Free R_i of VM_i (Free $R_i(C)$) and meet the rest of resource requirement N_Req_i using SVM_i but for resource allocation SVM_i there must be need one-fourth (R_i) of SVM_i for free resource.

The proposed Algorithm 1 and Algorithm 2 prove the efficiency and high performance during the resource allocation and maximize the utilization of resource.

Algorithm 2: Allocation of Virtual machine on SS_i Via MS_i

Input: List for MSi and SSi
Output: Improve the efficiency for VM_i allocation

Step 1. Repeat for {i = 0, 1, 2, 3, 4, 5, 6, …….n}
Step 2. For each SS_i in CS_i do
Step 3. For each VM_i in SS_i do
Step 4. For each list of SS_i on MS_i do
Step 5. For each N_Req_i do
Step 6. First MS_i check the condition for Free $R_i(C)$ for N_Req_i exe
If Free $N\_ Req_i$ <= Ri(C) then
MS_i will generate new SVM_i
Forward for Ref(SVM_i) to CSPi.
else
MS_i will check the condition for Free resource $R_i(C)$ at VM_i and Free resource for Ri(C) at $VM_i + 1$ for rest of remaining resources of $N\_ Req_i$.
If $N\_ Req_i$ <= (Free $R_i(C)$ at VM_i) + (Free Ri(C) at $VM_i + 1$) then
MS_i will generate new SVM_i after combining the Free $R_i(C)$ at VM_i and $VM_i +$ 1.
MS_i check for the Free $R_{i(}Z)$ on (SSi) for SVM_i resource alloc0ation
If Free 1\4(R_i)(SVM_i) <=R_i(Z)(SS_i)
then
MS_i resource allocates to SVM_i to SS_i
MS_i send request the Ref(SVM_i) to CSP_i
Update list SVMi on SSi
Forward Ref(SVMi) to user.
else
[Execute Algorithm-1]
[End of if Statement]
[End of if Statement]
[End of if Statement]
Step 7 Return

5 Results and Comparative Analysis

5.1 Cloud Simulation Tool (CloudSim)

We have proposed CloudSim toolkit architecture and used latest version to implement the proactive algorithm approach to execute all the tasks along with windows upgraded version. The computing resource will be provided on the basis of needs and their management. Heterogeneous environment can be created to simulate the workload in computing environment. The cloud simulator obtains the result from number of different algorithm such as scooter and considers different parameter such as minimum cost, reduced execution time and reduce utilization of resource. There are various use case in which the different size with number of VM. On the basis of VM, we have created workload in cloud computing.

Table 3 Result of Algorithm 1 execution

	Cloudlet identity	Exec time (s)	Exec cost (C$)	Wait. time (s)	Resc. consumption (utili) n (%)
Proposed approach	Cloud let zero	460.67	122.2	6.3	42
	Cloud let one	500.2	165.37	5	30.79
	Cloud let two	480.6	180.29	4.45	50.12
	Cloud let three	497.9	186.78	4.51	41.1
	Cloud let four	475.35	200.89	4.12	31.94
	Cloud let five	1090.9	458.9	5.84	46.66

5.2 *Cloud Simulation Result*

Comparison analysis of the proposed algorithm better than existing algorithm based on four parameters and the cloudlet are deploying on number of virtual machine which is the host on server. Firstly, the cloudlet is executing from server, and the based parameters are concluding the following result 460.67, 500.2, 480.6, 497.9, 475.35, 1090.9 s as shown in Table 3. The existing scooter algorithm gives comparatively slightly less result than proposed algorithm. In the execution of cloudlet on different virtual machine, the execution cost of server is measured, and we are getting the total execution cost as 122.2, 165.37, 180.29, 186.78, 200.89, 458.9 and it is bit less compared to the scooter algorithm. In the other scenario, the cloudlet has been executing on host from server, and execution time is slightly less compared to existing algorithm. The cloudlet has execution time on different host is getting less from server 6.3, 5, 4.45, 4.51, 4.12 and 5.84 s [14].

For the comparison from the existing algorithm of scooter, it is taking less time, i.e., 7.526, 6.13, 5.265, 4.96, 4.42, 6.62, executing during cloudlet execution from server. In the fourth number of cloudlet, the execution on different host during execution time individual from server is 42, 30.79, 50.12, 41.1, 31.94, 46.66% of resource utilization. It is less resource utilization. The proposed algorithm is giving less resource utilization compared to scooter algorithm such as 82.466, 84.569, 85.941, 86.792, 88.41 and 89.68. We can derive that there is no more difference, the result is slightly good compared to the existing algorithm result, and our Algorithm 1 utilization of resource is less compared to the existing algorithm.

The cloudlet is executing on different number of virtual machine on host in which it is allocated virtually to the virtual machine. The manner in which the cloudlet is requesting for the VM allocation then its check the available free resource and provided to the allocation of resource and improved the optimization in Algorithm-1. We have implemented the result, and it is bit better with additional parameters but less utilization of the resource compared to the existing algorithm. In Table 4,

Table 4 Result of Algorithm 2 execution

	Cloudlet identity	Exec time (s)	Exec cost (C$)	Waiting time (s)	Resc. consumption (utili) n (%)
Proposed approach	Cloud let six	495.16	59.72	4.99	78.41
	Cloudlet seven	461.66	90.39	3.92	82.15
	Cloudlet eight	1188.63	255.61	8.14	79.77
	Cloudlet nine	1250.7	214.55	11.22	75.34
	Cloudlet ten	1230.23	306.37	11.7	90.57

compared to existing algorithm, the proposed Algorithm 2 is representing the better performance [14].

The main focus is utilization of resource as compared to the scooter algorithms in Table 4. The utilization of resource using cloudlet on various host execution time individual in server is 78.41, 82.15, 79.77, 75.34 and 90.57%. It provides slight bit compare with the existing algorithm scooter, i.e., 80.554, 81.854, 82.65, 80.33 and 81.56. The derivation and comparative analysis result are from Fig. 3 that represents enhancement, and Fig. 4 also shows better result. So, we are considering the Algorithm 1 for better results implementation and same for the Algorithm 2. The proactive approach on resource allocation gives better result compared to the existing algorithm, and comparative analysis gives better result to the existing.

The analysis represented in Algorithm 2 derived the enhancement in the existing algorithm and hence it shows better results in Fig. 3. The comparison of proactive algorithm shows enhanced result compared to the existing scooter algorithm.

The comparative analysis of Figs. 3 and 4 is always giving the better on to those parameters compared to the scooter algorithm. The resource utilization is a very key concern for any allocation policy and defines the better performance as shown in the result.

Fig. 3 Comparative analysis on Algorithm 1

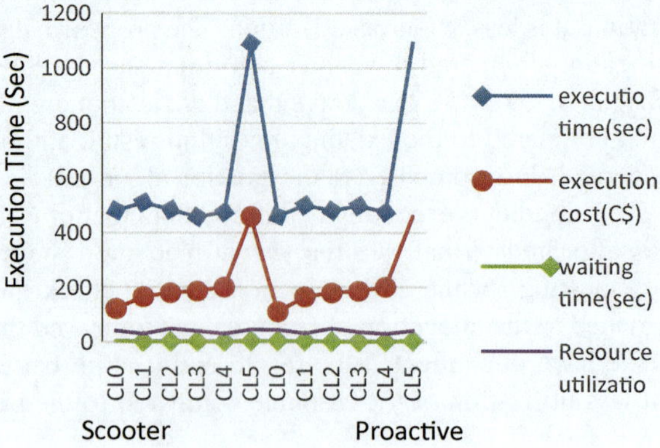

Fig. 4 Result of comparative analysis of Algorithm 2

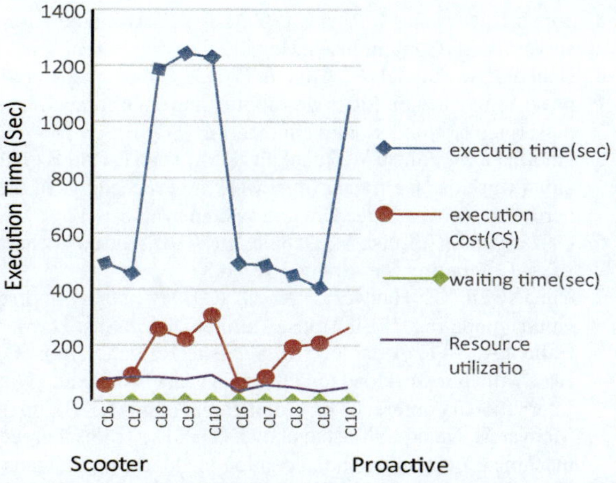

6 Conclusion

The proactive approach has been proposed for resource utilization in this paper. There are various optimization schemes have been used for heterogeneous resource allocation in cloud environment for fast response for VM provisioning and maximum utilization of resource. In data center, the resources allocation policies, there was very decreasing optimization of resource and reduced performance and efficiency. In the proposed algorithm, the dynamic allocation of resource is not affecting and violating the service level agreement as well as quality of service in this paper. The proposed approach is shows better results for direct VM allocation using VM serve site and its give enhanced utilization of resources. In analysis section, the proposed proactive approach has some limitations, if user is requesting for their remaining resource, then master server at cluster will be additional to find the free resource which will increase in time for waiting, and in such scenarios if the master server is not receiving any request of free resource, then it will directly send to the cloud service provider and the allocation of resource will be done by cloud service provider. There will be a possibility to resolve the waiting time of mentioned scenarios and will be getting better result to overcome the existing unresolved problems.

References

1. Fox A, Griffith R, Anthony J, Randy K, Andrew K, Gunho L, Patterson D, Ariel R, Ion S (2009) Above the clouds: A Berkeley view of cloud computing. Department Electrical Engineering and Computer Sciences, University of California, Berkeley, Rep. UCB/EECS 28, no. 13 (2009)
2. Chen Z, Yuanchang Z, Yanqiang D, Shaochong F (2015) Self-adaptive prediction of cloud resource demands using ensemble model and subtractive fuzzy clustering based fuzzy neural network. Comput Intell Neurosci

3. Sapan K, Nicholas I, Ravi S (2009) Time series prediction using support vector machines: a survey. IEEE Comput Intell Mag 4(2)
4. Bankole A, Samuel A, Ajila A (2013) Cloud client prediction models for cloud resource provisioning in a multitier web application environment. In: IEEE 7th International symposium on service oriented system engineering (SOSE), pp 156–161. IEEE
5. Ghorbani, M, Yanzhi W, Yuankun X, Massoud P, Paul B (2014) Prediction and control of bursty cloud workloads: a fractal framework. In Proceedings of the 2014 international conference on hardware/software codesign and system synthesis 12
6. Calzarossa MC, Luisa M, Daniele T (2016) Workload characterization: a survey revisited. In: ACM Computing Surveys. vol 48 no 3
7. Yin JX, Lu XZ, Hanwei C, Xue L (2015) A bursty and self-similar workload generator for cloud computing. IEEE Transa Parallel Distrib Syst 26(3)
8. Eldin AA, Ali R, Amardeep M, Stanislav R, Sara S, Luna O, Seleznjev, Johan T, Erik E (2014) How will your workload look like in 6 years? Analyzing wikimedia's workload. In: 2014 IEEE International conference on cloud engineering (IC2E), pp 349–354
9. Mahyar M, Nejad L, Mashayekhy DG (2015) Truthful greedy mechanisms for dynamic virtual machine provisioning and allocation in clouds, IEEE Trans Parallel Distrib Syst 26(2)
10. Nagadevi S, Kasmir S (2019) Virtual machine provisioning and allocation in a cloud environment using improved auction based model. Int J Innov Technol Exploring Eng (IJITEE) 8(7S). ISSN: 2278–3075
11. Roy N, Abhishek D, Aniruddha G (2011) Efficient autoscaling in the cloud using predictive models for workload forecasting. In: 2011 IEEE international conference on cloud computing (CLOUD), pp 500–507. IEEE
12. Calzarossa MC, Daniele T (2015) Modeling and predicting temporal patterns of web content changes. J Netw Comput Appl 56:115–123
13. Wang K, Minghong L, Florin C, Adam W, Chuang L (2015) Characterizing the impact of the workload on the value of dynamic resizing in data centers. Perform Eval 85
14. Ankita J, Arun Y, Brijesh C (2019) A proactive approach for resource provisioning in cloud computing. Int J Recent Technol Eng (IJRTE) 7(5S3). ISSN: 2277-3878
15. Sharrukh Z, Daniel G (2013) A combinatorial auction-based mechanism for dynamic virtual machine provisioning and allocation in clouds. IEEE Trans Cloud Comput 1(2)
16. Kaur P, Shikha M (2017) Resource provisioning and work flow scheduling in clouds using augmented shuffled frog leaping algorithm. J Parallel Distrib Comput 101:41–50
17. Calheiros RN, Enayat M, Rajiv R, Rajkumar B (2015) Workload prediction using ARIMA model and its impact on cloud applications' QoS. IEEE Trans Cloud Comput 3(4):449–458
18. Islam S, Jacky K, Kevin L, Anna L (2012) Empirical prediction models for adaptive resource provisioning in the cloud. Future Gene Comput Syst 28(1):155–162
19. Jiang Y, Chang P, Tao L, Rong C (2013) Cloud analytics for capacity planning and instant VM provisioning. IEEE Trans Netw Serv Manage 10(3):312–325
20. Yang J, Chuanchang L, Yanlei S, Cheng B, Zexiang M, Chunhong L, Lisha N, Junliang C (2014) A cost-aware auto-scaling approach using the workload prediction in service clouds. Inf Syst Front 16(1):7–18
21. Liu C, Yanlei S, Li D, Shiping C, Chuanchang L, Junliang C (2015) Optimizing workload category for adaptive workload prediction in service clouds. In: International conference on service oriented computing, pp 87–104. Springer, Berlin, Heidelberg
22. Patel JK, Vasu J, I-Ling Y, Farokh B, Jie X, Peter G (2015) Workload estimation for improving resource management decisions in the cloud. In: IEEE twelfth international symposium on autonomous decentralized systems (ISADS), pp 25–32. IEEE
23. Gong Z, Xiaohui G, John W (2010) Press: predictive elastic resource scaling for cloud systems. In 2010 International conference on network and service management (CNSM), pp 9–16
24. Zia Ullah Q, Shahzad H, Gul M (2017) Adaptive resource utilization prediction system for infrastructure as a service cloud. In: Computational intelligence and neuroscience
25. Panneer S, Liu L, Antonopoulos N, Bo Y (2014) Workload analysis for the scope of user demand prediction model evaluations in cloud environments. In: 2014 IEEE/ACM 7th International conference on utility and cloud computing (UCC), pp 883–889. IEEE

26. Seth S, Nipur, S (2017) Dynamic threshold based dynamic resource allocation using multiple VM migration for cloud computing systems. In: International conference on information, communication and computing technology, pp 106–116. Springer, Singapore
27. Gill SS, Rajkumar B (2018) Resource provisioning based scheduling framework for execution of heterogeneous and clustered workloads in clouds: from fundamental to autonomic offering. J Grid Comput 1–33
28. Caron E, Frédéric D, Adrian M (2010) Forecasting for cloud computing on-demand resources based on pattern matching. Ph.D. diss., Inria

Blockchain-Based Intelligent Transportation System with Priority Scheduling

Nakrani Dhruvinkumar Janakbhai, Maru Jalay Saurin, and Minal Patel

Abstract Intelligent transportation system (ITS) extracts efficiently and reliably and ensures security of information exchange between information infrastructures residing whether on roadside or vehicles on road by incorporating with blockchain. Also ITS includes vehicular ad hoc network (VANET), roadside units (RSU), priority scheduling, blockchain and some other vital components, for combining the according to characteristics, provides information exchange upon blockchain technology, and will give the best ITS. Basic requirements can be sent emergency, life-saving and legal information which may have nationwide critical data. To achieve the requirements for the transportation system, priority-based scheduling protocol and algorithm will be merged with efficient and reliable blockchain-based transportation which will reduce the network traffic and will not affect the service ratio and reliability of transmission channel.

Keywords Blockchain · Vehicular ad hoc network · Intelligent transportation system · Priority scheduling

1 Introduction

ITS is the major objective here with the help of various components mentioned above as blockchain, VANET and priority scheduling, and various analysis based on the used the technologies and tools for implementations of the particular systems.

N. D. Janakbhai (✉) · M. J. Saurin · M. Patel
Department of Computer Engineering, Gandhinagar Institute of Technology, Gandhinagar, Gujarat, India
e-mail: dhruvin.nakrani.99@gmail.com

M. J. Saurin
e-mail: jalay.maru@gmail.com

M. Patel
e-mail: mppatel.adit@gmail.com

K. Kotecha et al. (eds.), *Data Science and Intelligent Applications*, Lecture Notes on Data Engineering and Communications Technologies 52,
https://doi.org/10.1007/978-981-15-4474-3_34

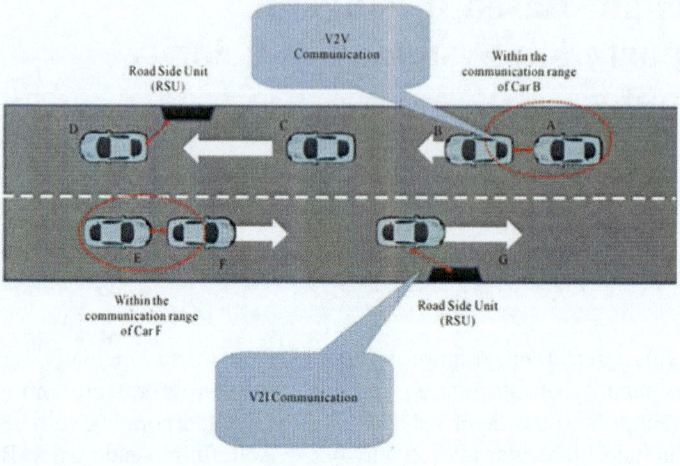

Fig. 1 Generic image embracing RSU, V2V and V2I

Vehicle-to-vehicle (V2V) and vehicle-to-infrastructures (V2I) [1] are classified elements for transmission of messages whether related to traffic guide, vehicle condition guide or emergency notifications [2]. In such cases, communications are done between close vehicles and nearby RSUs (infrastructure). As shown in Fig. 1, VANET as a part of ITS also supplies some other important utilities like Global Positioning System (GPS), traffic observations, managing the traffic and other applications [3].

By using the above available routing protocols, new class of application is used in ITS that requires time critical and very high data transfer rate which is enabled by VANET [4].

Due to the fact that security is main concern in vehicular communication systems (VCSs), blockchain technology [5] is used in providing solution for removing the central control and management entity and inserting the concept of transmission between individual nodes in the vehicular network infrastructure or network system. In general, it can be defined as a synchronous and distributed ledger which is ultimately data blocks. Block saves records of user information, link receipting previous block and some header information which will be combined in the sparse network of information structure of transportation. With this implementation in VANET using priority scheduling algorithm, information exchange between security domains or co-existing in same domain can be then accelerated as the messages will directly be sent to the decided destination node rather then routing through intermediate nodes which actually enhances weakness in whole communication network. So, the distributed network of blockchain performs better robustness under singular point of failure.

Scheduling and protocols are used in order to get control over transmission of information rates and limitations. One of the efficient protocol used with VANET is DOVE [3]: data will be disseminated to a specific number of receivers in network.

Number of receivers can be controlled accurately; also, small communication overhead is incurred with very low dissemination delay. One distinct scheduling technique used is SCORE: data scheduling in VANETs [2] which provides different time slots. In one particular time slot, determined user is assumed to be connected. Once, one is connected with its given time slot, information exchange begins with the help of other components of ITS. There are various algorithms available for scheduling like Most Requests First (MRF), Longest Total Stretch First (LTSF), First Deadline First (FDF), Maximum Quality Increment First (MQIF), Least Selected First (LSF), etc. [6].

2 Proposed Structure and Flowchart Model for Priority Scheduling in VANET with Blockchain-Distributed Network

As shown in above figure, in which blockchain-based structure of how communication is done in real-time, unlike the traditional structure, Internet serves here indirectly network managing services by allowing dynamical transmission of data. V2V and V2I kinds of data transmission are done with reliability, security and at higher transfer rate. Various components provide functionality as implemented before. One central entity can be implemented accessible to all the authenticated users for isolated environment which gives the facility to store and produce vehicle cryptographic information or materials like vehicle identities, repositories of issuing it, database of licence of users on road, pseudonyms and its certificates, etc. These materials should be kept with higher privacy and security for particular network-related operations. These whole data are generally accessed under three situations as (I) initial registration, (II) change the identity-related information and (III) adversary revocation. When new vehicles leave the manufacturer, its registration should be done, and all the information should be kept securely. Also, in case, if vehicle's identity changes, the according to data also should be changed. Blockchain also maintains data about malicious acts of users which is recognized by using blockchain look up. After confirmation of malicious activity, identity of adversary is made public (Fig. 2).

The model proposed here concerned to send the messages from the RSU to the vehicle according to priority set earlier for various kinds. RSU buffer holds priority queue, and the count number represents the buffer in which messages are stored. The more the count number, the higher the priority is. These messages will be sent to vehicles before any other block of information for gaining better service ratio. Both the flowcharts represent how the general flow of proposal will be maintained. In both the cases, either training mode or running mode, blockchain is set to be used by ITS also with scheduling techniques.

As from the figure, in training mode, parameters are set firstly, and the RSU with help of blockchain functionality is responsible for keeping track of the vehicle, interval time and speed of the vehicle, by which, packets will be distributed among

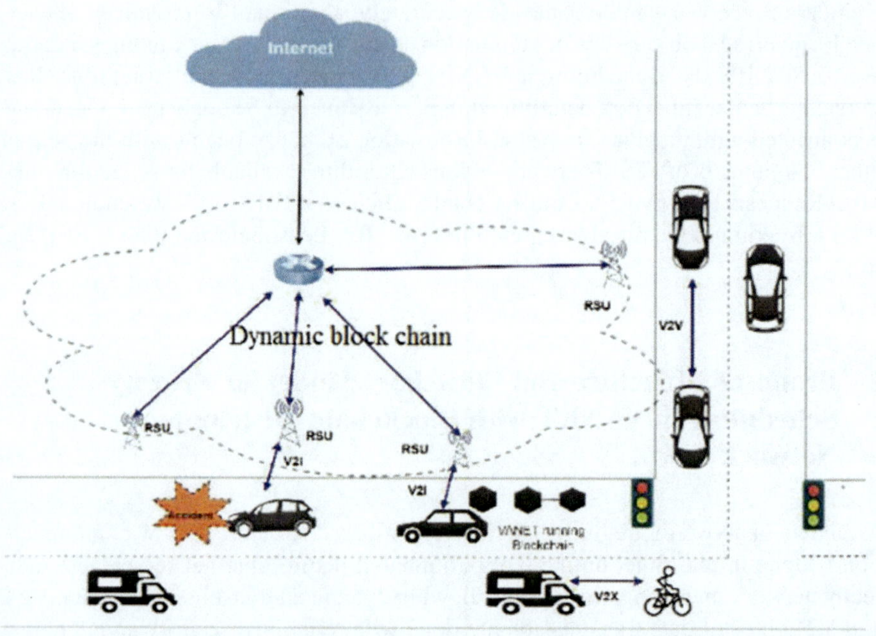

Fig. 2 Proposed structural image of the thesis

same network domain. RSU will only cache the data temporarily on blockchain nodes if the mobility movements of the vehicle are regular, otherwise it will avoid the caching. Hello packets are received at OBU on vehicle sent by RSU for performing registration process. When training mode is expired, the RSU will be changed to running mode process model (Figs. 3 and 4).

When RSU is switched to running mode, it will then be broadcasting the messages to the vehicles in blockchain manner. If On Board Unit (OBU) gets duplicate messages, it will inform RSU about duplication for maintaining reliability. Running mode is switched to training mode once it gets completed. The number of messages received in specific time is determined while training.

3 Proof-of-Work: Algorithm

1. Initialize RSU in VANET with registration data of vehicles
2. VANET is initiated with priority scheduling technique
3. While vehicle is in the VANET, it does step 4–8 for each vehicle
4. RSU retrieves process time and vehicle velocity
5. RSU assumes vehicle leaving time from VANET
6. Gives time slot as defined priority for communication
7. Vehicles get connected with RSU

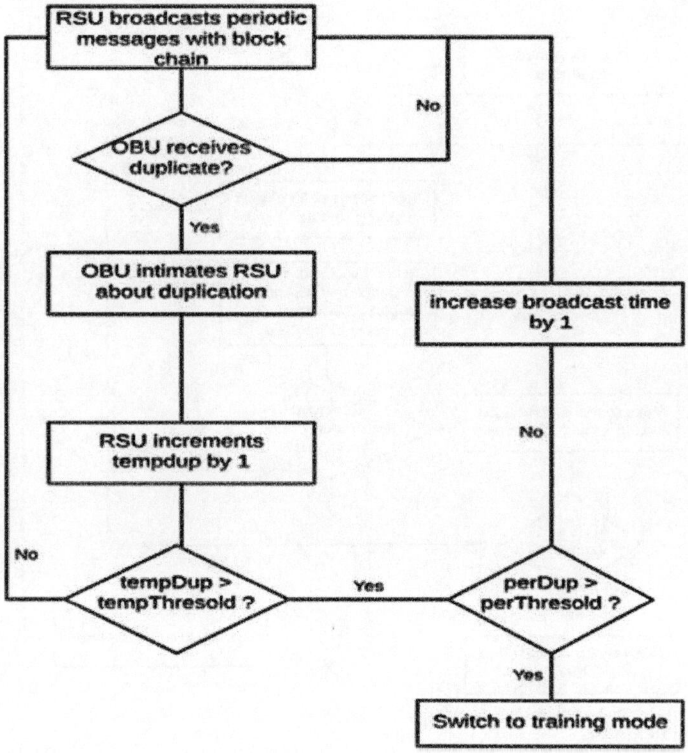

Fig. 3 Training mode

8. Blockchain-based information exchange is continued till interval time
9. Vehicles connection is handovered using block chain when they leave one network domain and enters another network domain.

Proposed model of this thesis is described with proposed algorithm. Initialization of RSUs and VANET with priority scheduling technique is done earlier. During the time of actual communication, RSU will communicate with vehicles on road based on priority scheduling which also recognizes communication process time and vehicle velocity for determining vehicle's network leaving time. Then, the higher priority vehicle will be processed first than other. The technique gives time slot for various classified vehicles based on their priority, and information exchange is done with decentralized blockchain data dissemination. Vehicles will never be disconnected while residing in the VANET blockchain-based distributed network, rather it will get handover when they leave one and enter another network domain. So, the security and privacy are maintained using blockchain, and performance is enhanced by priority scheduling.

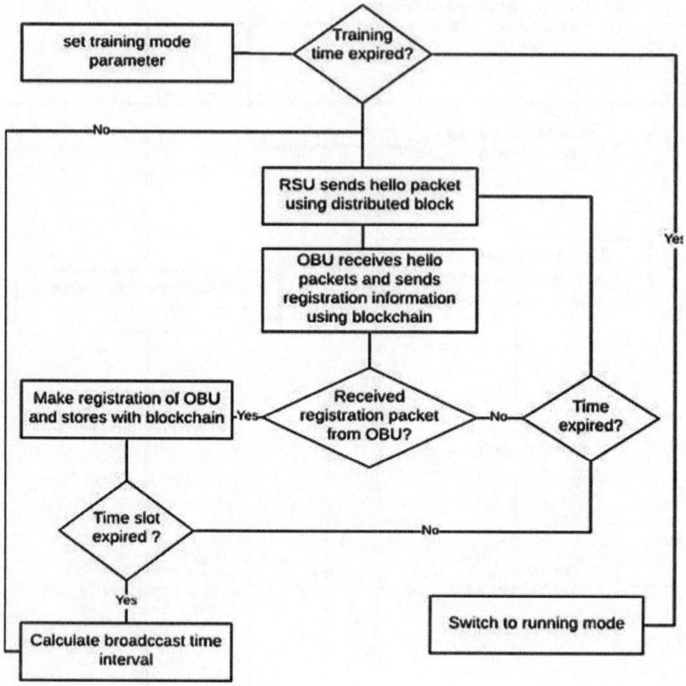

Fig. 4 Running mode

4 Conclusion

If the implementation of the proposed thesis model is performed and the given functionalities are tested, these may enhance the overall performance of the current V2I communication in VANET which is distributed based on blockchain and performs priority scheduling technique named SCORE. Whenever the vehicles' connections are established, information exchanged within this decentralized network will be under high security and privacy to corresponding users. Also, vehicle users will be monitored by this proposed VANET model.

References

1. Decker C, Wattenhofer R (2013) Information propagation in the bitcoin network. In: IEEE P2P 2013 Proceedings, pp 1–10
2. Kumar V, Chand N (2010) Data scheduling in VANETs: a review. Int J Comput Sci Commun 1(2):399–403
3. Yan T, Zhang W, Wang G (2012) DOVE: data dissemination to a fixed number of receivers in VANET. In: 2012 9th Annual IEEE communications scocirty conference on sensor, mesh and ad-hoc communication and networks (SECON) (2012)

4. Mershad K, Artail H (2012) SCORE: data scheduling at roadside units in vehicle ad hoc networks. In: 19th International conference on telecommunications (ICT 2012), IEEE. ISBN: 978-1-4673-0747-5
5. Nithya P, Darsini N, Kumari S (2013) A survey of routing protocols for VANET in urban scenarios In: IEEE conference on pattern recognition, informatics and mobile engineering. ISBN: 978-1-4673-5845-3
6. Nakamoto S (2008) Bitcoin: a peer-to-peer electronic cash system

Survey Paper on Automatic Vehicle Accident Detection and Rescue System

Utsav Chaudhary, Army Patel, Arju Patel, and Mukesh Soni

Abstract There are two sides of every coin, it happens with technology and there are pros and cons of it. In the current scenario, vehicles have provided a convenient commute, but the increasing number of vehicles on roads creates heavy traffic and road accidents. To overcome the problem of road accident, our project on automatic vehicle accident detection and rescue system will help detecting the accident and call in the emergency medic with sending current location. Our survey on this topic is made to construct such a system which is efficient and reliable. Moreover, we analyze different approaches on which work has been done to create a successor system which overcomes false detection of an accident.

Keywords GPS module · GSM modem · Gyroscope · Sensors · Accelerometer · Accident detection · Crash · Vehicle · Arduino

1 Introduction

The increasing number of vehicles leads to a higher number of accidents. If we go through statistical data, over 1.25 million people die in a road accident every year. On average 3424 deaths per day are noted (W. H. Organization, n.d.). The most obvious reason for one's death in an accident is an inadequate source of first aid. This happens because responders are not brought to the knowledge immediately [1]. It is extremely vital when it involves incidents involving vehicle accidents. The analysis shows that

U. Chaudhary (✉) · A. Patel · A. Patel · M. Soni
Smt. S. R. Patel Engineering College, Unjha, Gujarat, India
e-mail: utsavchaudhary108@gmail.com

A. Patel
e-mail: patelarmyj@gmail.com

A. Patel
e-mail: patelarju3003@gmail.com

M. Soni
e-mail: soni.mukesh15@gmail.com

K. Kotecha et al. (eds.), *Data Science and Intelligent Applications*, Lecture Notes on Data Engineering and Communications Technologies 52,
https://doi.org/10.1007/978-981-15-4474-3_35

if we decrease just 1 min in accident response time that can increase the chances of saving an individual's life up to 6%. To reduce response time, it is needed to implement advanced traffic technologies which will ultimately reduce fatalities. Our survey is to design and implement such a system that detects the event of an accident and send a rapid response to an emergency helpline for reducing response time significantly. Emergency services or helplines like fire brigade, ambulance services and police department will receive the location of incident and response accordingly.

2 Literature Survey

See Fig. 1.

2.1 *Image Processing*

This system is based on video monitoring from traffic surveillance cameras. Video from a surveillance camera is taken as input and as much as possible frames are analyzed for vehicle detection. Here, every vehicle is of different shapes, different colors and viewed from a different angle. The weather effect is also accountable for false tracking of the vehicle. In this work, more important has been given to vehicle detection and tracking stage, since they are essential to extract suitable vehicle features and vehicle parameters that can be used as factors for determining crashes at intersections. In this work, a weighted blend for further calculation is utilized where low-level highlights are obtained from vehicles and low-level examination is done on vehicle area [2]. Using the low-level features and vehicle velocity of the correctly tracked vehicles, crashes were detected by the system using an accident index calculated from speed, area, orientation and position indexes of the tracked vehicles.

2.2 *Machine Learning*

This study takes an account of accidents experienced in past. To identify accident, knowledge is obtained from passing by automobiles and machine learning tools which helps us to observe usual and unusual actions of automobiles [3]. This study focuses on strange driving actions which could lead to a possible crash. Clustering algorithms are able to identify accidents accurately. Vehicular ad hoc networks (VANETs) are broadly used in interaction with automobile to automobile and automobile to pathways interactions. Applications of VANET are vehicle crash alert, distance alert, self-driving car, parking assist, map access and Internet access [4].

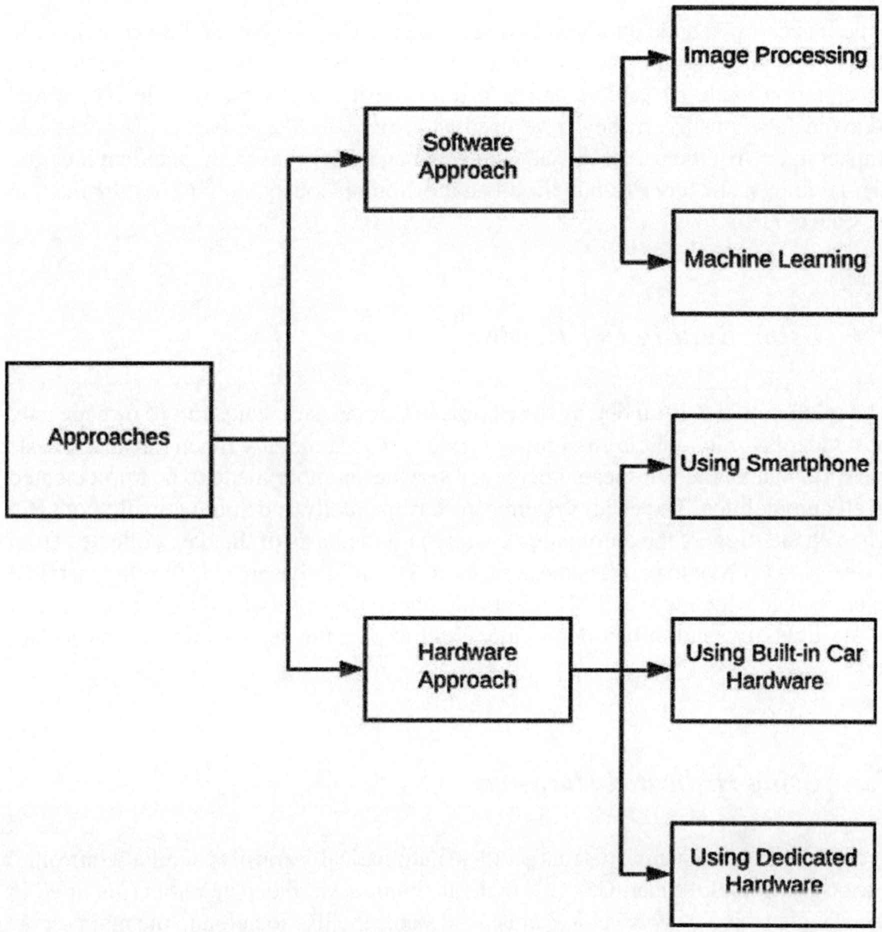

Fig. 1 Classification of different approaches to solve problem

Drivers will be warned about a crash in advance and they will have time to avoid a collision. Machine learning tools are used to forecast the collision that may occur.

2.3 Using Smartphone

Detecting accident based on a smartphone has both advantages as well as disadvantages. As smartphones are independent, it is hard to detect an actual crash and on advantage, smartphones provide rich data to analysis accidents. Developing this kind of system can be problematic as one needs to take care of false positives [5]. They provide an approach in which smartphone is used for detecting accident through forces experienced by vehicle and automatic emergency alert system (White et al. [5]). In this approach, sensor like accelerometer is used to measure sudden acceleration and deceleration made by vehicle and g-forces experienced by the vehicle. Therefore, to avoid false positives, they have used acoustic data like airbag deployment and impact noise. By the use of this acoustic data and GPS sensor, the incident location can be known. In order to handle this data simultaneously, an android application was developed.

2.4 Using Built-in Car Hardware

The eCall initiative is funded by the European Commission who aims to provide a set of standards for in-vehicle emergency services. This generally functions on the basis of E112, that is, the European emergency services number made to be implemented in all automobiles. The eCall system can start manually and automatically both [6]. The vehicle triggers the automatic activation on the basis of the data collected from the sensors in order to perform the activation. This initiative provides a setup installed in all vehicle which will call 112 automatically on the event of a vehicle crash. It will also wirelessly send airbag deployment and impact the sensor information (eCall, n.d.).

2.5 Using Dedicated Hardware

A dedicated system which detects accident automatically consists of microcontroller, various sensors, GPS and GSM. A dedicated unit as an independent part is fitted in the vehicle which detects vehicle crash and sends the live location to the main server by which ambulance will know the current location of the accident and assist the patient to the concerned hospital as well as it will also monitor important things of patient like temperature and pulse rate [7]. In their work, they also mentioned that traffic light signals can be controlled via RF communication to provide a clear path for an ambulance to reach its destination earlier.

3 Proposed Methodology

In this work, we will be using a hybrid mechanism in which sensors like gyroscope and accelerometer are used for detecting an accident and to reduce false positive we will use output of sound level meter and airbag sensor (if available in car) and sending location through GSM modem with the help GPS module to emergency responders (Fig. 2).

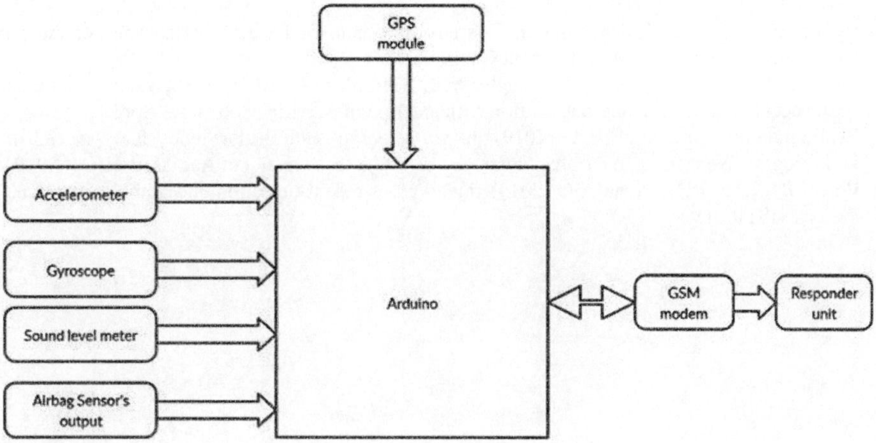

Fig. 2 Block diagram of accident detection module

On the other hand, the responder will have the live location of the victim in a device having Google Map services by which rescue and emergency help can be provided to the victim quickly. In order to process data, which is received simultaneously from a different location, we will develop Web application and short message service is also provided at the responder end.

4 Conclusion

To conclude, the proposed system will significantly decrease the response time which makes it more reliable and efficient for the people. Countries like India should use this system as traffic accidents are the major problem in these types of highly populated countries. False detection rate can be decreased by using some machine learning techniques though it is not featured in our system, it can be considered as future work.

References

1. Sadaphal I, Sangameshwarkar R, Singh S, Patil S, Kale S (2019) Accident detection and alert system using android application. Int J Res Appl Sci Eng Technol 7(V)
2. Vasu L (2010) An effective step to real-time implementation of accident detection system using image processing. Submitted to the Faculty of the Graduate College of the Oklahoma State University
3. Dogru N, Subasi A Traffic accident detection by using machine learning methods. International Burch University, Sarajevo, Bosnia and Herzegovina

4. Boukerche A, Oliveira H, Nakamura E, Loureiro A Vehicular ad hoc networks: a new challenge for localization-based systems. Comput Commun 2838–2849
5. White J, Thompson C, Turner H, Dougherty B, Schmidt DC (2011) WreckWatch: automatic traffic accident detection and notification with smartphones. Artic Mob Netw Appl
6. Ecall mobility and transport (2019) https://ec.europa.eu/transport/road_safety/specialist/knowledge/esave/esafety_measures_unknown_safety_effects/ecall_en. Accessed 21 Oct 2019
7. Prachi B, Kasturi D, Priyanka C (2014) Intelligent accident-detection and ambulance-rescue system. IJSTR 3(V)

Blockchain-Based Mechanisms to Address IoT Security Issues: A Review

Ochchhav Patel and Hiren Patel

Abstract Internet of things (IoT) is seen as combinations of various communication technologies and embedded equipment such as sensors, micro-controllers, radio-frequency identification, wireless devices, etc., which work as a single unit to achieve a specific task in a smart way, i.e., with less human intervention, and decisions are made automatically. IoT is used in many sectors such as smart home automation, smart grids, smart cities, vehicular networks, etc. Due to easy accessibility and reachability to such devices or networks, there are severe security and privacy issues in IoT. User authentication, access control mechanism, confidentiality and data correctness are a few of the major concerns among them. Though it looks like conventional security concerns, but due to (i) the dynamic nature of IoT, (ii) the resource-scarce nature of the devices and (iii) different standards and communication stacks involved, traditional network security countermeasures may not be applied directly to IoT. In recent years, usage of blockchain technology to address various security issues, specifically data correctness, has opened up a facet for IoT to explore the option of using blockchain in IoT. Researchers have started investigating the same. Primitively, blockchain is a non-editable, decentralized and cryptographically secured ledger that is tolerant against byzantine failure. This nature of blockchain makes it an attractive alternative to address security issues such as data integrity, entity authentication, etc. Also, in an IoT environment, where devices are spread across a vast geographical area, it requires a decentralized validation mechanism that a blockchain provides. In this paper, we intend to make an exhaustive survey of various researches being carried out to secure IoT networks using blockchain technology. We also aim to provide a summarized view for various prospective research directions in IoT which can be addressed by blockchain technology.

O. Patel (✉)
LDRP Institute of Technology and Research, Kadi Sarva Vishwavidyalaya,
Gandhinagar, Gujarat 382015, India
e-mail: ochchhavpatel@gmail.com

H. Patel
Vidush Somany Institute of Technology and Research, Kadi Sarva Vishwavidyalaya,
Gandhinagar, Gujarat 382015, India
e-mail: hbpatel1976@gmail.com

K. Kotecha et al. (eds.), *Data Science and Intelligent Applications*,
Lecture Notes on Data Engineering and Communications Technologies 52,
https://doi.org/10.1007/978-981-15-4474-3_36

Keyword Blockchain · Internet of things · Security · Privacy · Confidentiality

1 Introduction

The Internet of things is the community of physical objects, more than a few devices, buildings, vehicles and other objects which are embedded with electronics, software, sensors, network connectivity, permitting these objects to acquire and interchange data [1]. The Internet of things (IoT) has blasted in recent years, and it is not slackening down any time soon. Gartner forecasts that 20.4 billion connected things will be in use worldwide by 2020, and a report of Gartner (April 10, 2019) predicts that the IoT market will raise to over $3 trillion annually by 2026. The IoT involvement did not stopped to various use cases, but is widely adopted in many other areas such as health care, military, agriculture and smart domain also [2].

IoT finds application in many different fields [3, 4] such as patients remote monitoring, energy consumption control, traffic control, smart parking, inventory management, production chain, transportation and logistics domain and smart health care. Applications of IoT are enormous and user devices in IoT connected with the worldwide Web are keeping on growing, prevention of unauthorized access to IoT data is essential. Since sharing the medical facts of a patient is unethical, the data collected from such environment should be maintained securely. In application of healthcare system, authentication and authorization [5] are key challenges in Internet of things. Authentication and access authorization in IoT, user devices use logging mechanism with username and password. But that kind of mechanisms are often susceptible to attacks. If anybody gets the logging details, then can try to access the healthcare data of a patient purposely and may try to regenerate the medical devices which may downfall the entire system [6]. Smart environment using IoT that making its service easy and comfortable to the intelligence [7] of contained objects. Various projects like smart cities, smart health [8], smart home, smart office, smart shopping, smart mobility and smart transport are running. Butler is a European Union FP7 project [9], its primary persistence about enabling the development of secure and smart life assistant applications in terms of a context and position aware. After the entry of IoT in transportation domain communication, control and data distribution have been improved. Smart components are playing the important role as they carry out the important operations and make work more efficient in heavy loaded vehicles. In transportation domain include personal vehicles, commercial vehicles, trains, UAVs and other equipment. The IoT in smart transportation has rapidly changed the heavy vehicle industries with the help of mobile and connectivity advancements. Geo-fencing that captures the location of devices is most benefitted for transportation industry using IoT [1]. There are three layers [10] in IoT architecture, viz. application layer on the top, network layer in middle and perception layer in the bottom, as shown in Fig. 2. The perception layer plays the role of collection and transmission of data from/to the other layers. The network layer is responsible for steering and conveying the data collected by the perception layer via network technologies such as Wi-Fi, Bluetooth,

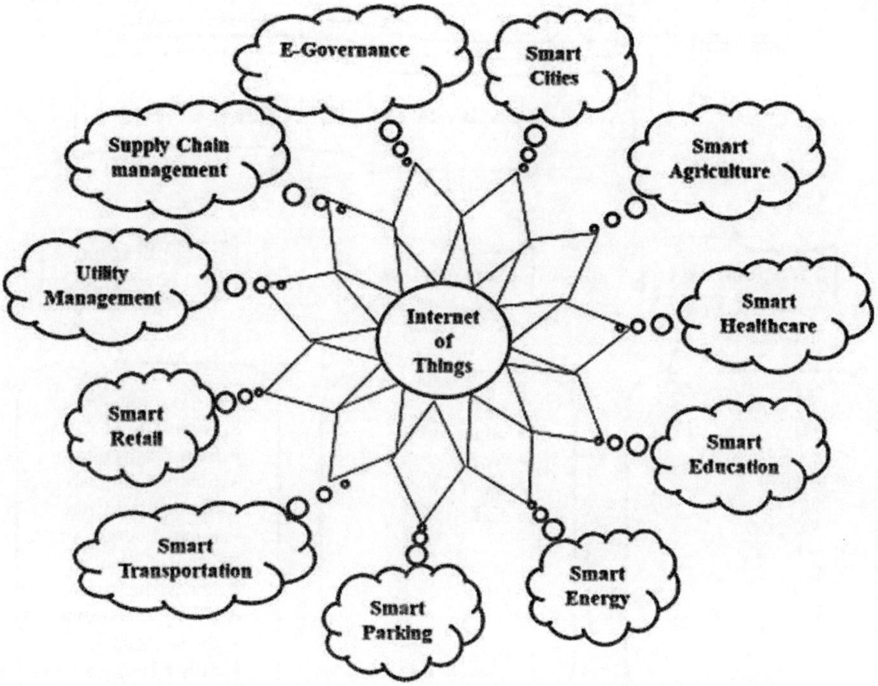

Fig. 1 Internet of things paradigm as a result of the convergence of different visions

3G/LTE and Zigbee. Top layer of the IoT framework is application layer where user interface is created for various application domains as mentioned in Fig. 1.

As mentioned in Fig. 2, the perception layer in IoT architecture mainly consists of sensors and hardware devices so some of the threats are possible by intruder such as spoofing, denial of service (DoS), timing attacks and reply attack. Second layer of the IoT is network layer related to network, communication technologies and corresponding hardware and protocols where traditional security problems of communication networks such as man in middle attack and Sybil attack. Third layer of IoT is the application layer provides interface for kind of services like smart home, smart health care, smart vehicle and smart grids where security threats for the application is dependent on particular domain [8].

Though at first glance, IoT seems to be very attractive options for many of the recent applications, there are few issues yet to be addressed before widely adopting IoT. This paragraph discusses such open issues being faced in IoT. The main issues [1] of IoTs are naming, transport protocol, QoS support, authorization, data integrity, privacy and mobility support. In mobility management, we know several features in IPv6 over IPv4 in mobility application due to efficient routing is achieved by using flexible address and fragmentation at source host and discover the destination path of a maximum of transmission unit. Improved security by using the IPsec protocol

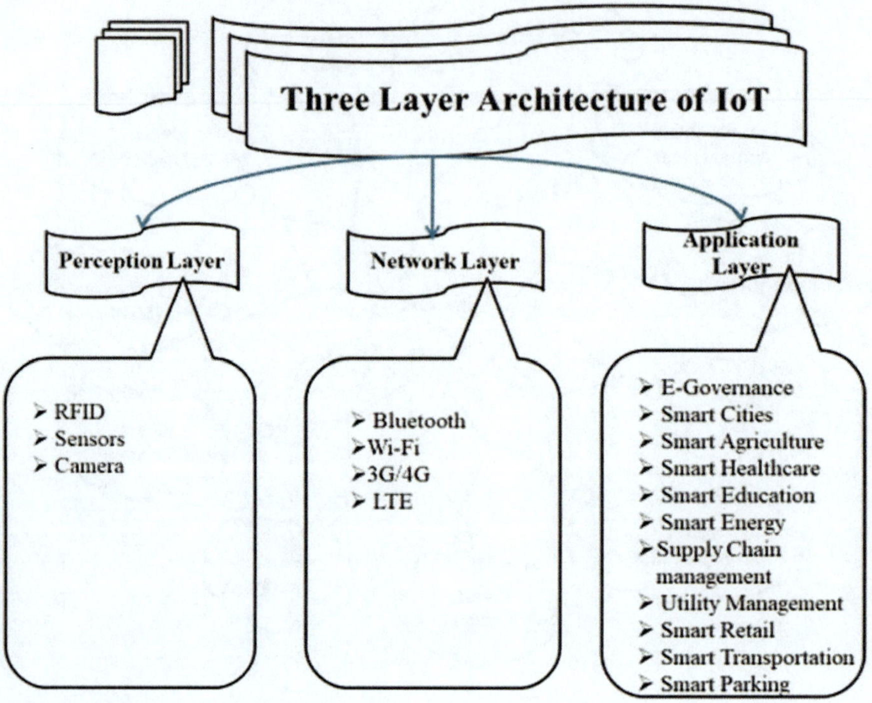

Fig. 2 Three-layer architecture of IoT

achieves better security than IPv4 protocol. Without guarantees in terms of system-level security applicable participants are dubious to adopt IoT solutions on a large scale. The IoT is vulnerable to attacks for several reasons where IoT components spend most of the time unattended so, it is easy to physically attack them. Another reason is the communications are wireless, which makes snooping extremely simple. Most of the IoT components are characterized by low capabilities in terms of both energy as well as computing resources [8]. Main security goals [11] of IoT may include the confidentiality, integrity, availability (CIA), energy efficiency and heterogeneity also. In this section, we mentioned major security challenges to be addressed to turn Internet of things technology into a mainstream and extensively deployed one. In a specific way mentioned here, three key issues [11] requiring innovative approaches that is data confidentiality, privacy and trust.

Due to its nature of immutability, decentralization and avoidance of central authority for decision making, blockchain has become a popular approach for many applications. Few of such applications are as mentioned in Fig. 1. Blockchain used in payment processing and money transfer [12] where blockchain technology provides fast, secure and low-cost universal payment processing services. Use of encrypted distributed ledgers blockchain provides trustful transaction [13] without the need for intermediaries such as correspondent banks. Blockchain provides digital ID that can replace traditional systems with a highly trusted mechanism of managing identities. Blockchain can empower users to have greater control over their own identity using blockchain. Blockchain provides decentralized security and privacy in IoT, involve energy, delay, computational task, etc., that is not suitable for most resource-constrained IoT devices. Blockchain provides secure and private [14] network using mining process for scalability also. Property transfer that is complex and worthy work where blockchain provides important role using of a decentralized application that is smart contract in property transfer process, the entire process, from signing lease agreements to managing cash flow to filing maintenance requests, can be conducted in a secure and transparent manner. In a blockchain-based smart home system, smart home entity can be considered as gateways and main object of homes. Similarly, they all are treated as nodes in the home-based blockchain network. These nodes not only store data but also can be worked as miners in a blockchain-based smart home and perform scalability with security and privacy. IoT issues can be addressed by blockchain architecture and protocols for data access, using smart contracts and a publisher—subscriber [15] mechanism. IoT devices grieve from impersonation and data tampering attacks due to their architectural and computational limitations, which are unable to provide tolerable level of security. Blockchain plays the role and provides unique hardware fingerprints to establish data provenance while Ethereum provides a decentralized digital ledger which is able to withstand data tampering attacks [16].

Asymmetric encryption wants a centralized key administration system, which cannot meet the wishes of an efficient developing IoT system. If the key management system is attacked, a large quantity of IoT units is in all likelihood to be affected. Also, traditional protection methods have a tendency to be expensive for the IoT in terms of strength consumption and processing overhead because sensors are lightweight, of slow processing and of much less memory. Blockchain technology can overcome that kind of problem where proof of work (consensus) calculation is particularly computationally intensive and time-consuming, so IoT devices are restricted and most IoT applications need low latency. Blockchain protocols create significant network traffic flow, which is a catastrophe for the communication of IoT devices.

In IoT, information is added to the network with regular interval means information added to blockchain network where blocks are added and need to be validated in a blockchain network, so mining is required that should be taken into consideration because it demands to high computational power. Alternative matter related to scalability, when the number of nodes increases, the number of transactions increases and the mining and validation process takes lengthier time to be completed [17]. IoT-based Raspbian Pi sensor node can be connected directly to the blockchain as a full node that can validate other transactions or a lite but partial node can only keep a track of its own transactions. The temperature sensor senses the environment and its value is extracted via a Web UI or mobile application. The Web UI or mobile app connected to the blockchain node send the sensor reading to the blockchain environment by smart contract only so Web application or Web app becomes interface between IoT devices and blockchain [18]. This script is organized as follows: Sect. 1 introduces the applications, security threats in IoT and how to address IoT security. Also include introduction about blockchain and issues of IoT addressed by blockchain and challenges of blockchain with IoT when integration performed between these two technologies. Section 2 includes all kind of security issues of IoT and mechanism which have been used to solve IoT security problems using blockchain. Section 3 describes challenges and opportunities of IoT and blockchain domain. Section 4 covered tools and technologies which have been used by existing researcher for their experimentation. Section 5 describes future directions so someone can perform their research in IoT security using blockchain. Section 6 that is final section is concluded with entire paper.

2 Background Theory

A series of blocks is contained in blockchain, in such a way that every new block is cryptographically connected to the previous block so provide tamper proof data storage and transaction. The process of validating and adding transaction to a block and then spreading that block on the blockchain network, to be known by all other the nodes in blockchain is termed as mining [17]. Miner nodes are used to do the mining, and the selection of a node to mine a new block is done based on certain criteria. There are so many types of nodes [18] available in blockchain. Simple node is the most basic, which can only send and receive transactions but cannot store the complete copy of the blockchain. Full node is the node which can maintain a complete copy of the blockchain, but they do not mine a block. A block validation mechanism performed by consensus algorithms such as proof of work in which a miner must perform some predefined work, which is often a mathematical puzzle or challenge which is hard to compute but easy to verify. PoW has the advantage of protecting transactions and blocks from being altered. But one disadvantage of that algorithm is required high computational resources. Another algorithm is proof of stake where there is no mining required by computational solving approach. There are other mechanisms to validate blocks like delegated proof of stake, proof of hold, proof of use, proof of time, proof of minimum aged stake and proof of importance [2].

Blockchain can be permission that is called private or permission—less (public) network. Permission network makes boundary on the consensus contributors and only the chosen trustful nodes have the rights to validate transactions. It does not require a lot of computation to reach a consensus, thus, it is not time and energy consuming. Another type is public blockchain, uses an unlimited number of anonymous nodes, based on the cryptography, each actor can securely communicate. Each node is represented by a pair of private and public keys. Public blockchain is time and energy consuming. We know cloud platform provided various services under the centralized control of one trusted entity, in cloud, it is possible to the single point of failure concerning security and privacy also availability of cloud services. Whereas, blockchain provides a way that all the miner and full nodes in the blockchain network maintains a same copy of the blockchain state and the trust is distributed among all the network nodes. Even though, one of them among blockchain node, want to try to alter data, then system will be reject it, and the blockchain state will be remain untampered [18].

3 Related Work

There are various traits of IoT such as connectivity, heterogeneity, ubiquity, self-organization, mobility, and scalability. Even though, to fulfill full deployment, the security requirements have to be addressed. Confidentiality means to protect data from unauthorized entity, in other words, only the persons can access data who are authorized to access sensitive data. And most important element that is helps to build and develop trust among existing entity in a network. In the IoT context not only users, but also sanctioned objects may access data so important aspects are access control mechanism and object authentication process [8]. Confidentiality should be providing so that the data collected or transmitted are protected and available to authorized users only. As we know collected data from the IoT devices or a sensor should not be transmitted to any other devices without proper encryption applied. Only encoded messages should be delivered to neighboring devices so that malicious entities cannot access the collected data [12]. Confidentiality [19] is very essential, especially in IoT as it makes sure that data and routing information are transferred to only authorize users that can access and alter securely. Hence, speaking forthrightly, the exchanged data within IoT applications should be secure from intermediate and unauthorized entities. The functionality that guarantees that data and directing information have not been changed in transit by a midway or a malicious node is termed as integrity. Hence, any change of exchanged data is imperatively detected.

The functionality in IoT that make sure that entity identity is extremely protected from third party, by defining the rules under which data referring to individual entity may be accessed is termed as privacy. Privacy is the unique considerations required to guard the data of humans from publicity in the IoT environment, in which IoT entities or object can be given a special identification and the capacity to speak autonomously over the Internet. Privacy defines the guidelines under which data referring to individual users may be accessed. Some of the precautions are used to maintain the privacy like set the password, disable universal plug and play and create a separate network or update firm ware [6]. In fitness care, purposes lack of gorgeous mechanism for ensuring privateness of private or sensitive facts has harnessed the adoption of IoT technologies. Health care purposes characterize the most outstanding application field, whereby the lack of fantastic mechanisms for making sure privateness of private or sensitive facts has harnessed the adoption of IoT technologies. In addition, in the IoT vision, a prominent function will be performed via Wi-Fi communication technologies. The ubiquitous adoption of the Wi-Fi medium for replacing facts may additionally pose new issue in time period of privacy violation. In fact, wireless channel will increase the risk of violation due to the remote access capabilities, which potentially expose the gadget to eavesdropping and overlaying attacks. Hence, the privateness represents actual open difficulty that may additionally restrict the improvement of the IoT [11]. Privacy needs to be addressed in the IoT device itself, in storage, during communication and at processing.

The concept of trust is used in a giant range of different contexts and with diverse meanings. Trust is a complex notion about which no consensus exists in the computer and facts science literature, even though its importance has been widely recognized. Mayer, Davis, and Schoorman (1995) define trust as the willingness of a party to be vulnerable to the action of another party based on the expectation that the other will perform a particular action important to the trustor, irrespective to the ability to monitor or control that other party. Another view of trust by Kimery and McCord (2002) about online transaction is online trust is a customer's willingness and enables to accept an online transaction according to their positive and negative expectations on future online shopping behavior [11]. Trust has been seen and interpreted in many different methods and in distinctive contexts which make it is a very elaborate concept. Because trust is influenced with the aid of many attributes that can be related or no longer to security, authentication, confidentiality, integrity, availability and identity management ought to be regarded a trust principle or have confidence components. Identification consists of providing special identifiers to the entities within the IoT service and correlating these digital identities to real world, legally binding identities. In a cellular network for IoT service, endpoint units are recognized the use of IMSI or IMEI. Networks are identified the use of community codes and country codes where each method of providing identification has various ranges of impenetrable assurance related with it. Blockchains are immutable digital ledger systems implemented in a distributed fashion [20] (i.e., without a central repository) and usually without a central authority. Blockchain applications in multiple use cases and use them as a secure way to create and manage a distributed database and maintain records for digital transactions of all types.

There is a need of architecting the IoT in a reliable manner allowing the ability to automatically adapt to the unexpectedly security break; hence, trust in IoT is very essential for the working. Various solutions in security occur to ensure the different requirements in a various IoT applications. Transport layer security mechanisms or virtual private network used for ensuring confidentiality can be used as an example. To provide integrity, message integrity codes could be used. To secure end-to-end communication between IoT entities, certificate-based authentication could be used. Various IoT issues such as confidentiality, privacy and integrity [18] by configuring and managing IoT devices using blockchain smart contracts are addressed via blockchain. Trust-free distributed architecture [21] is provided by blockchain, where IoT devices are proposed to be configured and managed through Ethereum or other blockchain platform with smart contracts. Bubble of trust [2] has been proposed for IoT devices on decentralized environment to achieve (a) identification, (b) authentication, (c) integrity and (d) availability. Using C++ and Ethereum, it uses blockchain for identification and trust among IoT devices by making use of a concept called bubble. Authors claimed their proposal to be secure, efficient and cost-effective. Authors plan to work upon certain issues such as reducing number of miners and revoking already granted access rights. IoT devices have less memory and low computational capability so it may be easily vulnerable to network attacks so researchers have been introduced integrating technology [19] which is combined with IoT and blockchain. Allocation of the resources in static form to prevent rogue devices accessing from the server, identify trusted and untrusted devices from the network and address DDoS attack also. Researcher claimed new abstract architecture [23] based on service centric networking (SCN) and blockchain provide trust, security and encouragement IoT in a decentralized manner. Proposed decentralized architecture (software stack) [24] based on Blockchain + IPFS and IoT, for retrieving IoT data in a secure manner, peer to peer file storage and abolish the need of centralized IoT data management. Furthermore, researcher enhances their work with hyperledger platform with advanced IoT applications.

4 Challenges and Opportunities

There are various drawbacks that the researchers feel about the IoT security domain such as client privacy, denial of service, secure communication and double spending. It is essential to the information and data of individuals from the exposure in the IoT environment, which makes the privacy in IoT a very consideration. There are times when any particular service is not available, this situation whenever arises, is termed as denial of service [21]. There could be multiple reasons for such inaccessibility, but on a higher probable cause, it refers in most cases to the infrastructure that cannot manage due to capacity overload. There are three types of DOS attacks; (1) application layer attack, (2) protocol and (3) volume-based attack. The biggest challenge in an IoT is the secure communication using symmetric and asymmetric encryption. Public and private keys are complimentary to each other and both are asymmetric keys. Any message that is encrypted by a public key will be decrypted by a private key. Same follows the other way around. For generating a shared key using asymmetric elliptic curve keys [19] is done using elliptical curve Diffie–Hellman.

As there has been an over-exhaust in the recent past, with the improvement, development of science and accessibility of the wireless devices has increased. Cloud computing seems to be the possible solution for such large number of wireless devices connected to Internet, but security challenges could not be satisfied by it. For private and sensitive data in an IoT will continue to face security problems. Thus, the need for a completely decentralized peer to peer and secure technology as a blockchain can have the sustainability to overcome these problems [15]. Due to the architectural and computational limitations, many IoT devices suffer from data tampering and impersonation attacks. BlockPro have been implemented by many researchers which is a blockchain-based system for data provenance and data integrity [16].

5 Tools and Technologies

Though blockchain has the potential to evolve various industries, like health care, logistics and supply chain, insurance, financial and many more reputed companies are also adopting blockchain, but the blockchain is well known for cryptocurrencies such as Bitcoin and Ethereum. There are multiple networks for blockchain such as Ethereum, Hyperledger Fabric and Bitcoin [2]. Ether is the cryptocurrency provided by Ethereum which is a public blockchain. Ethereum virtual machine is the operating system used in Ether where smart contacts are executed. Ethereum takes only 14s to validate a block in a network, whereas the Bitcoin takes 10 min. Ethash is the consensus mechanism for Ehtereum for block validation mechanism. Hyperledger is the other option, which is an open-source blockchain created by the Linux Foundation, more specifically by IBM and it does not provide a cryptocurrency. There are so many blockchain tools available to create blockchain [25] environment such as remix IDE, solc, geth, embark, ganache, baas, mist and metamask. One of the easiest and browser-based tools to use for the creation and deployment of smart contracts is the remix. For writing, debugging, testing and deploying smart contracts is the solidity language. It is syntax is a loosely typed programming language and has syntax similar to European Computer Manufacturer's Association (ECMA) script. Though one needs something to convert solidity script into a format readable by the EVM, this purpose is fulfilled by solidity compiler. Solidity compiler can be divided into two parts one is solc coded in C++ and second is solc-js that uses Emscripten for cross-compiling from solc C++ code to JS. In Go Programming, Geth is an Ethereum client used for Ethereum client. Logically, Geth [25] is a program which works as a node for the Ethereum platform. There is a development framework for Ethereum-based dApps. It is used for give permission to the developers to develop and deploy dApps on decentralized technology. It has the eligibility to permit you to create smart contracts which can be made available in Javascript code. It also provides with the application of migration if the application has multiple contracts. With the Javascript developers and the support of the test-driven development of smart contracts can handle contracts on different blockchains like testnet, live network and private net.

6 Future Directions

The applications of IoT and its importance in daily life in increasing to a very large extent, and hence it should require secure communication among devices in a network. The researchers working in this wide field have proposed the concept bubble of trust [2] where the virtual zones which are proven secured are created where devices can communicate in a way that is completely secure. The ultimate purpose is to enhance the work to establish a revocation mechanism for the devices that are compromised. The design protocol is the one that aims the optimization of the miner's number in an already defined system as well as how the miners that are selected can be placed. Credibility verification method for IoT devices have been implemented by the researchers and are getting the advantage in the aspects of the storage space and response time. It has also been mentioned by them that there is a lot of work in the near future for IoT environment and in determining how to choose the number of blockchain structure node (BCS) [26]. Not all researchers have implemented service centric networking (SCN) to provide reliable connectivity and the global scalability in IoT network. It is considered as a combined amalgamation of SCN and blockchain to empower the IoT [23]. A new combination of the architecture of OSCAR [27] and ACE [28] is the authorization framework to provide the end-to-end solution for the secure authorized access to IoT resources. Their enhancement is based upon the updation of private Ethereum blockchain network to use the PoS version of the ledger.

7 Conclusion

With the advancement in technology to develop smart applications such as smart cities, smart transportation, smart health care, etc., the Internet of things (IoT) has become a topic of buzz as it is one of the basic building blocks behind it. IoT equipments are often used in a coarse way as they are to be installed in any territory with/without observations. Unobserved devices are prone to fall in security breaches which may cause severe damages. Hence, it is of utmost importance to understand all such security problems and propose solutions for the same. Some of the network security issues can be addressed through traditional cryptographic primitives, however, due to the distinct nature of IoT, few security concerns are to be addressed in a diverse way. One of such solution provider to security predicament is blockchain technology. Due to a few attributes such as non-editability, transparency, avoidance of central arbiter and distributed environment, researchers have commenced exploring the use of blockchain technology to find out the solutions to IoT security problems. In this research, we have studied various existing mechanisms proposed in recent times with their detail analysis. We have discussed the fundamental terminologies related to blockchain technology along with all the challenges and opportunities. We also have investigated various tools and technologies pertaining to blockchain along with future research work which can be carried out in this domain. In the upcoming time, we wish to propose a novel security mechanism for IoT using blockchain technology.

References

1. Atzori L, Iera A, Morabito G (2010) The internet of things: a survey. Comput Netw 54(15):2787–2805
2. Hammi MT, Hammi B, Bellot P, Serhrouchni A Bubbles of trust: a decentralized blockchain-based authentication system for IoT. Comput Secur 78:126–142

3. Yuan R, Shumin L, Baogang Y (2007) Value chain oriented RFID system framework and enterprise application. Beijing: Sci (2007)
4. METRO Group Future Store Initiative. http://www.futurestore.org/
5. Vilamovska AM, Hattziandreu E, Schindler R, Van Oranje C, De Vries H, Krapelse J (2009) RFID application in healthcare—scoping and identifying areas for RFID deployment in healthcare delivery, RAND Eur
6. Miorandi D, Sicari S, De Pellegrini F, Chlamtac I (2012) Internet of things: vision, applications and research challenges. Ad Hoc netw 10(7):1497–1516
7. SENSEI FP7 Project, Scenario Portfolio, User and Context Requirements, Deliverable. http://www.sensei-project.eu/
8. Shantha JR (2016) A neoteric authentication scheme for Iot healthcare system. https://doi.org/10.5281/zenodo.192911
9. BUTLER Project. http://www.iot-butler.eu
10. Shin H, Lee HK, Cha HY, Heo SW, Kim H (2019) IoT security issues and light weight block cipher. In: International conference on artificial intelligence in information and communication (ICAIIC), IEEE, pp 381–384
11. Nabil D, Tandjaoui D, Romdhani I, Medjek F (2018) Trust management in internet of things. https://doi.org/10.4018/978-1-5225-5736-4.ch007
12. https://www.fool.com/investing/2018/04/11/20-real-world-uses-for-blockchain-technology.aspx
13. Guo, Y, Liang, C (2016) Blockchain application and outlook in the banking industry. Finan Innov 2(1):24
14. Yaga, D, Mell, P, Roby, N, Scarfone, K (2018) Blockchain technology overview (No. NIST Internal or Interagency Report (NISTIR) 8202 (Draft)), National Institute of Standards and Technology
15. Rifi, N, Rachkidi, E, Agoulmine, N, Taher, NC (2017) Towards using blockchain technology for IoT data access protection. In: 2017 IEEE 17th international conference on ubiquitous wireless broadband (ICUWB), IEEE, pp 1–5
16. Javaid, U, Aman, MN, Sikdar B (2018) BlockPro.: blockchain based data provenance and integrity for secure IoT environments. In: Proceedings of the 1st workshop on blockchain-enabled networked sensor systems, ACM, pp 13–18
17. Mining in Bitcoin. Available at: https://en.bitcoin.it/wiki/Mining/
18. Makhdoom, I, Abolhasan, M, Abbas H, Ni W (2019) Blockchain's adoption in IoT: the challenges, and a way forward. J Netw Comput Appl 125:251–279
19. Dorri A, Kanhere SS, Jurdak R, Gauravaram P (2017) Blockchain for IoT security and privacy: the case study of a smart home. In: 2017 IEEE international conference on pervasive computing and communications workshops, PerCom workshops, IEEE, pp 618–623
20. Zheng Z, Xie S, Dai HN, Wang H (2016) Blockchain challenges and opportunities: a survey. Work Pap
21. https://www.cloudflare.com/en-in/learning/ddos/what-is-a-ddos-attack/
22. Ming Z, Yang S, Li Q, Wang D, Xu M, Xu K Blockcloud: empowering IoT through a service-centric blockchain. Blockcloud Technical White Paper
23. Ali MS, Dolui K, Antonelli F (2017) IoT data privacy via blockchains and IPFS. In: Proceedings of the seventh international conference on the internet of things, ACM, pp 14
24. https://hackernoon.com/top-12-blockchain-development-tools-to-build-blockchain-ecosystem-371a1b587248
25. Qu C, Tao M, Zhang J, Hong X, Yuan R (2018) Blockchain based credibility verification method for IoT entities. Secur Commun Netw
26. Vučinić M, Tourancheau B, Rousseau F, Duda A, Damon L, Guizzetti R (2015) OSCAR: object security architecture for the internet of things. Ad Hoc Netw 32:3–16

27. Seitz L, Selander G, Wahlstroem E, Erdtman S, Tschofenig H (2017) Authentication and authorization for constrained environments (ACE). Int Eng Task Force Int Draft Draft Ietf Aceoauth Authz 07 work Prog. Available at: https://datatracker.ietf.org/doc/html/draft-ietf-ace-oauth-authz-07
28. Alphand O, Amoretti M, Claeys T, Dall'Asta S, Duda A, Ferrari G, Zanichelli F (2017) IoT Chain: a blockchain security architecture for the internet of things. In: IEEE wireless communications and networking conference (WCNC), IEEE, pp 1–6

Experimental Analysis of Measuring Neighbourhood Change in the Presence of Wormhole in Mobile Wireless Sensor Networks

Manish Patel, Akshai Aggarwal, and Nirbhay Chaubey

Abstract Wireless sensor networks are different from mobile ad hoc networks and vehicular ad hoc networks in terms of resource capacity and applications. Sensor nodes are deployed densely in the given area. Nodes are not personally monitored. Once deployed, they are unattended. Therefore, primary requirement for sensor network is security. Possible attacks for wireless sensor networks include jamming, selective forwarding, wormhole, Sybil, etc. Wormhole is very harmful attack for sensor network, and it creates many more attacks. For dynamic wireless sensor networks, two nodes P and Q located in different area can become one hop neighbours after some amount of time. Nodes connected through tunnel will increase the number of neighbouring nodes. In this research, we present the results of an experimental analysis of measuring neighbourhood change in the presence of wormholes in dynamic WSNs with varying speeds and varying time interval.

Keyword Mobile ad network · Unattended · Dynamic · Tunnel · Mobility

1 Introduction

Wireless sensor networks are useful for sensing related applications [1, 2]. Sensor nodes having low resources are easily hacked by malicious entity. Compared to wired communication, wireless is not reliable where packet loss occurs [3, 4]. Considering

M. Patel
Smt S.R.Patel Engineering College, Gujarat Technological University, Ahmedabad, India
e-mail: it43manish@gmail.com

A. Aggarwal (✉)
School of Computer Science, University of Windsor, Windsor, Canada
e-mail: akshai.aggarwal@gmail.com

N. Chaubey (✉)
Dean, Faculty of Computer Science, Ganpat University, Mehsana, India
e-mail: nirbhay@ieee.org

K. Kotecha et al. (eds.), *Data Science and Intelligent Applications*, Lecture Notes on Data Engineering and Communications Technologies 52,
https://doi.org/10.1007/978-981-15-4474-3_37

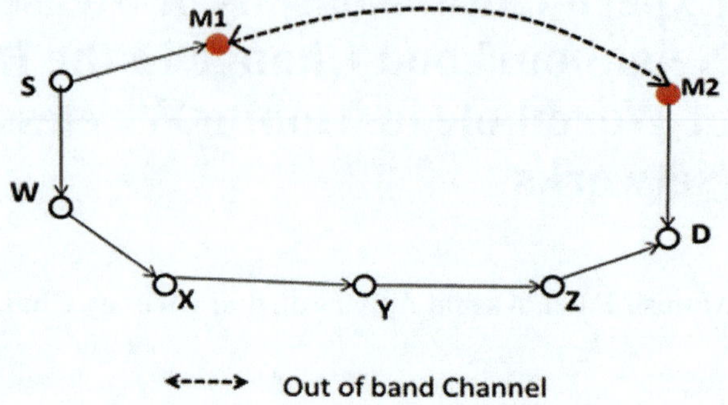

Fig. 1 Wormhole tunnel constructed between nodes M1 and M2 [29]

the fundamental features of sensor nodes, security is primary and necessary requirement [5, 6]. Various security approaches are presented in [7–13]. Once high-speed tunnel is created, then attacker can launch traffic analysis attack, selective forwarding attack, packet dropping attack and many more [14–18].

Figure 1 shows high-speed tunnel which is created between two unauthorized nodes M1 and M2. M1 lures traffic from one location and sends it to unauthorized node M2. Due to this, nodes S and D believe that they are neighbours. But they are not real neighbours. They are fake neighbours. Wormhole attack detection and prevention approaches are presented in [19–26].

In this study, we have measured the neighbourhood change effect after creating wormholes in dynamic WSNs. This analysis is based on our previous work [27, 28]. Impact due to nodes mobility in wireless sensor networks is discussed in Sect. 2. Rate of change of neighbours is calculated in Sect. 3. Experimental results and analysis are presented in Sect. 4. Finally, conclusions are presented in last section.

2 Impact of Mobility and Problem Formulation

Figure 2 shows the impact of node mobility. At t time, the location of node i is denoted by $\text{Loc}_i(t)$, and the velocity is denoted by $V_i(t)$. At $(t + \Delta t)$ time, the updated location of node i is $\text{Loc}_i(t + \Delta t) = \text{Loc}_i(t) + V_i(t) \cdot \Delta t$. Connectivity between two nodes is dependent on relative location $\text{Loc}_{(i,j)}(t) = \text{Loc}_i(t) - \text{Loc}_j(t)$ and $V_{(i,j)}(t) = (V_i(t) - V_j(t))$.

Therefore, the calculated distance between two nodes at time Δt is

$$\text{Dis}_{(i,j)}(t + \Delta t) = \left|\text{Loc}_i(t + \Delta t) - \text{Loc}_j(t + \Delta t)\right| = \left|\text{Loc}_{(i,j)}(t) + V_{(i,j)}(t) \cdot \Delta t\right|$$

If the distance between two nodes is greater than the transmission range, then they are not considered as neighbours. Figure 1 shows nodes S and D are connected through wormhole. They are not true neighbours. For dynamic wireless sensor networks, two nodes A and B located in different area can become true one hop neighbours after some amount of time, and it creates wormhole attack illusion.

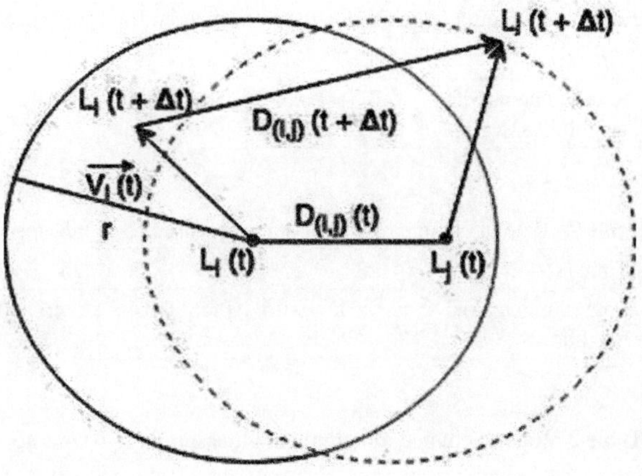

Fig. 2 Relative locations of node i at time t and $(t + \Delta t)$

3 Computation of Neighbourhood Change

Due to the tunnel creation, the neighbouring nodes are increased. The list of neighbouring nodes contains both true and fake neighbours. Suddenly, increasing the neighbours indicates there might be the presence of wormhole. The change in the neighbours is calculated by every node at certain time intervals. At certain time t, the change in the neighbours of node P is computed as.

Rate of change of neighbours (RCN) at time $t = 1 -$ (No. of neighbours at time $t2$ of node $P -$ No. of new neighbours at time $t2$ with respect to time $t1$)/max (No. of neighbours at time $t2$ of node P, No. of neighbours at time $t1$ of node P)].

For example, at time $t1$, node P had seven neighbouring nodes. At time $t2$, it had ten neighbouring nodes, and seven of them were not present during the previous time $t1$. For this situation, rate of change of neighbours is calculated as follows:

Rate of change of neighbours $(t) = 1 - ((10 - 7)/10) = 0.7$.

4 Experimental Results and Analysis

Experiments are performed by using Network Simulator Version 2. Different scenarios are created, and neighbourhood change is measured without an attack and after creating an attack.

4.1 Rate of Change of Neighbour Without Creating an Attack

The rate of change of neighbours depends on how to choose the value of time interval. Longer time duration results in high RCN value. Therefore, it is better to choose an

Table 1 Rate of change of neighbours measured for 1 m/s speed without wormhole attack

Time (s)	52	51	50	49	48	47	46	45	44	43
Rate of change of neighbours	0.51	0.46	0.46	0.46	0.46	0.46	0.40	0.40	0.40	0.37

Table 2 Rate of change of neighbours measured for 5 m/s speed without wormhole attack

Time (s)	28	27	26	25	24	23	22	21	20	19
Rate of change of neighbours	0.51	0.46	0.46	0.46	0.46	0.46	0.40	0.40	0.40	0.37

Table 3 Rate of change of neighbours measured for 10 m/s speed without wormhole attack

Time (s)	14	13	12	11	10	9	8	7	6	5
Rate of change of neighbours	0.59	0.59	0.52	0.46	0.46	0.46	0.39	0.39	0.37	0.34

optimal value of time interval for getting good detection accuracy. The optimal value for rate of change of neighbour should be between 0.40 and 0.50, not much high and not much less. Increasing the value of the speed will result in decreasing the value of time interval. Tables 1, 2 and 3 indicate different neighbourhood change rate values measured with varied speed and at different times in the absence of an attack.

4.2 *Rate of Change of Neighbour After Creating an Attack*

Tables 4, 5 and 6 show the rate of change in the neighbourhood value measured at different times with different speeds in the presence of a wormhole attack.

In previous work [28], it was found that, if the RCN value is greater than the upper threshold, then it indicates the presence of an attack in the network. If the RCN value is below the lower threshold, then no attack is present in the network [28]. If the RCN

Table 4 Rate of change of neighbours measured for 1 m/s speed after creating an attack

Time (s)	52	51	50	49	48	47	46	45	44	43
Rate of change of neighbours	0.79	0.79	0.79	0.79	0.79	0.74	0.74	0.74	0.67	0.67

Table 5 Rate of change of neighbours measured for 5 m/s speed after creating an attack

Time (s)	28	27	26	25	24	23	22	21	20	19
Rate of change of neighbours	0.80	0.76	0.76	0.76	0.76	0.72	0.72	0.72	0.65	0.65

Table 6 Rate of change of neighbours measured for 10 m/s speed after creating an attack

Time (s)	14	13	12	11	10	9	8	7	6	5
Rate of change of neighbours	0.87	0.82	0.82	0.78	0.78	0.78	0.70	0.70	0.65	0.62

value is in between the upper and lower thresholds, then the new neighbouring nodes are considered to be suspicious nodes, and further verification is done, as discussed in a previous study [28]. For RCN = 0.46, we have fixed the time interval. For 1 m/s speed, the time interval is 52 s. For 5 m/s speed, the time interval is 27 s. For 10 m/s speed, the time interval is 11 s. In [28], 0.46 is taken as lower threshold.

5 Conclusion

Experimental results indicate that wormhole attack increases number of neighbours. In the absence of wormhole, the rate of change of neighbourhood value is very less. Due to node's mobility, it is difficult to accurately detect an attack. Our future work includes finding impact of variants of wormhole in dynamic wireless sensor networks.

References

1. Misra S, Singh S (2012) Localized policy-based target tracking using wireless sensor networks, ACM Trans Sens Netw 3(27):1–30
2. Misra S, Chatterjee S (2014) Social choice considerations in cloud-assisted WBAN architecture for post-disaster healthcare: data aggregation and channelization. Inf Sci 284:95–117
3. Kar P, Roy A, Misra S, Obaidat MS (2015) Energy-efficient connectivity re-establishment in WSN in the presence of dumb nodes. In: Proceedings of IEEE ICC 2015—4th IEEE international workshop on smart communication protocols and algorithms (SCPA 2015), London, UK, pp 8–12
4. Roy A, Kar P, Misra S, Obaidat MS (2015) D3: distributed approach for the detection of dumb nodes in wireless sensor networks. Int J Commun Syst. https://doi.org/10.1002/dac.2913
5. Manish. Patel., Akshai. Agarwal.: Security Attacks in Wireless Sensor Networks: A Survey, Proceedings of the IEEE International Conference on Intelligent Systems and Signal Processing (2013)
6. Misra S, Goswami S, Taneja C, Mukherjee A (2014) Design and implementation analysis of a public key infrastructure-enabled security framework for zigbee sensor networks. Int J Commun Syst. https://doi.org/10.1002/dac.2893
7. Zhang Z, Li P, Wu F, Deng J (2015) Double edge-based trackback for wireless sensor networks, Int J Commun Netw Distrib Syst 15(1):107–126
8. Boukhlouf D, Kazar O, Kahloul L (2016) Network security: distributed intrusion detection system using mobile agent technology. Int J Commun Netw Distrib Syst 17(4):335–347
9. Nitesh K, Jana PK (2016) Distributed fault detection and recovery algorithms in two-tier wireless sensor networks. Int J Commun Netw Distrib Syst 16(3):281–286

10. Das AK (2015) An efficient and novel three-factor user authentication scheme for large-scale heterogeneous wireless sensor networks. Int J Commun Netw Distrib Syst 15(1):22–60
11. Chatterjee S, Das AK (2015) A secure and effective access control scheme for distributed wireless sensor networks. Int J Commun Netw Distrib Syst 14(1):40–73
12. Tripathy S (2013) ESF: an efficient security framework for wireless sensor network. Int J Commun Netw Distrib Syst 10(2):176–194
13. Yan L (2009) Game theoretic packet relaying model for wireless ad hoc networks. Int J Commun Netw Distrib Syst 3(1):68–89
14. Misra S, Singh R, Mohan SVR (2010) Information warfare-worthy jamming attack detection mechanism for wireless sensor networks using fuzzy inference system. Sensors 10(4):3444–3479
15. Misra S, Dhurandher SK, Rayankula A, Agrawal D (2010) Using honeynodes for defense against jamming attacks in wireless infrastructure-based networks. Comput Electr Eng 36(2):367–382
16. Misra S, Krishna P, Abraham K, Sasikumar N, Fredun S (2010) An adaptive learning routing protocol for the prevention of distributed denial of service attacks in wireless mesh networks. Comput Math Appl 60(2):294–306
17. Misra S, Ghosh A, Sagar AP, Obaidat MS (2010) Detection of identity-based attacks in wireless sensor networks using signalprints. In: Proceedings of the 2010 IEEE/ACM international conference on green computing and communications (GreenCom 2010), Hangzhou, China, pp 35–41
18. Dhurandher S, Misra S, Agrawal D, Rayankula A (2008) Using honey nodes along with channel surfing for defense against jamming attacks in wireless networks. In: Proceedings of the 3rd international conference on systems and networks communications (ICSNC 2008), Sliema, Malta, 26–31, pp 197–201
19. Singh R, Singh J, Singh R (2016) WRHT: a hybrid technique for detection of wormhole attack in wireless sensor networks. Hindawi Publishing Corporation. Mobile Information Systems, Volume, Article ID 8354930, pp 13
20. Chiu HS, Lui KS (2016) DEPHI: wormhole detection mechanism for ad hoc wireless networks. In: Proceedings of the IEEE 1st international symposium on wireless pervasive computing, pp 1–6
21. Ji S, Chen T, Zhong S (2015) Wormhole attack detection algorithms in wireless network coding systems. IEEE Trans Mob Comput 14(3):660–674
22. Mukherjee S, Chattopadhyay M, Chattopadhyay S, Kar P (2016) Wormhole detection based on ordinal MDS using RTT in wireless sensor network. J Comput Netw Commun, Article ID 3405264:15
23. Hai TH, Huh E-N, Jo M (2009) A lightweight intrusion detection framework for wireless sensor networks. Wirel Commun Mob Comput 10:559–572
24. Anwar RW, Bakhtiari M, Zainal A, Abdullah AH, Qureshi KN (2015) Enhanced trust aware routing against wormhole attacks in wireless sensor networks. Int Conf Smart Sens Appl
25. Khalil I, Bagchi S, Shroff NB (2008) MOBIWORP: mitigation of the wormhole attack in mobile multihop wireless networks. J Ad Hoc Netw 6:344–362
26. Qazi S, Raad R, Mu Y, Susilo W (2013) Securing DSR against wormhole attacks in multirate ad hoc networks. J Netw Comput Appl 582–593
27. Patel M, Aggarwal A, Chaubey N (2016) Two phase wormhole attack detection in dynamic wireless sensor networks. In: IEEE international conference on wireless communications signal processing and networking, 23–25 March, Chennai, India
28. Patel M, Aggarwal A, Chaubey N (2018) Detecting wormhole attack in mobility based wireless sensor networks. Int J Commun Netw Distrib Syst 21(2)
29. Patel MM, Agarwal A, Chaubey N (2017) Wormhole attacks and countermeasures in wireless sensor networks: a survey. Int J Eng Technol 9(2):1049–1060

Review of Blockchain Technology to Address Various Security Issues in Cloud Computing

Parin Patel and Hiren Patel

Abstract Being distributed in nature, Cloud is an in-use technology over the last decade where the data is stored on the Cloud service provider's site where the data owner does not have direct access to the storage where her data is stored. This leads to severe security issues. Out of various such issues, data correctness or integrity is one of the major aspects of Cloud data storage security. Researchers have been exploring the options of using blockchain technology to address the data storage correctness concern in Cloud computing. Blockchain technology has provided a new dimension to address the traditional security issues in computing technologies, where every transaction is treated as an atomic operation, and once performed, it becomes immutable. The transactions are stored in a block, and such blocks are linked together to form a chain through a cryptographic hash code. Such chains are distributed in nature in the sense that there is no single central administrative authority to manage them, and any node in the network can verify the correctness of data or transaction. In this paper, we intend to make an exhaustive study on various alternatives to solve the data integrity issues in Cloud computing using blockchain technology. Being distributed, immutable, transparent, cost effective and efficient to use, blockchain is definitely a solution to many issues including Cloud data storage security problems.

Keyword Cloud computing · Blockchain technology · Security · Storage · Smart contracts integrity

---

P. Patel (✉)
Research Scholar, Kadi Sarva Vishwavidyalaya, Gandhinagar, Gujarat 382015, India
e-mail: patelparinv@gmail.com

H. Patel
Vidush Somany Institute of Technology and Research, Kadi Sarva Vishwavidyalaya, Kadi 382715, India
e-mail: hbpatel1976@gmail.com

K. Kotecha et al. (eds.), *Data Science and Intelligent Applications*, Lecture Notes on Data Engineering and Communications Technologies 52,
https://doi.org/10.1007/978-981-15-4474-3_38

1 Introduction

The Cloud computing is utilized by numerous individuals in their everyday lives. Cloud computing is extremely helpful in business advancement as it gets surprising outcomes in an opportune way. Figure 1 shows the architecture of Cloud computing. There are three different types of Cloud services viz. software as a service (SaaS), infrastructure as a service (IaaS) and platform as a service (PaaS). SaaS is the service which provides online software services on pay and use basis. Client can pay to use services or applications that are hosted on the Cloud. Microsoft has Cloud-based online services like Office, Gmail, Outlook.com and Salesforce [1]. IaaS is the service which provides all infrastructure to build their application on Cloud-like servers, storage, workspace, etc. Amazon Web Services (AWS), Cisco Metapod, Google Compute Engine (GCE), Joyent and Microsoft Azure are the examples of it [2]. User can deploy their own software using platform as services (PaaS). DigitalOcean, Google Apps, Salesforce, Workday, Concur, Citrix GoToMeeting and Cisco WebEx are the examples of PaaS [2, 3]. There are three deployment models of Clouds like private Cloud, public Cloud and hybrid Cloud. Private Cloud is operated and maintained by specific organization. Private Cloud can be hosted by third party service provider or onsite of the organization. It provides more security, scalability and flexibility. Private Cloud examples are Citrix, Cisco, CSC, Dell, EMC, HP, IBM, Mirantis, Rackspace [4]. In public Cloud, Cloud service provider (CSP) makes it available for public and commercial use. Cloud provider manages all supporting infrastructure, hardware and software. Clients can access all services through Web browser. Public Cloud provides many features like scalability, law maintenance, reliability and lower cost services. Public Cloud examples are Amazon Elastic Compute, Google App Engine, IBM's Blue, Microsoft Azure, Salesforce Heroku and others

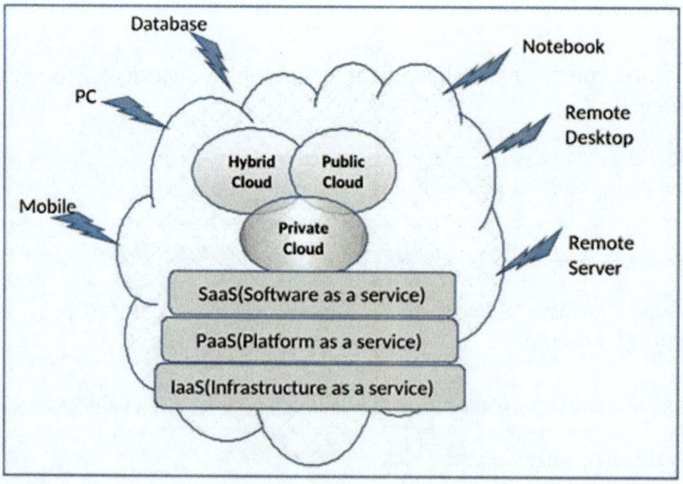

Fig. 1 Cloud architecture [3]

[5]. Community Cloud infrastructure is shared by the organizations which are having similar interest. Examples of community Cloud are Google Apps for Government, Microsoft Government Community Cloud [6]. Hybrid Cloud is the combination of the private and public Cloud. In this infrastructure, applications and data can easily move between private and public Clouds. Hybrid Clouds are more flexible, cost effective and provide control over the data. Management tools such as Cisco Cloud Center, Egenera RightScale Cloud Management and Scalr Enterprise Cloud Management Platform help businesses handle workflow creation, PAN Cloud Director service catalogs, billing and other tasks related to hybrid Cloud [7].

As the user data is stored on the Cloud service provider's site which is not under the control of the Cloud user, severe security issues may arise. Being a non-trustworthy element, CSP may play malicious and modify, destroy or use the data. Apart from this, several other issues that are mentioned beneath need also to be addressed. Multiple serious attacks like virus attack, Man-in-the-Cloud Attack, Denial of Service (DoS) Attacks, OpenStack Components Attacks and hacking of the client's data are the biggest Cloud computing data security issues [8]. When user is transferring important data on to the Cloud, it is important to ensure about security of the system. Choosing the ideal Cloud setup is also the major concern when you are adopting Cloud storage. If you are not choosing the correct Cloud, then it may be you have to face some serious hazards. Some companies have small data, so they prefer public Clouds, while huge data organizations usually use private Clouds. Cost barrier is also major obstacles for small organizations. For effective working of Cloud computing, you need to hold up under the high charges of the data transfer capacity. Data backup is the major issues faced by Cloud service in the case of data loss. There must be a proper backup policy for the recovery of data to deal with such kind of loss. Cloud management is the complex functionality of Cloud. It consists of lot of technical challenges.

At the point, when you store your information on Cloud, another person is having overall control on data. Data privacy and integrity are major issues in Cloud. You lose your privacy control on your data. Someone else is accessing your data without your permission, so it can lead to data leakage. Data is stored on the Cloud hacked or harmed by someone, so it may lead to data vulnerability. Devices are accessing your data that might not be safe. Storage gateways and API help to migrate data onto the Cloud. Data may be hacked when you store data on the Cloud.

There are multiple facets to address the security concerns of Cloud computing. Traditional cryptographic techniques are also widely used. However, researchers have started exploring the option of using recently introduced blockchain technology to solve the security problem of Cloud computing due to its decentralization and distributed nature.

Blockchain was fancied by an individual (or group of people) using the name Satoshi Nakamoto [9] in 2009 to serve as the general public transaction ledger of the cryptocurrency bitcoin. Blockchain is the distributed ledger which can be accessed by everyone on the network. In blockchain technology, anyone can join the network. A blockchain is the chain which contains data in every block. There are blocks in the blockchain, and every block consists of list of transaction, timestamp, nonce, previous hash value and Merkle root as shown in Fig. 2. Hash is the unique mathematical

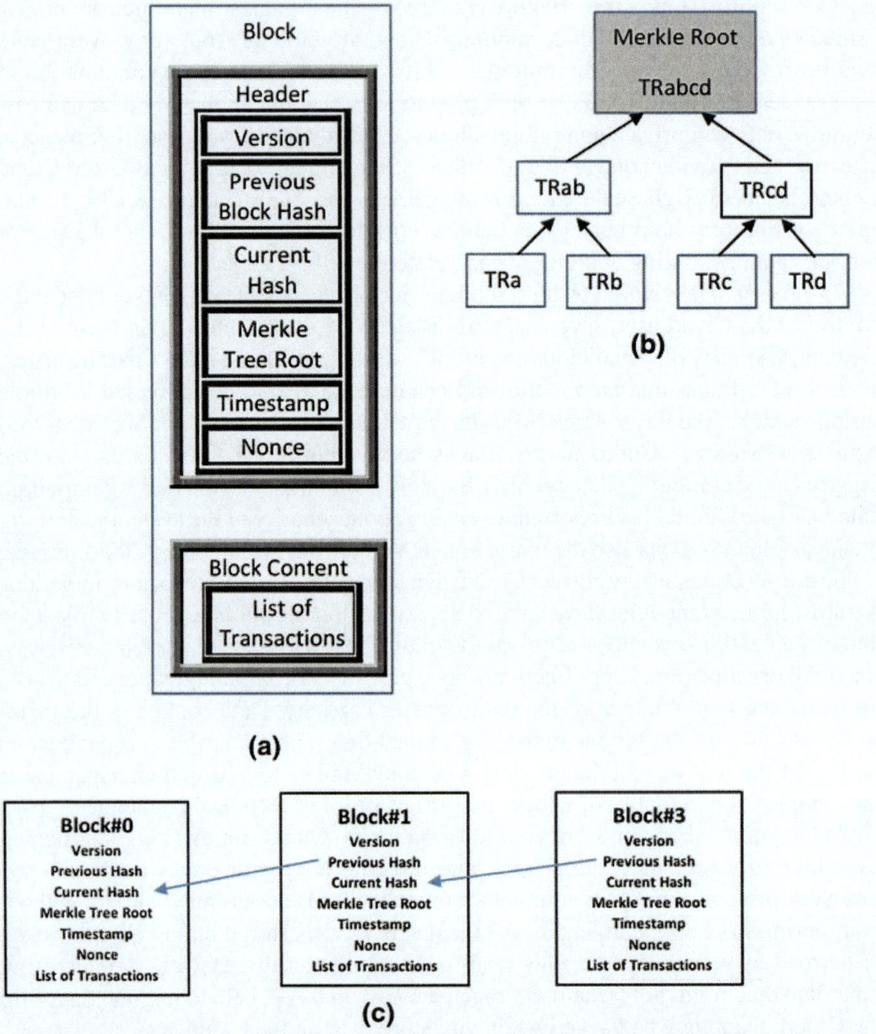

Fig. 2 **a** Blockchain block structure. **b** Merkle tree. **c** Blockchain structure [10]

code which is store with every block generated in the chain. Merkel tree is built continuously by hashing pair of nodes until only one hash left. Every leaf node is a transaction of data, and every non-leaf node is a hash of a previous hash values. Merkle tree requires even number of leaf that is why it is called as binary tree. After creation of the block, it is broadcasted through network, and then, validation process starts. Miners solve mathematical puzzles and lend their computing power. The process of solving puzzle and creating a hash is called mining. Miners required huge computational power to solve puzzles [10]. All nodes in the network check whether the calculation is correct or not. Once the majority of nodes in the network

Table 1 Blockchain platform comparison

Platform	Ethereum [12]	Hyperledger [13]	Bitcoin [14]
Smart contract	Yes	Yes	No
Data access	Public/private	Private	Public
Consensus	Proof-of-work	PBFT	Proof-of-work
Performance scalability	Low	High	Low

come to consensus and agree to the solution, the block is added to the chain. When new block is added to the chain, it updated in all existing copies into network.

Nonce value must be solved by miners. It is a complicated algorithm which is solved by miners in the network. Timestamp value is the time when block is added to the chain. The previous hash value is the hash value of the previous block. Using this previous hash value, new hash value is generated for new block. Transaction must be added when it was approved by the miners.

In a blockchain, every single node holds the copy of each and every transaction. If someone decides to alter the transaction than everyone in the network can verify that altered transaction. If someone is trying to cheat the system, then everyone in the network will not allow to perform the transaction. Noting down transaction in the ledger makes sure that double spending is not allowed. Moreover, it removes the trust we put into centralized system. Trust is developed through consensus where everyone is connected to network and making sure that the system is working fruitfully. Blockchain provides secured and tamperproof platform [11]. Comparison between blockchain platforms is described in Table 1. Ethereum and Hyperledger support smart contracts where bitcoin does not support. Ethereum supports both private and public data accessibility. Hyperledger supports private and bitcoin supports public data access. Ethereum and bitcoin use POW. Hyperledger uses PBFT. Ethereum and bitcoin have low scalability, and Hyperledger has high performance scalability.

In this paper, we study about blockchain technology and solution provided to security issues of Cloud storage. From the study, researcher can address various security issues by detailed study and understanding of this paper.

The rest of the paper is organized as follows. In Sect. 2, we discuss related work which addresses security issue in Cloud storage using blockchain technology. Section 3 focuses on the current security challenges and issues. Section 4 provides knowledge about the current tools available to implement blockchain using Cloud. We conclude our study in Sect. 5.

2 Related Work

Cloud storage uses Internet to store data over the network. Traditional Cloud storage technology handles information encryption, access control, data duplication and network management, etc. Cloud storage services cannot handle the issue of security and cannot be able to deal with the demands of users. Public key cryptography and identity-based cryptography play an important role for the security in Cloud. In public key cryptography, public–private key pair is used to encrypt and decrypt the data. Identity-based encryption used user identification, such as mobile number and email address for public key generation.

In the aspect of Cloud storage security for health data, authors of [15] have proposed a blockchain-based solution. In this proposed work, they used elliptic curve cryptography. All encrypted patients' data is stored on the blockchain using private accessible unit. Authors in [16] have proposed an architecture that protects data from unauthorized access using blockchain technology. In this system data stored in a distributed manner and manage access permission of files. Holt [17] has proposed architecture where any attempt to modify data cannot be accepted. In this paper, they have implemented log verification process using cryptographic solutions. Authors [18] have presented the work on privacy of audit logs using blockchain technology. They have used linked database technique which provides privacy and immutability to logs. Authors in [19] have discussed about the issue of log storage and proposed a secure log storage using blockchain which helps to crate immutable logs. It uses blockchain consensus algorithm for validation for logs.

Authors in [20] used genetic algorithm for file block replica placement problem between multiple users and multiple data center. Use blockchain to store metadata like URLs and hash values. In the proposed work [21], authors have implemented smart contract to store file of owner's data on blockchain network. They have also implemented verification and missing file recovery using smart contract on Cloud. Using blockchain technology, authors in [22] record unalterable timestamp and generate blockchain receipt for each and every data block for validation of provenance data. Blockchain-based provenance system enhances the integrity and availability of data. Authors in [23] proposed a secure solution for data provenance using two-folder encryption method to enhance security. SPROVE [24] architecture is implemented using digital signature and encryption to provide confidentiality and integrity to provenance data. Dstore [25] allows data owners to hire the local disks of other peers to save their data in a distributed manner without using third party. In this solution, periodically, audit happens with challenge verification solution based on Merkle hash tree. It uses smart contract to maintain lease relationships. Paper [26] has proposed an architecture named as Storj which provides end-to-end encryption of data without using third party. In blockchain-based P2P, Cloud storage network user can share and transfer date without relying on third party service providers.

3 Current Challenges and Opportunities

Secure Cloud storage has implemented using blockchain technology for different domains. From the study, we have identified some security issues that must be still addressed for security of the Cloud storage. As per the current data breaches, blockchain technology still needs more security concern. As per the report from Coindesk within the first 9 months of 2018, over $927 million were stolen by hackers from cryptocurrency exchanges [27]. In 2018, numerous exquisite cryptocurrencies such as ZenCash, Verge and Ethereum Classic fell victim to 51% attacks [28]. As per the current survey, there are many data breaches which occur due to bugs in the system. System does not provide integrity unless there is a problem of key theft or loss by hackers [29]. An attacker tries to access user's key by doing various attempts on user's smartphone or personal computer. There is still key theft and loss which much be addressed. Outsourcing of data security is significant key issue. Management of access control is also one important key issue. At the point, when somebody needs to get information on Cloud, then how to give access to that clients is a likewise challenge in Cloud and blockchain. Access control violations are likewise should be tended to. In the current system, transactions are scattered; it will increase the chances of security attacks [29]. Despite the fact that the security of blockchain is ceaselessly improved, issues have proceeded to be reported.

4 Tools and Technologies

Solidity is the language which was intended to target Ethereum Virtual Machine (EVM). Developers can execute business logic in smart contract with the use of solidity. Ethereum node can be implemented using go programming language in Geth. To implement different tasks on Ethereum blockchain, Geth is used [30]. Truffle is a framework which is used to develop Ethereum-based application [30]. To convert solidity program into EVM readable format, Solc compiler is used. Ethereum-based dApps can be developed using Embark framework [30]. Metamask is a wallet which used to give connection between Ethereum blockchain and Internet browsers like Firefox or Chrome and so forth [30]. Hyperledger provides blockchain platforms and networks for software developers. It is open-source software development approach that gives transparency, longevity and interoperability [31]. Remix IDE is one of the easier browser-based tools which is used for creation and deployment of smart contracts. It provides writing, debugging, testing and deploying smart contract written in programming language called solidity [32]. Solium tool makes sure that the code is formatted and resolves security and vulnerability issues in your code. It ensures that code is free from security holes [33]. Ganache is the tool which allows to create private Ethereum blockchain and test dApps. It allows to test without paying any gas. It also manages mining speed and gas cost [34]. Mist provides you the place where you can store ether tokens and run smart contract. It is an official Ethereum wallet which is available in Linux, Mac and Windows operating systems [35].

5 Conclusion

Data security in the Cloud environment has been a burning issue for quite a while. Apart from traditional network security approaches, the issue can also be addressed using blockchain technology. Very few researchers have plunged into this domain. We intend to address the issue of data storage security in Cloud computing with blockchain implementation. Usage of the third party between cloud user and service provider can be eliminated (due to severe concern of trust) by the notion of the smart contract in Blockchain. Another property of blockchain viz. immutability has affirmative outcomes in terms of preserving storage correctness through maintaining every transaction intact. The distributed nature of blockchain technology allows flexibility in different ways viz. (i) any node can verify correctness of transaction (ii) malicious nodes cannot add transaction to network as it has to be verified by selected number of legitimate nodes (iii) to add a block containing valid transaction in a network, the node (miner) has to prove its credibility through consensus mechanisms such as PoW and PoS. In this research, we have explored various recently proposed mechanisms to address the issue of Cloud data storage correctness through blockchain technology. We have studied such approaches in detail and analyzed them with their pros and cons. We further have discussed the current challenges and opportunities in this domain along with tools and technologies which can be used for further implementations. In the future, we intend to propose a security model for Cloud data storage with the use of blockchain technology.

References

1. Avoyan H (2019) 3 types of cloud computing services—monitis blog. Monitis blog. Available at: https://www.monitis.com/blog/3-types-of-Cloud-computing-services/. Accessed 30 Nov 2019
2. IaaS P (2019) SaaS (Explained and compared). Apprenda. Available at: https://apprenda.com/library/paas/iaas-paas-saas-explained-compared/. Accessed 30 Nov 2019
3. Lifewire (2019) Take 5 minutes to learn the basics of cloud computing. Available at: https://www.lifewire.com/what-is-Cloud-computing-817770. Accessed 30 Nov 2019
4. Anon (2019) Available at: https://www.networkworld.com/article/3007991/3-types-of-private-Clouds-which-one-s-right-for-you.htm. Accessed 30 Nov 2019
5. SAM Solutions (2019) 4 best cloud deployment models (An overview with examples). SAM Solutions. Available at: https://www.sam-solutions.com/blog/four-best-Cloud-deployment-models-you-need-to-know/. Accessed 30 Nov 2019
6. Techno-pulse.com (2019) Techno-Pulse. Available at: https://www.techno-pulse.com/. Accessed 30 Nov 2019

7. Search Cloud Computing () What is hybrid Cloud? - Definition from WhatIs.com. [online] Available at: https://searchCloudcomputing.techtarget.com/definition/hybrid-Cloud. Accessed 30 Nov 2019
8. Jabir RM, Khanji SI, Ahmad LA, Alfandi O, Said H (2016) Analysis of cloud computing attacks and countermeasures. pp 1–1. https://doi.org/10.1109/icact.2016.7423295
9. Bitcoin.org (2019). Available at: https://bitcoin.org/bitcoin.pdf. Accessed 30 Nov 2019
10. Zheng Z, Xie S, Dai H, Chen X, Wang H (2017) An overview of blockchain technology: architecture, consensus, and future trends. Available at: https://doi.org/10.1109/bigdatacongress.2017.85
11. Cheng S et al (2017) Research on application model of blockchain technology in distributed electricity market. IOP Conf Ser: Earth Environ Sci 93:012065. https://doi.org/10.1088/1755-1315/93/1/012065
12. Ethereum GitHub implementation. Available at: https://github.com/ethereum/goethereum. Accessed 13 Feb 2018
13. Hyperledger GitHub implementation. Available at: https://github.com/hyperledger/fabric-sdk-py. Accessed 13 Feb 2018
14. Bitcoin GitHub implementation. Available at: https://github.com/bitcoin/bitcoin. Accessed 13 Feb 2018
15. Al Omar A, Bhuiyan MZ, Basu A, Kiyomoto S, Rahman MS (2019) Privacy-friendly platform for healthcare data in Cloud based on blockchain environment. Futur Gener Comput Syst 95C:511–521. Available at: https://doi.org/10.1016/j.future.2018.12.044
16. Zyskind G, Nathan O et al. (2015) Decentralizing privacy: using blockchain to protect personal data. In: Security and privacy workshops (SPW), IEEE, pp 80–184
17. Holt JE. Logcrypt: forward security and public verification for secure audit logs. In: Proceedings of the 4th Australasian workshops on grid computing and e-research (ACSW'06), Tasmania, Australia, pp 203–211
18. Sutton A, Samavi R (2017) Blockchain enabled privacy audit logs. In: d'Amato C et al (ed) The semantic web—ISWC 2017. ISWC 2017. Lecture notes in computer science, vol 10587. Springer, Cham
19. Kumar M, Singh AK, Kumar TVS (2018) Secure log storage using blockchain and cloud infrastructure. In: 9th ICCCNT, IISC, Bengaluru, India, IEEE
20. Li J, Wu J, Chen L (2018) Block-secure: blockchain based scheme for secure P2P Cloud storage. Inf Sci 465:219–231. https://doi.org/10.1016/j.ins.2018.06.071
21. Li J, Wu J, Chen L, Li J (2018) Deduplication with blockchain for secure cloud storage. In: 6th proceedings of CCF conference, big data, China. Available at: https://doi.org/10.1007/978-981-13-2922-7_36
22. Liang X, Shetty S, Tosh D, Kamhoua C, Kwiat K, Njilla L (2017) ProvChain: a blockchain-based data provenance architecture in cloud environment with enhanced privacy and availability. In: 2017 17th IEEE/ACM international symposium on cluster, cloud and grid computing (CCGRID), Madrid, pp 468–477. https://doi.org/10.1109/ccgrid.2017.8
23. Asghar MR, Ion M, Russello G, Crispo B (2012) Securing data provenance in the cloud. In: Camenisch J, Kesdogan D (eds) Open problems in network security. Lecture notes in computer science, vol 7039. Springer, Berlin, Heidelberg
24. Hasan R, Sion R, Winslett M (2009) Sprove 2.0: A highly configurable platform-independent library for secure provenance, ACM, CCS, Chicago, IL, USA
25. Xue J, Xu C, Zhang Y, Bai L (2018) Dstore: a distributed cloud storage system based on smart contracts and blockchain. In: 18th international conference, ICA3PP 2018, Guangzhou, China, Part III. https://doi.org/10.1007/978-3-030-05057-3_30
26. Andrew A (2019) Blockchain security: how far have we come in 2019? Forbes, Forbes Magazine, 27 Mar 2019. Available at: https://www.forbes.com/sites/andrewarnold/2019/03/27/blockchain-security-how-far-have-we-come-in-2019/#26feb29b2457
27. Chandhok A (2019) Top five blockchain security issues in 2019. LedgerOps, LedgerOps—Elite Cybersecurity, 28 Mar 2019. Available at: https://ledgerops.com/blog/2019/03/28/top-five-blockchain-security-issues-in-2019

28. Wilkinson S, Boshevski T, Brandoff J, Buterin V (2014) Storj a peer-to-peer cloud storage network
29. Park JH, Park JH (2017) Blockchain security in cloud computing: use cases, challenges, and solutions. Symmetry 9:164. https://doi.org/10.3390/sym9080164
30. Blockchain Technology Explained: Introduction, Meaning, and Applications. Hackernoon. Available at: https://www.hackernoon.com/blockchain-technology-explained-introduction-meaning-and-applications-edbd6759a2b2 Accessed 30 Nov 2019
31. Hyperledger (2019) About—Hyperledger. Available at: https://www.hyperledger.org/about Accessed 30 Nov 2019
32. SitePoint (2019) Remix: develop smart contracts for the ethereum blockchain—SitePoint. Available at: https://www.sitepoint.com/remix-smart-contracts-ethereum-blockchain/ Accessed 30 Nov 2019
33. Dzone.com (2019) 10 tools for blockchain development—dzone security. Available at: https://dzone.com/articles/10-tools-for-blockchain-development Accessed 30 Nov 2019
34. Codementor.io (2019) Developing for ethereum: getting started with ganache. Codementor. Available at: https://www.codementor.io/swader/developing-for-ethereum-getting-started-with-ganache-l6abwh62j Accessed 20 No 2019
35. Ltd, Attores Pte (2019) Step-by-step guide: getting started with ethereum mist wallet. Available at: https://www.medium.com/@attores/step-by-step-guide-getting-started-with-ethereum-mist-wallet-772a3cc99af4 Accessed 17 Nov 2019

Comparative Analysis of Statistical Methods for Vehicle Detection in the Application of ITS for Monitoring Traffic and Road Accidents Using IoT

Diya Vadhwani and Devendra Thakor

Abstract In this paper, the intelligent transportation systems application an accident alert system was considered as scope for research. The vehicle detection analysis in the ITS application of accident alert is one of the major tasks. As there are many accidents occurring in construction area along roadside, so there should be safety for workers on roadside. Road safety and workers safety are must; so to avoid the accident, the prototype was developed using Internet of things to evaluate the distance of vehicles crossing near and far from prone area. Flow rate of vehicles and vehicle count has some statistics so this paper gives an immense idea to statistical evaluate analyzed the vehicles using statistical Poisson's, binomial and negative binomial methods.

Keywords Intelligent transportation system · Internet of things · Poisson's method · Binomial · Negative binomial · Accident alert · Vehicle detection

1 Introduction

1.1 Introduction to ITS, Accident Alert and IoT

Many applications of intelligent transportation system such as vehicle detection, accident notification, toll collection, smart parking, integrated multimodal transport and traffic analysis are useful for the safety purpose of peoples. The main aim of this paper was to how to use and compare the statistical methods in the vehicle detection of accident alert application of ITS. In the application of accident alert, the vehicle detection is one of the crucial tasks to be completed, so flow rate of vehicles and count of vehicles (here distance from prone area) were calculated using Poisson's, binomial and negative binomial method.

D. Vadhwani (✉) · D. Thakor
Uka Tarsadia University, Bardoli, Gujarat, India
e-mail: dnvadhwani@gmail.com

K. Kotecha et al. (eds.), *Data Science and Intelligent Applications*,
Lecture Notes on Data Engineering and Communications Technologies 52,
https://doi.org/10.1007/978-981-15-4474-3_39

In this paper, the novel approach was developed to evaluate the distances of vehicles crossing near and far from prone area using Internet of things technology. This distance will allow us to find some statistics to evaluate the count and flow rate of vehicles. Count and flow rate of vehicles are used in analyzing the traffic measurements [1].

2 Literature Review

The literature review of various statistical methods used in vehicle analysis was done to use an appropriate model for flow rate of vehicles and vehicle count. In the following section, various statistical methods give an appropriate study of parameters and different function evaluations to find the probability of vehicle arrivals. The below comparative analysis table gives the study of different statistical methods to evaluate the parameters like count of vehicles, flow rate, volume of traffic, density, speed, etc. To detect the vehicles in accident, notification is an important task. In this paper, the distance parameter is considered for parametric evaluation. Since the parameters calculated in [1–3] were count of vehicles, but in order to avoid the accident at construction site, there should be information (Table 1).

1. The main contributions of this paper are as follows:
2. The Internet of things technology was used in collection of datasets.
3. In the proposed method, the distance is calculated from prone area.
4. The distance parameter is the random value of vehicle distance which is used in the statistical evaluation to find the number of vehicles crossing the prone area which is calculated using Python Poisson's method, binomial method and negative binomial method (SCIPY-MODULE, NUMPY-MODULE).
5. Comparative analysis between Poisson's, binomial and negative binomial methods gives us to choose the appropriate statistical method for future prediction methods for the evaluation of vehicles and accident avoidance.

3 Proposed Method

The proposed method for the evaluation of vehicle detection and traffic flows is taken from my literature study and analysis papers [1, 4].

Assumption for vehicle detection and analysis from [1] was taken, and algorithm was made to find the number of vehicles using Poisson's, binomial and negative binomial methods.

Let us consider that after the prototype [1] installation, the datasets are collected. Let the vehicles crossed in the near prone area is the random number. So, to apply different statistical methods, there should be some characteristics which are as follows: It is a discrete distribution; each occurrence is independent of the other occurrences,

Table 1 Comparative analysis of statistical methods about the vehicles which comes nearer to prone area. So, our contribution is the distance calculation of vehicles from prone area

Paper title	Statistical methods	Parameters	Comments/remarks
Probability approach for ranking high-accident locations [2]	Poisson's and binomial distribution	Accident rate, hazardous location	Estimates the probability of locations which is hazardous
Mathematical study for traffic flow and traffic density in Kigali roads [3]	Poisson's distribution	Traffic flow, speed, density	Number of accidents and number of vehicles are analyzed
Statistical analysis of vehicle detection in the ITS application for monitoring the traffic and road accident using Internet of things [1]	Poisson's method	Count of vehicles, flow rate	Evaluate the count of vehicles and find the probability of accident after using the prototype in the construction area
Applying the random effect negative binomial model to examine traffic accident occurrence at signalized intersections [5]	Random effect negative binomial (RENB)	Traffic volume, numbers of phases per cycle	The geometric elements, traffic factors and traffic control measures, total accident frequency at signalized intersections in Singapore were evaluated

and the random number describes discrete occurrences over an interval, and the occurrences in each interval can range from zero to infinity.

Algorithm steps for system model for vehicle detection can be installed externally (from [1]): (for vehicle detection and analysis).

Step 1: Find the distance of vehicle crossed in construction/prone area.
Step 2: When the vehicle detected is nearer to the prone area, i.e., distance less than 4 m, then the warning will be sent to workers using a buzzer or LED; here, LED is considered as output. So, the LED output will be HIGH.
Step 3: When the vehicle is detected is far from the prone area, i.e., distance more than 4 m, then no warning will be sent to workers. So, the LED output will be LOW.
Step 4: Let the number of vehicles is 100 as per Poisson's method in Python (vehicle count = n/here random number = 96 as per datasets collected).
Step 5: Let the number of vehicles is 100 as per binomial and negative binomial in Python (vehicle count = n/here random number = 96 as per datasets collected).
Step 6: Check the distance value (If distance < 4 m, find the number of vehicles).

Step 7: Apply the Poisson's distribution method, binomial and negative binomial method on random values (vehicle count/vehicle detected).
Step 8: Find the count of vehicles whose distance is less than 4 m and greater than 0 using (Poisson's, binomial and negative binomial methods).
Step 9: Find probabilities using Poisson's method as: *P*(*n*) = Probability of exactly '*n*' vehicles arriving in given time.

$$P(n) = \lambda^n e^{-\lambda}/n!$$

where
P(*n*) = Probability of exactly '*n*' vehicles arriving in given time, λ = Average arrival rate (vehicles/unit time), *n* = Number of vehicles arriving in specific time interval.
e = 2.71828 (constant) ('*e*' is the base of the natural logarithm system).
Step 10: Find the Probability using binomial method:
The binomial distribution formula is:

$$b(x;\ n,\ P) = nCx * Px * (1 - P)n - x$$

where *b* = Binomial probability, *x* = Total number of 'successes' (here, vehicles crossing near prone area) *P* = Probability of a success on an individual vehicle, n = Number of vehicles.
Step 11: Find the probability using negative binomial method.
Let the negative binomial experiment consists of x trials, i.e., number of vehicles and results in r successes, i.e., number of vehicles whose distance is less than 4 m. If the probability of success (vehicle detected near prone area) on an individual/each trial is P, then the negative binomial probability is:

$$b * (x;\ r,\ P) = x - 1Cr - 1 * Pr * (1 - P)x - r,$$

where *x* is the number of vehicles required to produce *r* successes in a negative binomial experiment, *P* is the probability of success.
Step 12: Compare the probability and statistics collected from all the methods.

4 Implementation and Comparative Analysis of Statistical Methods

In this paper, the algorithm is implemented in Python language to evaluate the probability, number of vehicles in the accident alert method using Poisson's, binomial and negative binomial methods. The dataset was collected for 1 h to check the distances. Number of vehicles = 96, number of vehicles near prone area = 29, probability of success = 0.3. The following were the results generated from all three statistical distributions (Figs. 1, 2 and 3):

From Poisson's Distribution

To calculate one hour data collected total of 96 vehicles per hour, λ = I/3600 * t [1]

λ = 96 * 60/3600 = 1.6, λ = 1.6 vehicles per second.

Probability of exactly arrivals = 1.89, mean = 1.6, variance = 1.6, standard deviation = 1.264, $P(n) = 0.20189$ for $n = 0$, $P(n) = 0.\ 3230$ for $n = 1$ and $P(n) = 0.000382$ for $n = 9$.

From Binomial Distribution

X (Binomial (n, p)) $n = 96, p = 0.3, x = 29$, probability using binomial method: $P(X = x) = 0.0883$, $P(X \leq x) = 0.5671$, $P(X < x) = 0.47931$, $P(X \geq x) = 0.5206$, $P(X > x) = 0.4328$, mean = 28.8, variance = 20.16, standard deviation = 4.49.

From Negative Binomial

X (NB(r, p)) $n = 96, p = 0.3, x = 29$, $P(X = x) = 0.00461$, $P(X \leq x) = 0.96318$, $P(X \geq x) = 0.4143$.

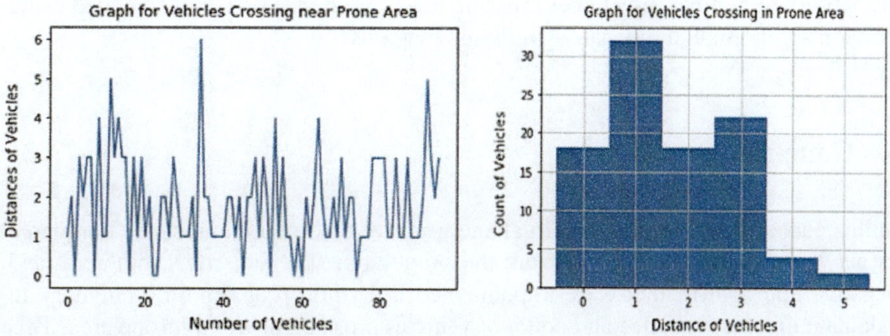

Fig. 1 Graph of poisson (vehicle count less than 4 m)

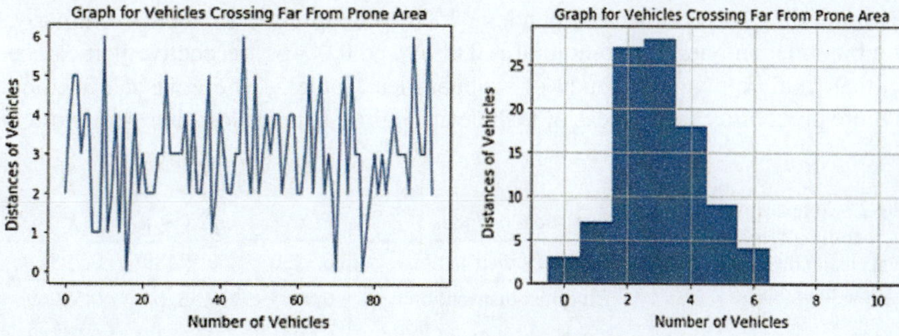

Fig. 2 Graph of binomial method (vehicle count less than 4 m)

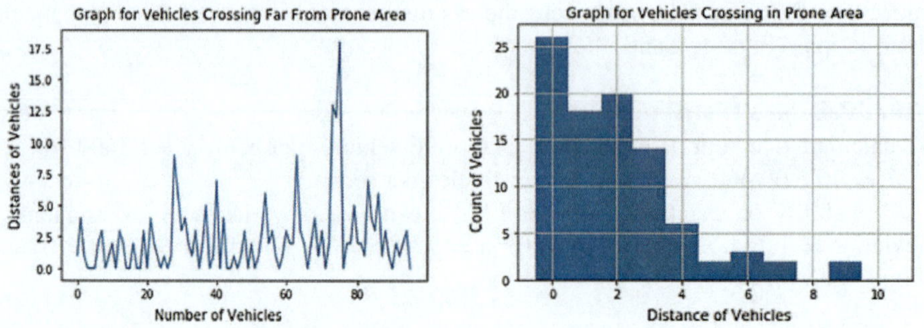

Fig. 3 Graph of negative binomial (vehicle count less than 4 m)

Probability of exactly arrivals = 1.89, mean = 67.667, variance = 225.556, standard deviation = 15.019

The following is the comparative analysis for probability distributions:

The analysis of Poisson's, binomial and negative binomial method is done as per the number of vehicles crossing nearer to prone area; so for safety of workers and to avoid the accident at construction site, the new methods can be implemented, and probability of number of vehicles crossing nearer prone area can be calculated using Poisson's, binomial and negative binomial methods.

5 Conclusion

In this paper, the Poisson's, binomial and negative binomial methods are compared as per the distances calculated from the prone area. The Poisson's, binomial and negative binomial methods are implemented in Python programming language to calculate the total vehicles and count of vehicles crossing nearer to prone area. This work is useful in finding the distances of vehicles nearer to prone area, which helps in evaluation of probability of accidents per unit time. As per Poisson's method, the probability of vehicles crossing nearer in the construction area was 0.3230, which helps in finding the number of vehicles which can cause the accidents. Similarly, using binomial and negative binomial is 0.0883 and 0.00461, respectively. It can be seen that from comparative study of probabilities (Table 2), the Poisson's method has more probability for success of vehicle arrivals at distance less than 4 m. Future

Table 2 Comparative study of probability of vehicle arrivals using different statistical methods

Statistical methods	$P(X = x)$	$P(X \leq x)$	$P(X \geq x)$
Poisson's distribution	0.3230	0.52489	0.7981
Binomial distribution	0.0883	0.56761	0.52069
Negative binomial distribution	0.00461	0.96318	0.04143

work is to calculate the traffic flow parameters and evaluate more number of statistical methods and compare with the implemented methods. Also by considering the traffic parameter, the prediction methods can be evaluated.

References

1. Vadhwani DN, Thakor D (2019) Statistical analysis of vehicle detection in the ITS application for monitoring the traffic and road accident using internet of things. In: Paper accepted and presented in international conference on advances in VLSI and embedded systems (AVES-2019). Springer, Surat, 20–21 Dec
2. Al-Ghamdi AS (2000) Probability approach for ranking high-accident locations. WIT Trans. Built Environ 49
3. Idrissa K (2017) Mathematical study for traffic flow and traffic density in kigali roads. Int J Math Comput Sci 11(3):104–108
4. Vadhwani DN, Buch S (2019) A novel approach for the ITS application to prevent accidents using wireless sensor network, IoT and VANET. In: 2019 IEEE international conference on electrical, computer and communication technologies (ICECCT), IEEE, pp 1–7
5. Chin HC, Quddus MA (2003) Applying the random effect negative binomial model to examine traffic accident occurrence at signalized intersections. Accid Anal Prev 35(2):253–259

An Extensive Survey on Consensus Mechanisms for Blockchain Technology

Jalpa Khamar and Hiren Patel

Abstract Blockchain technology is a cryptographic technique that enables users to maintain their data in a decentral way with a non-editable form without an arbitrator. Originally motivated from the underlying platform of bitcoin cryptocurrency, researchers found blockchain to be very useful in many application domains, including smart contracts, insurance, banking, finance, and many other sectors. As multiple stakeholders try to add a block (containing the data or transactions) into the chain, there is a need for a mechanism to come to an agreement on which stakeholder's block should be added. Such a mechanism is known as consensus mechanism which is a primary factor of significance because it decides the correctness of data to be added as well as trustworthiness of the node which is attempting to add the block. An efficient consensus mechanism achieves security, higher precision, and better performance. In this research, we aim to study various consensus mechanisms in blockchain technology with their strengths and challenges. We discuss factors affecting the consensus mechanism and provide our suggestions for drafting an effective consensus mechanism.

Keywords Blockchain · Consensus · Bitcoin

J. Khamar (✉)
LDRP Institute of Technology and Research, Research Scholar, Kadi Sarva Vishwavidyalaya, Gandhinagar, India
e-mail: khamarjalpa7@gmail.com

H. Patel
Vidush Somany Institute of Technology and Research, Kadi Sarva Vishwavidyalaya, Gandhinagar, Gujarat 382015, India
e-mail: hbpatel1976@gmail.com

K. Kotecha et al. (eds.), *Data Science and Intelligent Applications*, Lecture Notes on Data Engineering and Communications Technologies 52, https://doi.org/10.1007/978-981-15-4474-3_40

1 Introduction

The primitive concepts of carrying out some computation work or burn some of your wealth in any form before you add the block to the chain was first introduced by a group of researchers in 1991 [1]. However, the idea of blockchain was mostly not in use until it was first referred by Satoshi Nakamoto [2] in 2009 which was the birth of bitcoin cryptocurrency.

The blockchain is a decentralized, distributed ledger uses to record transactions between two parties in a secure, transparent, and immutable manner [3]. Data is recorded in blocks that are needed to be verified by network participants (also known as peers or nodes) before it is finally added on to the chain. The dynamically run-time updated blockchain ledger is available on a distributed peer-to-peer network across all stakeholders, which makes it possible for all participants to verify; hence, a group of malicious people cannot just add any random data block to the chain. Once fixed numbers of participants (not knowing each other and situated geographically apart, i.e., located decentralized) verify/confirm the transactions in the block, the block is ready to be added into the chain. Once a block is added to the chain, it is immutable, i.e., it cannot be altered, tampered, or deleted. The added block is visible to every node of the network, i.e., the block is transparent to the entire network. Any node of the network can verify the integrity of any block which results in auditability property of the blockchain. The blockchain, therefore, enables us to securely record, store, and transfer data without the need for a central authority to process and verify it. All the transactions occur in a scattered way that expels the need to authorize the transactions by mediators [4]. Blockchain has some key characteristics (as mentioned above), such as decentralization, transparency, immutability, and auditability [5].

In the case of bitcoin [2], individuals can make electronic money transactions without the need for a bank or involvement of any trusted third party; enabling each user to directly interact with others without any extra cost to be paid on the transactions. Although bitcoin is the most distinguished application of blockchain, the primeval concept can be applied to disparate application areas beyond cryptocurrencies.

In recent times, blockchain has also gained its popularity among researchers as the notion behind blockchain can be used in variety of domains. Many researchers have been working in different application areas such as study of digital currency based on blockchain [6], some are working on blockchain applicability in non-digital currency such as application of blockchain in smart city [7], medical information security management [8], and also in underlying blockchain technology such as scalability of consensus algorithms [9], smart contracts [10], difficulty control in mining [11] to name a few.

The rest of the paper is organized as follows. Section 2 discusses the background theory of blockchain technology. Section 3 depicts various types of blockchain. Different consensus mechanisms are studied in Sect. 4 followed by challenges and opportunities conversed in Sect. 5. Different tools and technologies related to blockchain technology have been conferred in Sect. 6. The paper has been concluded by a list of references used.

2 Basics of Blockchain

Blockchain technology has taken the world by surprise, with the hundreds of businesses and flourishing projects around the world. Although it seems a relatively new concept, blockchain, as well as decentralization, have been around for quite some time; these technologies have received attention in recent years. To explain how blockchain works, we must define the fundamental concept of it. The blockchain is a distributed ledger (data structure) which contains information about transactions or events. This is replicated and shared among network participants [12]. The blockchain can also be used as a generic term to refer to purely distributed peer-to-peer systems as a whole instead of referring to a software unit that is part of a purely distributed system [13]. Blockchain is an sequential list of blocks where each block contains a small list of transactions. Each block in a blockchain is "chained" to the previous block, by containing a hash of the representation of the previous block as seen in Fig. 1. So, historical transactions in the blockchain cannot be deleted or modified without invalidating the chain of hash. Combined with computation and incentive constraints block creation schemes, this can in practice prevent falsification and revision of the information stored in the blockchain. The first generation of blockchains, like bitcoin, provided a large public book to store cryptographically signed financial data transactions [14].

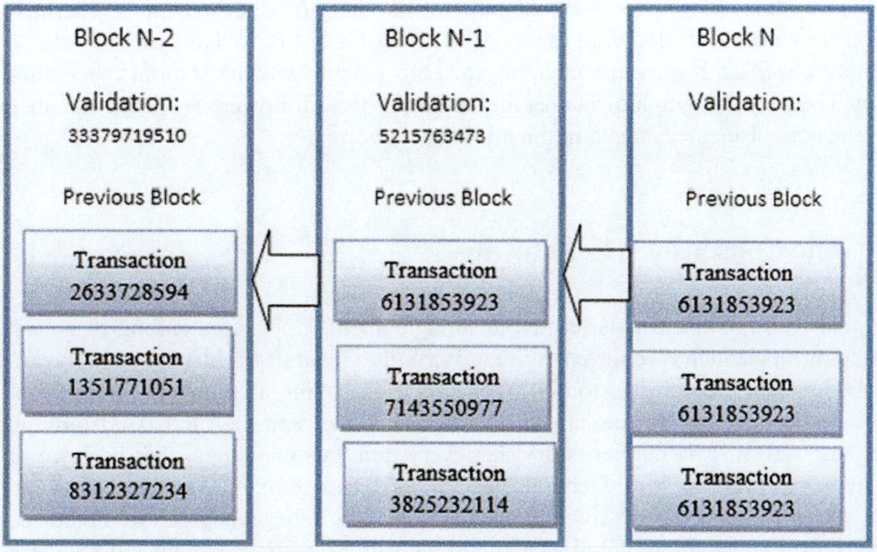

Fig. 1 Chain of blocks in blockchain

3 Types of Blockchain

Before we introduce the words such as permissioned blockchain or permissionless blockchain, we need to understand the specific application domain of the blockchain in form of centralized or decentralized monitoring mechanism. There are few applications in which a closed group of users are permitted to access the chain and hence it required a centralized approach to the chain. For such type of application, permissioned (or sometimes referred to as private) blockchain is used. Examples of permissioned blockchain are Ripple [15] Corda [16], Hyperledger Fabric [17]. In contrast to it, permissionless (sometimes referred to as public) blockchain mechanism is an approach where any number of participants may join the network. There is no central authority or restrictions for getting someone registered to the network. All the participants are able to access the database and store a local copy, and according to their computing power, they can update it. Bitcoin [2] and other cryptocurrencies such as Ethereum [18] are examples permissionless blockchain. There is also a hybrid category named consortium blockchains which is open to the public but not all data is available to all participants. Users have limited rights and blocks are validated based on predefined rules. Consortium blockchains are, therefore, "partly decentralized". R3 consortium [19] is an example of a consortium blockchain. In all such category, to validate transaction authenticity, to put the legitimate transactions in a block, to verify the block, and to add the verified block to the chain, there is a need for some work (in terms of computation or other) to be carried out. In a permissionless blockchain, the proof of work (PoW) [20] is one of such compute-intensive algorithms. And the process of adding such block to the network is often called mining. The next paragraph discusses the mining process followed by the significance of the consensus mechanism in the mining process.

4 The Consensus Mechanisms

In real world, consensus is termed as an agreement. Consensus among dishonest nodes in blockchain is achieved through Byzantine Generals Problem [21]. In Byzantine Generals Problem, a group of Generals each having a commando over them, decides to attack or to retreat. All the messages between then is passed through any messenger. It is not necessary that every time this messenger will be loyal; it may be betrayer. So, loyal Generals need to reach to a common consensus knowing the existence of betrayer. Since blockchain does not rely on any central authority, distributed nodes are required to agree to the validity of transactions and for these consensus mechanisms play a vital role in blockchain. We present a taxonomy of different consensus mechanisms in the blockchain by classifying them on the basis of the different environment of blockchain.

4.1 Permissionless Blockchain

4.1.1 Proof of Work

Proof of work (PoW) is a consensus algorithm which was firstly used in the bitcoin [2]. In PoW, by performing some form of computational work, consensus is achieved [22], For instance, in decentralized network, some value has to be selected to store or record the transactions. It can be done by selecting any random value. However, selecting random value is vulnerable to attacks, so to avoid attacks, nodes need to perform lots of computation in different forms (e.g., adding nonce to the original value). By using node's own computational power, it can be decided which transaction is to be added to the blockchain.

4.1.2 Proof of Stake

Proof of stake (PoS) consensus requires the node with larger amount of stake (e.g., coin, mileage, speed, etc.) to validate the data [23]. The stakeholders having more coins are more likely to add a new block to the existing blockchain [24]. PoS consumes less power as compared to PoW. In PoS, at the establishment of blockchain network, tokens are distributed to all the validating nodes. During regular time intervals, a specific node is selected to commit the new block. But if any node having more stakes, it has more power and it can commit the new block. Usually, PoS requires considerably less computational work, so the cost of executing PoS is substantially lower. Main problem with the PoS is "nothing at stake" problem. In the PoW blockchains, reward is given to the longest chain, so the miners are awarded to mine one single chain. But with PoS, there is little to prevent a miner from mining on numerous PoS chains and the cost of mining is very low. Therefore, a PoS miner operating on various chains can make it difficult for the network to reach consensus.

4.1.3 Delegated Proof of Stake

There are many peoples in delegated who have to vote for a delegation, which also includes some "witnesses" who are miners who review the transactions and maintain the chain [25]. A person having more stakes has more power to vote for the witness. After the review, the witnesses, therein check the transactions and create blocks, including the valid ones. The list of witnesses is always mixed. The witnesses will have to produce blocks consecutively in the list one at a time with a build up speed of 2 s per block. If a witness does not produce his or her block, it may be removed from the delegation. Whenever a witness creates a block to attach to the chain, he receives a reward.

4.1.4 Proof of Activity

The objective of the PoA protocol is to have a decentralized cyptocurrency network whose security is based on a combination of proof of work and proof of stake [26]. In general, PoW protocols give decision-making authority to entities performing computational tasks, while PoS protocols confer decision-making power to entities that hold a stake in the system. While we argue that proof of stake-based protocols offer significant benefits, proof of stake is neither problem-free nor effective in mitigating all the key risks that successful cryptocurrency faces.

4.1.5 Proof of Luck [27]

Proof of luck is an example of an Intel SGX-based consensus protocol, where all participants choose a random number (good luck) and a large number of wins (lucky ones). The lucky block is then used as the next block in the blockchain. Random number selection takes place in the SGX environment so it cannot be duplicated. Each CPU can only select one random number per block. To execute it, after all the ledgers from all the miners are synchronized, each miner will create for themselves a new block to append to their current chain, then a random number ranged from 0 to 1 will be assigned for each created block, which could be considered a lucky value. All of the nodes will have to agree that the chain with the total largest lucky value would be the main chain. As a result, proof of luck is considered to be fair for all the miners. Furthermore, it would be very difficult to make attacks, like double-spending attack, because the attacker should be very lucky to perform their illegal actions successfully.

4.1.6 Proof of Importance

The proof of importance is a modified version of the proof of stake, where instead of taking into account only the balances of the nodes, to determine the next winning node; to solve the upcoming block, it considers the aspects, including the reputation specified by a function defined by the particular system and the number of node transaction [28]. Therefore, this consensus method considers the productive network activity of the nodes that is more effective than the only equilibrium of nodes [29]. NEM is a cryptocurrency that uses PoI for consensus.

4.1.7 Proof of Elapsed Time

PoET is designed to achieve distribution consensus in a lottery type function. It was originally designed with the aim of create a fair mechanism to distribute mining rights within permits networks [30]. PoET complies with a four-stage process flow. First, each validator requests a timeout period randomly distributed from a reliable enclave.

Second, the validator with the shortest waiting time wins the elections and is given the leadership for the transactional block in question. A function is used to create a timer for the transaction block that is guaranteed to have been created by the enclave. Then, another function is used to verify the origin of the timer. The enclave comes in the form of a secure CPU instruction set, ensuring equity in the randomness of the selection among all participants. This is achieved through low-level implementation, using Intel software protection extensions (SGX) [31]. This facilitates the PoET algorithm by providing a random distribution of leadership across a population of participants fairly. A certificate of execution provides verification of a claim from participants to the leadership, providing a low cost of participation. This offers a strong incentive for participation in the network, since the algorithm is perceived as fair and accessible.

4.2 Permissioned Blockchain

Synchronous Mechanisms:

RAFT Raft algorithm is considered as a consensus algorithm for private blockchains [32], which is applied more to ad hoc networks such as the intranet. Compared to PBFT and Paxos, the Raft algorithm is highly efficient and simple, and it is widely used in distributed systems. A Raft is a leader-based algorithm that uses leader selection as an integral part of the consensus protocol. Book entries in a Raft-based system are transferred in only one direction from the leader to other servers. A network is called split when more than half of the nodes are outside the control of the current leader. Host failure and communication interruption caused by packet loss are the main reasons for network separation. If the network splits up, the blockchain network with the Raft algorithm will restart the new leader election process.

Byzantine Fault Tolerance Byzantine fault tolerance is commonly considered with respect to Byzantine General Problem [33]. With regard to Byzantine General Problem, Byzantine fault tolerance is reached when loyal generals reach a majority agreement on their strategy. Typical Byzantine defects are the most difficult to treat, as no restriction or hypothesis is made around the behavior that a node can exposure.

Asynchronous Mechanisms:

Practical BFT The Practical Byzantine Fault Tolerance (PBFT) [34] is one of the most well-established BFT algorithms. In PBFT, there are two kinds of nodes: A leader node, and some validating peers (nodes); and these peers will execute some rounds for appending a block to the chain. Specifically, it relies on three cycles of message exchange; pre-prepare, prepare and commit phase before reaching an agreement. This ensures that $3f + 1$ nodes can reach consensus also in the presence of Byzantine knots; this turned out to be optimal.

Delegated BFT DBFT is a variant of PBFT. This fault tolerant algorithm divides client within peer-to-peer system into two types; ordinary node and bookkeepers [35]. An ordinary node does not participate in consensus, and it just votes the nodes which are supported by bookkeeper nodes. The selected bookkeeper nodes participate in consensus. A random bookkeeper node broadcast its transaction to the entire network. If 66% of other bookkeeper nodes agree about the validity of transaction, then it is committed and next round of consensus is started.

5 Challenges and Opportunities

Blockchain is currently the subject of numerous research and development activities, both by academics and industry. However, there are still major challenges ahead of the mass penetration and adoption of the market. In this section, we highlight the major challenges in different domains which can be resolved by using blockchain.

Blockchain for enterprise [36]: In the enterprise, blockchain as participants are already known, the scalability issue is easy to solve as compared to public blockchain. However, to reach scalability, we must first keep in mind the usage context and the performance measures, we want to optimize: the transactions throughput, validation latency, and number of participating nodes, number of validation nodes, energy costs, calculation costs, storage fees or other criteria. In overall, scalability is an active area of research; we can mention some initiatives such as: fragmenting a global ledger into smaller sub-ledgers which are running by a subset of nodes, removing old transactions in order to optimize the storage using a blockchain hierarchy. Though we have techniques like ring signature, zero knowledge proof, homomorphic encryption which provide privacy in public blockchain, by incorporating these techniques, privacy can be achieved for enterprise blockchain.

Healthcare industry [37]: Healthcare industry has a unique concern with regard to privacy and security in order to protect the medical information of patients. Interoperability is also a major issue in health sector. The main limitation that prevents interoperability is the use of centralized storage of data in medical applications and institutions. Centralized storage of data is a problem for healthcare providers because they store all records in a central database. Data portability and mobility is a growing demand in the health sector as patients become more mobile. Due to smart devices, sensors and other Internet features, devices become more predictable so the ability to carry this data is also important.

Security threats [38]: The 51% attack is a technique that occurs when an attacker has possession of 51% of the hash power. Formation of chain of blocks, which is separated from real version of chain initiates this attack. This chain becomes the real chain for execution. This allows the double-spending attack [39]. Since the blockchain policy abides by the longest chain rule [2], if attackers can get 51% of the hash power or more, they will be able to manage the longest chain by persuading network nodes to follow their chain. However, for double-spending attack, it is not strictly necessary to obtain 51% of the hash power; it will work with the less than half of the hash power, but the success probability is less [40]. The stronger the blockchain network, the attack will become more expensive. Thus, cryptocurrencies with a high network hash are supposed to be more secure against the attack of 51%.

6 Blockchain Implementation

The expansion of blockchain platforms has introduced different architectures to satisfy the application requirements in an evolving communal ecosystem. In this section, we present a profane evolution of some of the blockchain architectures. We provide a retrospective analysis of these architectures and provide information on current problems for future areas of research.

Hyperledger fabric: Hyperledger fabric [41] (github.com/hyperledger/fabric) is an implementation of a distributed ledger platform for the execution of smart contracts, which exploits technologies, with a modular architecture allowing plug-in implementations of various functions. This is one of the many projects being incubated as part of the Hyperledger project. A developer version of the Hyperledger framework (called "v0.5-developer-preview") was released in June 2016. (Github.com/hyperledger/fabric/wiki/Fabric-Releases) The distributed ledger protocol is peerly managed. The fabric distinguishes between two types of peer: A validation peer is a node in the network responsible for executing the consensus, transaction validation, and general ledger. A non-validating peer is a node which serves as a proxy for connecting the clients (issuing transactions) to the validators. A non-validation the peer does not execute transactions but can verify them.

Ethereum: Basically, Ethereum is a "global computer" as a platform gives users the ability to run distributed applications in a decentralized environment way [42] This means that applications running on Ethereum are available anywhere and anytime. Ethereum blockchain has two types of accounts: Externally account owned and contract account to specify an authorized SHS person. During installation, external property account was created automatically by default. Smart contract account can be configured with the processing policies transactions. Ethereum can be developed by using Turing-complete language such as solidity.

Corda: Corda is a distributed ledger platform for the collection and processing of financial data agreements. The Corda platform supports intelligent contracts that meet the definition of Clack, Bakshi, Braine. Corda provides framework for running smart contracts. It is a unique approach for data distribution and transaction semantics while retaining the functions of distributed ledgers, who initially lured institutions to projects like R3, namely reliable execution of financial arrangements in an automatable and enforceable manner [43].

Quorum [44]: Quorum was developed by JP Morgan [45] as an authorized implementation of Ethereum in the general ledger. Ethereum is a public blockchain without permission that can be used across several areas to implement decentralized applications. It supports Turing-complete intelligent contracts and can be used to build general purpose blockchain applications in several areas. Being public and without authorization, its security is assured by the proof of work consensus algorithm. PoW consensus algorithm adds deliberate cryptographic difficulty to prevent Sybil attacks on Ethereum blockchain.

7 Conclusion

Blockchain technology has been a buzz in the computing era for the last few years due to its attractive characteristics such as immutability, distributed, transparent, and no central arbitrator. The data and transactions stored on blockchain are tamper-proof. Any node in the network can verify the correctness. The smart contract is an excellent application of blockchain, which is being used by many different domains. The transactions are packed into a block and every block is added to the chain after certain verification. The process of adding a block into the network is sometimes called mining where mutual agreement in terms of consensus mechanism plays a major role. In this research paper, we have studied the background theory for blockchain technology along with their types and classifications. Further, we have discussed various consensus mechanisms in detail with their detail analysis. We also have explored challenges and opportunities in the domain of blockchain consensus. In the future, we would like to propose a novel consensus mechanism, which would take criterion such as performance, a number of miners in the network, practical consideration (such as node/link failure), nodes credibility, etc., into consideration.

References

1. Haber S, Stornetta WS (1990) How to time-stamp a digital document. In: Proceedings of the 10th annual international cryptology conference on advances in cryptology, pp 437–455(11–15 Aug)
2. Nakamoto S (2008) Bitcoin a peer-to-peer electronic cash system. https://bitcoin.org/bitcoin.pdf
3. Helo P, Hao Y (2019) Blockchains in operations and supply chains: a model and reference implementation. Comput Indust Eng 136:242–251. ISSN: 0360-8352. https://doi.org/10.1016/j.cie.2019.07.023.(http://www.sciencedirect.com/science/article/pii/S0360835219304152
4. Litke A, Anagnostopoulos D, Varvarigou T (2019) Blockchains for supply chain management: architectural elements and challenges towards a global scale deployment. Logistics 3(1):5
5. Kouhizadeh M, Sarkis J (2018) Blockchain practices, potentials, and perspectives in greening supply chains. Sustainability 10(10):3652
6. Danezis G, Meiklejohn S (2015) Centrally banked cryptocurrencies
7. Biswas K, Muthuk kumarasamy V (2016) Securing smart cities using Blockchain technology. In: 18th IEEE international conference on high performance computing and communications, 14th IEEE international conference on smart city and 2nd IEEE international conference on data science and systems, HPCC/SmartCity/DSS, pp 1392–1393 (12, 14 Dec)

8. Liu PTS (2016) Medical record system using Blockchain, big data and tokenization. In: 18th international conference on information and communications security, ICICS (29 Nov–2 Dec, pp 254–261
9. Vukoli M (2016) The quest for scalable Blockchain fabric: proof-of-work vs BFT replication. In: IFIP WG 11.4 international workshop on open problems in network security, iNetS, pp 112–125
10. Idelberger F, Governatori G, Riveret R, Sartor G (2016) Evaluation of logic-based smart contracts for Blockchain systems. Cham, Switzerland, pp 167–83
11. Kraft D (2016) Difficulty control for Blockchain-based consensus systems. Peer-to-Peer Netw Appl 9:397–413 (201601-01)
12. Tama BA et al (2017) A critical review of Blockchain and its current applications. In: International conference on electrical engineering and computer science (ICECOS), IEEE
13. Drescher Daniel (2017) Blockchain basics. Apress, Berkeley, CA
14. Swan M (2015) Blockchain: blueprint for a new economy. O'Reilly, Sebastopol, US
15. Schwartz D, Youngs N, Britto A (2014) The ripple protocol consensus algorithm. Ripple Labs Inc White Pap 5:8
16. Brown RG, Carlyle J, Grigg I, Hearn M (2016) Corda: an introduction. R3 CEV 1:15
17. Sousa J, Bessani A, Vukolic M (2018) A byzantine fault-tolerant ordering service for the hyper ledger fabric Blockchain platform. In: 48th annual IEEE/IFIP international conference on dependable systems and networks (DSN), IEEE, pp 51–58
18. Baliga A (2017) Understanding Blockchain consensus models. Persistent
19. Guo Y, Liang C (2016) Blockchain application and outlook in the banking industry. Fin Innov 2(1):24
20. De Angelis S (2018) Assessing security and performances of consensus algorithms for permissioned Blockchains
21. Lamport L, Shostak R, Pease M (1982) The byzantine generals problem. In: ACM Trans Program Lang syst 4:382–401
22. Bach LM, Branko M, Mario Z (2018) Comparative analysis of Blockchain consensus algorithms. In: 41st international convention on information and communication technology, electronics and microelectronics (MIPRO), IEEE
23. Yuan Y, Wang F-Y (2016) Towards Blockchain-based intelligent transportation systems. In: IEEE 19th international conference on intelligent transportation systems (ITSC), IEEE
24. Ray J (2018) Proof of stake FAQ. https://github.com/ethereum/wiki/wiki/Proof-of-Stake-FAQ
25. Nguyen GT, Kyungbaek K (2018) A survey about consensus algorithms used in Blockchain. J Inf process syst 14(1)
26. Bentov I et al (2014) Proof of activity: extending bitcoin's proof of work via proof of stake. ACM SIGMETRICS Perform Eval Rev 42(3):34–37
27. Milutinovic M et al Proof of luck: an efficient Blockchain consensus protocol. In: proceedings of the 1st workshop on system software for trusted execution, ACM
28. Salimitari M, Chatterjee M (2018) An overview of Blockchain and consensus protocols for IoT networks. Preprint at arXiv:1809.05613
29. Proof of Importance. https://nem.io/technology/. Accessed 05 Apr 2018
30. Maple C, Jackson J (2018) Selecting effective Blockchain solutions. In: European conference on parallel processing. Springer, Cham
31. Costan V, Devadas S (2016) Intel SGX explained. https://eprint.iacr.org/2016/086.pdf
32. D Huang, X Ma, S Zhang (2019) Performance analysis of the Raft consensus algorithm for private Blockchains. IEEE Trans Syst Man Cybern: Syst
33. Castro M, Liskov B (1999) Practical byzantine fault tolerance. http://pmg.csail.mit.edu/papers/osdi99.pdf
34. De Angelis S et al (2018) PBFT vs proof-of-authority: applying the cap theorem to permissioned Blockchain
35. NEO White paper (2014) http://docs.neo.org/en-us. Accessed 10 Feb 2018
36. Hamida EB, Brousmiche KL, Levard H, Thea E (2017) Blockchain for enterprise: overview, opportunities and challenges. In: The thirteenth international conference on wireless and mobile communications, nice, France, (ICWMC 2017). https://hal.archives-ouvertes.fr/hal-01591859

37. McGhin T, Choo KK, Liu CZ, He D (2019) Blockchain in healthcare applications: research challenges and opportunities. J Netw Comput Appl 135:62–75. ISSN: 1084-8045. https://doi.org/10.1016/j.jnca.2019.02.027
38. Sayeed S, Marco-Gisbert H (2019) Assessing Blockchain consensus and security mechanisms against the 51% attack. Appl Sci 9(9):1788
39. Jimi S (2018) Blockchain: how a 51% attack works double spend attack. https://medium.com/coinmonks/what-is-a-51-attack-or-double-spend-attack-aa108db63474. Accessed 1 May 2018
40. Rosenfeld M (2014) Analysis of hashrate-based double spending. Preprint at arXiv:1402.2009
41. Cachin C (2016) Architecture of the hyperledger Blockchain fabric. In: Workshop on distributed cryptocurrencies and consensus ledgers, vol 310
42. Aung YN, Tantidham T (2017) Review of ethereum: smart home case study. In: 2nd international conference on information technology (INCIT), IEEE
43. Brown RG et al (2016) Corda: an introduction. R3 CEV 1:15
44. Baliga A et al (2018) Performance evaluation of the quorum Blockchain platform. Preprint at arXiv:1809.03421
45. Chase JPM (2018) A permissioned implementation of ethereum. https://github.com/jpmorganchase/quorum. Accessed 20 Feb 2018

Measuring IoT Security Issues and Control Home Lighting System by Android Application Using Arduino Uno and HC-05 Bluetooth Module

Gajendrasinh N. Mori and Priya R. Swaminarayan

Abstract Today's lot of domains have adopted the IoT-based applications in the home, agriculture, medical, industries, and so on where we find that the use of electronic devices becomes very common. McKinsey Global Institute predicted that the IoT covered more than $11.1 trillion market share by 2025. By looking at this prediction, we have to focus on the security of IoT applications. In this paper, first we will check security measures and then suggest IoT-based application that controls home lighting system using Arduino Uno microcontroller, breadboard, HC-05 Bluetooth module that will be operated by Android application. Smart phone has Bluetooth in-build so it will become very cheap, efficient, and preferable for the use. This paper gives the idea of how to control home lighting system and enable home lighting using a smart phone application. This system has five main components: an Arduino Uno as a microcontroller, breadboard, HC-05 Bluetooth module for transferring signal, n-channel DC 5 V relay for connection with lighting system, and Android application in smart phone for user interface.

Keywords Arduino Uno · Microcontroller · Breadboard · HC-05 Bluetooth module · N-channel DC 5 V relay module · MIT application inventor 2 · Home automation · Android application · IoT

1 Introduction

Today's Android is one of the leading and most preferable operating systems in smart phone, and it increases day by day due to their popularity, ease, and portability. Android application is also preferable to develop smart home appliances.

G. N. Mori
The Mandvi Education Society Institute of Computer Studies, Mandvi, Surat, India
e-mail: gajendranmori@gmail.com

P. R. Swaminarayan (✉)
Parul Institute of Computer Application, Parul University, Vadodara, Gujarat 391760, India
e-mail: swaminarayan.priya@yahoo.com

K. Kotecha et al. (eds.), *Data Science and Intelligent Applications*, Lecture Notes on Data Engineering and Communications Technologies 52,
https://doi.org/10.1007/978-981-15-4474-3_41

This project is based on an Android application which is used to control a home lighting appliance via Bluetooth module remotely [1]. Bluetooth technology works on short-range wireless communication and provides convenience, intelligence, and controllability [2]. This system designs nothing but personal area network (PAN) in any smart home environment where home light can connect and monitor by a single microcontroller. Home automation system provides computerized or automatic control to some electrical and electronic systems at home. This system will assist and provide support to the needs of elderly and physically disabled people in the home [3].

2 Related Works

First, home automation was introduced in 1970 but failed to meet all the expectations of the people because of not supporting full functionality [4]. Then in 2003 after the improvement of network, K. Y. Lee and J. W. Choi suggest that a smart home becomes a unit where all the appliances can connect, communicate, and monitor remotely [5] and also suggest that IoT uses IPv6, and it has large address space; also, each object becomes a communicating device so that there can be possible security threats [6]. Also by the researcher proposed a block chain technology that uses IoT architecture with lightweight, decentralized security, and privacy also. Here used architecture is hierarchical for smart homes to provide security and privacy [7] and also suggests that home automation system has full functionality that remotely controlled via wireless communication between Arduino and cell phone [8]. Home automation system also implements with Arduino Uno and Digilent chipKIT that is low-cost system design but one issue again standing is that Bluetooth is expensive for the same [9]. In Bluetooth supported home automation, all home appliances are easily connected to Arduino Uno board and appliances are connected with i/o ports of relay. Also, the program is written using high-level interactive C programming language for microcontrollers. Here, if we tried out Wi-Fi rather than Bluetooth, then the Wi-Fi-based home automation is also possible, and it has mainly three modules that are the server, the h/w interface module, and the s/w package for communication [10]. Then after, H. Kanma proposed a home automation via Bluetooth that can possible to access remotely using GPRS. The researchers use cell phone that enables Bluetooth connectivity and GSM modem for Internet connectivity [11]. In 2013, S. V. A. Syed Anwaarullah proposed an inexpensive home automation system that monitors using an Android device, and the proposed system uses RESTful Web services for communication between home automation system and home appliances [12].

3 Security Issues of IoT

As far as security is concerned, Mr. Andrew Meola identified below security issues that are as follows [13].

3.1 Public Perception

In IoT, it may possible to access private information so many people afraid by using it because in one survey in 2015 concludes that 44% Americans were very concerned, and they feel that their information was stolen.

Vulnerability to Hacking: Vulnerabilities were also introduced by hackers to hack personal information by controlling devices such as smart lighting system, smart locks, smoke detection system, and so on.

Confidence of Organizations: AT&T reported for IoT security that 85% organizations use IoT technology but out of them only 88% have lack of full confidence in security concern so they conclude that IoT implementation with significant risk.

Large amount of Data: Every day normally less than 10,000 household's devices generate more than 150 million data points, and with this, it also opens many doors for the hackers to hack sensitive information of the users.

Eavesdropping: Hackers attack on the smart home devices and continue eavesdropping all moments by capturing unencrypted data.

4 System Architecture and Working Principle

This topic briefly gives an idea of how the system architecture of the proposed system looks like, Fig. 1, and how it works.

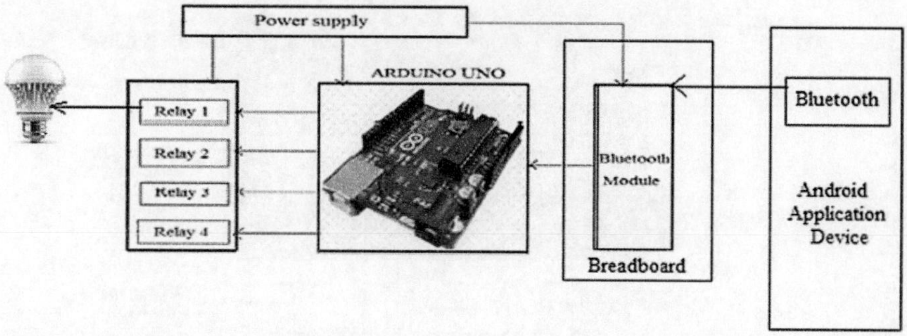

Fig. 1 System architecture

The above system architecture mainly consists of Arduino Uno microcontroller board, HC-05 Bluetooth module, n-channel DC 5 V relay, and Android application in Android smart phone and follows below steps to know the flow of system.

1. Connect Arduino Uno board with laptop and then mount Bluetooth module and relay on Arduino Uno board.
2. Develop an Android application in MIT application inventor 2 for user interface.
3. After getting command, HC-05 Bluetooth module sends signal to the Arduino Uno microcontroller.
4. As per logic application gives a command to the relay via board, and finally, light works as per command like ON or OFF.

5 Hardware Descriptions

5.1 Arduino Uno

The Arduino Uno microcontroller is the best board to developed initial system in IoT. Here, Uno stands for one in Italian also chosen on the introduction of Arduino S/W (IDE) 1.0. It is the most usable and popular board in the whole family of Arduino. It is based on the ATmega328P with 14 digital i/o pins and 6 analog inputs pins. It has 16 MHz quartz crystal USB connection, AC-to-DC adapter with power jack, and ICSP header with a reset button for resetting uploaded code [14]; see Fig. 2.

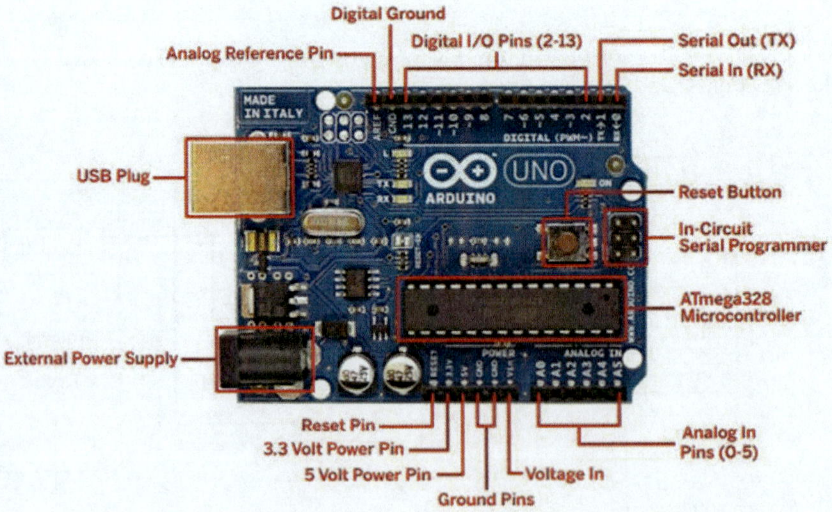

Fig. 2 Arduino UNO microcontroller

5.2 HC-05 Bluetooth Modules

Bluetooth is an open-standard short range in wireless data and voice communication technology [15]. It is based on Bluetooth module and designed for wireless communication. It is designed and used for one master or maximum eight slave configuration and can enhance data rate up to 3 Mbps modulation with 2.4 GHz radio transceiver and baseband. It uses CSR Bluecore 04 External single chip of Bluetooth module that embedded CMOS technology and also uses adaptive frequency hopping feature (AFH). See Fig. 3.

5.3 N-Channel DC 5 V Relay

It is relay interface board and uses and controls by microcontrollers that are as follows: Arduino, AVR, ARM, PIC, and so on. Here, this relay uses low-level 3.3–5 VDC-triggered control signal normally and operates on open and closed contacts. It is mainly used in auto-control circuit for high and low current signals. It signals input voltage range in-between 0 and 5 V [16]; see Fig. 4.

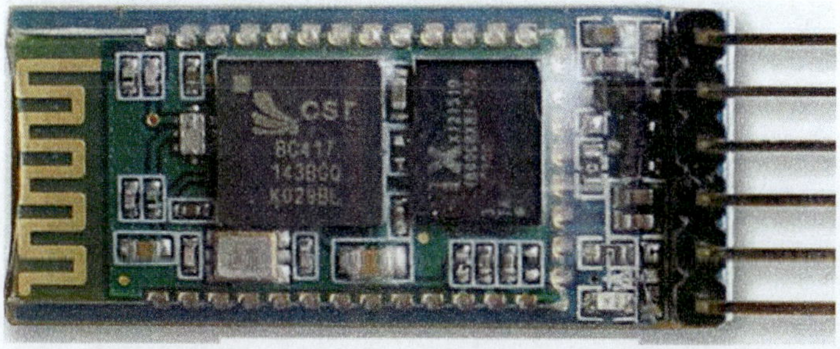

Fig. 3 HC-05 Bluetooth module

Fig. 4 N-channel DC 5 V relay

6 Software Descriptions

6.1 MIT Application Inventor 2

It is a free, easy, drag, and drop; blocked and visual programming languages that give easy access to the people regardless of coding experience for creating mobile applications for Android devices. New people on MIT application inventor 2 can develop the first application <30 min. Here developed project gives freedom of software and empowering all interested people especially young enthusiastic people to move from technology consumption to technology creation [17].

7 Implementation and Working

To implement home lighting system, we have to connect all the hardwares and then mount Bluetooth module and relay on Arduino Uno microcontroller board; see Fig. 5.

After design circuit, we need to develop an application in MIT application inventor 2 for user interface to light ON and OFF and so create design as shown in Fig. 6.

Now let us see block code of this design in Fig. 7.

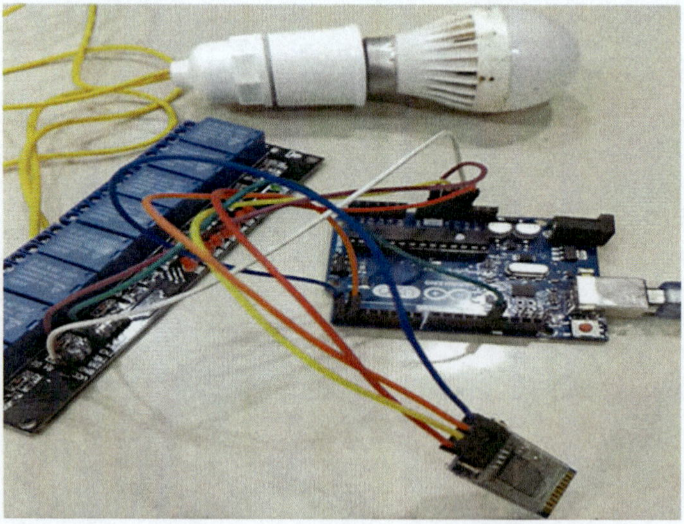

Fig. 5 Schematic of connection

Fig. 6 MIT application inventor-based application

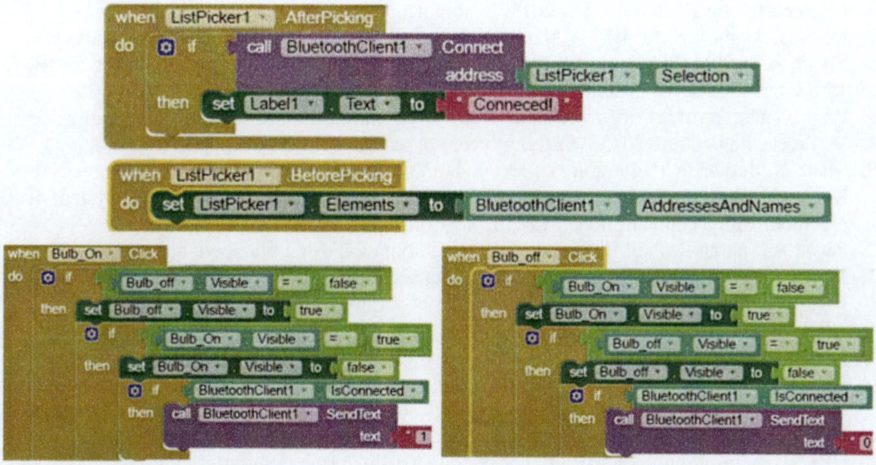

Fig. 7 Block code in MIT application inventor 2

8 Conclusions

In this paper, first some IoT security issues are identified and then circuit design is created with actual implementation that enhances easy, low-cost, reliable, and flexible lighting system for the home. It provides an understanding of communication methods by device through Bluetooth technology and develops an easy user interface by MIT application inventor 2 website. Designing circuit, developing user interface, and communication through Bluetooth module are interesting challenges. Overall, this work will help the needy people who are aged and physically disabled. As far as future enhancement is concerned, we would give the facility of autoconnection and would pair it when the phone comes into the range of the Bluetooth. Also, we tried out this concept with other sensors and actuators to develop an IoT environment.

References

1. Piyare R, Tazil M (2011) Bluetooth based home automation system using cell-phone. In: IEEE 15th international symposium on consumer electronics
2. David N, Chima A, Ugochukwu A, Obinna E (2015) Design of a home automation system using Arduino. Int j Sci Eng Res 6:795–801
3. Piyarel R, Leel SR (2013) Smart home-control and monitoring system using smart phone. In: Independent computer consultants association 2013, ASTL, vol 24, pp 83–86
4. Chassiakos A, Aghnatios W, Ary J Home automation via bluetooth using the arduino uno microcontroller
5. Lee K, Choi J (2003) Remote-controlled home automation system via bluetooth home network, vol 3, pp. 2824–2829
6. Riahi A, Challal Y, Natalizio E, Chtourou Z, Bouabdallah A (2013) A systemic approach for IoT security. In: IEEE. DCOSS, 2013, Boston, United States, pp. 351–355
7. Dorri A, Kanhere S. Jurdak R (2016) Blockchain in IoT: challenges and solutions, article
8. Piyare R, Tazil M (2011) Bluetooth based home automation system using cell phone. In: IEEE 15th international symposium on consumer electronics (ISCE), Singapore, pp 192–195
9. Raghavan S, Tewolde G (2015) Cloud based low-cost home monitoring and automation system. In: Proceedings of the 2015 ASEE north central section conference, pp 1–10
10. Malik N, Bodwade Y Literature review on home automation system
11. Kanma H, Wakabayashi N, Kanazawa R, Ito H (2003) Home appliance control system over bluetooth with a cellular phone. IEEE Trans Consum Electron 49(4):1049–1053
12. Pawar A, Sharan R, Patil R, Chavan S Home automation using bluetooth and IOT
13. Meola A How the internet of things will affect security & privacy. Tech Insider, Bus Insider. URL http://www.businessinsider.com/internet-of-things-security-privacy-2016–8?IR=T
14. ARDUINO UNO REV3. URL https://store.arduino.cc/usa/arduino-uno-rev3
15. Interfacing of AT command based HC-05 serial bluetooth module with minicom in linux
16. URL https://www.rhydolabz.com/hardwares-relays-c-170_182/2channel-5v-relay-moduleblack-p-2580.html
17. URL https://appInventor2.mit.edu/explore/about-us.html
18. URL http://humboldtmcu.blogspot.com/2014/07/programming-arduino-uno-hardware-memory.html

Advanced 3-Bit LSB Based on Data Hiding Using Steganography

Kinjal Patani and Dushyantsinh Rathod

Abstract Nowadays, security plays a main role for data hiding and makes it confidential from unauthorized access. So steganography and cryptography techniques are used for data security. In this paper, 3-bit least significant bits technique is used for embedding secrete image into the cover image. The resultant image after embedding the secrete image into cover image is a "Stego Image." Using Vvisual Cryptography, transferring the stego image over the wireless channel or Internet becomes more securable because it divides the encrypted message and transfers it into two shares based threshold. In this paper, ECC algorithm is used to kept data more secure and secrete which makes very hard to get secrete message even if the secrete message is disclose. These proposed systems will work on large size of image and result with high performance and accuracy with high PSNR for color image.

Keywords Visual cryptography · LSB · Steganography · ECC · Data hiding

1 Introduction

Steganography is the method of covering text within some form of cover file. It originates from the Greek words Steganós (Covered) and Graptos (Writing). There are two different concepts cryptography and steganography. In cryptography, plaintext is encrypted whereas in steganography, embedding of text or image into cover file. The cover image could be a text, image, audio, or video. The general approach is LSB insertion. Among all these methods, LSB is the very simple method to implant the data in image. In LSB method, the information is embedded into bit stream of image [1].

K. Patani (✉) · D. Rathod (✉)
Department of Computer Engineering, Alpha College of Engineering and Technology, Gandhinagar, Gujarat 382721, India
e-mail: kinjalpatani2004@gmail.com

D. Rathod
e-mail: Dushyantsinh.Rathod@gmail.com

K. Kotecha et al. (eds.), *Data Science and Intelligent Applications*, Lecture Notes on Data Engineering and Communications Technologies 52, https://doi.org/10.1007/978-981-15-4474-3_42

Image steganography is the most common form of steganography that used in various applications for hiding text in image, video in image, and audio in image.

Now, in this paper, steganography and cryptography are used for the security and hiding of data. Steganography (hiding information) and cryptography (protecting information) are totally different from one another. Due to invisibility or hidden factor, it is difficult to recover information without known procedure in steganography [2].

In this proposed work, we implement the 3-bit LSB algorithm for embedding the secret image into the original image so that the secret image cannot be visible to anyone from their naked eyes. After that, we implement the cryptography algorithm to hide the information securely using ECC algorithm. From this hybrid approach, it will provide highly security for data hiding over the network. Peak signal to noise ratio (PSNR) and MSE parameters are used for quality measurement and the performance of the system.

Steganography: Cryptography and steganography are both methods used to hide or protect secret data. However, they differ in the respect that cryptography makes the data unreadable, or hides the meaning of the data, while steganography hides the existence of the data. The main features of a steganography system are to get high load capacity and visual quality and decrease the amount of noise [3].

Techniques of Steganography: In steganography, the cover object used can be in text, image, audio, video, etc. Image steganography is the most popular one because of the Internet multimedia transmission. Image steganography is the process of embedding the secret information in an image file known as image steganography. Steganography is divided into two parts:

Spatial Domain Methods: In this method, the data can be hidden by changing some bits in the image pixel values. It includes LSB, pixel value differencing, edge-based data embedding method, random pixel embedding method [4].

Frequency Domain Technique: This is a more complex way of hiding information in an image various algorithms and transformations are used on the image to hide information in it frequency domain embedding can be termed as a domain of embedding techniques. Frequency domain is classified into: Discrete Fourier transformation technique: The discrete Fourier transform to get frequency component for each pixel value. The discrete Fourier transform (DFT) of spatial value $f(x, y)$ for the image of size $M \times N$ is defined in equation for frequency domain transformation. Discrete cosine transformation technique: The discrete cosine transform (DCT) is a technique for converting a signal into elementary frequency components. It is widely used in image compression. Discrete wavelet transformation technique: A discrete wavelet transform (DWT) is any wavelet transform for which the wavelets are discretely sampled [4].

Least Significant Bit: It replaces least significant bits of cover object with secret message. It is most popular and simple technique when dealing with images. It has low computational complexity and high embedding capacity. It works best when the image file is larger than the message file [5]. LSB insertion is a fundamental approach for embedding data in cover file. Digital images which are used as cover file are majorly of two types: 8-bit images and 24-bit images. 1 bit of data can be concealed in 8-bit images and 3 bits of data can be stored in 24 bit images [6]. Each pixel in image is made from the combination of RGB pixel which is represented by 3 bytes in 24-bit bitmap. Three bytes are used to represent the red, blue, and green color values of each pixel. It is best to use 24-bit bitmap because when an image is of high quality and resolution, it is easier to hide information inside it. Using LSB steganography, one can embed 3 bit data in each pixel of 24-bit BMP image and replace those LSB's with 3 message bit [7].The primary advantage of LSB technique is that itis easy to implement and chance of degrading quality of original image is reduced [6].

Stego-3 bit LSB: In this method 3 MSb's of secret image is stored in 3 LSb's of cover image. This will also rarely affect the cover image. The stego image appears nearly same as cover image. The retrieved image quality is degraded but it is better than Stego-2bit LSB and 3 bit LSB provide stego image less degrade than 4-bit Stego image [8]. The key advantage of color image is their ability to hide large amount of data in their pixel definition because each pixel is defined using three color composition and a small change in these three definitions in last bits is not effect much in the actual image definition [9].

ECC (Elliptic Curve Cryptography): Elliptical curve cryptography (ECC) is a public key encryption method depending on elliptic curve theory used to generate quick, smaller, and effective cryptographic keys. ECC is a well-known technique till date, equipped with the expertise to explore spectacular results in complete exponential interval. The technology used in existence of majority of public key encryption technique, like RSA, and Diffie–Hellman. As per several researchers, ECC can produce a level of security with a 164-bit key, which other systems need a 1024-bit key to achieve. ECC contribute to maintain equivalent security with minimized computing power and battery usage. The use of ECC within cryptography was first proposed in 1985, individually by Neal Koblitz from University of Washington, and Victor Miller at IBM. An elliptic curve is not oval in shape (ellipse); it represents a looping line intersecting two axes. ECC works on properties of specific equation created by a mathematical group and extracted from points where line intersects the axes. Multiplying the point on curve with a number to generate another point on the curve. It is hard to discover which number was used, even knowing the original point and the result. Equations based on elliptic curves have valuable feature for cryptography purposes: that are easy to perform, extremely difficult to reverse [6].

Elliptic curve cryptography authentication scheme offers considerably greater data security for a given key size. If the key size is smaller, it is also possible to implement for a given level of security so that it consumes less power and less heat production. The smaller key size makes faster cryptographic operations, running on smaller chip and on more compact software [3].

Visual Cryptography: The Visual Cryptography used for sharing visual secrets is developed and pioneered by Moni Noar and Adi Shamir in 1994. Visual Cryptography is a secret sharing scheme that takes secret image as input(i.e., Printed, Handwritten) encrypts the input image into a set of other images called shares in such a way, if shares are printed on transparencies and superimposed or staked over one another original secret is revealed. Simplest form of visual cryptography or visual secret sharing scheme considers binary image as input and deals with each and every pixel independently. To encode a pixel of the secret image, we split the secret pixel into n versions in such a way that if all n versions are printed on transparencies and superimposed, the original secret pixel is revealed. This process has to be applied for entire secret image [10]. Visual data is encrypted using visual cryptography. It is achieved with dividing the representation into two shares based on a threshold. The both shares of the similar picture are desired to recover the authentic data. Since the encrypted information is spread in both the pictures, it is not possible for anybody for obtaining the information by obtaining just a single part of the picture. Therefore, both chunks are necessary for getting unique data. The information cannot be recovered from an indistinct picture. Thus. it is extremely needed to coat the picture correctly [11].

2 Proposed Work

In this paper, they will be use of 3-bit LSB Steganography method along with shifting and adding algorithm which is used for embedding the secret image into the another cover image (Original image) and for more security, elliptic curve cryptography (ECC) is used for transferring the secret data over the wireless channel.

2.1 Proposed Model

See Fig. 1.

2.2 Steps Description

Embedding Phase: LSB is considered to be the most common technique in the spatial domain image steganography, because of rather than its simplicity. Changing the values of 3-LSB bits of image's data will make the image's alteration not perceptible for any human eyes, because the slight difference of colors. Each pixel in color image is specified by three values, one each for red, blue, and green color components. The high embedding rate will lead us to increase the payload capacity within the color image without sacrificing the imperceptibility

Encryption Phase: In the encryption phase, the secrete information to be ciphered is provided as input by the user of the software. After encryption, the resultant data will be confidential data so it cannot be understood by any unauthorized person. The encryption algorithm used here is ECC cryptography. The encryption key used is known to the dispatcher and the recipient. Apart from that the encryption key and decryption used are the different. The result of encryption would be cipher data.

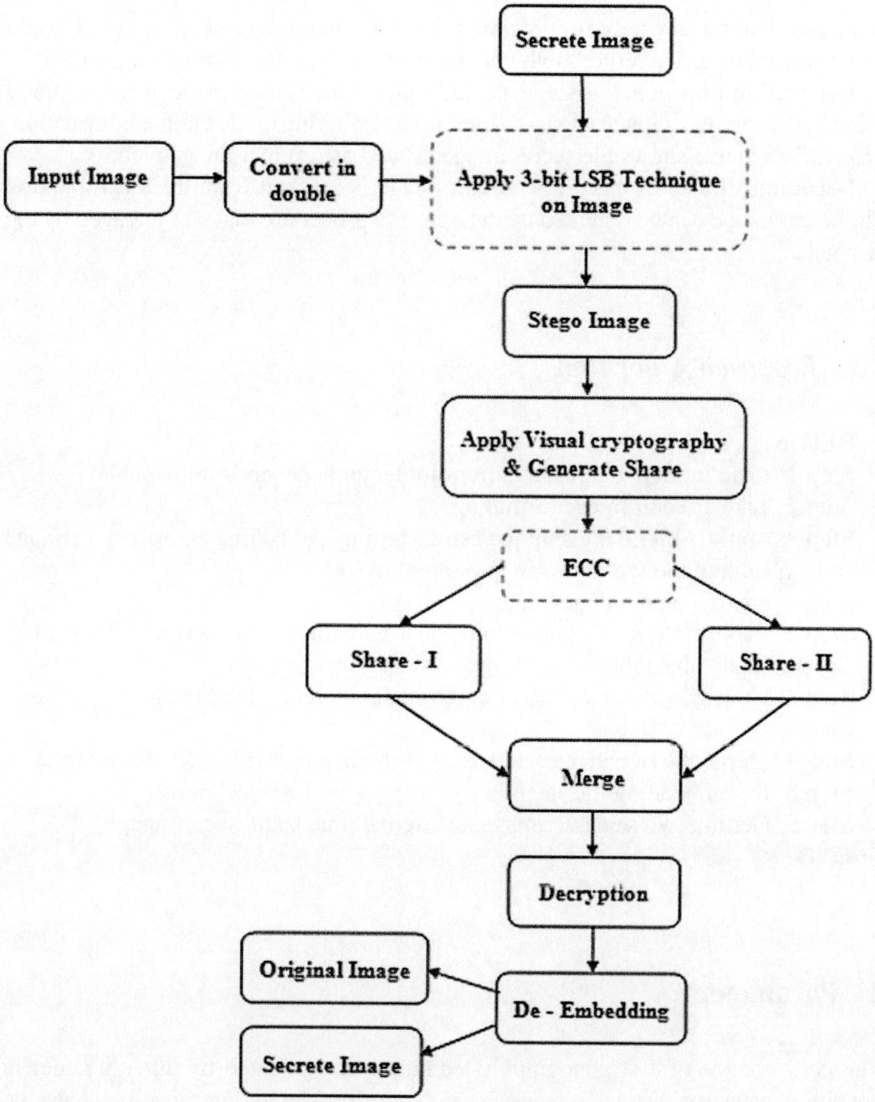

Fig. 1 Flowchart for proposed method

Cipher information is worthless information that cannot be understood extremely with any trouble.

Now, apply Visual Cryptography on this cipher data and generate the two share of similar picture, here the encrypted information will divided into two parts and which will embed into the two similar pictures.

Overlapping Phase: The both shares of the similar picture are desired to recover the authentic data. Since the encrypted information is spread in both the pictures, it is not possible for anybody for obtaining the information by obtaining just a single part of the picture. Therefore, both chunks are necessary for getting unique data.

Decryption Phase: In this stage, the ECC algorithm is used in the reverse method using the different enciphering key as used for enciphering of the unique information. Lastly, the scheme shows the secret image as well as original image picture.

Decoding Phase: In this stage, secret image is cracked from the stego picture. The secret image can be regained by the opposite procedure that was engaged by the dispatcher.

2.3 Proposed Algorithm

BEGIN
Step 1: Read input image from current folder and convert it into double.
Step 2: Take a secret image for hiding.
Step 3: Apply 3-bit LSB technique using shifting and adding algorithm on image and embedding the secret image into input image.
Step 4: Stego image will generate.
Step 5: Apply Visual Cryptography on stego image and encrypt it by ECC Cryptography algorithm.
Step 6: Now, generate two shares and transfer the encrypted image into the two shares.
Step 7: Merge the two images and decrypt the image using ECC algorithm.
Step 8: De-embedding the stego image using 3-bit LSB technique.
Step 9: Getting two separate image as original image and secret image.
END

3 Parameters

The performance of a steganographic technique can be rated by three parameters are hiding capacity, distortion measure, and security. The hiding capacity means the maximum amount of information that can be hidden in an image. It can also be represented as the number of bits per pixel.

3.1 Mean Square Error (MSE)

The mean squared error (MSE) of an estimator is one of many ways to quantify the difference between values implied by an estimator and the true values of the quantity being estimated. LSB technique contains the highest value of the PSNR and lowest value of MSE. Due to these values the quality of image has become better. If MSE attains a better quality of image if there is a low value of MSE (Mean square error). Means low value low Error.

3.2 Peak Signal-to-Noise Ratio(PSNR)

The PSNR measures the peak signal-to-noise ratio between two images. This ratio is often used as a quality measurement between the original and a compressed image. Higher the PSNR means better the quality of the compressed or reconstructed image [9]. The PSNR value can be calculated as:

$$\mathrm{PSNR} = 10\log_{10}\frac{R^2}{\mathrm{MSE}}$$

4 Conclusion

The proposed technique uses 3-bit LSB algorithm for embedding the secrete image in original image which is stego image. The steganography using LSB is made for the color image. The security is increased in steganography by ECC method. Because of this, no Attackers can access or manipulating the secret message which is hidden behind the cover image. This algorithm proved good for security using both steganographic and cryptographic techniques. Stego image obtained with good quality is trustworthy, protected and possesses a high PSNR and accuracy.

References

1. Yadav S, Tripathi A, Yadav P (2017) Image steganography on color image using SVD and RSA with 2-1-4-LSB technique. In: International conference on computation of power, energy, information and communication (ICCPEIC), IEEE
2. Biradar R, Umashetty A (2016) A survey paper on steganography techniques. Int J Innov Res Comput Commun Eng 4. ISSN: 2320-9801
3. Gahan A, Devanagavi G (2019) A empirical study of security issues in encryption techniques. Int J Appl Eng Res 14(5). ISSN: 0973-4562
4. Singh N (2017) Survey paper on steganography. IRJES 6(1):68–71. ISSN: 2319-183X
5. Kaur S, Bansal S, Bansal R (2014) Steganography and classification of image steganography techniques. In: IEEE
6. Bharti S, Behal S, Sharma V (2018) Security enhancements for high quality image transaction with hybrid image steganography algorithm. In: Proceedings of the international conference on computing methodologies and communication, IEEE

7. Lokhande U (2014) An effective way of using LSB steganography in images along with cryptography. Int J Comput Appl 88(12):0975–8887)
8. Thakur R, Saravanan C (2016) Analysis of steganography with various bits of LSB for color images. In: International conference on electrical, electronics, and optimization techniques, IEEE
9. Yadav R, Nigam S (2018) An imporve color image stgnography using ECC algorithm. IJRRETAS 4(4). ISSN: 2455-4723
10. Jain A, Soni S (2017) Visual cryptography and image processing based approach for secure transactions in banking sector. In: International conference on telecommunication and networks, IEEE
11. Chavan S, Gurav Y (2018) Lossless tagged visual cryptography scheme using bit plane slicing for image processing. In: Proceedings of the international conference on inventive research in computing applications (ICIRCA), IEEE I.SBN: 978-1-5386-2456-2

Improving Energy Efficiency and Minimizing Service-Level Agreement Violation in Mobile Cloud Computing Environment

Pandya Nitinkumar Rajnikant and Nimisha Patel

Abstract In mobile cloud computing, high energy consumption and service-level agreements (SLAs) violation are the challenging issues considering that the demand for computational power is growing rapidly, which requiring large-scale cloud data centers. The mobile cloud computing enables customers to outsource the large and complex tasks to the cloud data centers for the execution and result generations. To provide resources such as CPU, RAM, memory, and bandwidth cloud data centers host incoming tasks, due to the large data centers that provide the basic resources for the host operations, and it consumes a huge amount of energy, which results in very high operational costs. In order to improve the energy efficiency and SLA violation of a cloud data center, energy-aware methods for overloaded host detection and VM selection from an overloaded host are needed. Cloud computing research community felt the need to provide or improve the energy-efficient solutions that reduce the impact of the issues. In this paper, we have shown many existing algorithm for energy efficiency and SLA violation reduction.

Keywords Mobile cloud computing · Data center · Energy efficiency · Green computing · Service-level agreements · Virtual machine · Virtualization · Consolidation

1 Introduction

Mobile cloud computing has radically transformed the traditional model of ownership (in the past, IT resources were acquired) into the current subscription model. It provides on-demand access to flexible infrastructure in the form of pay-as-you-go

P. N. Rajnikant (✉)
Sankalchand Patel University, Visnagar, Gujarat, India
e-mail: nitin.pandya85@gmail.com

N. Patel
CE Department, Sankalchand Patel College of Engineering, Visnagar, Gujarat, India
e-mail: nimishapatelce@gmail.com

K. Kotecha et al. (eds.), *Data Science and Intelligent Applications*, Lecture Notes on Data Engineering and Communications Technologies 52,
https://doi.org/10.1007/978-981-15-4474-3_43

model services based on actual resource usage to meet the demand for large-scale data center IT resources. Data centers consume approximately 1.3% of the world's total electricity supply; this figure is expected to increase by 8% in 2020. Server resource usage data collected over a six-month period from more than 5000 production servers indicates that most time servers are running at 10% to 50% of their full capacity, resulting in a waste of energy [1].

One of the main challenges of mobile cloud computing research is energy efficiency, and researchers are focusing on this challenge. Energy efficiency not only minimizes overall cloud data center (DC) spending, but also has a positive impact on the environment. CO_2 and greenhouse gases (GHGs) play an important role in the pollution of the environment, and reports published in 2007 highlight the fact that the information and communication technology (ICT) industry has produced around 2% of total CO_2 emissions. On the other hand, large ICT companies like Google already host hundreds of thousands of servers in 2010; these companies consumed around 271.8 billion kWh of electricity. In addition, recent reports suggest that the ICT industry's energy consumption will reach around 1963 billion kWh in 2020 [2].

Mobile cloud computing has become a new paradigm for service delivery over the Internet, using a dynamic pool of virtualized computing resources. In order to meet growing customer demands, many cloud service providers (e.g., Amazon, Microsoft, and Google) continue to rapidly expand their data centers, resulting in huge energy consumption and high volume of emissions of CO_2. The electricity consumed by data centers has increased by 56%, and experts expect the annual energy bill paid by data center operators to exceed the cost of equipment, if the current trend continues. Due to the economic and ecological impact of data center energy consumption, there is increased interest in the development of energy-conscious cloud computing approaches. Such approaches result in significant reductions in operating costs for data centers (due to reduced energy costs) and minimize the negative environmental impact associated with CO_2 emissions [3].

Modern society is increasingly dependent on large-scale distributed systems, in areas such as business, government, and defense. A distributed system is essentially a collection of independent computers, sharing resources and interacting with each other to achieve a common goal, via a computer network. The shared resources in a distributed system include data, storage capacity, and computing power, and the distributed computing infrastructure can be used to serve a variety of purposes, ranging from running resource-intensive applications (applications related to processor and memory) for scalable Internet applications (processor and network applications). Distributed systems have become a powerful means of supporting the evolution of civilizations, providing business services, the development of research and science, and the well-being of populations [4].

2 Adaptive Energy-Aware Algorithm

Most existing approaches take into account the threshold of use of the host processor for the host overload detection process and assume that the available bandwidth is always equal to the basic bandwidth, thus causing performance degradation. This article presents regression-based algorithms called Gdr and MCP to set a higher dynamic utilization threshold for the CPU to detect overloaded hosts [1].

A new dynamic algorithm called bandwidth dynamic VM selection strategy (Bw) is introduced in VM selection to balance the trade-off between power consumption, number of migrations, host performance, and number of total stop hosts.

- Adaptive heuristic algorithms Gdr and MCP are proposed to define a dynamic upper utilization threshold of the processor for an optimal use of host resources and to detect overloaded hosts, thus minimizing the number of active hosts, which means a reduced energy consumption as well as an ALS violation.
- A VM selection algorithm sensitive to dynamic bandwidth (Bw) has been introduced to select VMs from an overloaded host. The basic idea of this algorithm is to minimize the migration time of virtual machines from an overloaded host to an appropriate host, thus minimizing the overall violation of the SLA.
- The efficiency of the Gdr, MCP, and Bw algorithms is evaluated using the CloudSim simulator and demonstrates its superiority by comparing them to the different algorithms discussed in the literature.

3 SLA-Aware Energy-Efficient Resource Management

Two workload consolidation techniques try to reduce energy consumption and the resulting ALS violations. The available capacity and power (PCA) and the required capacity and power (RCP) are based on the optimal best-fit decreasing algorithm (BFD) which is traditionally used for the packaging of in-line trays. The BFD algorithm is considered a good VM placement algorithm because it places VMs with higher resource requirements first. The remaining virtual machines with lower resource requirements are placed on the servers with the remaining resources.

The proposed techniques focus on the available CPU capacity and the energy that have yet to be consumed rather than on the overall CPU capacity and maximum energy consumption. This allows you to clearly see the amount of energy that will be consumed by a certain virtual machine. The proposed techniques also guarantee the agreed QoS level using threshold mechanisms [2].

- Detailed analysis of the heuristics for managing selected energy resources is provided using the same environments and assumptions.

- Two new techniques for managing energy-efficient resources sensitive to SLA, ACP, and RCP are proposed to optimize energy, SLA violations, and network load simultaneously.
- An analysis of the temporal and spatial complexity of the proposed and selected solution is presented.

4 Energy-Efficient and SLA Compliant Approach for Resource Allocation and Consolidation

In this article, the authors suggest an approach and placement of multi-objective VMs sensitive to energy and QoS which aims to achieve a compromise between energy efficiency, system performance, and quality of service. Two complementary algorithms are a multi-objective VM ACO placement algorithm and a multi-objective VM ACO consolidation algorithm. Based on the results of the placement algorithm, our VM consolidation algorithm optimizes the use of resources through VM migration, focusing on finding optimal solutions to minimize energy consumption total data center, in addition to minimizing the number of VM migrations, minimizing SLA violations, and minimizing the number of active PMs [3].

- The proposal of a multi-objective energy and QoS-sensitive cloud resource management approach strikes a balance between energy efficiency and the system's ability to meet QoS and SLA requirements.
- Carrying out an in-depth performance analysis, comparing the performance of the proposed approach with six other single-objective and multi-objective approaches with regard to eight performance parameters. This benchmarking performance provides interesting insights into the effect of taking into account different targets on the energy efficiency, performance, and SLA compliance of the system.

5 Energy-Efficient and SLA-Based Resource Management in Cloud Data Centers

5.1 Power- and SLA-Based Management of Resources

Strongly enhanced by virtualization technologies that facilitate the simultaneous execution of multiple applications on a single physical resource, the dynamic method of activating and deactivating servers can be applied to reduce overall consumption of inactivity, thereby improving efficiency energy. In this regard, some research simply uses migration to aggregate the load on certain servers and deactivate those that are inactive, in order to improve energy efficiency [4].

5.2 *Power- and Failure-Aware Management of Resources*

In this article, the authors propose two alternative algorithms in the scheduling of work driven by deadline (a job is defined as a set of tasks) which take into account both the capacity and the reliability status of the compute nodes in the process of making selection decisions when provisioning resources. The first scheduling alternative, the optimistic best-fit (OBFIT) algorithm, schedules jobs in nodes that should not fail before the job deadline and cumulatively provides the minimum resources required to get the jobs done on time. The second alternative, called the pessimistic best-fit algorithm, calculates the average of the average available capacity level C among the calculation nodes which will not fail before the work deadline, and from a set of MPs which will fail before the user work deadline, he selects the PM with Cp capacity such that C average + Cp results in the minimum capacity available to run a virtual machine [4].

5.3 *Interference-Aware Management of Resources*

The VM arrangement minimizes performance interference and the number of active physical servers, thereby reducing power consumption by disabling redundant servers. The proposal involves co-locating a virtual machine that has a large cache demand with a virtual machine that has a small cache demand, in order to improve overall performance [4].

6 A Joint CPU-RAM Energy-Efficient and SLA Compliant Approach for Cloud Data Centers

They introduce two new strategies to improve the consolidation of energy-sensitive VMs in data centers. In the proposal, they call their CPU and RAM energy-aware (CREW) and under-load detection (UD), in which CREW improves the placement of VMs by using a power model that takes into account the energy consumed by the CPU and RAM. To improves detection of under-loaded (UD) servers by taking into account host processor usage history to establish a minimum threshold to shutdown the servers [5].

From the above algorithms, Table 1 describes the comparison with different parameters.

Table 1 Comparison of existing algorithm with different parameter

Parameter	Adaptive energy-aware algorithm	SLA-aware energy-efficient resource management	Energy-efficient and SLA compliant approach for resource allocation and consolidation	Energy-efficient and SLA-based resource management in cloud data centers	A joint CPU-RAM energy-efficient and SLA compliant approach for cloud data center
Energy consumption	12% reduced	29% reduced	30.85% reduced	11.2% reduced	Better than as mention in cloudsim
AVG SLA violation	22% reduced	15% reduced	21% reduced	18% reduced	Better than as mention in cloudsim
Performance metrics	Better than as mention in cloudsim	Degradation due to migration	Better than as mention in cloudsim	Better than existing algorithm in the cloud	Better than as mention in cloudsim
No. of host and VM migration	22%	31.80%	27%	20%	10%
Experiment time	34% improved	32% improved	23% improved	10% improved	12% improved

7 Conclusion

In this article, we observed that energy-aware adaptive algorithms have the potential to improve energy efficiency and minimize SLA violation in a mobile cloud environment. From the comparison of existing algorithm for energy-efficiency using minimizing SLA violation we conclude that, energy-efficient and SLA compliant approach for resource allocation and consolidation is a better approach than other algorithms, in which energy consumption is reduced and the average violation of SLA is also reduced.

References

1. Rahul Y et al (2018) Adaptive energy-aware algorithms for minimizing energy consumption and SLA violation in cloud computing. IEEE Access 6:52923–55936
2. Saad M et al (2018) SLA-aware energy efficient resource management for cloud environments. IEEE Access 6:15004–15020
3. Mohammad-Hossein M, Kara N, Barachi M (2018) An energy efficient and SLA compliant approach for resource allocation and consolidation in cloud computing environments. Sustain Comput: Inf Syst 17:9–24

4. Sampaio Altino M, Barbosa J (2016) Energy-efficient and SLA-based resource management in cloud data centers. Adv Comput 100:03–159
5. Pedro C et al (2016) A joint CPU-RAM energy efficient and SLA-compliant approach for cloud data centers. Comput Netw 94:1–13

Access Control Mechanism for Cloud Data Using Block Chain and Proxy Re-Encryption

Umangi Mistry and Rajan Patel

Abstract This paper studies block chain and proxy re-encryption methods in our system. Block chain provides transparent property and serves privacy and confidentiality to the system. The storage of block chain can be a barrier for implementing the records in the system. Moreover, it is difficult to store the large amount of data exceeding into the practical storage capabilities of the current block chain usages. The most important issues of block chain are the performance, limited storage, privacy, energy consumption and scalability. Using normal public-key encryption, User A (data owner) needs to download and decrypt the requested data and then re-encrypts data using the target user's public key. This process requires extra computation cost and communication to the data. Another solution for this is to allow data owners to define access policies and encrypt the data with **encryption** under the access policies; only authenticated users whose keys are matching their policies can decrypt the ciphertext. But the thing is here data owner needs to download, decrypt, and re-encrypt the requested data in case data access policies change dynamically and frequently. To overcome these drawbacks, in this paper our system proposed a proxy re-encryption scheme that gives solutions for securing data and sharing in cloud computing environment. This system will reduce communication cost and extra computational cost.

Keywords Access control · Block chain · Proxy re-encryption · Cryptography · Cloud computing · Security

---

U. Mistry (✉) · R. Patel
Department of Computer Engineering, Gandhinagar Institute of Technology, Kadi-Kalol High way, Moti Bhoyan, Gandhinagar, India
e-mail: 180120702002@git.org.in

R. Patel
e-mail: rgpce21@gmail.com

K. Kotecha et al. (eds.), *Data Science and Intelligent Applications*, Lecture Notes on Data Engineering and Communications Technologies 52,
https://doi.org/10.1007/978-981-15-4474-3_44

1 Introduction

One of the most challenging issues for any system is preserving the privacy and security to the user's data. System contains sensitive data, the most important requirements of the preserving privacy to the access control mechanism to restrict the users to perform the certain operations. Generally, an access control mechanism can be implemented by three processes: identifying the user, authorizing the user, and authenticating the user. An access control also needs to support accountability and revocation—the ability to support the data records [1–3]. Block chain is a newly introduced technology. Nowadays, it can be used in developing many applications such as health care, insurance, and finance. Block chain stores the records of transactional. Block chain tracks the information flow. Block chain can be used for storing the data records. It can provide an on-demand access services to a specific information anywhere and anytime. The records may be stored as a transaction in the form of block or as a normal database record. The block chain services can be used to combine and to verify the information from different sources and to respond to the data access request. Block chain can support a cryptographic approach for identifying and authenticating the users and also support a static storage [2, 4, 5].

The proxy re-encryption is applied to support valid access control on secret data. A block chain-based secret-data sharing model that gives permissions to the system users to access the data records from a cloud storage after their identities have been verified is explained in our proposed system [5]. Applications based on block chain of any data management system offers different solution for different conditions. In this work, three types of block chain systems are analyzed. Then, a secret data-sharing model can support data confidentiality. To keep sensitive user data confidential against untrusted servers, different cryptographic methods are used to provide security, authentication, and access control in clouds. For sharing the data over the network, the shared data must be encrypted. There are many encryption schemes that provide security and access control over the network. Cloud services and applications may require all standard security functions including integrity, data confidentiality, privacy, access control, and robustness. Providing security to the cloud data is the challenging task. There are some cryptographic methods to secure the data stored in cloud storage systems [6].

2 Related Theory

2.1 Cloud Computing and Cloud Data

Cloud computing is a popular option for people and businesses for many reasons like cost-savings, increased productivity, speed and efficiency, performance. Companies which provide cloud services enable users to store files and applications on remote servers, and then all the data can be accessed by the Internet. Cloud computing delivers the different services through the Internet. These resources include tools and services like servers, databases, data storage, networking, and software. Rather than storing files on a hard drive or local storage device, cloud-based storage is more convenient to save them on a remote database. Cloud computing can be both public and private. Cloud data is the digital data which are stored in cloud data storage. The responsibilities of cloud data storage providers are to keep data safe, accessible, and available.

Proxy Servers: The most important role of proxy server is to providing security. A proxy server is a server that is in between a client application such as a Web browser and a real server. It intercepts all requests to the real server. It forwards the request to the real server. A proxy server verifies the user requests and forwards incoming client requests to other servers for further communication. A proxy server works as an intermediator between the Web browser and a Web server, and it is located between a client and a server. A proxy server is a computer system or router that applications are relay between client and server. The meaning of proxy word means "to work or act as a behalf of another." Here a proxy server acts on behalf of the user.

Proxy Re-encryption: Proxy re-encryption is a cryptosystem which allows third parties or also called as a proxy to convert a ciphertext which has been encrypted for one party and then decrypted by another party. Proxy re-encryption schemes are similar to symmetric or asymmetric encryption schemes. Proxy re-encryption is a cryptographic primitive which translates ciphertexts with use of encryption key to another encryption key. It can be used to forward the encrypted messages without showing the clear texts to other users. Advantage: Proxy re-encryption schemes are unidirectional. A proxy re-encryption algorithm converts a ciphertext using the public key public key of user A (PKA) to ciphertext public key of user B (PKB) by using the re-encryption key RK A $\rightarrow$ B. The server does not know the original text. PKA and PKB can only be decrypted by different keys Key A and Key B. Applications of proxy re-encryption are forwarding an e-mail, securing network file storage, and performing cryptographic operations on storage devices.

Access Control: Access control mechanism makes the restrictions on selective users in the field of physical and information security. To give the permission to access different resources called as an authorization. Access control mechanism assures users identity. Access control is a way of limiting accessing processes to a system or to physical or virtual resources. In cloud computing, access control is a process by which users are granted permissions to access the resources or information.

2.2 Block Chain

A block chain is an implementing or can be said as growing list of records which is called as blocks that are linked together by using cryptography. Each block contains cryptographic hash function of the previous block hash value. A block chain is restricted to modify the data. The main advantage of block chain is user cannot edit, delete, and modify the data, so that user can see the original form of the real data. A block chain is managed by a peer-to-peer (P2P) network. Block chain is a collection of many nodes. The IP address is used for identifying the node. That node can be a physical or a virtual machine. Block chain nodes can be linked together by a peer-to-peer (P2P) communication network. The block chain can use its own layer of protocol for a node communication. The user public key is used as an address to identifying such users. The user private key is used to convert cryptographic message and transaction. Theoretically, the user can login from any node within the block chain network [4]. The block chain technology can be a suitable technology for privacy-preserving model.

The Challenges of Block chain: *Privacy*—block chain transparently shows all the transactions, so it preserves the privacy using the public key. *Performance*—the processing speed of block chain is slow because the block chain has to perform some tasks such as signature verification, redundancy in addition to traditional database processes. *Scalability*—the larger the requirements on storage, bandwidth, and computational power when the larger the block chain grows. Scalability is a major issue for a block chain. *Energy Consumption*—block chain consumes a large amount of computational power and a large amount of electricity [4].

Cryptography: Cryptography is the study of techniques for secure communication in the presence of third parties. To prevent third parties or the public from reading private messages, cryptography constructing and analyzing the protocols. It provides information security such as data confidentiality, data integrity, and authentication. Cryptography focused on message encryption to provide data confidentiality. Encryption maintains the secrecy in the communications.

Encryption–Decryption: In cryptography technique, the process of encoding a message or information in a way that only authorized parties can access it who has encryption key and those who are not authorized cannot access it is called the encryption. In an encryption scheme, the targeted information or message, which is referred as a plaintext, is encrypted by an encryption algorithm. A cipher generating ciphertext that can be read only after it is decrypted. An encryption scheme usually uses a pseudorandom encryption key generated by an encryption algorithm. Only an authorized user can easily decrypt the message by using the key provided by the original user, but not to access by unauthorized users. Decryption is the process of taking an encoded data or encrypted text and converting it back into the original form of data or text that you or the computer can read and understand. The term decryption is used to describe a method of unencrypting the data using the proper codes or cryptography keys.

The Proposed Model for Secret Data Access

Proxy re-encryption server uses identities in cryptographic operations. Proxy re-encryption serves as a solution that promising to secure the sharing the data in the cloud computing. It fulfills the drawbacks of general public-key encryption. Proxy re-encryption server allows to re-encrypt data from one key to another key without getting access. These techniques are used to protect the secret shared data in cloud computing. Each part of data is ciphered with its own encryption key. Cloud server is responsible for re-encryption in proxy re-encryption scheme. It reduces time of communication process and extra computational cost.

A secret-data sharing model is proposed for privacy to preserving block chain technology and proxy re-encryption. There are five entities in the system model, the data owner which is User A, the user which is User B, the proxy re-encryption server, the cloud data storage, and the private block chain as shown in Fig. 1. User A is an entity who has its own or owner of the secret data and wishes to read or write the data from the system. User has a full right to control all accesses on his/her own data. He/she is responsible for defining an access controlling with re-encryption key. User B gets the re-encrypted secret data and decrypts it by using his/her private key. The owner can check the private block chain as necessary if they want.

The cloud storage is responsible for storing the encrypted secret data and information. The proxy re-encryption server is responsible for verifying the authenticity of the user, and it re-encrypting the encrypted secret data and access control on block chain. The private block chain is responsible for storing meta-data and access log. The user B who wants to access the secret-data stored on the cloud server.

A block chain is used to cover the drawbacks like performance, energy consumption, and scalability issues. The main advantages of using block chain in system are providing transparency to the end-user. To ensure the authenticity, the keys are verified before storing the data on block chain.

Data Availability: In our model, only the meta-data is stored on block chain. The proposed system model is guaranteed the availability of the encrypted real data, and a cloud data storage is provided user to store the encrypted the real form of data. The users can search the data on block chain and then query the data through the proxy re-encryption server. The authentication mechanism is for the authentic user to access the data.

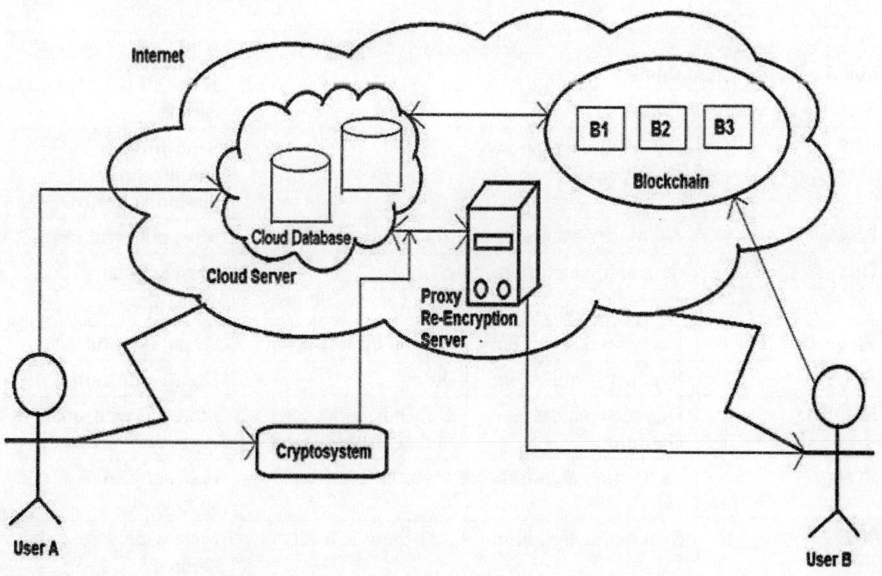

Fig. 1 Architecture of secret data access model

Data Privacy: The privacy of the data owner is maintained by supporting and safeguarded by the access control on his/her data. The data owner can revoke the access to his/her data by using the re-encryption key. In our model, the servers are assumed to be semi-trusted. The encrypted real data which is stored on the cloud storage is encrypted with the use of public key of the data owner. Only the data owner can create the re-encryption key by using BBS proxy re-encryption algorithm. The proxy re-encryption server can know all re-encryption keys, and it can try to perform the re-encryption on the encrypted secret data.

Data Security: In this section, malicious access attacks and replay attacks on the proposed model and how the secret data access model can handle such attacks will be discussed below.

Malicious Access: A malicious user wants to access secret data without authenticity. Our proposed model is protected against malicious attackers by using the proxy re-encryption server. The proxy server stores the meta-data on the block chain only after it successfully verifies by the proxy re-encryption keys. Then the server re-encrypts the data for only the authorized users who have re-encryption key. By using these mechanisms on the proxy re-encryption server, the secret data are protected from malicious attackers.

Replay Attack: A replay attack would happen when the malicious attacker copies the transaction information of some authorized users from the block chain and send the request to the proxy re-encryption server. The proposed model protects against this attack as it uses the user request lifetime on the proxy re-encryption server.

List of cryptography algorithm schemes as shown in Table 1:

Blaze, Bleumer, Strauss (BBS), Rivest–Shamir–Adleman (RSA), advanced encryption standard (AES), data encryption standard (DES), triple DES, Blowfish

Table 1 Security algorithms

Security algorithm	Type	Key size	Ability
BBS	Proxy Re-encryption algorithm	128-bit	Semi-trusted and atomic, three dissimilar keys
RSA	Asymmetric algorithm	2048-bit and 4096-bit	Most powerful forms
DES	Symmetric algorithm	64-bit	Most popular
Triple DES	Symmetric algorithm	56-bit	3 bundle of keys
Blowfish	Encryption algorithm	128-bit up to 448-bit	Faster algorithm
AFGH	Symmetric algorithm	56-bit	Highly influential
ECDSA	Digital signature algorithm	Depends on its size and algorithm	Powerful with smaller keys
RC4	Encryption algorithm	40-bit to 1024-bit	Fastest by JAVA supported
AES	Symmetric algorithm	128-bit to 256-bit	Most widely used and secured

(now known as Twofish), RC4, AFGH algorithm, elliptic curve digital signature algorithm (ECDSA).

3 Conclusion

In this paper, a block chain-based secret data access model is proposed using proxy re-encryption to solve user access control on cloud data. The proposed model explained the properties of block chain such as transparency. In cloud computing, security is an important factor of quality of service. Proxy re-encryption techniques are used to keep the sensitive user data confidential against untrusted servers. This paper surveys different security algorithms for proxy re-encryption schemes used in cloud data storage system. In the extended paper, we will set up the test-bed for the simulation of our proposed work.

References

1. Jin Ho P, Hyuk P (2017) Blockchain security in cloud computing: use cases, challenges, and solutions. Symmetry 9:1–13
2. Li X, Jiang P, Chen T, Luo X, Wen Q (2017) A survey on the security of blockchain systems. Future Gener Comput Syst
3. Zyskind G, Nathan O et al (2015) Decentralizing privacy: using blockchain to protect personal data. In: Security and privacy workshops (SPW), IEEE, pp 180–184
4. Zheng Z, Xie S, Dai H, Wang H Blockchain challenges and opportunities: a survey. Work Pap
5. Verma D, Desai N, Preece A, Taylor I A blockchain based architecture for asset management in coalition operations. SPIE
6. Gierschner F, Beyond B (2018) Consumers and health information technology: a national survey. In: California Health Care Foundation

Privy Cloud/Web Server

Maitri Patel and Rajan Patel

Abstract A stolen tablet, a crushed hard drive, a broken telephone—these are the things that modernized terrible dreams are made of. Luckily, while we may lose our equipment, and we can refuse losing the data set away on them. Cloud storage empowers us to store our data freely from our devices, ensuring those profitable archives, chronicles, and reports from bad occurrences and allowing us to access them from other devices as well. As capacity limit is expanding each year passing, everyone is moving to cloud storage. In this project, we will deal with every one of the issues in regards to the storage capacity and loss of data with only a credit-card sized computer that is the Raspberry Pi. After you are effectively done by putting away your information, you can get to it from anyplace on the planet and from any self-assertive number of gadgets in the meantime while as yet shielding the information safe from being hacked.

Keywords Cloud server · Web server · Raspberry Pi · Cloud storage

1 Introduction

As nowadays there is boom in online industry all the small traders as well as retailers are getting their business online so for them it is very useful. Cloud computing is a quick-developing business standard. Enterprises think that it is advantageous in several ways [1, 2]. Accessibility can be simplified through cloud computing. It also provides virtual storage space, security against unauthorized access, data loss and report for backup issues. Key advantage of using cloud server is, "users have to emolument only for the used resources from the cloud and get rid of from data storage, software licenses, servers and hardware" [3]. Also, it is very user friendly

M. Patel · R. Patel (✉)
Gandhinagar Institute of Technology, Khatraj-Kalol Road, Gandhinagar 382721, India
e-mail: rgpce21@gmail.com

M. Patel
e-mail: maitru1487288@gmail.com

K. Kotecha et al. (eds.), *Data Science and Intelligent Applications*, Lecture Notes on Data Engineering and Communications Technologies 52,
https://doi.org/10.1007/978-981-15-4474-3_45

as it is type of just "plug and play." The user need to do is plug in the power cable and place his/her HTML page in a folder. After this, his/her cloud is up and running. The user has not any need of managing big computer, but just has to purchase his/her domain [4, 5].

Cloud computing is broad concept and it wraps each type of online service. It defines three types of cloud service models: software as a service (SaaS), platform as a service (PaaS), and infrastructure as a service (IaaS) [6]. Proposed work has used PaaS as cloud service model. PaaS suppliers abstract significant work of dealing with servers. Also, they provides clients an environment in which operating system, server software and underlying server hardware and network infrastructures are managed which reduces user burden and make users to focus on their work. Every one of the servers that are being utilized today are muddled and utilizes part of space additionally you have to join a PC to cooperate with it which costs all the more regardless of the possibility that the server should be little for small association or say school.

The main purpose of the proposed work is to provide user with affordable, portable and easy to carry, small in size, consuming less power, and making equal to no noise cloud/Web server. For this, Raspberry Pi model 3 is being used. It also has expansion capabilities for cloud storage. Purpose of the proposed work is to expel this garbage and make the server considerably more reduced, convenient, and helpful and that even devours less measure of power. Also, to provide the customer with easier to carry and portable personal cloud storage, which is accessible anywhere from the world at decreased cost. Proposed Web/cloud sever will run on a Raspberry Pi, i.e., an affordable and tiny computer which runs on 5v electricity (that our mobile phone charges on) which is way more affordable than customary PC we utilize. It also has expansion capabilities that mean the more amount of space that you need, the much bigger hard drive you attach to it (Fig. 1).

Fig. 1 Raspberry Pi with expansion capabilities

2 System Requirements and Design

2.1 System Requirement Specification

2.1.1 Hardware Requirements for Proposed Work

Raspberry Pi Model 3 (B).

- Memory card (For Raspbian OS).
- External storage device.

Power adapter.

- USB cable (micro USB).

2.1.2 Software Requirements

Raspbian OS.

- Next cloud service for Linux.
- SSH (PutTy).
- MySql.

2.1.3 Costing and Budget for Proposed Work

See Table 1.

Table 1 Cost and budget details

S. No.	Item	Cost (in ₹) (approx.)
1.	Raspberry Pi 3	3000
2.	Memory card	400+
3.	External storage (PD, HDD, etc.)	1000+
4.	Power adapter	500+
5.	Micro USB cable	300+
Total		5200+

Note Here (+) symbol means that the cost may vary depending upon the client's requirements

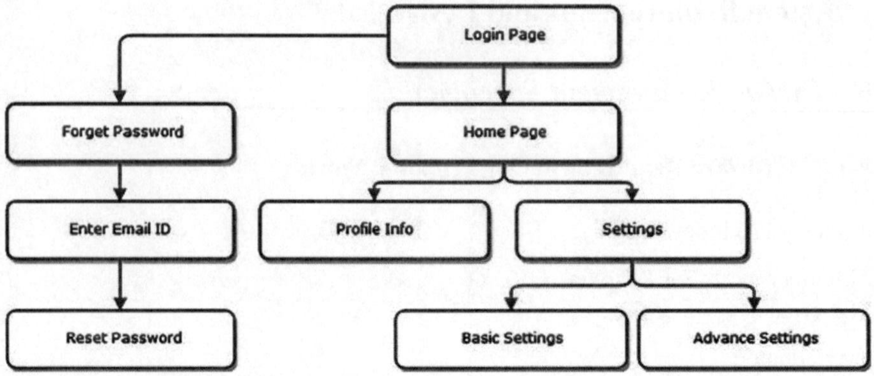

Fig. 2 Application navigation

2.2 *System Design*

2.2.1 User Interface

User will make use of the existing Web browsers such as Microsoft Internet Explorer or Google Chrome or Mozilla Firefox.

2.2.2 Application Navigation

See Fig. 2.

3 Implementation

3.1 *Implementation Environment*

Privy Cloud Server is developed in Linux and NextCloud framework is used as GUI in our proposed work. It is an open-source framework forked from own cloud. Any database like MySQl, SQLite, MariaDB, etc., can be chosen to work with the proposed work.

3.2 *Security Feature*

Proposed work is a Web application which will be accessed through URL. The important security feature is to make it accessible through HTTPS instead of HTTP.

Hyper text transfer protocol secure (HTTPS) is the secure version of HTTP in which 'S' stands for secure channel. Using this protocol, data is sent between user's browser and the Web site that he/she is connected to in encrypted format.

Proposed work provides security even when the product is offline. Also, it is secure against brute force attack with the use of fail2ban open-source code. Proposed application keeps user detached from outside world as it totally runs offline. Also, it makes hard for any hacker to attack directly on server. For more protection and security, a UFW open-source firewall is also added into it. Also, for Web application firewall, ModSecurity firewall is used.

4 Testing

4.1 Testing Plan

To find any kind of error, program gets executed is called as testing. A good test finds undiscovered error with high probability. The test which discovers undiscovered error is known as successful test. For any system, a test is vital. To perform system test, a logical assumption "if all parts of the system are correct, then goal will be successfully achieved" is made. The candidate system undergoes for various online tests like responsiveness, its value, stress and security. For user acceptance testing, a series of tests are performed.

4.2 Testing Strategy

4.2.1 Load Testing (Freezing)

Practice of demonstrating the projected utilization of a software program by simulation of multiple users that access the program concurrently is referred as load testing. Though other types of software systems can also be load tested, this testing is more relevant to the systems which contains multiple users and constructed using client/server model like Web servers. For example, a word processor or graphics editor can be forced to read an extremely large document; or a financial package can be forced to generate a report based on several years' worth of data. The utmost precise load testing put on actual usage, as contrasting to test using theoretical or analytical modeling.

4.2.2 Stress Testing (Click Test)

The procedure used to determine the stability of a given system or entity by performing entire or mordant testing is known as stress testing or torture testing. In order to analyze the results, sometimes it includes testing further than the normal operational potential to a breaking point. Following are reasons that are included:

- To confirm mathematical model is accurate enough in predicting breaking points or safe usage limits.
- To conclude breaking points or safe usage limits.
- To endorse intended specifications are being met.
- To identify modes of failure (how exactly a system fails).
- To test stable operation of a part or system outside standard usage.

4.3 Testing Methods

- Cpuburn (temperature).
- Cpuburn-a53.
- Click test.

4.4 Test Cases

4.4.1 Stress Test

This test was carried out to see if the Web application freezes but it worked correctly and the temperature remained stable too (Fig. 3).

4.4.2 Cpuburn-a53 Test (Overclocking)

This test was carried out to see if the temperature exceeds the threshold (80c) or not, stopping the server to continue working (Fig. 4).

4.4.3 Stress Test (Click Test) (Using Tool)

This was the test to check the load balancing and at a time how many users the server can handle. We gave it 10 users and each user clicked 200 times in one second sending heavy traffic to the server but it passed this test too (Fig. 5).

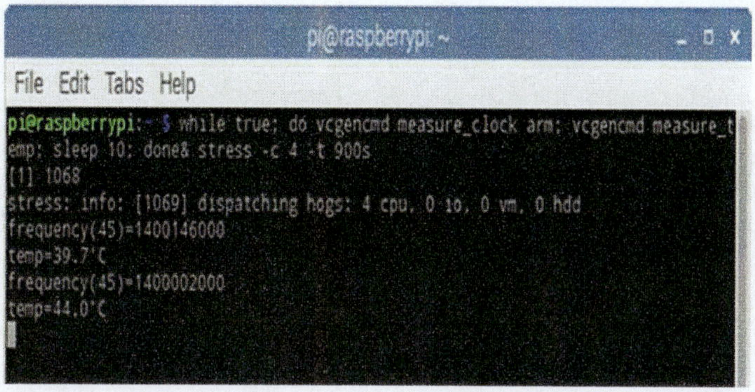

Fig. 3 Result of stress test

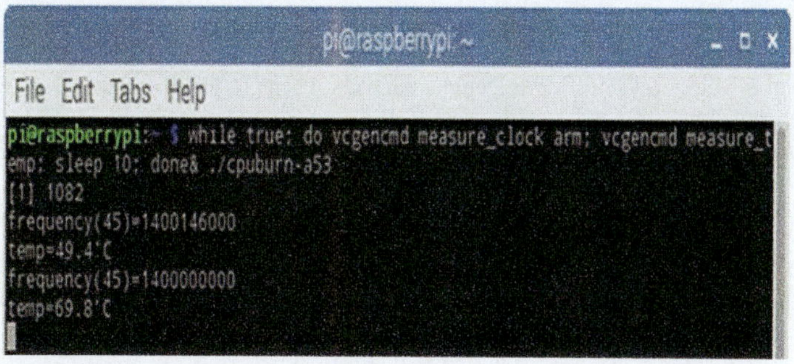

Fig. 4 Result of Cpuburn-a53 test (overclocking)

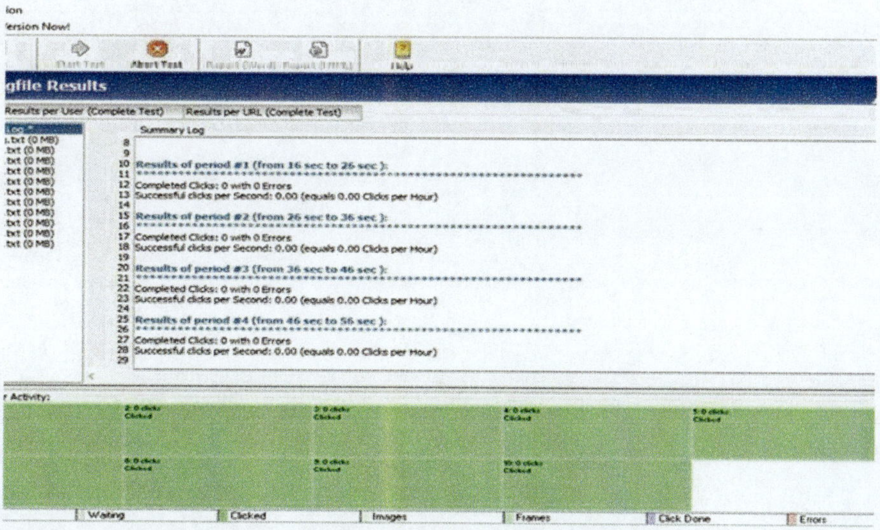

Fig. 5 Result of stress test (click test) (Using tool)

5 Conclusion

All the plans and the tasks related to storage capacity will be functioning efficiently through the system. The proposed system will fulfill all the problems faced due to security breaches by making this product offline. We solve major issues which help the attackers or hackers who try to breach our server through Internet. It will also be cost efficient, so that buying huge servers will get out of trend. Future work involves to remove any bugs if found and providing post assistance.

References

1. Kharat K, Pawar V, Wagh N (2019) Implementation of stable private cloud using Openstack with virtual machine results. IJCET
2. Homayoun S, Teing Y, Choo K, Ali Dehghantanha A, Parizi R, Epiphaniou G, Hammoudeh M (2019) Private cloud storage forensics: seafile as a case study. Springer, pp 73–127
3. Kiskani M, Sadjadpour H, Rahimi M, Etemadieh F (2018) Low complexity secure code (LCSC) design for big data in cloud storage systems. IEEE
4. Singh S, Mathur S, Khatri S (2018) Analysis of scaling in cloud infrastructure. IEEE, pp 1375–1379
5. Verma D, Tyagi R (2018) Cloud computing security: a review. ResearchGate
6. Chilamkurti N, Manogaran G, Varatharajan R, Vijayakumar V, Rawal B (2018) Secure disintegration protocol for privacy preserving cloud storage. Springer

Fault Tolerance in Cloud and Fog Computing—A Holistic View

Yash Shah, Ekta Thakkar, and Sejal Bhavsar

Abstract Cloud computing is paramount in information technology, and cloud computing has revolutionized the traditional ways of dealing with IT resources. Fog computing can be used in both large cloud systems and big data structures which is used for solving growing difficulties in information technology. Cloud computing and fog computing are used for large amount of storage, data and applications for the end users. Compared to cloud, fog computing is near to the end users. The major difference between cloud computing and fog computing is that cloud computing is centralized system and fog computing is distributed decentralized infrastructure. Many faults occur in cloud and fog computing, and there are various methods for fault tolerant techniques.

Keywords Cloud · Failure · Fault tolerance · Methods of fault tolerance · Fog computing

1 Introduction

Cloud computing and fog computing have been a wide evolution in information technology. Cloud computing is widely used nowadays, but there are several limitations in cloud computing. The fundamental limitation of cloud computing is connectivity between cloud and end user. This limitation can be removed by fog computing by placing fog nodes between cloud and the end user which would increase the connectivity with the end user. The data which is frequently used is kept in fog nodes, while the data which is kept for a longer amount of time is sent to cloud. In the

Y. Shah (✉) · E. Thakkar · S. Bhavsar
Gandhinagar Institute of Technology, Gandhinagar, Gujarat, India
e-mail: yashshah15199@gmail.com

E. Thakkar
e-mail: ektathakkar.2407@gmail.com

S. Bhavsar
e-mail: sejal.bhavsar@git.org.in

K. Kotecha et al. (eds.), *Data Science and Intelligent Applications*, Lecture Notes on Data Engineering and Communications Technologies 52,
https://doi.org/10.1007/978-981-15-4474-3_46

given method, a method is proposed fault tolerant techniques using collective network of fog and cloud computing in the smart building. A watchdog is connected with the smart building to supervise the fog nodes, and in case of failure and error, it is connected with the backup node. Moreover, the smart building data is sent to cloud which can be displayed to the end user using dashboard. Load balancing is also used to solve the problem of data congestion. For fault tolerance in cloud computing, self-healing [1] method is used.

2 Literature Survey

According to Table 1, more research gap is towards fault tolerance. So, next section contains holistic approach of combined fog and cloud computing.

3 Fault Tolerance

3.1 Fault Tolerance in Cloud

Fault Tolerance Metrics
Fault tolerance techniques involve various parameters for their evaluation. These parameters are helpful to find out how efficient the system performs if fault or failure occurs [2]. The parameters included are

1. Throughput: Throughput defines the number of tasks of the work flow that has completed the execution per unit time [2]. Higher the throughput, the better is the system performance [1].
2. Scalability: It defines the ability of an algorithm to tolerate the faults with the predetermined number of nodes [1].
3. Reliability: It aims to provide correct, accurate results within a certain specified period of time [1].
4. Response Time: It may be defined as the time taken by a particular algorithm to respond. This parameter should be the minimum [2].
5. Usability: It is the scope up to which a product can be effectively used by the user to achieve objectives with efficacy, proficiency and contentment [2].
6. Availability: The possibility that a particular job will function adequately at a given amount of time can be considered in terms of availability of resources [2]. As the system's reliability increases, the system's availability also increases.

Fault Tolerance Techniques
Fault tolerance helps in making a system more powerful. There are mainly two fault tolerance techniques in cloud computing:

Table 1 Literature survey on fault tolerance in cloud and fog computing

Author	Synthesis of research	Summary of contribution	Future of research
Kaur et al. [1]	Cloud Computing has gained a large amount of attention from the government and enterprises. With the increased use of cloud computing, the chances of fault occurrence also increase which are a challenging issue, thereby providing architectures and the existing techniques for fault tolerance of cloud computing	Different types of fault tolerance Metrics for fault tolerance Comparison of fault tolerance in cloud computing Cloud computing architectures and their comparison	To explore MPI architecture for more reliable and less costly fault tolerance techniques Implementation of fault tolerance techniques which can be done over cloud supported simulator so that it can be more reliable
Prathiba et al. [2]	Basically focuses on different types of fault occurring in cloud computing. Various fault tolerance models and mechanisms are also provided to recognize and reduce the amount of fault tolerance in cloud computing	Different types of fault tolerance metrics Brief details of existing techniques of fault tolerance Various fault tolerance models which can be used in cloud computing	With rapidly increasing use of cloud, cloud needs to be made fault tolerant, and methods and architectures contributing to fault tolerance and reliable and scalable architecture in cloud computing should be implemented
Ganesh et al. [3]	Brief description about different types of fault so to have a better understanding about the fault tolerant (FT) system of cloud computing. Fault tolerant systems can be used for continuous reliability	Key characteristic of cloud computing Overview of fault tolerant Various fault tolerant (FT) systems	To understand various types of fault tolerant (FT) systems in order to built an efficient FT system which can deal with different faults in diverse aspects
Ataallah et al. [5]	Discusses about various fault occurring in cloud computing. Major focuses on hardware failures which can reduce the efficiency in the cloud computing. Survey done for various faults in cloud computing	Fault tolerance requirements Various cloud fault tolerance policies Cloud tolerant developed techniques Efficient fault tolerant algorithm	Fault tolerance in cloud can be achieved by some policies and fault tolerance methods which includes active and proactive methods, such methods should be developed and implemented for better fault tolerant cloud systems

(continued)

Table 1 (continued)

Author	Synthesis of research	Summary of contribution	Future of research
Mohamed et al. [6]	Fog computing can be used to build IoT-based smart city applications, and there are various fog services and some methods and considerations for making fog tolerant fog-based systems	IoT-based smart city applications Fog services types Fault tolerant fog computing considerations Fault tolerant fog computing services	More reliable fault tolerant IoTSCA (IoT-based smart city applications) can be developed and various services that can be used and implemented as smartcityware in IoTSCA
Lyu et al. [4]	Systems having cluster of computers are essential but have complex software models. They have separate hardware for error detection and dedicated software for fault tolerance. Reliable clustered computing (RCC) systems, their techniques, architecture are areas focused on	RCC principle techniques RCC architecture assumptions Reliability Techniques Reliability Modelling and analysis	The manual recovery of faults is costly as well as time consuming in a reliable clustered computing (RCC), so building of architectures in which faults can be monitored and avoided in RCC needs to be implemented

- Proactive fault tolerance
- Reactive fault tolerance

(a) **Proactive Fault tolerance**: Proactive fault tolerance mechanism involves identifying the doubted constituents and proactively taking preventive measures like replacing the suspicious components before it could cause any faults or errors.

i. **Preemptive Migration**: This task is based on the feedback loop control mechanism. In this technique, the system is adamantly supervised and evaluated [1].

ii. **Self-Healing**: Self-healing system detects and responds to faults that occur during runtime. It involves recognizing and solving the failure immediately. In this, a larger task is broken down into smaller parts, and those smaller parts run on virtual machines as a single application making it handle failures efficiently [1].

iii. **Software Rejuvenation**: This approach involves resetting of the system state and making a fresh start whenever any failure occurs [1].

(b) **Reactive Fault Tolerance**: Reactive fault tolerance comes into action after a failure is encountered. It deals with reducing the impact of failure that has occurred.

i. **Checkpointing/Restart**: Checkpoints act as a marker, in this, whenever a system failure occurs, it will restart the process from the place that is marked [3].
ii. **Replication**: In replication, a job is replicated a number of times and is executed on different machines using distinct resources so that execution of the application is done successfully [2].
iii. **Task resubmission/Jjob migration**: In task resubmission, whenever a fault is detected for a job running on a specific machine, it will be shifted to another machine which is appropriate for that job [3].
iv. **Retry**: This method is used when we want to execute the task on the same virtual machine, task is retried many number of times until the success [2].

3.2 Fault-Tolerant Fog Computing

These are the considerations that can prevent faults or make our system fault tolerant.

- All the IoT devices in the network should be connected to more than one fog node, such that if one node fails, others can work as a backup node.
- Fault tolerance can be enhanced by overlapping the coverage of these nodes across the IoT distributed areas.
- In case of failure of a fog node, the 'watchdog' technique is used, where a fog node monitors another fog node.
- Replacing a faulty node must ensure

  (a) The capability of the replaced node must be equal to the faulty node.
  (b) The accessibility of the replaced node must be with all the IoT devices as well as the nodes.
  (c) It must maintain real-time services.

Fault Tolerance services: These are some services that are required for fault-tolerant fog computing.

Watchdog: For monitoring a node by another node.

Service Replication: To start a stateful service on another node (replacement node).

Replace Main Broker: Changing the fog node an IoT device uses to another node.

Find Fog Nodes: Finding a list of neighbouring nodes.

4 Proposed Work

Implementation of fault tolerance in collective network of cloud and fog is given in Fig. 1.

The architecture consists of four components namely

1. Watchdog: It supervises the smart building fog node, and in case of failure, fault or error, it switches over to backup/spare node and thus make system fault tolerant.
2. Smart Building Fog Node: They are the IoT device connected with the network of fog nodes send data to cloud, and in case of data congestion and overload, load balancing techniques are used.
3. Cloud: Massive storage and computation is done here.
4. Dashboard: Real-time monitoring of the system is done here.

Proposed implementation is done with the combination of watchdog in fog with load balancing and proactive fault tolerant method like self-healing in cloud.

Here, over load and data congestion problems causing faults are solved by load balancing, and rest of faults like corruption of fog node/node failure, etc., are solved by watchdog.

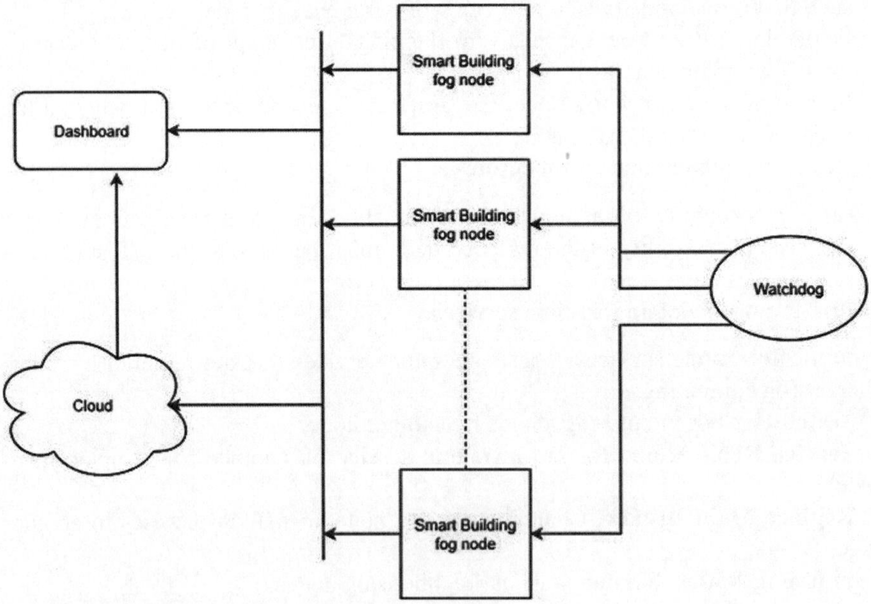

Fig. 1 Demonstrates working of proposed system

4.1 Fault Tolerance in Fog Devices

Fault tolerant fog network is prepared by the use of watchdog, a watchdog is used for detection of errors and faults, and switching over to a backup processor, if there is failure in the active processor. However, the application software is responsible for error detection, consequent recovery actions and data backup [4].

4.2 Load Balancing in Fog Nodes

Load balancing can help in performing small computations that do not require much compute capacity in fog devices itself, thereby saving the data transfer overhead towards the cloud and then fetching the eventual result from the cloud. Once the computation is done, the data that needs to be persisted can be stored on the local persistent storage devices, and thereby, it can be asynchronously replicated to cloud (Fig. 2).

4.3 Fault Tolerance in Cloud

This can be solved by 'self-healing' technique, a proactive fault tolerance technique used for making cloud systems fault tolerant.

5 Conclusion

Cloud and fog computing are emerging needs and essential, to use this technologies efficiently, faults and errors should be prevented. To prevent fault occurring in smart building a watchdog is used and to solve the problem of congestion, load balancing methods are used, along with proactive method self-healing in the cloud to make cloud fault-tolerant. For a system involving use of fog node, IoT devices and cloud computing, fault at any level may lead to failure in system. Holistic and robust fault tolerance method which predicts and nullifies fault at every level needs to be developed and implemented for better smart building or smart city applications. Comparison of different fault tolerance algorithms can be done in fog and cloud, and their different combinations can be used.

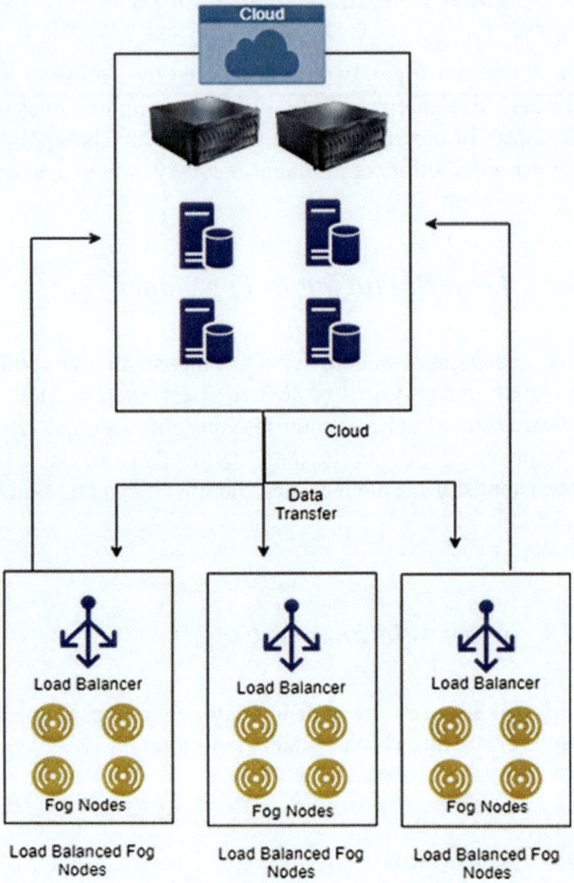

Fig. 2 Load balancing

References

1. Kaur P, Priya K (2015) Fault tolerance techniques and architectures in cloud computing—a comparative analysis. In: 2015 International conference on green computing and internet of things, pp 1090–1095
2. Prathiba S, Sowvarnica S (2016) Survey of failures and fault tolerance in cloud. In: 2nd International conference on computing and communications technologies (ICCCT). IEEE, pp 169–172
3. Ganesh A, Sandhya M, Shankar S (2010) A study on fault tolerance methods in cloud computing. In: IEEE international advance computing conference. IEEE, pp 844–849
4. Lyu MR, Mendiratta VB (1999) Software fault tolerance in a clustered architecture: techniques and reliability modeling. In: IEEE aerospace conference. Proceedings (Cat. No. 99TH8403), pp 141–150
5. Ataallah SMA, Nassar SM, Hemayed EE (2015) Fault tolerance in cloud computing-survey. In: 2015 11th International computer engineering conference (ICENCO), pp 241–245
6. Mohamed N, Al-Jaroodi J, Jawhar I (2019) Towards fault tolerant fog computing for IoT-based smart city applications. In: IEEE 9th annual computing and communication workshop and conference, pp 0752–0757

Automating Container Deployments Using CI/CD

Sejal Bhavsar, Jimit Rangras, and Kirit Modi

Abstract The biggest challenge faced by information technology (IT) industry is of the deployment and release, traditional methods used to release the final build as a part of deployment process, along with deployment of the database and code on the on premise infrastructure which was a long and cumbersome process, then with advent of cloud computing, things became easy, and the infrastructure was shifted from on premises to cloud, this made the on demand access of resources just a matter of clicks, but still deployment was a problem due to a major reason of "Code working on developer's machine but not working properly in staging or production environments". Primary reasons included versioning problems, software patches or bug fixes. It also had various problems such as scalability issues, the infrastructures were not fault tolerant as well as lack of disaster recovery strategy in place. With advent of containerization, the platform dependency issues were resolved, but again deployment release and rolling updates are a major issue with them as well. This paper focuses on high level overview about how CI/CD could solve the problem of deployment and rolling upgrades by automating the entire build and deployment process, as well as to handle the rolling updates process in an effective manner. It also provides a way of implementing the disaster recovery strategy regardless of the underlying cloud platform as well as fault tolerance mechanism if in case any failure occurs in staging or production environments.

Keywords Cloud computing · Containers · DevOps · CI/CD

S. Bhavsar (✉) · J. Rangras
Computer Engineering, Gandhinagar Institute of Technology, Gandhinagar, Gujarat 382721, India
e-mail: sejal.bhavsar@git.org.in

S. Bhavsar
Ganpat University, Mehsana, Gujarat 384012, India

K. Modi
Computer and Science Engineering, Parul University, Vadodara, Gujarat 391760, India

K. Kotecha et al. (eds.), *Data Science and Intelligent Applications*, Lecture Notes on Data Engineering and Communications Technologies 52,
https://doi.org/10.1007/978-981-15-4474-3_47

1 Introduction

Deployment is the most important phase in the entire software development lifecycle, as if the deployment is not done in the proper manner, then the whole product is of no use as it will not be used by its intended users. Traditionally, the deployment process was a manual process, wherein the infrastructure folks who were responsible for deployment of the product performed the deployment steps manually, which included transfer of the build files to the testing/staging server and transfer of database files to the server, then finally updating the manifests for pointing the business logic to the database, which then eventually resulted in the completion of the deployment process, this was a long process as well as cumbersome task, wherein support staff had to be present for maintaining the production/staging servers which were residing on premise. This also bumped up the costs as there was additional manpower required just to handle the deployment process, and it also included the capital costs to maintain the servers for hosting the product. Then eventually, with the advent of cloud computing, the whole scenario changed, companies started shifting toward cloud environments for performing deployments rather than deploying on the on premise environments. Cloud computing solved the major problem of capital costs as it was purely based on pay-as-you-go model and also offered on demand delivery of resources via Internet. Initially, there were limited options, but nowadays, there are many vendors in this space such as Amazon AWS, Google Cloud Platform, Microsoft Azure, IBM Cloud [1]. All these platforms offer numerous amount of services in various domains. While cloud computing solved the problem of capital costs, it had various other questions/concerns to take care of so as to make the application deployed on their environment more durable, secure and robust. Thus, various challenges with cloud computing are deployment, scalability, security as well as fault tolerance. Deployment is still a major concern as only the resources have been shifted to cloud, so the overhead of maintaining the server has been transferred to them, the deployment process is still the same old procedure of uploading the build files onto the server and then mapping it with the database to make it available in production environments. Performance issues were also a major concern as the servers were not performing up to the mark, and there was always a requirement to enhance them with high configuration which again bumped up the infrastructure costs. Maintaining backups of the servers was also a major issue as it was a very important part as a part of disaster recovery strategy as well as to handle the failovers.

2 Containers

2.1 Introduction

Traditionally, virtual machines were used for the purpose of server virtualization so as to host multiple applications on the same server in a virtualized environment, hence multiple applications used to access the same underlying hardware for the purpose of compute operations, even today the same concept, i.e., server virtualization, network virtualization and storage virtualization is being used in cloud environments and is also the heart of cloud computing paradigm. Even with virtualization, resource utilization is not efficient as well as the infrastructures suffer from performance issues. Basically, virtual machines partition the hardware resources and allocate them to the virtual machines so as to create an isolated environment, thereby allowing the application to access the allocated resources, but in many situations, when the application is not performing any compute operation, the resources allocated to that application are sitting idle and of no use. Thus, it leads to wastage of the resources. Containers on the other hand do not partition resources beforehand, but it allocates the required resources on the runtime.

2.2 *Containers Versus Virtual Machines*

According to Docker, containers are defined as follows:

"A container is a standard unit of software that packages up code and all its dependencies so the application runs quickly and reliably from one computing environment to another" [2]. Docker is an open source container engine which provides the base engine for running and maintaining containers on it. Docker also provides a better way to ship the software, i.e., via Docker images, these images are nothing but a package which contains multiple layers prepackaged into a lightweight bundle. This bundle contains base operating system, application dependencies, application code as well as network-related drivers and information. Using this image, containers can actually execute the application in a very efficient manner making it lightweight and highly scalable.

2.3 *Architecture*

Virtual machines (VMs) generally use hypervisor as a base engine to partition the hardware resources beforehand, thereby consuming the required resources. On the other hand, containers directly interact with the kernel of the operating system, hence there is no need to partition and assign the hardware resources of the system beforehand, and it can be consumed on runtime as per the requirements (Figs. 1 and 2).

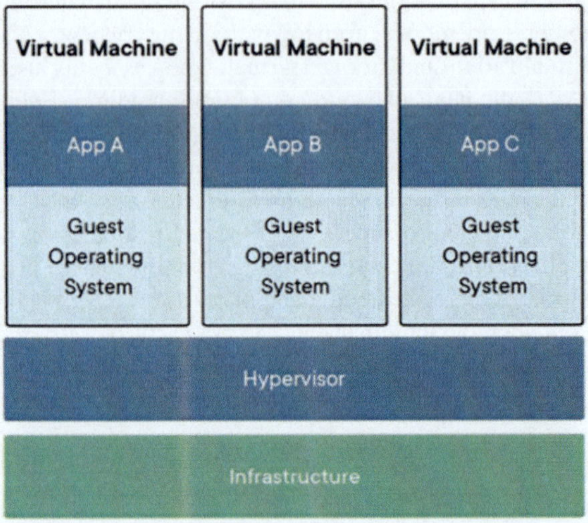

Fig. 1 Virtual machine architecture [2]. *Image source* https://www.docker.com/resources/what-container

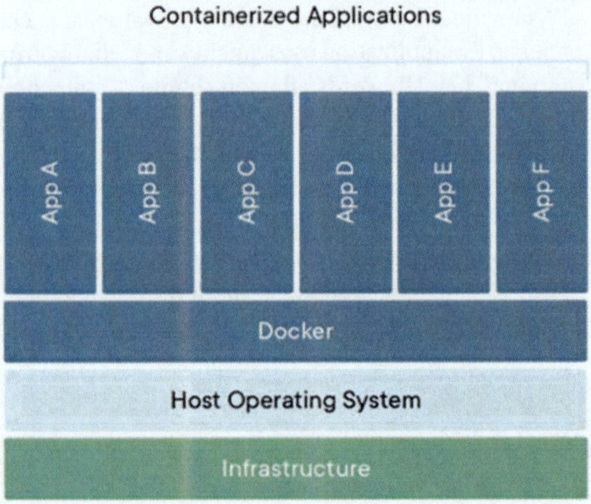

Fig. 2 Containerized application architecture [2]. *Image source* https://www.docker.com/resources/what-container

3 Deployment Process

Containers are powered by images, and for the deployment of the same, Docker images have to be created, Docker images are nothing but a package that contains application code, required dependencies as well the networking information for that application. Docker images are created from the manifests named as Docker files and can be used to create Docker images, once the Docker images are created, it can be deployed on the Docker engine from where the application can be managed.

Rolling updates are also easier with this approach as images can be tagged with specific versions, and once the latest version is ready to be rolled out, the running container can be updated with the latest version.

Multiple replicas of the containers can also be maintained easily using container orchestration tools such as Kubernetes or Docker Swarm, and using such tools, the services can be replicated onto other nodes so that multiple copies of any service exist, which make the entire infrastructure to be fault tolerant.

Major problem with using containers is that it has to be managed manually at every step and that issue can be resolved by CI/CD using which the entire workflow can be automated.

4 Continuous Integration and Continuous Deployment

Continuous Integration is a practice that helps multiple teams to integrate their code in one place so as to automate the build process [3].
Continuous Deployment is a practice that enables deployment of the application in real time in an automated manner with zero or minimal downtime, thereby pushing the small upgrades into production environment on frequent basis so that the final deployment does not become a long and cumbersome process [3].

The entire process starts from building the code, and once it is successfully built, it is deployed to the testing environment. After being deployed to testing environment, it is tested, once the testing phase is completed, it is then deployed to staging for further testing upon completion of which the changes are moved to production, and if there find any bugs, then they will be fixed, and the whole process is repeated again [4].

Various tools are available for CI/CD, one such popular open source tool is Jenkins, and it can be used to seamlessly integrate the infrastructure and create CI/CD pipelines to automate the deployment and testing process.

5 Jenkins

Jenkins is an open source tool which can be used for CI/CD, using Jenkins entire build, deployment and testing process can be automated, and it can also be integrated with containerized applications to automate the deployment process into containers. Using Jenkins as the CI tool, it can solve the problem of manually building the Docker images and pushing them to the registry and then spinning containers from them. Various components in the entire CI/CD workflow are as follows [5]:

Jobs: The entire workflow for the build and deployment process can be divided into various jobs in Jenkins, a job contains the entire workflow which will build the code, create the Docker image, push the image to the Docker registry, get into the deployment environment, pull the image from the registry and spin container from the image into the production environment.

Triggers: Triggers are responsible for invoking the job in Jenkins, when an event occurs such as a push in Github, Jenkins will invoke a job from that trigger which will then complete the remaining process.

Groovy: It is the syntax followed by Jenkins for writing the job script.

Stages: Jenkins also supports multistaged builds, for example, we can divide the job into multiple stages which can be used to divide the job into various parts.

CI/CD Using Jenkins

The entire process will start executing when the developers will commit their code on a version control system, and in this scenario, we consider that developers are using Github as their version control system. As soon as the code is pushed to Github, a Jenkins job will be trigged which will then build a Docker image from the code, this Docker image will then be pushed onto Docker trusted registry which is the central hub for storing the delivering Docker images across the globe, then Jenkins will perform an SSH connection in the server and clone the Docker image, after cloning the Docker image, it will then spin up a container from that image, if the container already exists, then it will simply update the existing container by using the latest version of the image. This can be further enhanced to integrate automation testing after the container has been started, and it can also deploy the application to multiple environments and can be used for container orchestration as well as distributed environments also (Fig. 3).

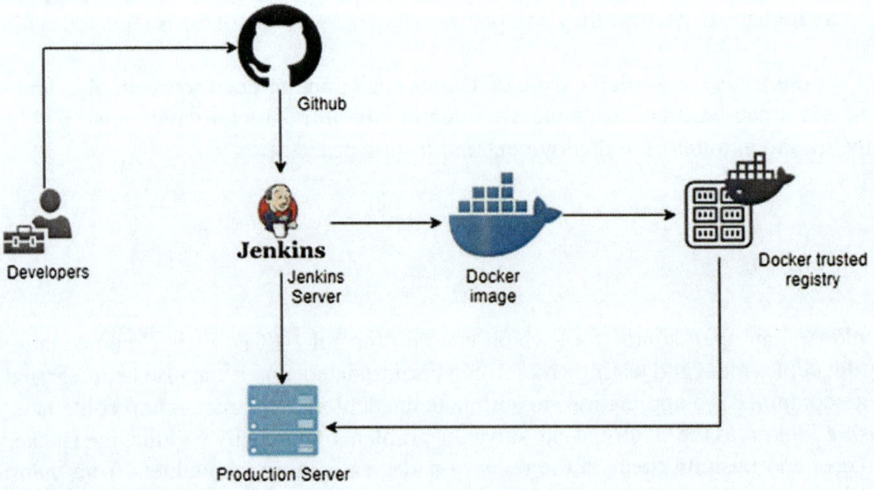

Fig. 3 CI/CD architecture

6 Future Work

This architecture can be integrated with multiple cloud environments and can be created as infrastructure as a code [6] in a way that it is completely cloud agnostic, this can be achieved by using open source tools such as Terraform by Hashicorp [6], and this tool can be used to create infrastructure as a code for almost all the cloud platforms. Furthermore, it can be enhanced to support container service offerings by multiple cloud platforms such as Amazon AWS offers "Elastic Container Service" [7] which uses same underlying Docker infrastructure but is managed and hosted by AWS [8].

7 Conclusion

Thus, by using CI/CD tools such as Jenkins, build and deployment processed can be automated with the best possible performance, and greater efficiency can be achieved with the same. Other tools such as Ansible, Chef and Puppet can be used for the similar use case, but Jenkins is preferred due to its efficiency as well as large community base.

References

1. Stackify Blog. https://stackify.com/continuous-delivery-vs-continuous-deployment-vs-continuous-integration/
2. Docker's Official Documentation. https://www.docker.com/resources/what-container
3. Bobrovskis S, Jurenoks A (2018) A survey of continuous integration, continuous delivery and continuos deployment. In: BIR workshops, pp 314–322
4. Sapana Kejadiwal Connuous integraon using Jenkins. Whitepaper lntinfotech. https://www.lntinfotech.com/wp-content/uploads/2018/01/LTI-Continuous-Integration-Using-Jenkins-WP.pdf
5. Jenkin's Official Documentation. https://jenkins.io/doc/
6. Terraform Documentation. https://www.terraform.io/
7. AWS ECS Documentation. https://docs.aws.amazon.com/AmazonECS/latest/developerguide/Welcome.html
8. DevSecOps Realities and Opportunities, Whitepaper by 451 Research. https://www.synopsys.com/content/dam/synopsys/sig-assets/reports/devsecops-realities-opportunities-451.pdf

Fault Tolerance and Detection in Wireless Sensor Networks

Hetvi Sheth and Raxit Jani

Abstract The dependableness of wireless sensor network (WSN) is tormented by faults that will occur due to numerous reasons like malfunctioning hardware, computer code glitches, dislocation and environmental hazards. The WSN that is not ready to agitate such things may suffer from some serious issues or can lead to hazardous consequences in important application contexts. The faults occurring in the sensor nodes are due to the sensor itself and the harsh environment. So as to confirm the network quality of a service, it is necessary for the WSN to detect the faults and to take actions to avoid more degradation of the service. The goal of this paper is to find the faults within the wireless sensor networks and provide the fault tolerance techniques.

Keywords Wireless sensor networks · WSN · Energy efficiency · Fault · Detection · Tolerance · Latency

1 Introduction

Advances in wireless communications have enabled the development of wireless sensor networks (WSN) which are expected to be one of the most encouraging technologies in the near future. Wireless sensors are tiny, cheap and battery-operated devices whose sensing capabilities change them to retrieve data. They can be deployed in a very sizeable amount to observe events of interest or send periodic reports regarding the surroundings [1]. WSN are key parts of system since they will be integrated in complicated assemblies so as to retort to current problems like the aging of the population, the improvement of natural resources device networks also can be accustomed notice foreign chemical agents within the air and also the water. They will

H. Sheth (✉) · R. Jani
Gandhinagar Institute of Technology, Gandhinagar, Gujarat, India
e-mail: hetvisheth9999@gmail.com

R. Jani
e-mail: janiraxit86@gmail.com

K. Kotecha et al. (eds.), *Data Science and Intelligent Applications*,
Lecture Notes on Data Engineering and Communications Technologies 52,
https://doi.org/10.1007/978-981-15-4474-3_48

facilitate to spot the sort, concentration and site of pollutants. In essence, detector networks can offer the tip user with the intelligence and a higher understanding of the setting we tend to authorize that in future wireless detector networks are associate degree integral part of our lives over the current personal computers [2]. The planning of energy-efficient wireless sensing element networks is, however, a really difficult issue. On the other hand, energy-constrained sensors measure resource-limited in terms of memory and procedure capacities that stops the utilization of advanced algorithms and makes the ancient protocols of the TCP/IP stack not like minded. On the opposite hand, in contrast to different networks, WSN measures designs for some specific applications. These applications vary from little size attention management systems to giant scale setting observance. Thus, any WSN reading should satisfy a collection of needs that differs from one application to a different [3]. As a consequence, energy saving techniques typically trade off lifetime maximization with different optimization objectives. The major source of power consumption at device nodes measured the communication tasks, followed by computation and sensing operations through totally different sensing platforms that have different energy consumption profile [4]. The subsequent remark totally holds.

The communication system has associate energy consumption abundant over the consumption system. Therefore, communication ought to be listed for computation. The radio energy consumption is of identical order of magnitude within the reception, transmission and idle state, whereas the ability of consumption drops to minimum of one order of magnitude within the sleep state. So, the radio should be turned off whenever potential. Counting on the precise application, the sensing system could be another important supply of energy consumption, so its power consumption should also be reduced [1] (Fig. 1).

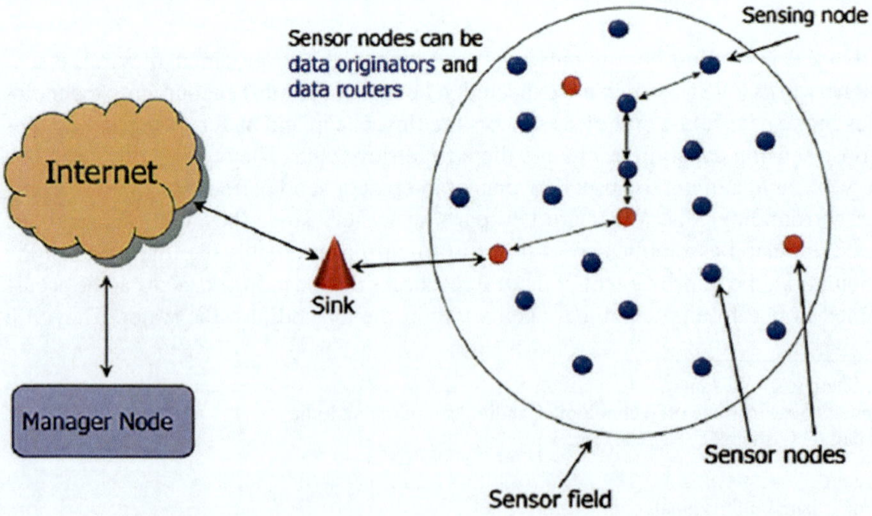

Fig. 1 WSN communication architecture

2 Energy Efficiency and Latency

Energy efficiency and latency are probably the most important issues in wireless sensor networks (WSNs). Since future sensor nodes are unreal and disposable, it is very vital to develop the techniques that prolong battery period as much as possible. Unwanted energy consumption needs to be avoided by attentive hardware design along with low level (operating system and support middleware) and high level (application) software program. Recent commercial sensor devices provide a high level of flexibility allowing programmers to selectively turn on/off the various hardware components. However, achieving a desired data rate and maximizing the lifetime of individual sensors are often conflicting goals and are subject of ongoing research. The first dichotomy is between the beacon-enabled and the beacon-less modes. One of the most important energy efficiency features is the possibility of turning the transceivers off most of the time and activating them only when required. Studies show how such sleeping mode affects transmission delay. In slotted CSMA-CA, a packet might be delayed by several sleep periods when a node fails to access the channel due to contention (especially near data convergence points like the sink node) although overall traffic in the network is relatively low. One of the proposals follows from the observation that in group networks many close-by nodes are equivalent from a routing point. The concept of Connected Dominating Set (CDS) approach is to select some of the nodes to constitute a network backbone and be active all the time providing network connectivity and temporarily storing messages for neighbouring non-backbone nodes. Non-backbone nodes sleep most of the time (saving energy) and periodically wake up to exchange messages with their backbone node neighbour. Since backbone nodes consume more energy than the other nodes, CDS protocols require nodes to alternate between backbone and non-backbone status. Two examples of CDS protocols are GAF (geographic adaptive fidelity) and Span. The first option is to fuse data generated by different sensors. The second option focuses on the paths of data routing phase. Energy efficiency is used as the optimization metric [3].

3 Fault Tolerance and Detection

Some sensor nodes may fail or get blocked because of lack of power, have physical harm or environmental interference. The failure of sensor nodes should not affect the overall activity of the sensor network. This is the responsibility of fault tolerance. Fault tolerance has the ability to sustain sensor network functionalities without any interruption due to sensor node failures. If the environment where the sensor nodes are deployed has minimum interference, then the protocols can be more at ease. A system failure occurs when the delivered service differs from the specified service. Hardware and software faults affect the system state and the operational behaviour, such as memory or register content, program control flow, communication links, etc.

Faults can be caused because of hardware failure or energy depletion. Communication can be disrupted due to environmental conditions. Hardware faults can even disrupt radio communication, ending all the communication to and from the gateway. A fault in transmitter can prevent the gateway to transmit tasks to the sensors furthermore as relay the information to the command node. Data sent by the sensors are lost if receiver of a gateway fails. As a result, the fault tolerance level depends on the application of the sensor networks investigated according to the parties involved in the process. Through self-diagnosis, the node itself can establish faults in its elements. With group detection, several nodes monitor the behaviour of another node. Finally, in hierarchical detection, the fault detection is performed employing a detection tree where a hierarchy is outlined for identification of unsuccessful nodes. Often in a hierarchical detection, the detection is shifted to a very powerful device such as the sink. The error is the state of the service when attempting to scan the device knowledge, and therefore, the failure occurs once the application does not send the data within the desired time interval. Two actions are performed that are: Fault Detection and Fault Recovery [5].

Fault Detection: To continue any function, the first step is to check whether the functionality is or will be faulty [6].

Fault Recovery: After detection of the fault, the next step is to recover it. The faults that can happen in sensor nodes are.

Node Fault: Under the harsh environment conditions, a variety of extreme conditions can cause issues such as battery leakage, antenna failures, circuit failures, cluster issues which can lead to weak performance of WSN. Due to battery failures at a certain stage, reading of sensors may become incorrect. Data acquisition application will not perform if sensor provides incorrect reading. Even the hardware failure can cause destruction in software. There are some hardware failures that do not harm the services of the sensor node. For example, though the sensor node does not provide correct readings, it can be used to route the packets at the desired location. To extend the lifetime of network cluster is used [7]. The small cluster combines to form a cluster heads. These cluster heads coordinate the nodes in their respective clusters, to perform the actions such as forwarding the data to the base station. If the failure occurs in the cluster heads, the data cannot be passed to the base station, and this can even occur due to the software bug which forwards the incorrect information [8]. The deviations may occur in data aggregation due to these failures.

Network Fault: Routing plays an integral role in WSN. Mainly, the routing protocols are used to collect data and coordinate among the nodes and even to track the data. The faults caused due to routing can led to transfer of improper data or can cause delay in transferring the data. The timing of the messages sent to desired location is not always same. The instability in the links causes variations in the delivery of message [9]. The error is even caused when during routing the nodes move to different location and are unable to reach to the exact place, which causes loss in the data. Another cause of fault is collision of messages which occurs when the phase changes and due to which huge amount of data is misplaced. This only occurs when the proximity is less. Similar to the node fault, software bug can cause fault, when it is present in the routing layer, it can deliver the message at incorrect location [10].

Table 1 Probability of no faulty sensor diagnosed as good

p	Average number of sensor	Column 1	Column 2	Column 3	Column 4
	4	6	10	15	20
0.05	1	1	1	1	1
0.1	1	1	1	1	1
0.15	1	1	1	1	1
0.2	0.9999	1	1	1	1
0.25	0.9994	1	1	1	1

Table 2 Probability of no good sensor diagnosed as good

p	Average number of sensors	Column 1	Column 2	Column 3	Column 4
	4	6	10	15	20
0.05	8.93E−66	8.88E−66	0.00E+00	0	0
0.1	1.04E−50	1.00E−50	0	0	0
0.15	7.17E−42	6.39E−42	0	0	0
0.2	1.43E−35	1.14E−35	0	0	0
0.25	1.16E−30	8.14E−31	0	0	0

Sink Fault: This type of fault occurs when the data are collected and are to be sent to the back-end system and suddenly the device fails. If the fault tolerant measurements are not available at that moment, massive failure of network is caused, and even the data cannot be accessed. This fault may occur where proper energy or power is not supplied. Even the network infrastructures give rise to this type of fault. In some areas, where network issues are found, an alternative to that should be kept, due to which fluctuations in the back-end system do not occur and loss can be recovered [6]. Even here, software bug arises faulty conditions. When the software stores the data, processes it and sends back to the back-end system, the bug present there can lead to loss of data [11].

4 Analysis of the Probability Algorithm

See Tables 1 and 2; Figs. 2 and 3.

5 Conclusion

A fault tolerant technique in which every device identifies its own standing, either its good or faulty and therefore the claim is then supported or reverted by its own neighbours as they additionally measure the node behaviour. This formula is analysed

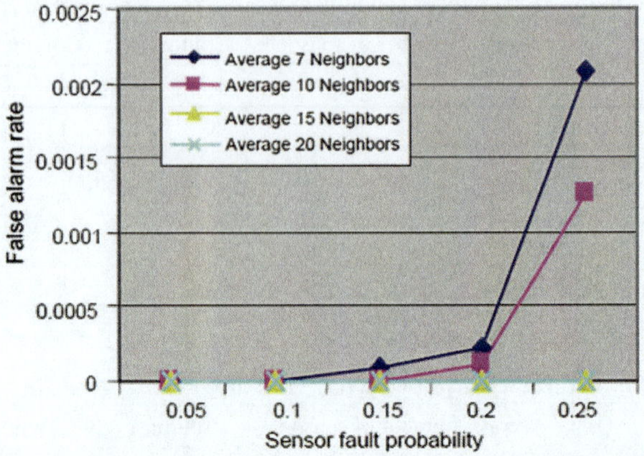

Fig. 2 Faulty sensor detection accuracy

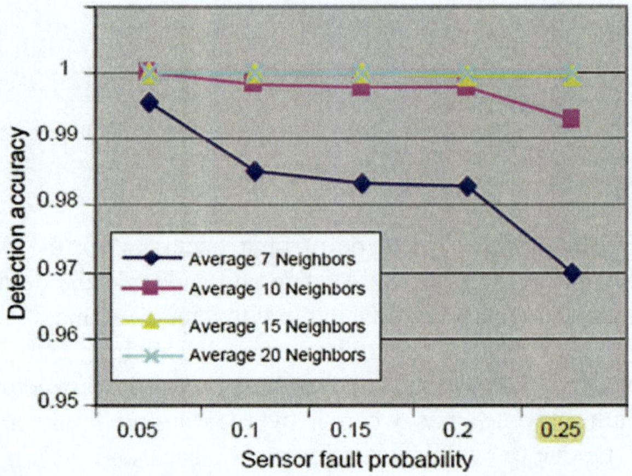

Fig. 3 False alarm rate in faulty sensor detection

using a probabilistic approach. We deeply examined the faults that occur in the sensor node. Although extensive works are done on fault tolerance in every layer of WSN, so many solutions are expected in near future. Use of resources could be lot of economical if resources are properly integrated and scheduled in different layers. So, cross-layer solutions are expected to own higher performance than current solutions. A new trend of WSN is to integrate with different wireless devices like RFID system and mechanism network. For instance, there are increasing range of applications that need the network system to move with the surroundings through actuators. It needs the utilization of device network along with mechanism to create wireless device and actuator network (WSANs). Fault tolerance techniques can be reused in WSANs.

References

1. Somani AK, Agarwal VK (1992) Distributed diagnosis algorithms for regular interconnected structures. IEEE Trans Comput 41(7):899–906
2. Cheraghlou MN, Khadem-Zade A (2017) Increasing lifetime and fault tolerance capability in wireless sensor networks by providing a novel management framework. Wireless Pers Commun 92:603–622
3. Mahmood A, McCluskey EJ (1988) Concurrent error detection using watchdog processors—a survey. IEEE Trans Comput 37(2):160–174
4. Banerjee I, Chanak P, Rahaman H et al (2014) Effective fault detection and routing scheme for wireless sensor networks. Computer Electr 40(2):291–306
5. Ding M, Chen D, Xing K, Cheng X (2005) Localized fault-tolerant event boundary detection in sensor networks. In: Proceedings of IEEE INFOCOM 2005, Miami
6. Koushanfar F, Potkonj M (2003) On-line fault detection of sensor measurements. In: Sensors
7. de Souza LMS, Vogt H, Beigl Bericht N, Fakultät (2007)
8. Liu H, Nayak A, Stojmenović I (2009) Fault-tolerant algorithms/protocols in wireless sensor networks. In: Guide to wireless sensor networks. Springer
9. de Souza LMS (2007) FT-CoWiseNets: a fault tolerance framework for wireless sensor networks. In: International conference on sensor technologies and applications (SENSORCOMM 2007)
10. Mishra S, Jena L, Pradhan A (2012) Fault tolerance in wireless sensor networks. International Journal
11. Titouna C, Aliouat M, Gueroui M (2016) FDS: fault detection scheme for wireless sensor networks. Wirel Pers Commun 86:549–562
12. Venkataraman G, Emmanuel S, Thambipillai S (2008) Energy-efficient cluster-based scheme for failure management in sensor networks. IET Commun 2(4):528–537

Design of Framework for Disaster Recovery in Cloud Computing

Jimit Rangras and Sejal Bhavsar

Abstract Cloud computing is the most widely used technique which is being used by IT organizations worldwide, and it was a major disruption in IT industry and changed the landscape of the entire industry when it came into the picture as it changed the way how software was developed and distributed among their users. Cloud computing enables us to use remote resources such as network, compute, and storage over the Internet without the need to acquire physical hardware such as network cables, and servers which was the case before cloud computing came out as a concept where companies had their own on-premise data centers. Basically, cloud computing is also powered by data centers that contain a collection of several servers/systems, and systems are prone to failure which can be due to several reasons such as natural calamities, network failures, and famines which can be termed as a disaster. Disaster recovery is a technique which is used to recover the data and applications and perform a failover in such a way that the users experience the least possible downtime as well as maximum data is recovered while performing the failover. This research paper focuses on crafting an ideal solution for handling disaster recovery in cloud with minimum possible downtime for containerized microservices.

Keywords Cloud computing · DevOps · Disaster recovery · Monitoring

1 Introduction

Cloud computing is an integral part of IT ecosystems these days, and it can also be considered as a backbone of many enterprise systems as it mainly drives how IT and software companies develop enterprise systems these days. Cloud computing resulted in a paradigm shift in this space as it changed the way of developing software. Prior

---

J. Rangras · S. Bhavsar (✉)
Gandhinagar Institute of Technology, Gandhinagar, Gujarat 382721, India
e-mail: sejal.bhavsar@git.org.in

J. Rangras
e-mail: rangrasjimit@gmail.com

K. Kotecha et al. (eds.), *Data Science and Intelligent Applications*, Lecture Notes on Data Engineering and Communications Technologies 52,
https://doi.org/10.1007/978-981-15-4474-3_49

to cloud computing, companies had to invest in their on-premise data center, and they had to maintain a support staff to look after certain operations such that keeping the network, servers up and running, maintaining, updating, and administering such systems and troubleshooting as and when required. It required a huge amount of investment in terms of physical resources which resulted in high capital expenses. Also if at anytime higher capacity or there was a requirement to scale the systems up for handling more traffic, a request had to be made to support/operations team, and they would increase the capacity of the system with a certain downtime, hence there was no elasticity and ease of scalability with systems.

Cloud computing solved majorly all such problems wherein companies had to invest in physical resources in the form of building their own data centers and maintaining them. It offers all such resources over the Internet which can be accessed and used remotely for various purposes, this made it easy for enterprises to get up and running with their digital initiatives in no time also majorly being based on pay-as-you-go model companies started adopting cloud computing in no time, and it created a huge disruption in IT industry as it solved one of the greatest problems of that time.

Eventually, it became the center point around which the entire IT industry revolves around these days, almost all the enterprise graded softwares that are developed these days are dependant on cloud in many ways. Reason behind such popularity is the elasticity and the agility which is offered by cloud, it also has many other advantages such as cost effectiveness as we only need to pay for the resources that are being used, and there are no upfront investments required since it is in a remote data center where the resources can be accessed over the Internet via Web browser or via APIs and desired infrastructure can be configured in a matter of clicks. Basically, it removed the entire hassle of getting hold of IT support folks, asking them to configure hardware, performing the hardware upgrade with long downtimes ultimately losing the business. Cloud enables us to configure any amount of desired capacity at any point in time, this makes it easy for companies to get up and running with their digital initiatives, and it also makes it easy for companies to scale up gradually as their traffic increases.

Cloud computing can be defined as on-demand delivery of services such as compute, storage, and networking over the Internet [1]. It helps organizations develop scalable solutions that can be deployed on cloud and has various advantages such as on-demand, elasticity, cost effectiveness, and scalability. It is offered as a service via cloud computing service providers, and some of the popular cloud computing service providers include Amazon Web Services, Microsoft Azure, Google Cloud Platform, IBM Cloud, etc.

2 Basics of Cloud Computing

2.1 Types of Cloud

Cloud platforms can be categorized into three types as follows [3]:

Public Cloud: It is available to the general public and can be accessed via the Internet to craft solutions and deploy applications.

Private Cloud: It is not available publicly and it belongs to a particular organization; it mainly resides in the on-premise data center and is used to deploy critical systems such as databases.

Hybrid Cloud: It is a combination of public as well as private cloud; in this model, public cloud is used to deploy less critical systems such as frontend or Web application, private cloud which resides in the on-premise data center of the company is used to deploy critical systems such as database, and both the cloud platforms can communicate over a secure communication channel.

2.2 SPI Framework for Cloud Computing

SPI framework [4] is an essential categorization of services that are offered as part of a cloud offering; it stands for SaaS, PaaS, and IaaS which are also known as software as a service (SaaS), platform as a service (PaaS), and infrastructure as a service (IaaS).

Software as a Service (SaaS): As its name suggests, in this model a software is offered as a service which can be accessed by its users remotely over the Internet; in this model, the user does not need to install the software on their system; it can be accessed via Internet in a Web browser itself; nowadays, most of the enterprise software is developed and delivered in this manner to the customers, and examples include Gmail, Google Drive, Microsoft Outlook, Dropbox, etc.

Platform as a Service (PaaS): As its name suggests, in this model a platform is offered as a service which can be accessed by its users over the Internet to perform certain tasks or to develop a solution using that platform, and examples include popular services such as JIRA—a project management tool and Slack—workspace communication platform.

Infrastructure as a Service (IaaS): As its name suggests, in this model an entire infrastructure is offered as a service; using that infrastructure users can create some solutions out of it, for example, users using cloud infrastructure for deploying their services, and some popular examples include major cloud platforms such as Amazon Web Services, Microsoft Azure, Google Cloud Platform, and IBM Cloud.

3 Problems with Cloud Computing

Although cloud computing can be accessed over the Internet, it is still a physical system which is present at some remote location which is being used remotely by us. Hence, like any system, it can also face various problems as follows.

Privacy: Since the cloud is a remote environment, we need to make sure that the tenancy is managed in the right manner; also it becomes a bit difficult to trust the public cloud platforms as we do not have access to their underlying systems.

Security: It has been the biggest concern in IT industry since decades, there are many components in the cloud which are being used in conjunction, and security aspect needs to be taken care of from both the sides, i.e., from the cloud service provider as well as the customer; hence, it is a shared responsibility.

Compliance: Many companies need to meet certain compliances so as to make sure that they are following the right set of processes and make sure that their user's data is secure; examples include HIPAA, SOX, PCI-DSS, etc.

Systems Failure/Unexpected Downtimes: Ultimately, cloud is also powered by physical systems which are prone to failures which eventually cause unexpected downtimes.

Disaster Recovery: As the cloud is powered by physical machines residing in a data center, it can get impacted due to systems failure, natural calamities, network failure, etc. Hence, disaster recovery refers to recovering cloud systems from such a situation with least possible downtime.

Reliability: As the cloud is in a remote environment, we need to be sure about the reliability of the data.

Cost: It depends on many factors such as cloud services being used and optimization techniques being used or not.

Above problems suggest the areas that can still be worked on in order to suggest an appropriate solution to a specific problem; in this paper, we will be focusing on disaster recovery, what are the existing approaches to deal with it and their drawbacks, and will also see the suggested approach for the same.

4 Disaster Recovery

Disaster recovery in the cloud involves a set of policies, techniques to recover and enable continuity of the infrastructure supporting certain applications so as to avoid business loss and maintain continuity. It deals with getting the systems back up and running with minimum possible downtime.

It is very essential to have a disaster recovery strategy in place as mission-critical systems on which enterprises are heavily dependent on are deployed in the cloud, and it is very crucial to ensure that such systems are available all the time to ensure business continuity. Disasters can occur in any manner; it could be due to some human error or due to some natural calamity or any other such systems.

Hence, it is a very important aspect that needs to be taken care of in order to make sure that the services which are driving an enterprise graded mission-critical system are up and running and if a disaster strikes then they should be able to recover from the same in the minimum possible time, also with the maximum possible data persistency.

5 Literature Review

See Table 1.

6 Existing Disaster Recovery Approaches

6.1 Factors Affecting Disaster Recovery

RTO (Recovery Time Objective): It refers to the time taken to route traffic from the impacted site to the DR site, and it is used to calculate the downtime that users will experience. Ideally, it should be minimum.

RPO (Recovery Point Objective): It refers to the amount of data that will be recovered when a disaster occurs. Ideally, it should be maximum.

Cost of the Cloud Infrastructure: It is the cost of the cloud infrastructure including the DR site.

6.2 Hot Backup Site [6]

In this technique, an alternate infrastructure with exact capacity is maintained in another region as a backup, and it is the exact same copy of the production system.

Advantages: It has a very high recovery point objective (RPO).

Drawback: Infrastructure cost gets doubled as we need to maintain an exact copy of the infrastructure which is not being utilized.

6.3 Cool Backup Site [6]

In this technique, we maintain minimal infrastructure, as well as there are no real-time backups of the production system.

Advantages: Low cost for the backup site.

Table 1 Literature review

S. No.	Year	Title	Author	Research area
1	2017	Study on disaster recovery in cloud environment [6]	A. Arul Mary, Dr. K. Chitra	In this paper, the impact of disaster is discussed for cloud environments, furthermore a couple of approaches have been suggested such as warm and hot standby, and some further research opportunities which are open issues such as disaster monitoring, resource scheduling, confidentiality have been suggested
2	2017	Disaster recovery in single-cloud and multi-cloud environments: issues and challenges [7]	Mohammad M. Alshammari, Ali A. Alwan, Azlin Nordin, Imad Fakhri Al-Shaikhli	In this paper, authors have discussed overall concepts about DR in cloud and suggested approaches for single as well as multi-cloud environments; however, they have suggested that further research is required and suggested an opportunity to work on a framework for DR for multiple cloud platforms
3	2016	Disaster recovery as a service [8]	Hassen BEN REBAH, Hatem BEN STA	Here, the authors have shed some light on the traditional DR strategies in cloud, also they have suggested a solution which could be served as disaster recovery as service (DRAAS) but they have also discussed some open questions such as is it really feasible to have a third party to protect your data?

(continued)

Table 1 (continued)

S. No.	Year	Title	Author	Research area
4	2014	Cloud standby: disaster recovery of distributed systems in the Cloud [9]	Alexander Lenk, Stefan Tai	Here, the authors have suggested cloud standby as an approach for disaster recovery solution for systems in cloud, its performance analysis and discussed future scope of deployments of services in infrastructure
5	2014	Load balancing in cloud computing [10]	Foram F Kherani, Prof. Jignesh Vania	Here, the author has focused on comparing various load balancing algorithms and given conclusion that one of the major topics of research is dynamic load balancing algorithm
6	2014	Disaster recovery in cloud computing: a survey [12]	Mohammad Ali Khoshkholghi, Azizol Abdullah, Rohaya Latip, Shamala Subramaniam & Mohamed Othman	Here, the author has focused on a very major challenge in cloud infrastructure, i.e., disaster recovery, and they have suggested various mechanisms such as failover and failback as a scope of future work
7	2013	Fault tolerance model for reliable cloud computing [13]	Anjali D. Meshram, A. S. Sambare, S. D. Zade	Here, the authors have suggested an approach to incorporate fault tolerance in cloud computing and have suggested a model for the same along with some other suggested future enhancement opportunities
8	2010	Disaster recovery as a cloud service: economic benefits and deployment challenges [14]	Timothy Wood, Emmanuel Cecchet, K. K. Ramakrishnan, Prashant Shenoy, Jacobus van der Merwe, and Arun Venkataramani	Here, the authors have discussed the need for disaster recovery for applications in cloud, existing traditional methods for disaster recovery as well as suggested the research opportunities that are present in this space

Drawback: It has very high RTO and low RPO as the time required to get the infrastructure up is very high as well as it does not have real-time backups of the production system.

6.4 Warm Backup Site [6]

In this technique, a furnished hardware configuration is ready and warmed up to be used, and only the latest backup needs to be transferred.

Advantages: RTO is relatively low.

Drawback: Cost is more as the infrastructure capacity is provisioned. RPO is low as the backup is not in real time. There are three main problems with the above-specified approaches for handling disaster recovery as follows:

- High cost
- Very high RTO—causing huge downtimes
- Very low RPO—causing huge data losses.

7 Proposed Solution

7.1 Decrease Cost

To reduce cost, an effective strategy would be to bootstrap the cloud infrastructure instead of keeping it up and running all the time.

7.2 Decrease RTO

To reduce the RTO/downtime, an effective strategy would be to automate the deployment process of the containerized microservices such that it can be deployed and are ready to serve users in the least possible time.

7.3 Increase RPO

To increase the RPO, an effective strategy would be to take timely backups of essential mission-critical components such as databases/object storage. Here is the proposed architecture diagram of the overall solution (Fig. 1).

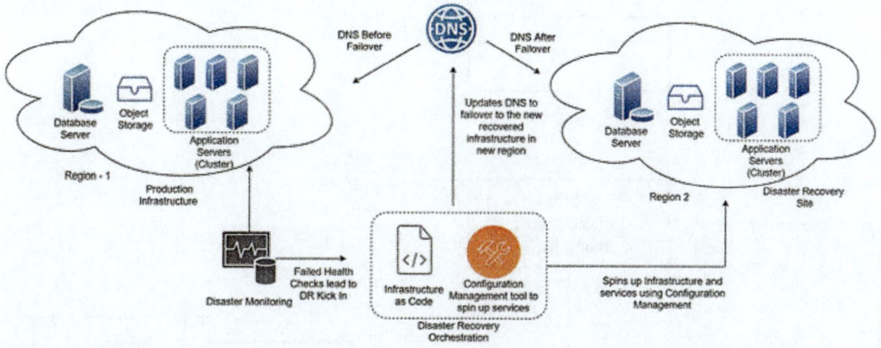

Fig. 1 Proposed solution for handling disaster recovery in cloud

Basically, if the increased cost is due to the complete replication of the cloud infrastructure, we can reduce that by using infrastructure as code (IaC) which is a very popular concept these days; in this method, we develop the entire cloud infrastructure as code which can be used to instantly spin up the cloud infrastructure as and when required; hence, there is no need to keep the infrastructure up and running all the time, thereby reducing the costs.

Secondly, the production infrastructure can be monitored constantly by establishing a disaster monitoring solution so as to check the application servers constantly, and as soon as the monitoring health checks start failing, it can trigger the execution of infrastructure as code so that the cloud infrastructure can be ready in the next few sec/min.

Once the cloud infrastructure is ready, all the services can be deployed in a multithreaded manner such that it takes the least possible time to deploy those services; once all the services are deployed, DNS can be modified to route the traffic to the updated copy of the infrastructure.

The proposed solution is more of a cloud-agnostic approach and hence can be applied to any cloud platform and can be crafted to handle disaster recovery in an efficient manner in the cloud. Here is the proposed flow of execution of the overall solution (Fig. 2).

8 Future Work

To work on the implementation of the above-proposed solution and to test it out on the real environments and enterprise graded systems for handling disaster recovery in cloud.

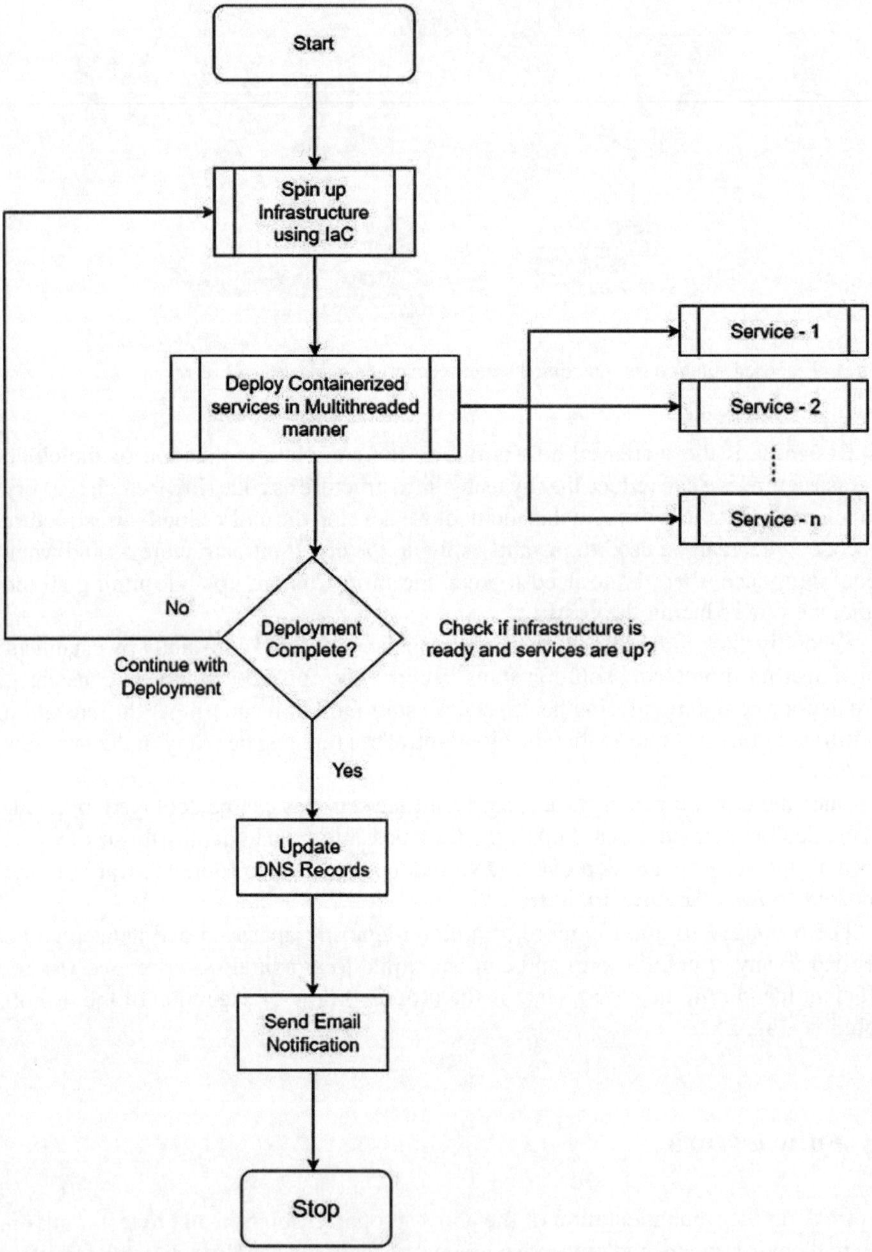

Fig. 2 Flowchart of the suggested approach for disaster recovery in cloud

9 Conclusion

Thus, an essential disaster recovery strategy is required to recover systems from failure with minimum possible downtime; some existing suggested approaches include warm standby, hot standby, and cold standby but they have their own challenges such as latency issues and cost aspects. The suggested approach would be to develop more of a cloud-agnostic solution in the form of a framework to perform disaster recovery of infrastructure as well as services. Suggested DevOps approaches that could be used with this solution could be infrastructure as code and leveraging existing CI/CD tools for automating the process.

References

1. What is cloud computing? Webpage at. https://aws.amazon.com/what-is-cloud-computing/
2. How is cloud computing different from traditional IT infrastructure? Webpage at. https://www.leadingedgetech.co.uk/it-services/it-consultancy-services/cloud-computing/how-is-cloud-computing-different-from-traditional-it-infrastructure/
3. Public cloud vs private cloud vs hybrid cloud, webpage at.https://azure.microsoft.com/en-in/overview/what-are-private-public-hybrid-clouds/
4. Naik N (2012) Cloud computing-SPI framework and it's challenges
5. What is IaaS? Webpage at. https://azure.microsoft.com/en-in/overview/what-is-iaas/
6. Mary A, Chitra K (2017) Study on Disaster Recovery in Cloud Environment. In 2017 World Congress on Computing and Communication Technologies (WCCCT) (pp. 165–167). IEEE (2017)
7. Alshammari., M. M., Alwan, A. A., Nordin, A., & Al-Shaikhli,.: Disaster recovery in single-cloud and multi-cloud environments: Issues and challenges. In 2017 4th IEEE International Conference on Engineering Technologies and Applied Sciences (ICETAS) (pp. 1–7). IEEE (2017)
8. Rebah, H. B., Sta, H. B.:Disaster Recovery as a Service: A Disaster Recovery Plan in the Cloud for SMEs. In 2016 Global Summit on Computer & Information Technology (GSCIT) (pp. 32–37). IEEE (2016)
9. Lenk, A., & Tai, S.: Cloud standby: disaster recovery of distributed systems in the cloud. In European Conference on Service-Oriented and Cloud Computing (pp. 32–46). Springer, Berlin, Heidelberg (2014)
10. Foram F K, Prof. Jignesh V.: Load Balancing in cloud computing, International Journal of Engineering Development and Research, Volume 2, Issue 1| ISSN: 2321-9939
11. Mahalle HS, Kaveri PR, Chavan V (2013) Load balancing on cloud data centres. Int J Adv Res Comput Sci Software Engineering, 3(1) (2013)
12. Khoshkholghi MA, Abdullah A, Latip R, Subramaniam S (2014) Disaster recovery in cloud computing: a survey
13. Meshram AD, Sambare AS, Zade SD (2013) Fault tolerance model for reliable cloud computing. Int J Recent Innov Trends Computing Commun 1(7):600–603
14. Wood T, Cecchet E, Ramakrishnan KK, Shenoy PJ, van der Merwe JE, Venkataramani A (2010) Disaster recovery as a cloud service: economic benefits and deployment challenges. HotCloud 10:8–15

Lightweight Vehicle-to-Infrastructure Message Verification Method for VANET

Mukesh Soni, Brajendra Singh Rajput, Tejas Patel, and Nilesh Parmar

Abstract We use vehicular communication technology to exchange varied data for different intentions such as traffic awareness, road assistance, and automatic toll plaza payment. Vehicles transmit information over a common communication channel and move on the road. Therefore, it is important to perform all required operations efficiently. Researchers proposed various vehicular communication schemes for rapid computation but they are not effective due to high computational cost. In this paper, we suggest a lightweight message verification scheme for vehicle-to-infrastructure (V2I) communication. After that, we implement the proposed scheme on Intel Core i7-7700 processor computer machine. Further, we compare the results of the proposed scheme with other relevant communication methods and it is noticed that the proposed protocol needs less computational resources compared to other vehicular communication schemes. Thus, the proposed scheme is efficient for VANETs.

Keywords Communication · Lightweight · Message · Verification · VANETs

M. Soni (✉)
Smt. S. R. Patel Engineering College, Unjha, Gujarat, India
e-mail: soni.mukesh15@gmail.com

B. S. Rajput
Government Polytechnic Narayanpur, Narayanpur, Chhattisgarh, India
e-mail: brajendrarajput89@gmail.com

T. Patel
ITM Universe, Vadodara, Gujarat, India
e-mail: tejas.patel@itmuniverse.ac.in

N. Parmar
U. V. Patel Collage of Engineering, Mehsana, Gujarat, India
e-mail: nileshparmar0355@gmail.com

K. Kotecha et al. (eds.), *Data Science and Intelligent Applications*, Lecture Notes on Data Engineering and Communications Technologies 52,
https://doi.org/10.1007/978-981-15-4474-3_50

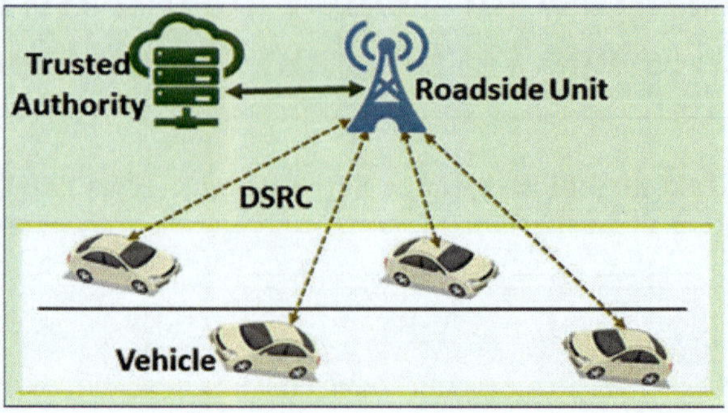

Fig. 1 Vehicle-to-infrastructure communication overview

1 Introduction

We use different advanced applications for better experience of the available services using network systems in the modern society. For better enhancement of ad hoc facility in the field of vehicular technology, scientists introduced the concept of the vehicular ad hoc network (VANET) to fulfill available opportunities in the fast-growing world. There are mainly two types of communications as vehicle-to-vehicle (V2V) and vehicle-to-infrastructure (V2I). Presently, V2I communication has gained notable attention due to the opportunity of investing novel and agreeable explications in the fields of status, moving details, traffic response, toll payment (e.g., FASTag), and safety [1]. Figure 1 shows the general structure of vehicular communications in which vehicles are connected to a roadside unit (RSU) to transmit significant packets within a reasonable period. Data is sent over a public channel using dedicated short-range communication (DSRC) and hence, it opens different opportunities for an adversary to intercept this data. If an adversary can perform different malicious activities such as modification, delay, and impersonation over vehicular communications, then it is a huge loss to vehicle users because it is necessary to maintain fundamental security requirements in the communication system. In addition, it is also important to minimize resource requirement during vehicular communications as vehicle are always in a moving situation. Therefore, a lightweight vehicular communication system is required to transmit vital information on the road [2, 3].

2 Literature Survey

Researchers [4] proposed SeRLOC skeleton using one-way hash and symmetric key to resist Sybil and wormhole attacks, but it suffers from advanced keys distribution, public and private keys generation, execution time, keys maintenance, private key protection, etc. Sun et al. [5] advised an identity-based vehicular data defending scheme without using certificate to preserve vehicle users' privacy. Bhoi et al. [6]

recommended transferring the data packets through the quickest path among the verified vehicles. Accordingly, the experiment outcomes have been reviewed for better secure path routing protocol, anchor-based street, greedy traffic-aware routing, path length, average end-to-end delay, traffic-aware routing, and network gap encounter.

A new threshold reputation-based scheme [7] was proposed to counter the node compromise attacks, stimulate the cooperation with intermediate nodes, increase vehicular nodes' attention and understand the truth of maintaining the probability of conveying datagrams for credits. Scientists [8] suggested a new vehicle management scheme to preserve privacy related to a present position of the vehicle. Moreover, they described the scenario for the adversary to realize the practical outcome of the proposed model. Ying et al. [9] proposed an authentication scheme using a smart card to verify the legitimacy of vehicle users, but it is vulnerable to offline identity guessing, location spoofing, and replay attacks. To solve these security issues, Chen et al. [10] suggested an authentication protocol for Internet of vehicles. However, the scheme [10] is designed using high-cost cryptographic such as exponential and symmetric key cryptography. Thus, it needs a high amount of time to execute all essential operations and hence, it is not effective for vehicular communications as vehicles move on the road.

Contributions To overcome performance and security issues for vehicular communications, we come up with a lightweight vehicle-to-infrastructure (V2I) communication system with the following contributions. Design a low-cost V2I message transmission protocol, which can be used to confirm messages (sent by a vehicle user) on the road. Discuss the performance analysis for the proposed system based on different performance measures and compare the results with other related communication methods.

Paper structure: Section 2 presents a lightweight vehicle-to-infrastructure communication protocol to improve performance results. Section 3 discusses performance analysis of the proposed system. Then, we do performance comparison of the suggested protocol with other related communication methods. Ultimately, we summarize our work in Sect. 4.

The Proposed Solution In this section, we suggest an authentication scheme to verify legitimacy of a vehicle user. If a server verifies a vehicle user correctly, then both (server and vehicle) can communicate with each other to share significant information in the vehicular ad hoc networks. In the proposed scheme, we systematically design two phases as (i) Registration (ii) Message generation and verification as follows.

Registration Process It is a one-time procedure for a vehicle user (V_i) as follows:

V_i chooses an identity (VID_i) as per vehicle identity number allocated by the respected authority and selects the password (PWD_i) as well as r_u (a random number) freely. Further, V_i computes $\text{RPW}_i = h(r_u||\text{PWD}_i)$ and sends $\{\text{VID}_i, \text{RPW}_i\}$ to the trusted authority (TA). After that, TA generates a secret key as x_i (random nonce) for VID_i and computes $A_i = \text{RPW}_i \oplus h(x_i)$ and $B_i = A_i \oplus \text{VID}_i$.

TA saves credentials B_i into a smart-chip (SC_i) and installs it in a vehicle. Further, TA stores VID_i and respective x_i in its secure database, and it is shared with all RSUs. Here, RSUs are deployed by TA.

V_i does $C_i = \text{VID}_i \oplus h(r_u||\text{PWD}_i)$ and updates SC_i with C_i and RPW_i.

Message Generation and Verification Process

This phase is performed between a vehicle user (V_i) and the smart-chip reader (SCR_i) as follows to get login into the system:

1. V_i enters VID_i and PWD_i. Now, SCR_i extracts B_i and C_i from SC_i to calculate $D_i = \text{VID}_i \oplus B_i \oplus C_i \oplus T_1$ and $E_i = h(T_1||M||D_i)$. SCR_i sends $\{\text{VID}_i, D_i, M, E_i, T_1\}$ to a nearby RSU. Here, T_1 is a current time-stamp.
2. An RSU receives $\text{VID}_i, D_i, M, E_i, T_1$ and checks its validity based on a time-stamp validation ($\Delta T \leq T_2 - T_1$), where T_2 is a current time-stamp and ΔT is a threshold time delay. If ΔT is not valid, then an RSU drops a request immediately. Otherwise, an RSU computes $h'(x_i) = \text{VID}_i \oplus D_i$ and matches it with $h(x_i)$. If both are same, then it calculates $E_i' = h(T_1||M||D_i)$ and does $E_i' \stackrel{?}{=} E_i$. If both are equal, then an RSU considers M as valid (=not modified) during a communication between V_i and an RSU.

3 Performance Analysis

We discuss different performance measures like computational cost, execution time, and communication cost. After that, we compare these performance attributes with other relevant vehicular communication methods.

Computational Cost Vehicular communication schemes perform different cryptographic operations during the message generation and verification process. This is known as the computational cost.

Execution Time Time is required to execute different cryptographic functions in the system, and it is called as the execution time.

Firstly, we calculate the computational cost for [9, 10], and the proposed scheme. Different cryptographic functions are used in these vehicular communication methods, and they are as one-way hash ($T_{\text{h}(\cdot)}$), asymmetric encryption (T_{AE}), asymmetric decryption (T_{AD}), asymmetric encryption/decryption (T_{SED}), and exponential (T_{EX}). The schemes [9, 10] are designed using T_{EX}, T_{AE}, T_{AD}, $T_{\text{h}(\cdot)}$, and T_{SED}. However, the proposed method is designed using only one-way hash operation. Besides, we have implemented all these operation on Intel Core i7-7700 processor computer machine with 8 GB 2400 MHz DDR4 RAM and 3.6 GHz frequency to calculate the average execution time for each vehicular communication protocol. The execution time is 0.0020 milliseconds (ms) for $T_{\text{h}(\cdot)}$, 0.0100 ms for T_{SED}, 0.0399 ms T_{EX}, 4.4063 ms for

Table 1 Comparison of schema

Schemes	Computational cost	Execution time (ms)
Ying et al. [9]	$1\,T_{EX} + 1\,T_{AE} + 1\,T_{AD} + 14\,T_{h(\cdot)} + 2\,T_{SED}$	12.2555
Chen et al. [10]	$6\,T_{EX} + 1\,T_{AE} + 1\,T_{AD} + 12\,T_{h(\cdot)} + 2\,T_{SED}$	12.4510
Proposed scheme	$7\,T_{h(\cdot)}$	0.0140

T_{AE}, and 7.7613 ms for T_{AD}. Based on this execution time, we calculate the total execution time for the communication system, and it is 12.2555 ms in [9], 12.4510 ms in [10], and 0.0140 ms in the proposed scheme. The computational cost and the execution time are described in Table 1.

Communication Cost Both (sender and receiver) transmit different variables during the communication phase, and it requires some amount of memory, which is known as the communication cost. It is measured in bytes. Here, one-way hash requires 32 bytes, random nonce takes 12 bytes, time-stamp needs 8 bytes, and symmetric encryption/decryption takes 32 bytes. After considering it, we have calculated the communication cost as 312, 312, and 96 bytes for [9, 10], and the proposed scheme, respectively. Hence, the suggested protocol needs less computational cost, communication cost, and execution time comparatively.

4 Conclusion

We have proposed a lightweight V2I message verification protocol. After that, we have analyzed performance results of the suggested scheme, and it shows that the proposed scheme performs better compared to other relevant communication methods in the computational cost and the execution time. Therefore, the suggested protocol can be implemented quickly to transfer messages between a vehicle and RSU on the road. Hence, the proposed scheme can be used to verify vehicle users in different vehicular communication applications such as traffic response, road assistance, automatic toll payment, and safety.

References

1. Chen W, Guha RK, Kwon TJ, Lee J, Hsu YY (2011) A survey and challenges in routing and data dissemination in vehicular ad hoc networks. Wirel Commun Mob Comput 11(7):787–795
2. Raya M, Hubaux JP (2007) Securing vehicular ad hoc networks. J Comput Secur 15(1):39–68
3. Eze EC, Zhang S, Liu E (2014) Vehicular ad hoc networks (VANETs): current state, challenges, potentials and way forward. In: 20th International conference on automation and computing. IEEE, pp 176–181

4. Zhang C, Lu R, Lin X, Ho PH, Shen X (2008) An efficient identity-based batch verification scheme for vehicular sensor networks. In: 27th Conference on computer communications. IEEE, pp 246–250
5. Sun J, Zhang C, Zhang Y, Fang Y (2010) An identity-based security system for user privacy in vehicular ad hoc networks. IEEE Trans Parallel Distrib Syst 21(9):1227–1239
6. Bhoi SK, Khilar PM (2014) SIR: a secure and intelligent routing protocol for vehicular ad hoc network. IET Netw 4(3):185–194
7. Zhou J, Dong X, Cao Z, Vasilakos AV (2015) Secure and privacy preserving protocol for cloud-based vehicular DTNs. IEEE Trans Inf Forensics Secur 10(6):1299–1314
8. Kang J, Yu R, Huang X, Jonsson M, Bogucka H, Gjessing S, Zhang Y (2016) Location privacy attacks and defenses in cloud-enabled internet of vehicles. IEEE Wirel Commun 23(5):52–59
9. Ying B, Nayak A (2017) Anonymous and lightweight authentication for secure vehicular networks. IEEE Trans Veh Technol 66(12):10626–10636
10. Chen CM, Xiang B, Liu Y, Wang KH (2019) A secure authentication protocol for internet of vehicles. IEEE Access 7:12047–12057

Security and Performance Evaluations of QUIC Protocol

Mukesh Soni and Brajendra Singh Rajput

Abstract QUIC protocol was proposed to carry out fast navigation of Internet, and it is a latest transport layer protocol build on UDP for Internet by Google developers. QUIC provides equivalent functionality to TCP, HTTP/2, and TLS as whole. In this paper, we do in-depth study of QUIC protocol from its implementation and application point of view. Further, we do testbed implementation using Amazon AWS services to evaluate the performance of QUIC protocol against TCP and UDP protocol. In this connection, we find different issues in QUIC protocol such as data loss recovery and forward error correction. Besides, we do security analysis on QUIC protocol to understand its security strengths.

Keywords Data · Performance · QUIC · Security · Transmission

1 Introduction

TCP is one of the most important transport layer protocols of TCP/IP protocol suite. It is widely used in data transmission and communication networks such as Internet. TCP is used for various network protocols such as HTTP, FTP, SMTP, and Telnet t transfer long messages. TCP/IP is a connection-oriented service with full duplex operation. TCP Header has fields such as source port, destination port, sequence number, acknowledgment number, header length, checksum, flags (URG, ACK, PSH, RST, SYN, FIN), the reserved bits, window size, urgent pointer, and options field. UDP is a transport layer protocol available in TCP/IP suite involving minimum communication. It is a connectionless protocol, and it is used for message delivery, but not for mutual data transmission. Therefore, UDP does not have any acknowledgment

M. Soni (✉)
Smt. S. R. Patel Engineering College, Unjha, Gujarat, India
e-mail: soni.mukesh15@gmail.com

B. S. Rajput
Government Polytechnic Narayanpur, Narayanpur, Chhattisgarh, India
e-mail: brajendrarajput89@gmail.com

K. Kotecha et al. (eds.), *Data Science and Intelligent Applications*, Lecture Notes on Data Engineering and Communications Technologies 52, https://doi.org/10.1007/978-981-15-4474-3_51

mechanism. Accordingly, the sender does not wait for ACK, but she/he continuously sends data [1].

QUIC is a new transport layer protocol, built on UDP by Google. QUIC has features like multiplexing, flow control, and security similar to TLS and reliability of TCP. QUIC protocol is developed to reduce latency for faster Internet/Web navigation faster. As per study, 7\% of Internet traffic is QUIC. It allows rapid experimentation and quick iterations for development. It has open-source development in Chromium [2]. The features of QUIC protocol are as follows.

Packets and payload sent by QUIC are fully encrypted and authenticated.

QUIC has a unique connection ID. When a connection is migrated, the request is retrieved using this connection ID.

It combines the crypto and transport handshakes which result in less round trip for connection establishment. QUIC is considered 0-RTT.

It uses XOR-based forward error correction method to recover lost packets.

Unlike TCP, QUIC provides plug-in control over congestion enabling gain information to control congestion.

QUIC transmits data in multiple streams instead of single stream which avoids head of line blocking.

Google services like YouTube, Google Duo, Gmail, Google Search Engine, Google Chrome have been using QUIC as a transport protocol [3]. QUIC protocol is being studied and employed by Uber to increase its app's performance. Extensive experimentation is being done before deploying QUIC in Uber App [4]. Ericsson is also working on employing QUIC in its 5G networks [5].

QUIC packets are of two types as (i) special and (ii) regular. Special packets are public reset packet and version negotiation packet. Regular packets are frame packet and FEC packet. The maximum QUIC packet size is 1370 bytes for IPv4 and 1350 bytes for IPv6. The header contains public flags of size 8 bits and connection ID is of 64-bit random number. This does not change during the connection. Figure 1 shows the QUIC packet header structure with different fields.

Now, we describe different steps to transfer data in QUIC, and they are as follows. (i) A client initiates the connection with CHLO, and the server responds with token in REJ packet if they have not communicated earlier. (ii) Version negotiation takes

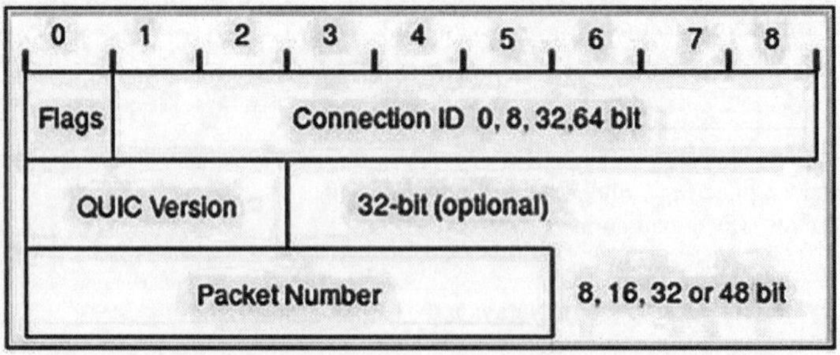

Fig. 1 QUIC packet header structure

place along with crypto handshake to reduce connection establishment latency and dedicated stream for connection and security handshake. (iii) QUIC uses streams for data transfer and provides ACKs for acknowledgment of QUIC packets. Streams can be terminated normally, abruptly, and with connection termination. Connection termination should become idle for pre-negotiated time as explicit shutdown and implicit shutdown [6].

We analyze various research works on QUIC protocol and discuss different observations on QUIC protocol in Sect. 2. We identify varied challenges in QUIC protocol based on the performance and security, which are described in Sect. 3. Finally, we conclude our observations on QUIC protocol in Sect. 4.

2 Literature Survey

Carlucci et al. [1] analyzed QUIC protocol to evaluate its experimental results considering QUIC version 21, and they found that QUIC shows higher good-put as compared to TCP CUBIC. Further, QUIC also decreases the overall page access time for HTTP/1.1 in case of a channel with random losses. Implementation in user space decreases the expected performance of the protocol because each message starts a context switch between kernel and user space. Therefore, to look into the better possible performance, QUIC was implemented in kernel space like TCP, UDP [7]. Authors found that QUIC always performs better than TCP, but it may have higher jitter rate for the low loss rate cases. It is also shown that QUIC is very much friendly with TCP as they share the throughput in equal while running concurrently. Also, QUIC is able to reach its peak data rate much faster than TCP in error-free environment. In addition, QUIC has faster loss recovery [7].

Wang et al. [8] did preliminary performance analysis of QUIC with bottleneck bandwidth and round trip propagation time (BBR) via GEO satellite Internet access on dedicated network emulation testbed. BBR-based congestion control algorithm is Google's contribution to make Internet faster. They mainly focused on evaluation of QUIC in Satellite Internet with long propagation time and found that QUIC with BBR helps to reduce the transmission delay and maintain the transmission rate even though the packet loss grows. Thomas et al. [9] evaluated Google QUIC performance on a public satellite communication, and they found that the page load time is twice longer in GQUIC as compared to TCP. And it is due to non-delegated congestion controller in GQUIC. Besides, GQUIC gives bad end-user experience on a high Bandwidth delay product (BDP) network.

3 Practical Challenges in QUIC Protocol

For the implementation, we have used Amazon AWS service, and server and client programs are hosted on AWS EC2 instance. In this, one instance is configured to work as a client and another as a server using Python program for QUIC, TCP,

Fig. 2 Data retrieval time comparison among QUIC, TCP, and UDP

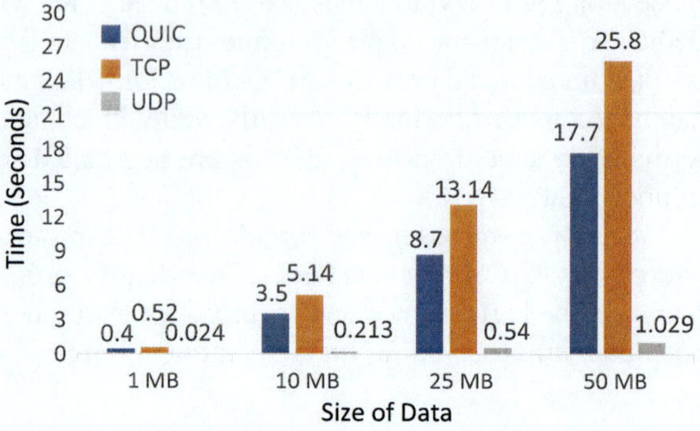

Fig. 3 Throughput comparison among QUIC, TCP, and UDP

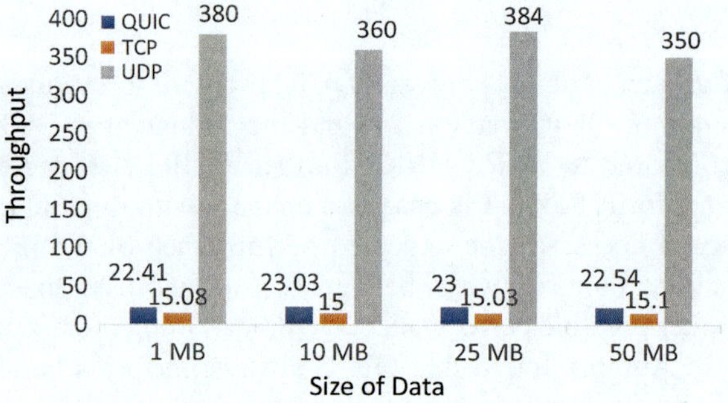

and UDP implementations. Specifically, for QUIC implementation, we have used a Python library, aioquic for QUIC transport layer protocol, and it supports IPv4 and IPv6 implementations with minimal TLS v1.3. In TCP, the server program uses standard Python socket library with a simple mechanism to read data using buffer and send it to a client. In UDP, the server program uses the same library as TCP, but here while creating socket object, it uses mention *SOCK DGRAM*. Based on the above configuration, we have performed different operations as (i) time to retrieve data (ii) throughput for different size of data such as 1 MB, 10 MB, 25 MB, and 50 MB. Figures 2 and 3 show the comparison among QUIC, TCP, and UDP for data retrieval time and throughput, respectively. QUIC performs better compared to TCP for throughput and data retrieval time. But, QUIC falls in between TCP and UDP for performance results.

4 Performance Issues

There are many implementations of QUIC, and not all implementations are standardized for real-time applications. Thus, it creates ambiguity in the implementation and posing a threat. Many Internet service provider firewall has blocked QUIC traffic for legal services, and it results in false blockage in a case of valid requests.

QUIC has its own encryption mechanism that integrates with TLS1.3, but it has to be analyzed with more specifications because it is not standardized.

5 Security Issues

QUIC does not provide forward secrecy. Further, a simple bit flipping and replay attack during handshake could prevent QUIC protocol from having minimal latency and then falling back to TCP. QUIC has duplicate packet detection, complete packet protection, retry mechanism, and anti-reflection mechanism which reduce the effect of a denial of service attack. But, these features are under analysis and it is yet to be shown whether it achieves its desired goals.

Forward erasure correction (FEC) is disabled in QUIC implementations. But if it is enabled, then false packet injection could be a possible attack on this newly developed protocol.

6 Conclusion

We have implemented QUIC protocol on AWS services to check its performance comparatively. From this, we identified that QUIC performs better compared to TCP for throughput and data retrieval time. But, QUIC lies in between TCP and UDP in terms of performance. Further, it provides fast data transfer of UDP and reliability of TCP. Besides, QUIC has some security concerns such as forward secrecy, replay attack, and denial of service attack.

References

1. Carlucci G, De Cicco L, Mascolo S (2015) HTTP over UDP: an experimental investigation of QUIC. In: Proceedings of the 30th annual ACM symposium on applied computing. ACM, pp 609–614
2. Langley A, Riddoch A, Wilk A, Vicente A, Krasic C, Zhang D, Yang F, Kouranov F, Swett I, Iyengar J, Bailey J, Dorfman J, Roskind J, Kulik J, Westin P, Tenneti R, Shade R, Hamilton R, Vasiliev V, Chang WT, Shi Z (2017) The QUIC transport protocol: design and internet-scale deployment. In: Proceedings of the conference of the ACM special interest group on data communication. ACM, pp 183–196
3. Mahindra R, Chandar V, Guo E (2019) Usage statistics of QUIC for websites. Online available https://w3techs.com/technologies/details/ce-quic. Last Accessed 10 Nov 2019
4. Mahindra R, Chandar V, Guo E (2019) Employing QUIC protocol to optimize Uber's app performance. Online available https://eng.uber.com/employing-quic-protocol/. Last Accessed 14 Nov 2019

5. Arkko J (2019) Gearing-up for modern internet transport. Online available https://www.ericsson.com/en/blog/2019/1/quic. Last Accessed 16 Nov 2019
6. Cui Y, Li T, Liu C, Wang X, Kuhlewind M (2017) Innovating transport with QUIC: design approaches and research challenges. IEEE Internet Comput 21(2):72–76
7. Wang P, Bianco C, Riihijarvi J, Petrova M (2018) Implementation and performance evaluation of the QUIC protocol in Linux Kernel. In: Proceedings of the 21st ACM international conference on modeling, analysis and simulation of wireless and mobile systems. ACM, pp 227–234
8. Wang Y, Zhao K, Li W, Fraire J, Sun Z, Fang Y (2018) Performance evaluation of QUIC with BBR in satellite internet. In: 2018 6th IEEE international conference on wireless for space and extreme environments (WiSEE). IEEE, pp 195–199
9. Thomas L, Dubois E, Kuhn N, Lochin E (2019) Google QUIC performance over a public SATCOM access. Int J Satell Commun Network 37(6):601–611

Biometric Fingerprint Recognition Using Minutiae Score Matching

Ronakkumar B. Patel, Dilendra Hiran, and Jayeshbhai Patel

Abstract Biometric fingerprint recognition is one of the difficult biometric pattern recognition problems. Fingerprints of human are unique and remain unchanged throughout a person's whole life. Biometric fingerprint minutia matching is one of the most important post-processing stages of biometric fingerprint authentication system. The fingerprint minutiae points are the most important features for comparing one fingerprint image with another. Usually, a fingerprint minutia matching is implemented after different preprocessing and post-processing steps like fingerprint image enhancement, fingerprint image binarization, fingerprint image thinning and fingerprint minutia extraction. An algorithm of fingerprint minutia score matching is proposed in this paper. The research work is implemented in Dot Net platform (c# language) using custom database of 100 fingerprint images from different 25 persons. Four different fingerprint images of same finger are used for biometric fingerprint matching experiment. The enrollment of total 25 fingerprint images is carried out, and other fingerprint images of users are matched with already enrolled fingerprint images. Similarity score is calculated for original fingerprint image with enrolled fingerprint image.

Keywords Biometric · Fingerprint · Recognition · Image · Enhancement · Thinning · Minutia · Matching

R. B. Patel (✉) · D. Hiran
PAHER University, Udaipur, India
e-mail: ronakcjp@gmail.com

D. Hiran
e-mail: sigmapawan72@gmail.com

J. Patel
Ganpat University, Kherva, India
e-mail: jayeshpatel_mca@yahoo.com

K. Kotecha et al. (eds.), *Data Science and Intelligent Applications*, Lecture Notes on Data Engineering and Communications Technologies 52, https://doi.org/10.1007/978-981-15-4474-3_52

1 Introduction

The term biometrics means bio and metrics. So, bio means living thing, and metrics means measurement. Biometric authentication is used for recognition of person based on his physiological or behavioral characteristics [1]. Fingerprint biometric recognition has been a widely used technique in both civilian and forensic applications. As compared to other biometrics modalities, fingerprint-based biometric modality is the most proven and most famous technique used for human recognition purpose.

The fingerprint authentication system is divided into four parts [2, 3]. First part is image acquirement part that captures the fingerprint image, and the next is preprocessing part for fingerprint image improvement, binarization and thinning of the image. The next part is minutiae extraction part to extract the feature from the thinning fingerprint image by using minutiae extractor methods to extract ridge ending and ridge bifurcation. The fourth part is fingerprint matching to match two minutiae points of fingerprint image by using minutiae matcher method.

1. Fingerprint image acquisition
2. Preprocessing stage
    - Image enhancement of fingerprint image
    - Binarization of fingerprint image
    - Thinning of fingerprint image
3. Post-processing stage
    - Minutiae extraction (ridge ending and ridge bifurcation) from fingerprint image
    - Matching and recognition of the fingerprint image.

A minutia matching of fingerprint is the last stage in post-processing. A minutia matching is used to find out that minutiae points of input fingerprint image and template fingerprint are from the same finger or not. In fingerprint matching, not only compare these two sets of minutiae points but also consider some problems like shifting of fingerprint, rotation of fingerprint and so on. The pattern-based fingerprint matching [4] and minutiae-based fingerprint matching approaches [5] are the famous approaches for biometric fingerprint matching. The proposed research work followed the minutiae-based fingerprint matching approach.

2 Literature Review

Malathi et al. 2016 [6] said that an efficient biometrics-based user recognition has been introduced. This includes the features like biometric system, fingerprint, iris, palm print, etc., for recognition of a person. The proposed system is most appropriate for human recognition, and it needs high security during online purchasing, net banking, etc.

Jain et al. 2016 [7] proposed a fingerprint-based recognition system for infants whose age range is between 0 and 12 months. Their research has pointed out the following issues: (i) design of comfortable, compact and high resolution greater than 1000 ppi fingerprint reader, (ii) fingerprint image enhancement algorithms to improve quality of infant fingerprint images and (iii) collection of longitudinal infant fingerprint image data to evaluate recognition accuracy over time. This collaboration between Michigan State University, Dayalbagh Educational Institute, Saran Ashram Hospital, Agra, India, and NEC Corporation has demonstrated the feasibility of recognizing infants older than 4 weeks using fingerprints.

Peralta et al. 2015 [8] reviewed and categorized the vast number of fingerprint matching methods like fingerprint minutiae-based matching, local and global minutiae matching, minutiae feature extraction techniques, etc., proposed in the specialized literature. In particular, the author focuses on local fingerprint minutiae-based matching algorithms, which give good performance. The results of matching are discussed in detail for future directions. In this paper, the author compiled the most relevant work in the scientific literature about fingerprint local minutiae-based matching.

Satya et al. 2014 [9] pointed out about an efficient approach for fingerprint recognition system, which will be implemented through many steps like fingerprint image acquisition, fingerprint isolation, fingerprint enhancement. The system implemented in two fingerprint images for the left and right thumbs. Various data of fingerprint images are collected (5 images of right thumb and 5 images of left thumb). These images are converted into digital forms for further processing.

Paraste and Shukla 2014 [10] represented a brief review of the field of fingerprint recognition. The purpose of this research paper is to review the different techniques like matching based on minutiae, matching based on correlation and other local, global methods for biometric fingerprint-based matching. The authors point out that automated fingerprint recognition systems can be divided into two parts: 1. minutiae-based matching and 2. pattern-based matching. The authors point out that the research on authentication of fingerprint will never stop because of its complexity and intractability.

The author describes lots of algorithms of fingerprint authentication systems in this paper. Zhang et al. 2011 [11] proposed high-resolution biometric fingerprint authentication systems which implements more security because they are using level three fingerprint features, like pores, that are not obtained in lower resolution lower than 500 dpi fingerprint images. In this paper, the author finds out the optimal resolution for an automated fingerprint authentication system. They carried out two most important fingerprint features: (1) minutiae points and (2) pores of fingerprint. The author uses high-resolution fingerprint scanner device for collecting images of fingerprint at various resolutions and finds out the minimum necessary resolution for biometric fingerprint authentication using minutiae and pores.

Choi et al. 2011 [12] proposed a new fingerprint matching algorithm which contains minutiae features and additional ridge features. The ridge features are made up of four elements: count of ridge, length of ridge, ridge curvature direction and type of ridge and minutiae features like type, orientation and positions. The proposed method gives higher matching score compared to the conventional minutiae-based approach. Still, the proposed method needs improvement for images which has the low quality.

Jain and Feng 2011 [13] proposed a system for matching latent fingerprint with rolled fingerprint using all the three levels of fingerprint features. The proposed system uses the reference point (location, direction and type), overall image characteristic (ridge quality map, ridge flow map and ridge wavelength map), secondary features (dots, ridge, pores), minutia details (x- and y-coordinates, minutiae direction, quality type). The author proposed baseline matching algorithm for minutia matching. The author also used additional features like reference point, ridge quality map, ridge flow map, ridge wavelength map, skeleton, etc., to improve performance for latent fingerprint matching.

Jain et al. 2010 [14] described an automated biometric fingerprint authentication system. They also discover research opportunities and challenges in this area of biometric. The author implements minutiae matching algorithm in this paper. A fingerprint matching module computes a match score between two fingerprints which should be high for fingerprint from same finger and low for those from different fingers. There are four approaches in most fingerprint matching algorithm: (1) image correlation, (2) phase matching, (3) skeleton matching and (4) minutiae matching.

3 Proposed Work

There are different fingerprint matching techniques like minutiae-based matching, correlation-based matching, ridge feature-based matching, etc., used. In our research, the minutiae-based matching is used as a matching technique. The minutiae-based matching technique contains five main steps [15].

- Orientation field estimation
- Ridge extraction
- Thinning
- Minutiae extraction
- Minutiae matching.

We have mentioned only last step in this research work. Other steps are described in our previous research work [16, 17]. The core point-based local matching technique is an efficient, easy to understand technique. The core point of a fingerprint is the highest curved point (Fig. 1).

The geometry of region (GR) technique is the most common and best result algorithm of core point detection.

The algorithm of GR is shown below:

Compute the smoothed orientation field $\theta'(i, j)$ by using the following equation.

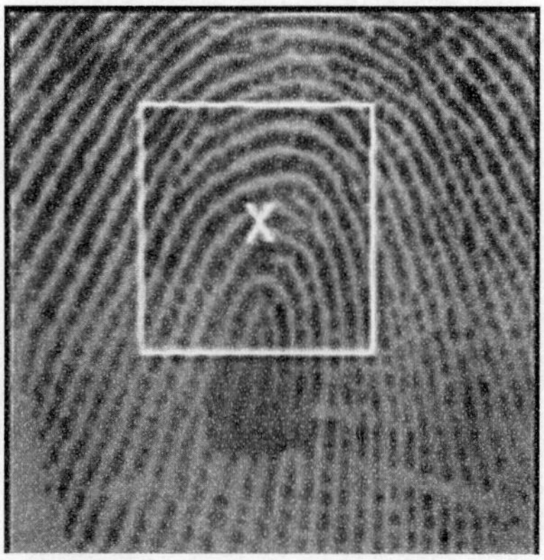

Fig. 1 X is the core point of a fingerprint [15]

$$\theta'(i, j) = \frac{1}{2}\tan^{-1}\left(\frac{\Phi'_y(i, j)}{\Phi'_x(i, j)}\right)$$

Compute ε (i, j), which is the sine component of $\theta'(i, j)$.

$$\varepsilon(i, j) = \sin(\theta'(i, j))$$

Initialize a label image A which is used to indicate the core point.

Assign the corresponding pixel in A, the value of the difference in integrated pixel intensity of each region.

$$A(i, j) = \sum_{R1} \varepsilon(i, j) - \sum_{R2} \varepsilon(i, j),$$

The regions $R1$ and $R2$ were determined empirically, and also their geometry is designed to capture the maximum curvature in concave ridges. In practice, the region is defined within the radius of 1015 pixels (should cover at least 1 ridge). In addition, $R1$ that is sandwiched $R2$ is expected to hold the maximum point.

Find the maximum value in A, and assign its coordinate as the core point.

If the core point still cannot be located, the steps (1–5) could be iterated for a number of times while decreasing the window size used in step (1) above. For instance, $w = 15$, 10 and 5 pixels, respectively.

After these six steps, the core point will be extracted. We save the core point into database.

To matching two fingerprint images: The minutiae information of these two images must be saved in the database. For each minutia, X and Y coordinate (row and column no. of the pixel), type (ridge ending or bifurcation) and the absolute distance to core point of the image are saved in database.

A pair of minutiae can be said a match pair if they are same type of minutiae and the absolute different values >= 0 and <= Er where Er is error rate value. A suitable error rate value is necessary because it is very common to get a nonzero absolute different value even two fingerprint images obtained from a same finger.

After matching each pair of minutiae, instead of giving Boolean value as a result of matching images or not, the match rate is introduced to give a reference of recognition result:

$$R = \sqrt{k * k / m * n}$$

where R is the match rate, k is the matched pair count, and m, n are the minutiae count from template and input fingerprint image. If the value of R is greater than a threshold value, then we say the two fingerprint images could be two copies of a same finger.

Pseudocode: Image Matching

1. Start
2. Input preprocessing image_1
3. Input preprocessing image_2
4. Find core and delta in image_1
5. Find core and delta in image_2
6. Rotate image_1 and image_2 based on their core and delta angle
7. Match minutiae in image_1 and image_2
8. If R > = threshold value
9. Matching
10. Else
11. Not matching
12. End.

4 Implementation

Twenty-five users' 100 fingerprints are used for implementation of research work using .NET platform. Implementation is done in two phases.

(1) Enrollment phase

  (a) Enhancement
  (b) Thinning
  (c) Extract minutiae

(2) Matching phase.

4.1 Enhancement

Image improvement is an important part for smoothening the image by removing noise from the image and also connecting broken ridges of the image. We use the composite algorithm of Gaussian mask and Sobel convolution for image enhancement. The Gaussian mask is carried out for smoothening images, and 3*3 Sobel convolution is used for edge detection [17] (Table 1).

4.1.1 Thinning

The thinned image is got by applying Zhang-Suen thinning algorithm [16] (Table 2).

4.1.2 Minutia Extraction and Core Point Detection

The following results are obtained by using minutia extraction algorithm [17] on thinned image which is extracted from Zhang-Suen thinning algorithm (Table 3).

4.2 Fingerprint Matching

A total of 100 fingerprints are used for experimental work. Four images of same fingers are used for matching experiments. A total of 25 fingerprint images are enrolled. The other fingerprint images are matched with the enrolled fingerprints, and some of the results are given below (Table 4).

5 Conclusion and Future Work

This matching technique, as we mentioned, is efficient and easy to understand or implement into code. The disadvantage of this method is the images must contain high quality. Here, only 25 enrolled fingerprints are used for experiment. Experimental results can be checked if numbers of enrolled fingers are increased to more than 500 fingerprints. Matching time can be reduced as specially in identification mode of recognition using classification of the database based on various features of the fingerprint pattern. The work can be extended in the future which can overcome the problem of low resolution of image quality. The security can be improved by storing the template in encrypted form. The work can be extended using extracting level three feature of fingerprint image for better recognition.

Table 1 Image enhancement [17]

Original image	Enhanced image
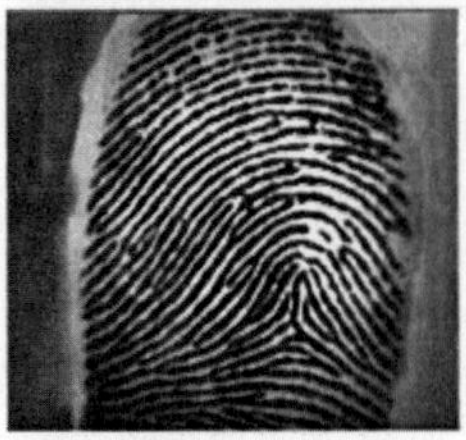 1_01_Original	 1_01_Enhanced
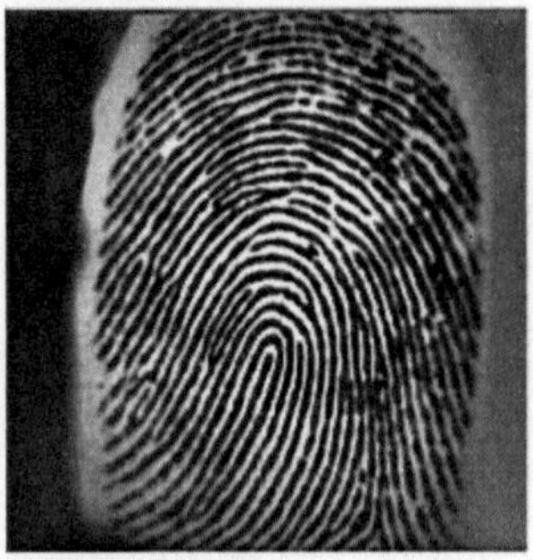 2_01_Original	 2_01_Enhanced
 3_01_Original	 3_01_Enhanced

(continued)

Table 1 (continued)

4_01_Original

4_01_Enhanced

5_01_Original

5_01_Enhanced

6_01_Original

6_01_Enhanced

Table 2 Image thinning by Zhang-Suen thinning algorithm [16]

Enhanced image	Thinned image

1_01_Enhanced	1_01_Thinned

2_01_Enhanced	2_01_Thinned

3_01_Enhanced	3_01_Thinned

(continued)

Table 2 (continued)

 4_01_Enhanced	 4_01_Thinned
 5_01_Enhanced	 5_01_Thinned
 6_01_Enhanced	 6_01_Thinned

Table 3 Minutia extraction algorithm on thinning image [17]

Thinned image	Minutia extraction
	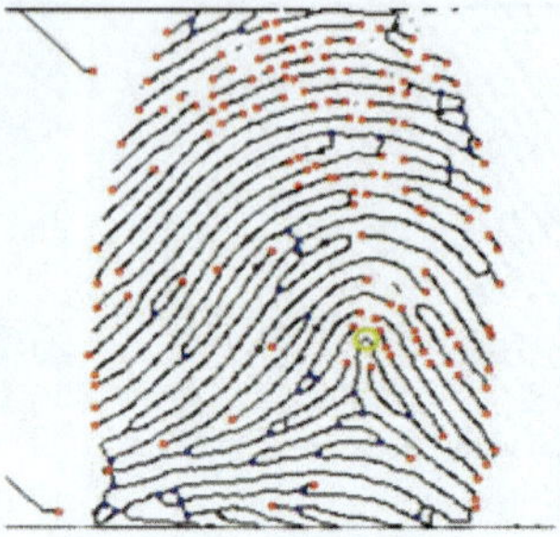
1_01_Thinned	1_01_Minutie_Extraction
	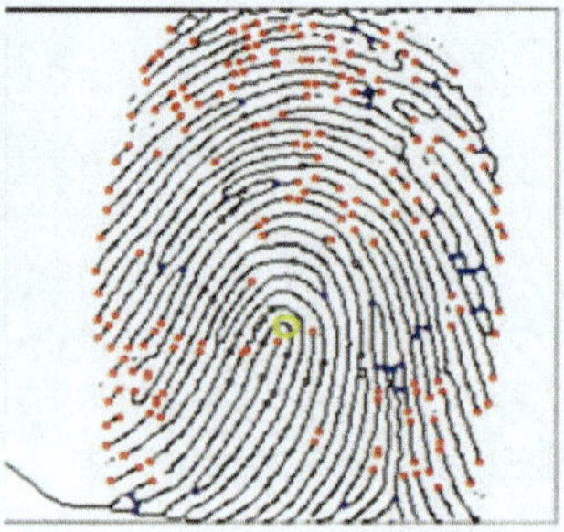
2_01_Thinned	2_01_ Minutiae_Extraction
	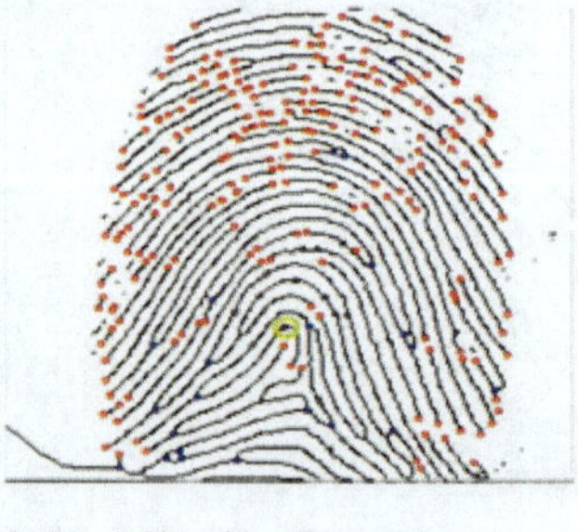
3_01_Thinned	3_01_ Minutiae_Extraction

(continued)

Table 3 (continued)

	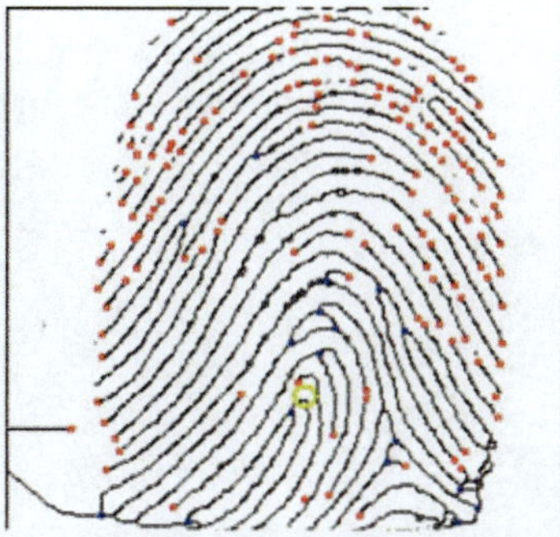
4_01_Thinned	4_01_ Minutiae_Extraction
	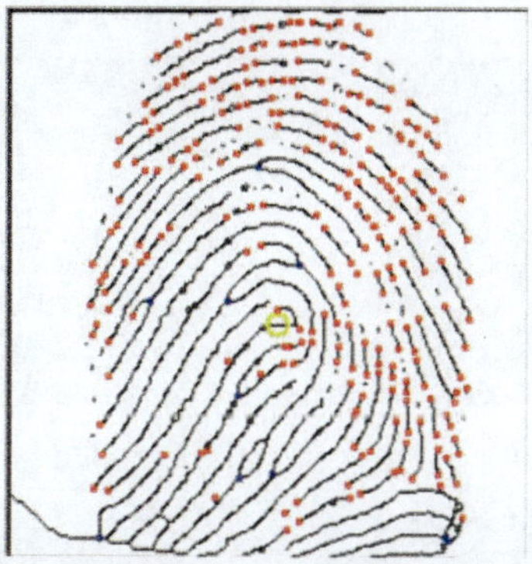
5_01_Thinned	5_01_ Minutiae_Extraction
	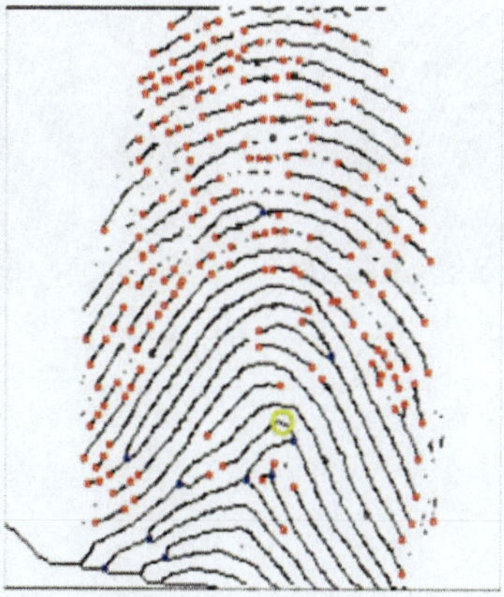
6_01_Thinned	6_01_ Minutiae_Extraction

Table 4 Fingerprint matching of original fingerprint image with enrolled image in database

Original image	Matched with following enrolled image in database	No of template minutiae	matched minutiae pairs	similarity score
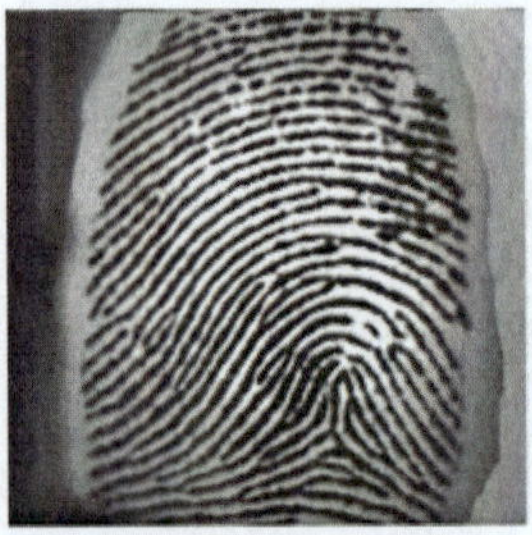 1_02_Original	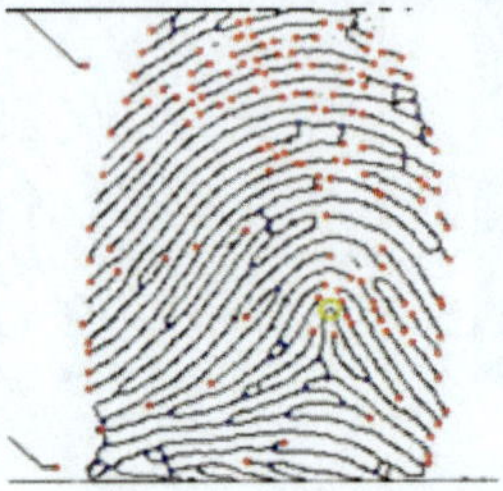 1_01_Enrolled	35	32	91.42
 2_03_Original	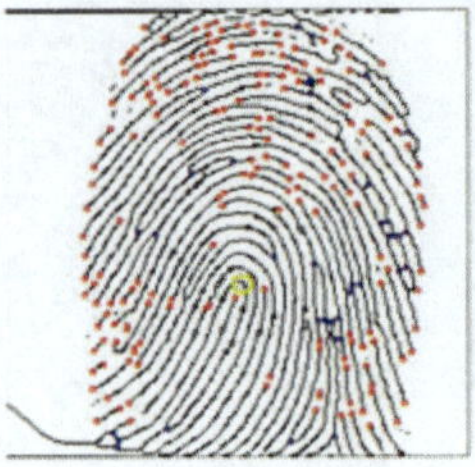 2_01_ Enrolled	28	23	82.14
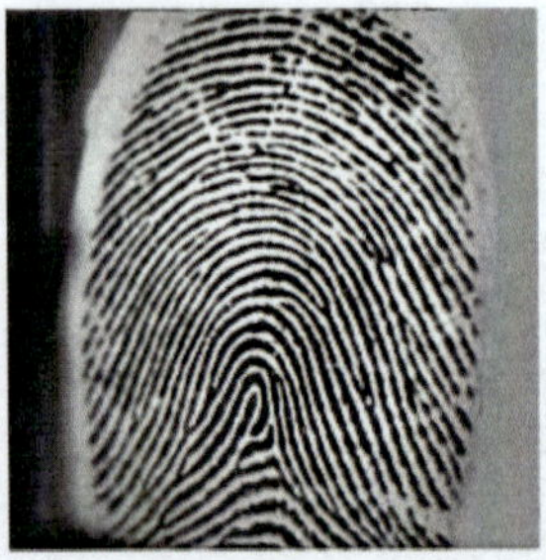 3_02_Original	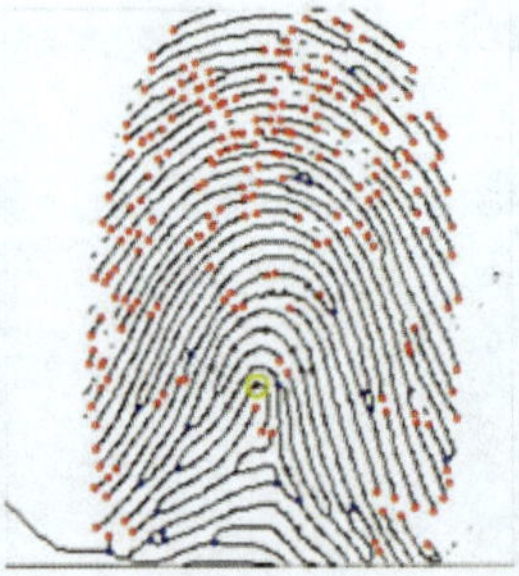 3_01_ Enrolled	26	21	80.77

(continued)

Table 4 (continued)

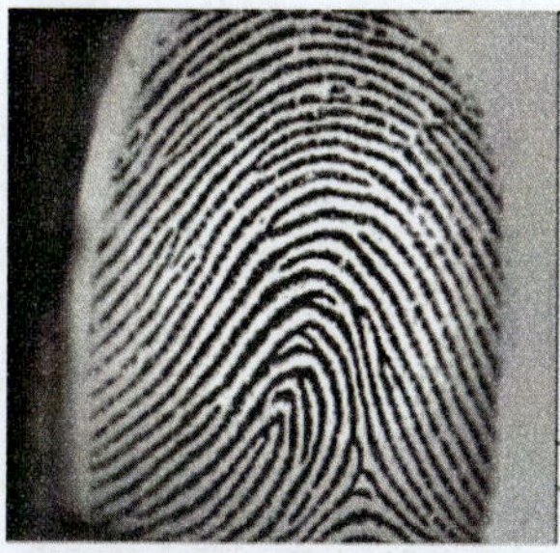 4_04_Original	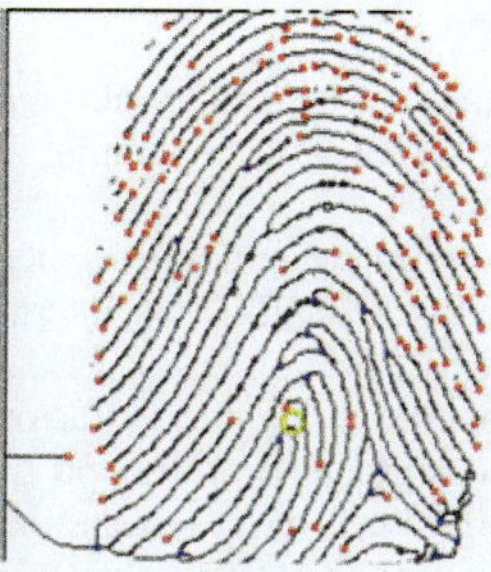 4_01_Enrolled	17	12	70.59
 5_01_Original	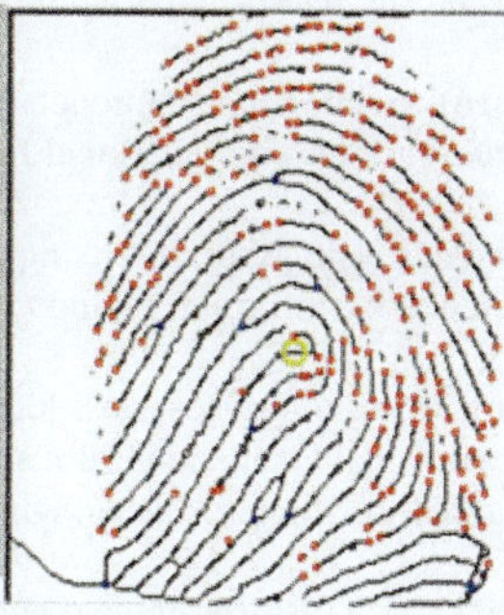 5_01_ Enrolled	46	46	100
 6_03_Original	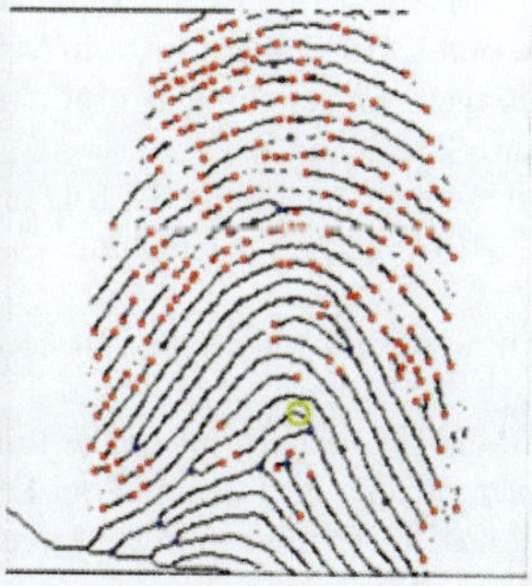 6_01_ Enrolled	28	26	92.86

References

1. Lavanya BN (2009) Minutiae extraction in fingerprint using gabor filter enhancement. In: 2009 International conference on advances in computing, control, and telecommunication technologies, pp 54–56. IEEE
2. Ali M, Mouad MH (2016) Fingerprint recognition for person identification and verification based on minutiae matching. In: 2016 IEEE 6th international conference on advanced computing (IACC), pp 332–339. IEEE
3. Chugh T (2017) Benchmarking fingerprint minutiae extractors. In: 2017 International conference of the biometrics special interest group (BIOSIG), pp 1–8. IEEE
4. Tudosa A, Costin M, Barbu T (2004) Fingerprint recognition using gabor filters and wavelet features. In: Proceedings of symposium of electronics and telecommunications, ETC, pp 328–332
5. Ravi J, Raja KB, Venugopal KR (2009) Fingerprint recognition using minutia score matching. Int J Eng Sci Technol 1(2):35–42
6. Malathi R, Jeberson R, Raj R (2016) An integrated approach of physical biometric authentication system. In: International conference on computational modeling and security, Published by Elsevier
7. Jain AK (2016) Giving infants an identity. fingerprint sensing and recognition. In: Proceedings of the eighth international conference on information and communication technologies and development, ACM, pp 29
8. Peralta D (2015) A survey on fingerprint minutiae-based local matching for verification and identification: taxonomy and experimental evaluation. Published by Elsevier Inc
9. Satya R (2015) An efficient approach for fingerprint recognition. Int J Eng Innov Res 4(2). ISSN: 2277-5668
10. Ajay Singh Paraste, Shailja Shukla (2014) A brief review paper on finger print identification method. Int J Electr Electron Res. ISSN 2348-6988
11. Zhang D, Liu F, Zhao Q, Lu G, Luo N (2010) Selecting a reference high resolution for fingerprint recognition using minutiae and pores. IEEE Trans Instrum Meas 60(3):863–871
12. Choi H, Choi K, Kim J (2011) Fingerprint matching incorporating ridge features with minutiae. IEEE Trans Inf Forensics Secur 6(2):338–345
13. Jain AK, Feng J (2011) Latent fingerprint matching. IEEE Trans Pattern Anal Mach Intell 33(1)
14. Jain AK, Feng J, Karthik N (2010) Fingerprint matching. Published by the IEEE Computer Society
15. Atipat J, Somsak C (2007) An algorithm for fingerprint core point detection. Comput Soc IEEE. 1-4244-0779-6/07
16. Ronak P, Dilendra H, Jayesh P (2019) Fingerprint image thinning by applying Zhang-Suen algorithm on enhanced fingerprint image. Int J Comput Sci Eng 7(4). ISSN: 2320-7639
17. Ronak P, Dilendra H, Jayesh P (2019) An algorithm for fingerprint minutiae extraction. Int J Comput Sci Eng 7(6). ISSN: 2347-2693
18. Thai (2003) Fingerprint image enhancement and minutiae extraction. The University of Western Australia, 8–10

Automatic Traffic Fine Generation Using Deep Learning

Shah Monil, Bhalerao Ketan, Patel Dhruvkumar, Patel Narendra, and Swadas Prashant

Abstract The aim of this paper is to fully automate the manual process of fine generation over crossroads. A video recorder needs to be set on crossroads and focused over the area which is not meant to be crossed when the red light is ON. Thus, the vehicles breaching this rule will be recorded. This video has to be browsed into the designed system and will be processed to various image processing algorithms for getting the region of interest, which is the number-plate of the vehicle breaching the rule. The detected number-plate would be cropped from the image and would be segmented internally by every character. The obtained string of number-plate would be queried to a database in order to obtain all the information about a vehicle's owner. Using this information, E-memo will be sent through the system to the vehicle's owner in the form of text message and Email.

Keywords Region of interest · Convolutional neural network · Number-plate detection · Character segmentation · Electronic memo · Edge detection · Fine generation

---

S. Monil · B. Ketan · P. Dhruvkumar · P. Narendra (✉) · S. Prashant
BVM Engineering College, Anand, India
e-mail: nmpatel@bvmengineering.ac.in

S. Monil
e-mail: shahmonil1989@gmail.com

B. Ketan
e-mail: head@scaai.siu.edu.in

P. Dhruvkumar
e-mail: dhruv814@gmail.com

S. Prashant
e-mail: pbswadas@bvmengineering.ac.in

K. Kotecha et al. (eds.), *Data Science and Intelligent Applications*, Lecture Notes on Data Engineering and Communications Technologies 52,
https://doi.org/10.1007/978-981-15-4474-3_53

1 Introduction

Computer science has developed significantly in recent years while expanding beyond the quadrilateral screens and is used to design high-quality real-life applications today. One of the important domains which can be drastically improved is the efficient management of traffic overcrowded crossroads. The traffic on Indian roads has been swelling since recent years. This has made the task of traffic management as well as the fine generation quite complex. Moreover, in countries like India, the tremendous population overcrowds the roads, and a lot of traffic rules are breached every day [1]. The overall rate of accidents across crossroads has also hiked up during the last 5 years. Due to these reasons, the Indian government has passed legal rules of collecting fines from the vehicle owners who break the rules on the crossroads as this would lower down the breaching of traffic rules by the public. Thus, the fine generation system holds a core place in controlling and managing traffic over crossroads. However, the entire system of the fine collection in several Indian cities is manual. The video on crossroads is manually recorded and then taken to the memo generation hub where all the vehicles are zoomed to obtain their number-plates, and then, a memo is generated after certain days. This reduces the efficiency of traffic management as well as control.

2 Existing Research

2.1 Literature Review

The past approaches involved detection of number-plate using naive algorithms of image processing which resulted in a certain degree of inefficiency in case of detection. [2] have discussed the recognition of number-plate using the n-fold pixel splitting approach involving the darkening of various pixels around the number-plate and then constructing a cropping window over it. [2] have also discussed the approach simplification via dilation of images in order to merge certain pixels of an image to get the required region of interest. [3] have explained the process of image edge detection through Canny's algorithm in order to draw a sharp line across the borders of number-plate. [3] also state the usage of many image processing algorithms like image binarization, thresholding, morphological transformation, etc., in order to get the entire focus over the number-plate. Thus, the earlier approaches suggested preprocessing the image containing the car and its number-plate in order to get the required region of interest.

Moreover, recent developments in the field of deep learning have given rise to a more efficient and flexible algorithm named convolutional neural network (CNN) [4, 5]. CNN is one of the most utilized algorithms for text extraction from images [4, 5]. Various applications have been designed using CNN in recent years, and the algorithm has terminated with correct results for majority cases [6]. An important feature of CNN is its ability to detect simple as well as complex features of object encompassed within an image [4, 6]. With every convolution, the text recognition algorithm keeps on becoming robust and resistant to noise in the images [5]. Owing to all the above reasons, CNN was chosen as the text recognition algorithm and was trained for recognition of 36 classes (0–9 and A–Z). Thus, every segmented character was iteratively given as an input to the CNN model in order to obtain the entire string of number-plates characters.

2.2 Obtained Insights

Image processing focuses on developing a computer system that is ready to perform processing on a picture. The input of that system could be a digital image, and also the system processes that image using efficient algorithms, and provides an appropriate desired output. The goal of image processing is to speed up or compress image/video information for the further implementation process. Moreover, Tkinter [7] follows the layered approach. As a result of this, it is inherited from the advantages of the GUI toolkit that had the desired time for mature. Also, it hides the complex and detailed calls and functions which make it painless to go with Tkinter. Furthermore, the connection to the database is very flexible in Python compared to Java. Portability is a major advantage of Tkinter over most other libraries which make user restrict from one to two platforms [7]. Additionally, it gives native look and feel for a particular platform where the program runs on.

3 Proposed Methodology

3.1 Initial Steps

The proposed system aims to detect the license plate from the images which can be generated from the video captured by the placed camera on crossroads. Firstly, a high-quality camera must be placed on crossroads in such a way that all the vehicles approaching from the front are clearly recorded. There is a possibility of getting a blurred video if the camera does not have advanced features of quality which turns out that the system cannot recognize the license plate. The following diagram summarizes the proposed methodology for the automatic generation of traffic E-memo (Fig. 1).

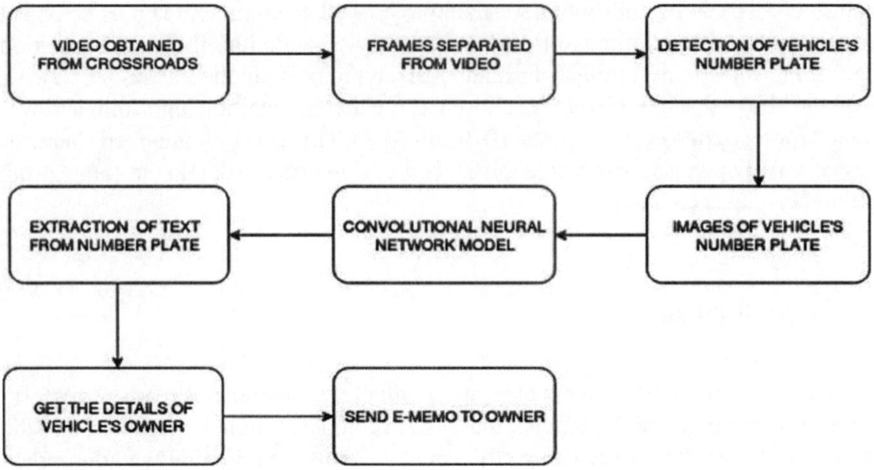

Fig. 1 Flow diagram

3.2 Detection of Number-Plate Through Image Processing Techniques

Mathematically, an image is a matrix of pixel values which is stored in the computer's memory. A raw image when directly passed for recognition can lead to inefficient results. In addition, a system slices video into the number of frames. Here, frames are separated at every 250 ms time interval and stored in the grayscale format.

The images having RGB form need more memory than grayscale form. Different operations of image processing will be applied to each image one by one. After converting the images from RGB to grayscale images, image enhancement has been performed. In enhancement operation, median blur filter is used to remove noise from the image. The next step is to use histogram equalization to enhance the contrast of an image. Subsequently, the enhanced image is converted into a binary form using the image thresholding technique. So, it classifies all the pixels with values above threshold as white and remaining pixels as black. Next operation is morphological transformation. Here, our goal is to identify the edges of the number-plate. So, the aim of the operation is to remove unnecessary white pixels and to determine the vivid white rectangular region where the license plate is located in an image. Moreover, this transformation has changed the shape of the object in a binary image. For that, a closing operation is applied, which is dilation followed by an erosion. Dilation increases the size of the white pixels, and erosion does reverse the process of dilation (Fig. 2).

After applying morphological transformations on the binary image, image contouring technique is performed to make bounding box around the enclosed regions (Fig. 3). It may happen that an unnecessary bounding box will be created around an

Fig. 2 Input RGB image

Fig. 3 Image after applying image contouring

enclosed region. Following properties of number-plates are used to omit unnecessary bounding boxes.

- Width is greater than the height.
- The count of white pixels is more in the region containing a number-plate.
- Height to width ratio is approximately 0.4.

3.3 Text Recognition from Number-Plate

From the previous phase, we got a cropped image of number-plate. Here, we have to identify the numbers written on a license plate. The CNN model expected a character image rather than an entire word. Hence, every number-plate stored in the folder was subjected to internal segmentation with respect to every character of number-plate. The contouring algorithm was utilized for accomplishing this task which detects the change in pixels and crops out the required region encompassing a single character. The character segmentation process is shown in Fig. 4.

Figure 5 shows the sample number-plate image on which character segmentation is performed. The input image is converted into gray scale to make the image suitable for image binarization (Fig. 6). After binarizing an image, image enhancement techniques are applied to separate each character from the number-plate. Here, we are using a flood fill algorithm to remove background noise (black pixels forming border in Fig. 6).

Image contouring process makes bounding boxes around each enclosed region. The character must form enclosed region so that it is correctly bounded. Image

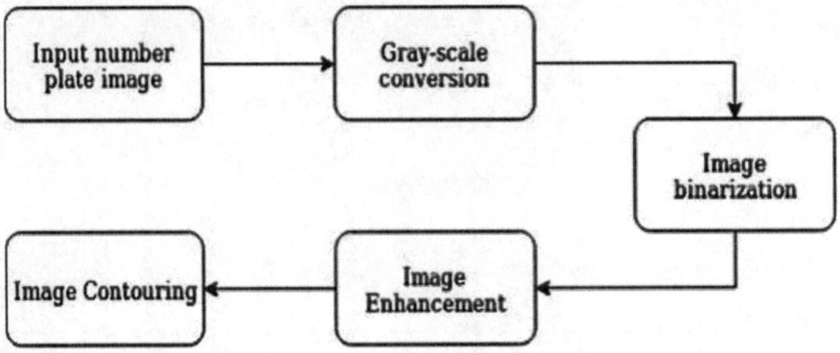

Fig. 4 Character segmentation process

Fig. 5 Cropped number-plate image

Fig. 6 Image binarization

Fig. 7 Image erosion in the vertical direction

Fig. 8 Contoured number-plate

erosion is applied to achieve this goal. It decreases the size of white pixels based on its width and height. Image erosion in direction of height overlaps black pixels each other and makes clear and separated black regions (Fig. 7). Now, image contouring is applied on the image shown in Fig. 7 so that characters are bounded easily (Fig. 8).

As shown in Fig. 8, the characters to be individually inputted to CNN model were contoured and later cropped off into separate images. However, utmost care was taken in order to remove blur and inconsistency from the images, and thus, the characters were efficiently segmented out from the number-plate image. The CNN model's features were applied on trial and test basis, and the overall accuracy varied by tweaking each of its internal parameters. Different flavors of CNN were worked upon, and every approach provided a certain difference in terms of accuracy of recognition (Table 1). The MNIST dataset was utilized in order to train a CNN model from scratch over the recognition of 36 classes (A–Z and 0–9).

Also, transfer learning was utilized for training the CNN appropriately, and the accuracy comparison was performed between self-trained CNN model and CNN model obtained through transfer learning. Following results were obtained after training CNN over various transfer learning models (Figs. 9, 10, 11 and 12).

It was observed that none of the transfer learning models were able to pull the accuracy of recognition beyond 90%. As a result, the CNN trained from scratch was

Table 1 Obtained accuracies from CNN trained from MNIST dataset

Filters	Strides	Kernel size	Accuracy
16	(2,2)	(3,3)	91.52
16	(3,3)	(3,3)	85.25
32	(2,2)	(3,3)	97.60
32	(3,3)	(3,3)	93.28
16	(2,2)	(2,2)	92.67
32	(2,2)	(2,2)	91.75
16	(2,2)	(4,4)	91.87
32	(2,2)	(4,4)	86.14

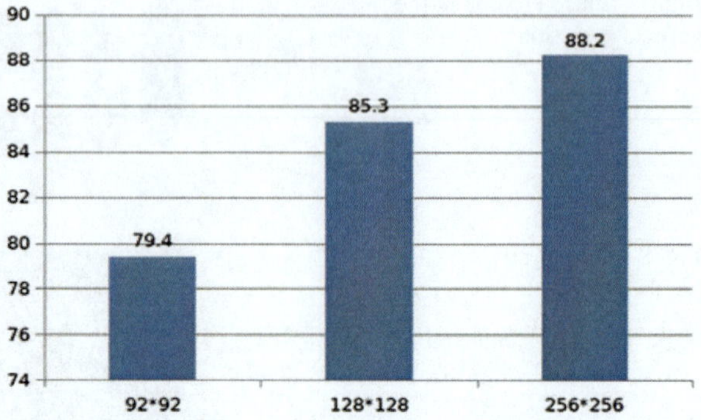

Fig. 9 Accuracies of Inception-V3 model

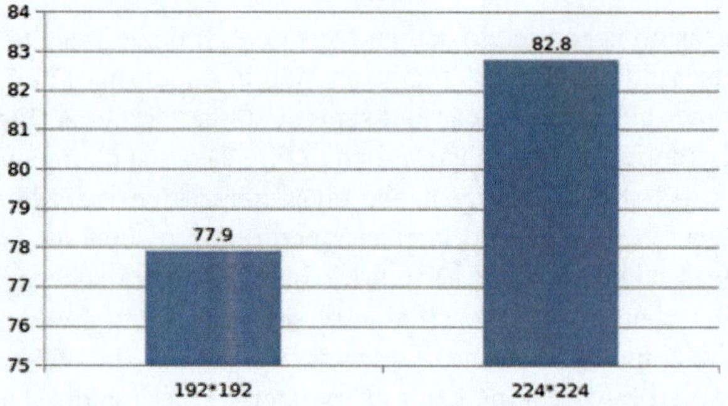

Fig. 10 Accuracies of VGG-16 model

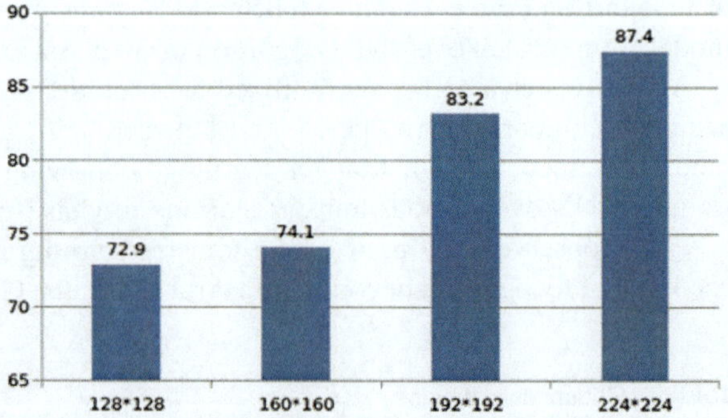

Fig. 11 Accuracies of MobileNet model

selected as the final model for recognizing the number-plate characters. Following were the properties of the final model which were utilized for character recognition:

- Initial accuracy: 91%
- Increased neural network's body by adding more layers
- Used LeakyReLU instead of ReLU
- Decreased learning rate to 1e-4
- Resulting accuracy: 97.6%.

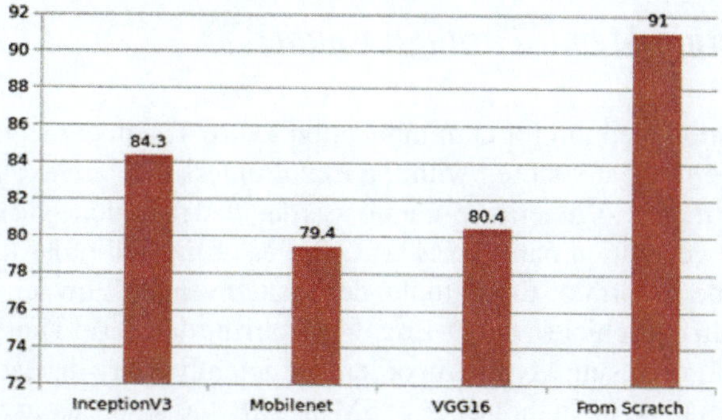

Fig. 12 Overall average accuracy comparisons

3.4 Designing GUI

Python's Tkinter library was utilized to design the GUI for the proposed system [7]. Some of the screenshots are shown below. "Select Video" button is used to select the video from a directory. System user can start analyzing a video by pressing the "Start Analysis" button. It takes some time to analyze. After completion, the system shows the detected number-plate with extracted text from it. The previous and next buttons are used for navigating if the video contains more than one number-plate (Fig. 13).

System user can generate a fine report by pressing the "Show Report" button. After pressing the "Notify All" button, the E-memo will be sent to all vehicle owners listed in the report. The upcoming sections discuss the methods used for E-memo generation.

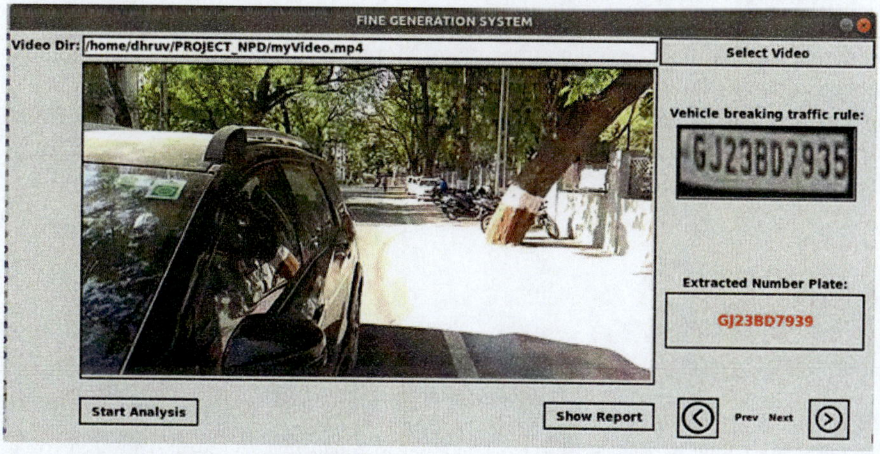

Fig. 13 Analysis of a video shown in GUI

3.5 Generating Memo Through Email

Email is a common medium for communication today. The documents mailed once to anyone are permanently stored with the Email unless they are explicitly deleted. Thus, one of the means to generate E-memo was decided to be done via Email. Python library for mail generation named SMTP-LIB was utilized in order to generate the required E-memo in form of Email to the designated vehicle's owner. The very first step was to obtain the vehicle owner's Email-Id from the database. Thus, a normalized schema was designed using MySQL in order to efficiently query the database in order to get the user details. The Python code of SMTP-LIB was stored as an event response block which got executed on the click of a button over the designed GUI. Thus, on the click of a button on the designed GUI, the Email will be sent to the vehicle's owner who is supposed to pay the fine. After retrieving the contact details of the owner of the vehicle who has broken the traffic rules, this system will send the text SMS within the time period of 12 h to that particular person.

4 Conclusions

The entire system rectified several defects of the existing traffic controlling and fine generation systems. Several insights were obtained owing to the proposed system. The overall quality of the video camera used to record the input video must be decent, and if possible sharp in order to efficiently extract the number-plate as well as the number on the number-plate. Also, various operations like morphological transformations perform better if performed with efficient and dynamic parameters. Edge detection is an efficient technique in order to detect the number-plate from a car's image. Convolutional neural networks are best-suited text recognition algorithms in order to get the number on the number-plate. Moreover, the accuracy of CNN highly depends upon its internal parameters and varies with variations in the input image's dimensions. An exception was also encountered during the design process of CNN for text recognition that CNN trained from scratch sometimes gives better accuracy than the CNN obtained from pre-trained transfer learning models.

References

1. Developing a license plate recognition system with machine learning in python. Online Available: https://blog.devcenter.co/developing-a-license-plate-recognition-system-with-machine-learning-in-python-787833569ccd
2. Lakshmi CJ, Rani AJ, Ramakrishna KS, Lantikiran M, Siddhartha VR (2016) A novel approach for Indian license plate recognition system. IJAEST 2(I):010–014
3. Swati B, Amit K, Sayali B, Jyoti C, Smita L (2017) Indian vehicle number plate detection using image processing. Int Res J Eng Technol (IRJET) 04(04)

4. Wang T, Wu DJ, Coates A, Ng AY (2012) End-to-end text recognition with convolutional neural networks. In: Proceedings of the 21st international conference on pattern recognition (ICPR), pp 3304–3308. IEEE
5. Gu J, Wang Z, Kuen J, Ma L, Shahroudy A, Shuai B, Liu T, Wang X, Wang G, Cai J, Chen T (2018) Recent advances in convolutional neural networks. Pattern Recogn 77:354–377
6. Krizhevsky A, Sutskever I, Hinton GE (2012) Imagenet classification with deep convolutional neural networks. In: Advances in neural information processing systems, pp 1097–1105
7. Tkinter. Online Available: https://pythontextbok.readthedocs.io/en/1.0/Introduction_to_GUI_Programming.html
8. Number plate detection on Indian car vehicles using YOLOv2. Online Available: https://medium.com/@ravidesetty/number-plate-detection-on-indian-car-vehicles-using-yo2l2ov2-8c99e1a259f5
9. License plate detection with OpenCV and Python. Online Available: https://cvisiondemy.com/license-plate-detection-with-opencv-and-python/
10. Gonzalez RC, Woods RE (1992) Digital image processing

Comparative Survey of Digital Image Steganography Spatial Domain Techniques

Chandani Navadiya and Nishant Sanghani

Abstract In the modern world, confidentiality is a major issue in communication, so to increase the security cryptography and steganography came into the picture. Steganography techniques provide security from attacker. In this paper, we study some different research papers on various image steganography techniques. Steganography is hiding information in some other information. Nowadays, steganography used for copyright ownership by watermarking on digital data and also used to share asymmetric algorithm key sharing and also contains a survey on theory of color based on human vision system (HVS).

Keywords Steganography · Image steganography · Encoding · Data hiding · LSB · EBE · PVD

1 Introduction

The advent of modern digital communication technology has made the transmission of messages quicker and more convenient, but it has also made it easier to track and secret messages. It is, therefore, a critical issue to maintain the confidentiality of messages during transmission. One method called cryptography is by encrypting or encoding secret information in such a way that it can be successfully decoded. Steganography is another approach to keep communications secret. A strategy hides secret information within a medium or carrier to keep the secret information unnoticed. They differ, however, in that cryptography makes the data unreadable, or hides the data's meaning, while steganography hides the data's existence.

C. Navadiya (✉) · N. Sanghani
Shantilal Shah Engineering College, Sidsar Campus, Vartej, Sidsar, Bhavnagar, Gujarat 364060, India
e-mail: chandaninavadiya@gmail.com

K. Kotecha et al. (eds.), *Data Science and Intelligent Applications*, Lecture Notes on Data Engineering and Communications Technologies 52,
https://doi.org/10.1007/978-981-15-4474-3_54

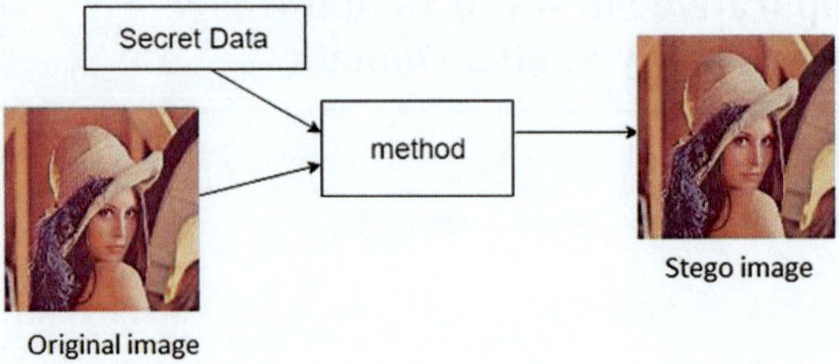

Fig. 1 Basic steganography process

Image steganography, as the name suggests, refers to the process of hiding data in an image file. The image chosen is called the cover image for this reason and the image obtained after steganography is called the stego image. Figure 1 shows the basic diagram of steganography process.

1.1 Technical Terms

1. Steganography is the practice of concealing a file, message, image, or video within another file, message, image, or video. The word steganography combines the Greek words steganos, meaning "covered," and graphe meaning "writing."
2. Steganography is hiding information into another information [1].
3. Digital steganography is hiding information on another digital data structure such that no one could track their information [2].
4. MSR: The mean-square error (MSE) and the peak signal-to-noise ratio (PSNR) are used to compare image compression quality [3]. The MSE is cumulative squared error between the two images which is original and the modified, whereas PSNR represents a measure of the peak error. The lower MSE's value, the lower the error.

$$\text{MSE} = \frac{1}{mn}\sum_{i=0}^{m-1}\sum_{j=0}^{n-1}[I(i,j) - K(i,j)]^2$$

5. PSNR: The peak signal-to-noise ratio (PSNR), in decibels between two images. This ratio is used to measure the quality of the original and the modified image. The higher the PSNR's value, the better the quality of the modified image [3].

$$\text{PSNR} = 10\log_{10}\left(\frac{\text{MAX}}{\text{MSE}}\right)^2$$

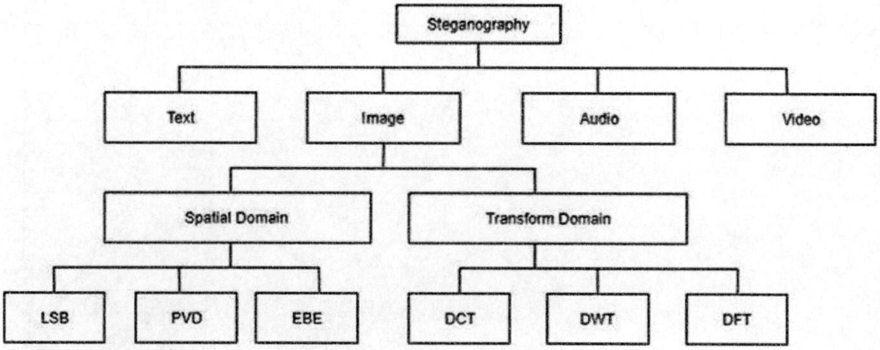

Fig. 2 Types of steganography methods

Data security is the most important thing in the digital world. Nowadays, steganography used for asymmetric key sharing and also for ownership copyright. Steganography is the technique of hiding secret data within an ordinary, non-secret, file or message in order to avoid detection.

2 Types of Steganography

2.1 Spatial Domain Techniques

Spatial steganography covers image data implicitly by changing the pixel values of the image to certain hidden message bits [4]. In Fig. 2,

1. least significant bit (LSB),
2. pixel value differencing (PVD),
3. edges-based data embedding method (EBE).

2.2 Transform Domain Techniques

The hidden message is converted into image coefficients that provide increased capacity for hiding and robustness of information [4].

1. Discrete cosine transform (DCT).
2. Discrete wavelet transform (DWT).

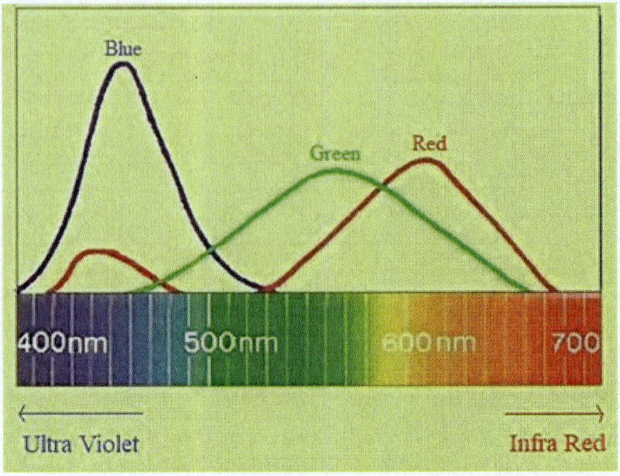

Fig. 3 Wavelength of color [3]

3 Theory of Color

A simplified theory of color vision is that there are four types of light receptors in the human eye: Rods: The rods which are sensitive only to white, black, and shades of gray. Three RGB cones: Cones are sensitive to different colors. The thousands of other shades that we can distinguish by various combinations of three types of cones (blue, green, red) [3]. Even computer screen uses a three-color system (RGB) in which the screen same number of red, green, and blue phosphor dots to represent each pixel. For example, Cyan color creates when blue and green are combined with correct ratio of intensities. White may be sensed when all three are lightened up. It would seem that all colors can be made by adding three primary colors in various proportions (Fig. 3).

4 Literature Survey

4.1 An improved LSB-based image steganography technique for RGB images [3]

The human eye has three vision cones which make the blue cones are less sensitive to light than the red and green cones. Means vision of human eyes strongly noticing color red and green. In the proposed method uses blue bits more than the other two color pixels. So, it seems less noticeable. Proposed methodology bit replacement strategy is 2-2-4 bot format for red, green, and blue, respectively.

4.2 Adaptive pixel value differencing steganography using both vertical and horizontal edges [4]

In this paper, given proposed methodology considers vertical and horizontal both pixels for find pixel value differencing. While in traditional PVD method only considers horizontal neighboring pixel value difference. Pros: Human eyes are less noticeable as compared with traditional LSB methods. Cons: The picture should not be plain color (or just a few colors) for better results.

4.3 Edge-Based Image Steganography: Problems and Solution [5]:

Cover image in this paper splits into three color components (red, green, blue). Apply edge detection algorithm on one of the components, depending on the size of the secret message select the threshold for finding the edge position. If the message size is small, then sharper edge positions are selected using high threshold value, but when the message size increases, by recording threshold value we pick more number of edge positions. Find edge pixels as threshold according to message size. And then extract data from the pixels at the edge. Embed the secret message bits on the other color component into the LSB of corresponding pixel positions. Pros: Less noticeable by human eyes as compared to other methods. Cons: Edge sharpness filter attack can destroy most of the edge pixel bits.

4.4 An Imperceptible LSB Image Hiding on Edge Region Using DES Encryption [6]:

Proposed method in this paper first encrypts a secret message using Data Encryption Standard (DES) symmetric key cryptographic type algorithm. Using LSB algorithm, It encrypted image was then inserted into the edge region of the cover image. Pros: Proposed method is more secure because it adds one more level of security by using cryptography algorithm. Cons: The main limitation of this paper's methodology is sender need to separately pass algorithm private key to the receiver side. DES key is a DES key (56 bit) need to be passed to the Receiver side.

Table 1 Comparison based on PSNR result

Image name	Method	PSNR
Lena.jpg	Traditional LSB	41.0053
Lena.jpg	Improved LSB [3]	55.9461
Lena.jpg	Pixel value differencing [4]	53.76
Lena.jpg	Edge-based embedding [5]	62.97 (when message bits 25,488)

Table 2 Comparison Analysis

Technique	Domain	Capacity	Visibility	Description
LSB	Spatial	H	L	Independent of image format
PVD	Spatial	M	L	Suitable for high contrast image
EBE	Spatial	L	L	Preferred for images with objects

4.5 A High Payload Steganographic Algorithm Based on Edge Detection [7]

In this paper, they distribute the cover image pixels into two parts: edge areas and plain areas. Then, edges area pixels are used to carry more secret bits. And plain area pixels carry less bits. Therefore, to increase the payload and to maintain good image quality, they use most significant bits (MSBs) of cover image for detecting edges of object from image. In the decoding side, the same edge information is obtained. Therefore, the secret data can be extracted. Pros: Data hiding capacity is variable. Cons: Cover image distortion detect on edges of image as message size increases (Tables 1 and 2).

5 Conclusion

In this survey, data hiding methods are described. The performance measurement of several methods is compared in terms of data hiding capacity and visibility. LSB method is nowadays known by all so it is not safe to use only that method. While EBE can be contain very less data. So, every method has its own advantages and limitations.

References

1. Lahiri S, Paul P, Banerjee S, Mitra S, Mukhopadhyay A, Gangopadhyaya M (2016) Image steganography on coloured images using edge based data hiding in DCT domain. In: IEEE 7th annual information technology, electronics and mobile communication conference (IEMCON), pp 1–8
2. Arora H, Bansal C, Dagar S (2018) Comparative study of image steganography techniques. In: International conference on advances in computing, communication control and networking (ICACCCN), pp 982–985
3. Singh A, Singh H (2015) An improved LSB based image steganography technique for RGB images. In: IEEE international conference on electrical, computer and communication technologies (ICECCT), pp 1–4
4. Swain G (2015) Adaptive pixel value differencing steganography using both vertical and horizontal edge. Springer Science+Business Media, New York
5. Rashid RD, Majeed TF (2019) Edge based image steganography: problems and solution. In: International conference on communications, signal processing, and their applications (ICCSPA), pp 1–5
6. Kusuma EJ, Indriani OR, Sari CA, Rachmawanto EH (2017) An imperceptible LSB image hiding on edge region using DES encryption. In: International conference on innovative and creative information technology (ICITech), pp 1–6
7. Bai J, Chang CC, Nguyen TS, Zhu C, Liu Y (2017) A high payload steganographic algorithm based on edge detection. Displays 46:42–51

WSN-Based Driver Cabinet Monitoring System for the Fleet of Long-Route Vehicles

Jyoti R. Dubey and Ankit R. Bhavsar

Abstract As per the Ministry of Road Transport and Highways, majority of accident happened in the Indian Highways out of total accidents occurs in India. Road accidents occur due to violation of lot many rules set by the Ministry of Road Transport and Highways. The comparative study of the accidents in 15 states, presented by Ministry of Road Transport and Highways, reveals that accidents have comparatively reduced in year 2017 as compared to 2016 but still the decrease is not that drastic. It still needs more attention and monitoring to decrease it to a minimal level. This paper presents a preliminary performance investigation of the WSN-based architecture model for driver cabinet monitoring system for the fleet of long-route vehicles. To developed prototype model, we use various sensors and integrate it using NodeMCU. For transmission of real-time sensor data to various stack holders, we used IOT-based online platform called ThingSpeak.

Keywords WSN · NodeMCU · ThingSpeak · IOT

1 Introduction

WSN is an evolving technology that has made users to access the data and services by electronic means regardless of their topographical location. WSN is a wireless communication system that facilitates transmission and reception of data by the use of air as medium from one place to another. Wireless sensor network includes a large number of portable, low powered, self-directed sensor nodes named as motes. These networks consist of a vast number of spatially distributed, tiny, battery-operated, embedded devices. These embedded devices are networked to collect, process and transfer data to the operators [1, 2].

J. R. Dubey · A. R. Bhavsar (✉)
FCAIT, GLS University, Ahmedabad, India
e-mail: ankit.bhavsar@glsuniversity.ac.in

J. R. Dubey
e-mail: jyoti.dubey@glsuniversity.ac.in

K. Kotecha et al. (eds.), *Data Science and Intelligent Applications*, Lecture Notes on Data Engineering and Communications Technologies 52,
https://doi.org/10.1007/978-981-15-4474-3_55

Table 1 Percentage share of total accidents during 2016 and 2017 for 15 states [3]

Driving rules violation	Number of accidents	Number of deaths
Overspeeding	327,448 (70.4)	98,613 (66.7)
Driving on wrong side	29,148 (6.3)	9527 (6.4)
Drunk and drive	14,071 (3.0)	4776 (3.2)
Use of mobile phone	8526 (1.8)	3172 (2.1)
Jumping red light	6324 (1.4)	1826 (1.2)
No violation and not known	79,394 (17.1)	29,999 (20.3)
Total	464,910	147,913

As per the Ministry of Road Transport and Highways, Table 1 represents that the total accidental fatalities in top 15 states in India for the year 2016 and 2017 which is accounted for total 89% share of total accidental fatalities occur on Indian highways. There is a drastic increase in number of accidental fatalities leading to death of people in Uttar Pradesh for the year 2017 as compared to 2016. It has been increased to 20,124 from 19,320. There is a considerable decrease in other states like Tamil Nadu, Maharashtra, Karnataka, Rajasthan, Gujarat, West Bengal and Telangana [3].

Figure 1 depicts percentage share of 15 states. States recorded with highest percentage of accidental fatalities are: Uttar Pradesh, Tamil Nadu, Maharashtra, Karnataka and Rajasthan. Madhya Pradesh, Andhra Pradesh, Gujarat and Telangana contribute to 7.2, 7.1, 6.9, 5.4 and 4.9 [3].

Fig. 1 Percentage share of total accidents during 2016 and 2017 for 15 states [3]

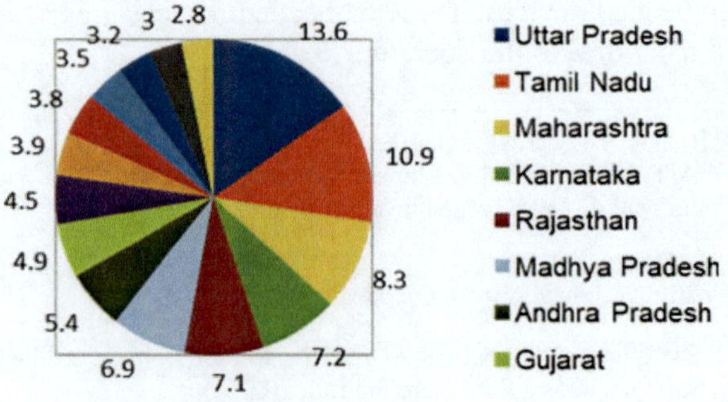

2 Causes of Accidents on Indian Highways

Table 2 depicts the various keen parameters responsible for the occurrences of accidents on Indian Highways along with number of accidental cases. Violation of rules like overspeeding and driving on the wrong side reported to 76.7% of total accidents and 73.1% of total deaths. Other parameters in totality reported to 17.1% of total accidents with 20.3% of total deaths. The other accidental parameters could be smoke detection in the vehicle, driver behavior changes, negligence of hurdles while driving on the highways and many more. As per the report of Ministry of Road Transport and Highways [3], it is a serious issue of concern to bring reduction in occurrences of road accidents on highways to overcome the life safety of the people and loss of property. Wireless sensor network which used Internet of thing (IOT) can provide an architecture that can be helpful in reduction or avoidance of fatalities on the highways. WSN-based accident avoidance architecture for the fleet of long-route vehicles [4] proposes a WSN-based architecture for the avoidance of accidents on highways. The entire architecture is divided into three major zones: activity zone, notification zone and control zone. The vehicles considered for the same architecture must be equipped with smart dashboards. Driver behavior monitoring is an important aspect of this architecture [3]. The driver behavioral parameters monitored include: head movement monitoring, drowsiness detection, fatigue detection, drunk behavior detection, eyelid movement and face monitoring [3]. The dashboard of the vehicle will be embedded with the various sensors for monitoring: fog detection, humidity detection, hurdle detection, alcohol detection, eyelid movement monitoring, head movement monitoring and so on [3]. The parameter that we are going to focus in our hardware prototype includes hurdle detection, smoke detection and humidity and temperature live tracking through sensors.

Table 2 Road accidents due to traffic rules violations—2017 [3]

Driving rules violation	Number of accidents	Number of deaths
Overspeeding	327,448 (70.4)	98,613 (66.7)
Driving on wrong side	29,148 (6.3)	9527 (6.4)
Drunk and drive	14,071 (3.0)	4776 (3.2)
Use of mobile phone	8526 (1.8)	3172 (2.1)
Jumping red light	6324 (1.4)	1826 (1.2)
No violation and not known	79,394 (17.1)	29,999 (20.3)
Total	464,910	147,913

3 Prototype for Proposed Driver Cabinet Monitoring Architecture

The proposed hardware prototype is designed to monitor various parameters like: smoke detection, hurdle detection while driving on highway, humidity and temperature detection what could be helpful in measuring fatigue level for the driver. Proposed architecture for the accident avoidance system will be composed of various components like: microcontroller NodeMCU with inbuilt Wi-fi module ESP8366, HC-SR046 ultrasonic sensor, MQ-5 smoke detection sensor, DHT-11 humidity and temperature measurement sensor, LED and USB cable.

3.1 Microcontroller and Sensors

3.1.1 NodeMCU

Figure 2 depicts NodeMCU. It is an open-source platform with inbuilt ESP8266 Wi-fi system on chip (SOC). ESP8266 module helps to transfer data collected from various sensors to be transferred on Web. The solitary reason behind using NodeMCU instead of other microcontrollers and Raspberry Pi is that it has got inbuilt Wi-fi module to help in remote access and it is much cheaper than Raspberry Pi. Its compact size made it easier to handle physically. The pins of NodeMCU are internally mapped with general-purpose input/output (GPIO) pins of ESP8266.

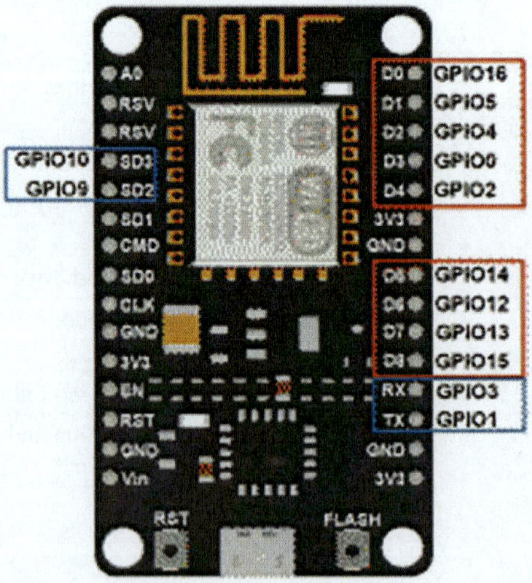

Fig. 2 NodeMCU with ESP8266 Wi-fi module [5]

Fig. 3 HC-SR046 ultrasonic sensor [6]

Fig. 4 MQ-5-smoke detection sensor [7]

3.1.2 HC-SR046 Ultrasonic Sensor

Figure 3 depicts ultrasonic sensor which can detect hurdles while driving on the highway. It can be fixed outside the cabinet, i.e., on the front topside of the vehicle. Features of HC-SR046 ultrasonic sensors measure distances based on transmitting and receiving ultrasonic signals. Four-pin module: VCC, trig, echo and GND and HC-SR046 ranges from 0 to 4000 cm (4 m). For example, D2 pin of NodeMCU is mapped to GPIO4 pin of ESP8266. One can connect number of sensors to the controller, the only thing required is Internet connection, and data can be transferred on the Web. Programming can be done in Arduino IDE.

3.1.3 MQ-5 Smoke Detection Sensor

Figure 4 represents MQ-5 smoke detection sensor. It has the following features: MQ-5 gas sensor module is useful for gas leakage detecting. It can detect smoke, LPG, i-butane, methane, hydrogen and so on. It can measure the values in parts per million (ppm). Threshold value as per prototype is >300. 1 ppm = 0.0001%. MQ-5 is a four-pin module: DOUT, AOUT, GND and VCC.

3.1.4 DHT-11 Humidity and Temperature Sensor

Figure 5 represents DHT-11 humidity and temperature sensor. It has the following features: DHT-11 is a commonly used temperature and humidity sensor. They

Fig. 5 DHT-11 humidity and temperature sensor [8]

are low-cost temperature and humidity sensors. It is composed of a basic chip that converts analog signals to digital signals. Humidity is measured in percentage, and temperature can be calculated in degree Celsius and Fahrenheit.

3.2 Block Diagram for the Prototype

Figure 6 represents the interfacing of sensors with microcontroller. Sensors like MQ-5, DHT-11 and ultrasonic sensors are interfaced with NodeMCU with inbuilt Wi-fi module ESP8266. Smoke sensor is connected with A0 pin of NodeMCU. Trig and echo pins of ultrasonic sensor are connected to D5 and D6 pins of microcontroller, smoke detecting LED at D0 pin and DHT-11 at D2 pin of the controller. Code that

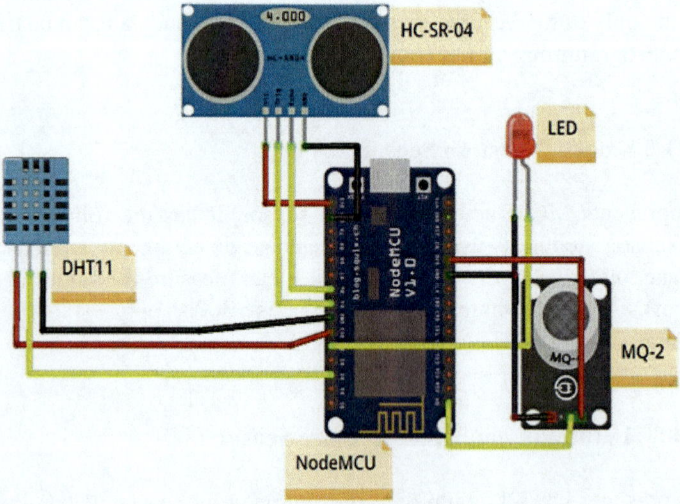

Fig. 6 Block diagram for the proposed hardware prototype model

to be dumped on controller is written on Arduino IDE. Here, LED will work as an indicator that will be ON/OFF as per the threshold value set for the sensors.

3.3 Working of the Hardware Prototype

Sensors like MQ-5, DHT-11 and ultrasonic sensors are interfaced with NodeMCU with inbuilt Wi-fi module ESP8266. Sensors are connected to different pins available on the controller board. Ultrasonic sensor gives the distance from the nearby obstacle. Humidity is in percentage, and temperature is displayed in degree Celsius and Fahrenheit. Threshold value is set for smoke sensor and for DHT 11 humidity parameter. Threshold value supposed for smoke is 300, and for humidity it is greater than 50%. LED which is connected at D0 pin of the controller is acting as indicator. If smoke detection is more than 300, the LED gets ‘LOW’ and if humidity is greater than 50% LED turns ON, giving two notifications, i.e., obstacle is very close to the vehicle, and humidity is HIGH. Figure 7 represents the real-time prototype implementation of the system architecture. The data from the different sensors is being sensed and can be displayed on serial monitor of the Arduino IDE. The data is also been sent to the cloud server ThingSpeak.

4 ThingSpeak

ThingSpeak fits in the cloud part of the diagram and provides a platform to quickly collect and analyze data from Internet-connected sensors. Display of sensor parameter values like: ultrasonic distance, humidity measurement, temperature and heat index on the serial monitor of the Arduino IDE is sent to cloud server ThingSpeak. ThingSpeak is an online open IOT platform with MATLAB analytics. Sensors are interfaced with NodeMCU. Connection is established with ThingSpeak, and Wi-fi connection is established.

Fig. 7 Prototype implementation model

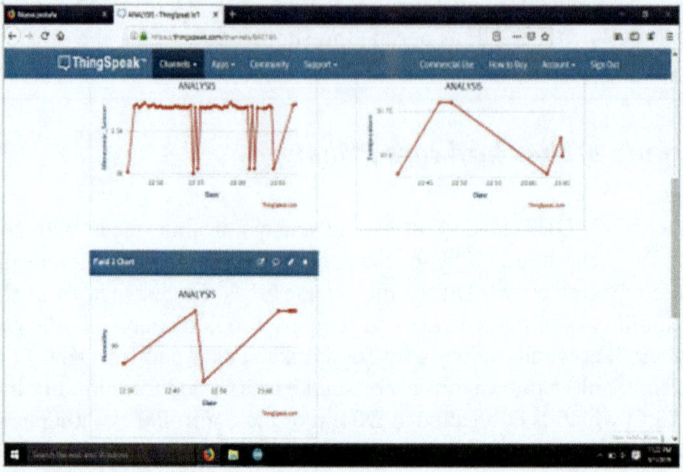

Fig. 8 MATLAB real-time representation of ultrasonic sensor distance, temperature and humidity on ThingSpeak

4.1 *Sensor Result Analysis*

MATLAB analysis allows real-time graphical representation of data on the Web. Figure 8 represents the real-time updation of the data collected from smoke sensor, ultrasonic sensor and DHT-11 sensor w.r.t time. Channel feeds that are generated are saved on ThingSpeak.

4.2 *ThingChart Mobile App*

ThingChart is a mobile app that can be linked to ThingSpeak using Channel_ID. For establishing connection, current channel id is 840166 which can be seen from Fig. 9. Plus symbol at the bottom of Fig. 9 can be used to establish new connection with a different channel id.

Result Generation on Mobile App ThingChart

Figure 10 represents the real-time feed generation on the mobile app 'ThingChart' with respect to date and time. It will display the latest value for the sensor.

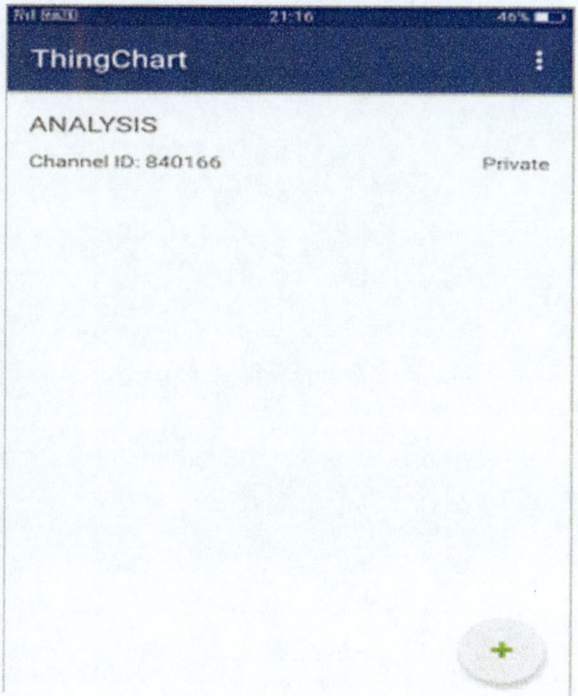

Fig. 9 ThingChart mobile app

5 Conclusion

In this paper, we have proposed a hardware prototype model for our WSN-based accident avoidance architecture for the long-route vehicles that can be helpful in monitoring certain parameters like smoke detection, hurdle detection, humidity and temperature detection. Collection, storage and visualization are done for the data collected from the sensor. This prototype model can be extended to send alert in terms of text message or a call to the driver of the vehicle with probable accidental condition.

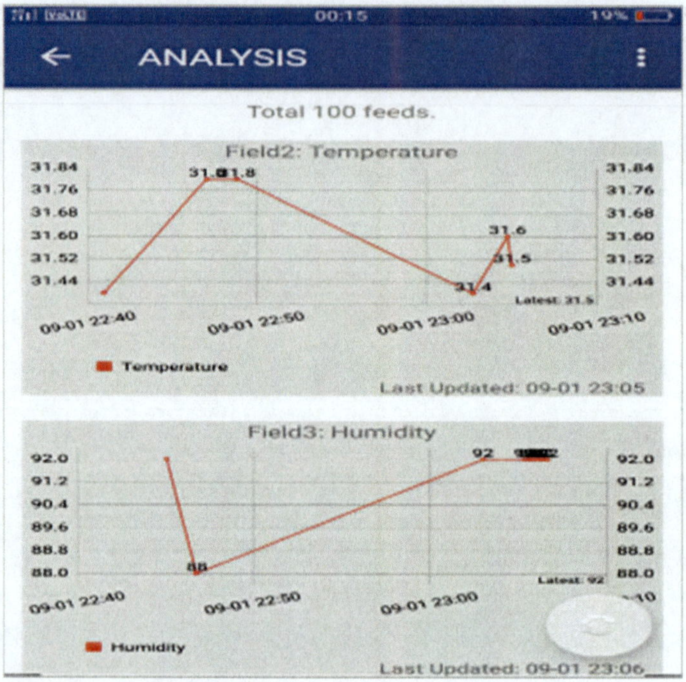

Fig. 10 Temperature and humidity feed record on ThingSpeak

References

1. http://shodhganga.inflibnet.ac.in/bitstream/10603/11663/12/11_chapter-2.pdf
2. WSN: https://www.intechopen.com/books/wireless-sensor-networks-technology-andprotocols/overview of-wireless-sensor-network
3. http://www.indiaenvironmentportal.org.in/files/file/road%20accidents%20in%20India%202017.pdf
4. Dubey JR, Bhavsar AR, Arolkar HA (2017) Wireless sensor network based accident avoidance architecture for the fleet of long route vehicles. EAI Endorsed Trans Scalable Inform Syst 4(15)
5. Arduino IDE: https://www.instructables.com/id/Quick-Start-to-Nodemcu-ESP8266-on-Arduino-IDE/
6. Dataset: https://components101.com/ultrasonic-sensor-working-pinout-datasheet
7. Sensor: https://www.waveshare.com/mq-5-gas-sensor.htm
8. Sensor Arduino: https://www.electronicshub.org/dht11-humidity-sensor-arduino/

An Approach for Privacy-Enhancing Actions Using Cryptography for Facial Recognition on Database

Arpankumar G. Raval and Harshad B. Bhadka

Abstract In the era of digital security, face recognition is highly demanded with other biometric strategies like fingerprints, iris recognition, and hand geometry for identification. In these existing ways, there are some disadvantages like time taken, pose, illumination problem, and also age effects. With this, it has been noticed that the stored images are not secured, and it is possible of cheating with the restriction in non-tolerable areas of security. Keeping these aspects in mind, this work has proposed mixture of two popular methods with cipher images to be saved in the database. Proposed method also focusing on fastest key generation for encryption and can deal with more subjects and less false match rate. Both principal component analysis and local binary pattern algorithms are utilized, and encryption system has been included for storing the images and for recognition in this proposed method. Reducing the measurement of image is the functionality of the PC. An algorithm and LBP describe the binary pattern for neighbor pixels and generate pattern for the mapping. So proposed work will increase the true matching rate and decrease the false match rate with cryptography. This proposed method is appropriate for real-time application.

Keywords Principal component analysis · Local binary pattern · Security · Cryptography · Encryption · Decryption · PCA · LBP · FMR · RR · Eigenvalue · Eigenvector

A. G. Raval · H. B. Bhadka (✉)
C. U. Shah University, Wadhwan, Gujarat, India
e-mail: harshad.bhadka@yahoo.com

A. G. Raval
e-mail: arpankumar.raval42088@paruluniversity.ac.in

A. G. Raval
FITCS Parul University, Vadodara, India

K. Kotecha et al. (eds.), *Data Science and Intelligent Applications*, Lecture Notes on Data Engineering and Communications Technologies 52,
https://doi.org/10.1007/978-981-15-4474-3_56

1 Introduction

The human body has some unique features to be used to identify each human being separately. The term ‘bio’ means the life of living organism, and ‘metric’ means the measurement. When we use some specific measurements of parts of human being for unique identification, it is called the ‘biometrics.’ The proposed work ensures matching with approved individual only from the database. The normal non-encrypted database gives chances to be cheated in non-tolerable fields like identifying the criminals [1, 2]. The biometric system is having major four phases—saving the images of part, feature extraction and identifying unique ones, database stores these features, and the program will match the retrieved features of inputted image with database image. But in this proposed method along with the image recognition, the image will go through the cryptographic method for the safety of the stored images [3].

2 Objective of Proposed Method

In large database, there are some possibilities of wrong matching images output. So to increase the recognition rate, one equation has been raised in this method for the calculation of the recognition rate and false match rate. It needs number of successfully matched images ***NS***, and number of total images ***NT***. The formula gives the general recognition rate $\boldsymbol{G_{RR}}$ for the particular database [1].

$$G_{RR=(\frac{100*NT}{NS})}$$

One more objective of this proposed method is to decrease the wrong match rate. After the implementation of the proposed technique in practical, the testing outputs of all the methods are combined and analyzed for checking how many times the result is matching with incorrect image from the database. One more equation has been used to find out for the same. Here, the general false match rate $\boldsymbol{G_{FMR}}$ is equal to the general recognition rate $\boldsymbol{G_{RR}}$ minus 100.

$$G_{FMR=100-}G_{RR}$$

One important module of this proposed method is the security of the saved images in the database called cryptography. This method is using symmetric key for the encryption and decryption of the images. Generation of the key for the cipher images is fast compared to other key generation techniques [1]. Both the steps of encryption and decryption are fast and secure.

3 Literature Review of Face Recognition Methods

3.1 Local Binary Pattern

The image is converted into 2D, the each pixel will have the binary value, and the pixel is having the neighbor values for the each pixel. In this method, it uses the eight pixels and finds the pattern of binary by comparing the values. If the value is equal or greater than compared pixel value, then it puts 1 and it puts 0 value if the pixel is having less number than the compared pixel value. This method is not feasible for complex and big scale image because it only works on 3×3 neighborhoods [4, 5].

3.2 Principal Component Analysis

This algorithm generates the eigenface for the measurement. It is also considered the eigenface-based algorithm: PCA is potential technique for the problem of face identification. This method uses lessen parts of the image and also finds the distinct verities of image and that is called eigenfaces. These defined points stored in eigenvectors are the main segment of image face information. This method is having less steps and also very strong and faster than LBP [6–9]. Following are the steps for the PCA.

Step1: Difference between each row of matrix with mean.

Step2: Produce the covariance of the matrix.

Step3: Find eigenvalues and eigenvectors.

Step4: Select the feature vector.

Step5: Select feature vectors and multiply these adjusted data with transposed feature vectors.

PCA is very strong, easy to implement, faster, and gives good performance with more dimensions and complex images also. But some disadvantages are as below which this proposed method will target to solve and reduce the false match rate with database security.

Disadvantages:

The Brightness Problem:

The different conditions of light can bring lot of problems in recognizing the face. It is not possible to fix the lighting conditions for each face. In some real-time face recognition, the lighting conditions may not be same. In this case, this method fails to compute true recognition as well as with stored images [7].

Different Pose Situation Problems:

The face poses and angles are always a hurdle for the methods to calculate the distance between features and for recognition. Pose can bring lots of changes in features, and it decreases the recognition rate of appeared faces with different poses. These situations can bring distinct outcomes [1, 7].

Fig. 1 FEI database

4 Tools and Technology

4.1 Training Data Set

To examine this face recognition work, FEI database has been used. It is Brazilian face database having exclusively face photographs for the testing. Here 200 peoples' 14 type of pictures in all the angles and poses. It is up to 180 degrees. Each image size is 640 × 480 pixels [10]. Following is one of the examples for one person subject [11] (Fig. 1).

4.2 Real-Time Images Database

To test Indian images also in this work, following are real-time images of same size. In this database, there are four pictures of 20 peoples. It is also up to 180°. Its size is also 640 × 480 [1] (Fig. 2).

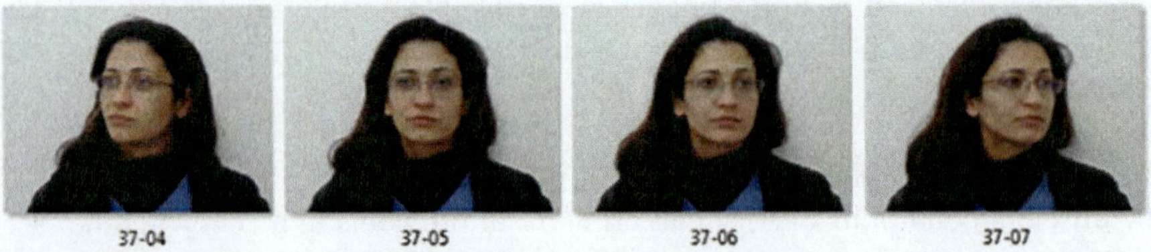

Fig. 2 Real database

5 Proposed Algorithm

5.1 *Steps to Insert Key and Image into Database*

1. Read the inputted color image.
2. Conversion of color image into gray scale.
3. Find mean of grayscale image.
4. Do subtraction of mean from each row.
5. Generate local binary pattern on found value of subtracted mean.
6. In this LBP matrix, find eigenvectors and eigenvalues.
7. Consider these eigenvectors as key.
8. Save this encrypted image and key into database.

5.2 *Planned Algorithm for Image Verification*

1. Read color image.
2. Read the inputted color image.
3. Conversion of color image into gray scale.
4. Find mean of gray scale image.
5. Do subtraction of mean from each row.
6. Generate local binary pattern on found value of subtracted mean.
7. In this LBP matrix, find eigenvectors and eigenvalues.
8. Consider these eigenvectors as key.
9. Apply this key on all the database images and decrypt.
10. Use Euclidean distance measurement technique to verify all images with selected image.
11. Give output as the image which has minimum difference between tested image and database.

5.3 *Planned Algorithms in Detail*

5.3.1 Euclidean Distance Method in Proposed Algorithm

Step 1: Read color image
Nowadays, only color images are available instead of black-and-white or grayscale images. In first step, input color image. It is difficult to use directly the color image.

In first step, we have input colored image for the process. Directly processing on color image is a difficult task. In this case, difficulties like visualization also create problems for coding. First also find each red, green, and blue values using imtool. It shows the pixel values for RGB (Figs. 3 and 4).

Step 2: Grayscale image conversion from color one

These color images should be converted into gray scale from 3D to 2D. It is conversion of shading to low-scale image as gray scale. Use MATLAB function for conversion rgb2gray (Figs. 5 and 6).

Step 3: Row-wise mean calculation

Finding the mean for each row means calculate all the values of gray scale and find the mean. Above figure shows the circled as mean of that row and also for each line.

Locate the mean of the image. Find the mean of the each row utilizing mean capacity (Fig. 7).

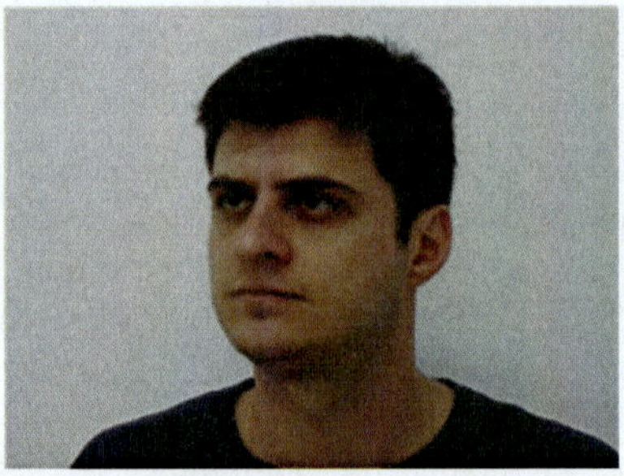

Fig. 3 Color image

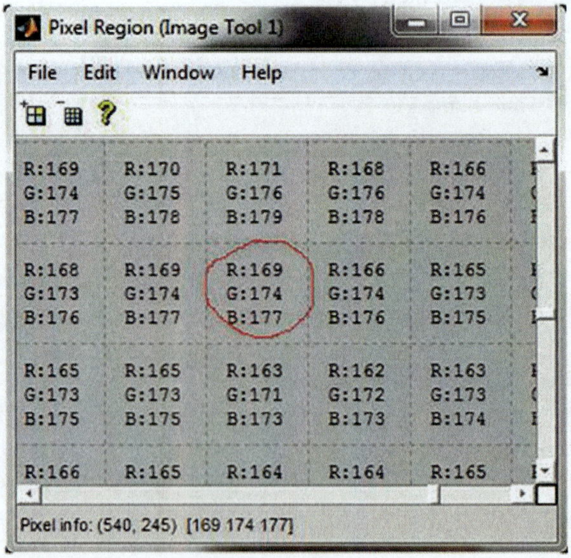

Fig. 4 Image's RGB pixel values

Fig. 5 Image in gray scale

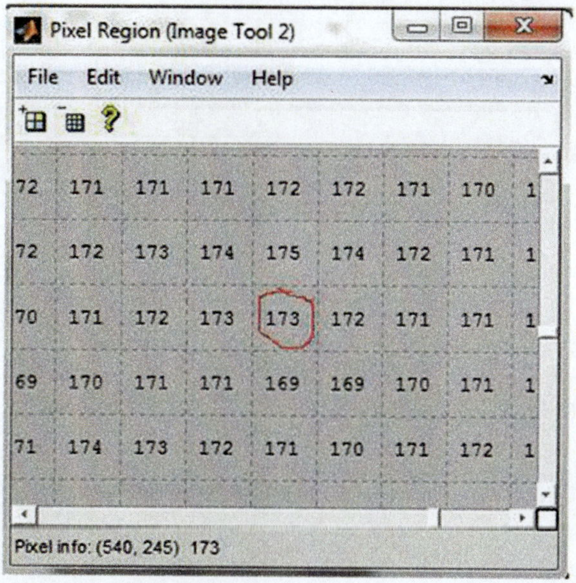

Fig. 6 Grayscale image's pixel value using imtool

Step 4: Subtract mean from all the rows of grayscale image
As it generates mean vector for each row, subtract it from each value of image pixels. Subtract the mean from each row in the grayscale image. It produces the image called shifted image. Then, find the deviation as the distance between each pixel value (Figs. 8 and 9).

Step 5: Use this array and find local binary pattern
Here at selected pixel, it has eight neighbors, and it is at the center of the neighbors. It is threshold for all values. To generate binary pattern, if the value of neighbor is greater than the value of center or threshold value, then put 1 else put 0. Even if the value is same as the threshold, then consider 0. Following is the example for highlighted cell in the figure. Consider it in 3×3 matrix for the binary pattern [12].

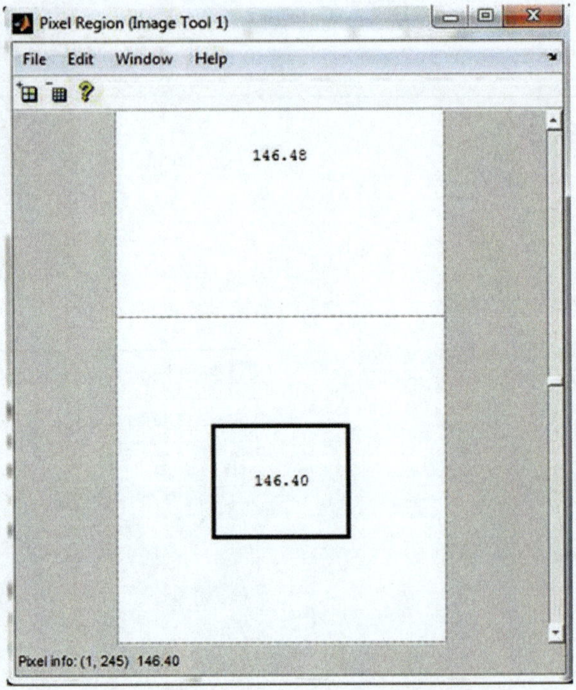

Fig. 7 Mean value of the image

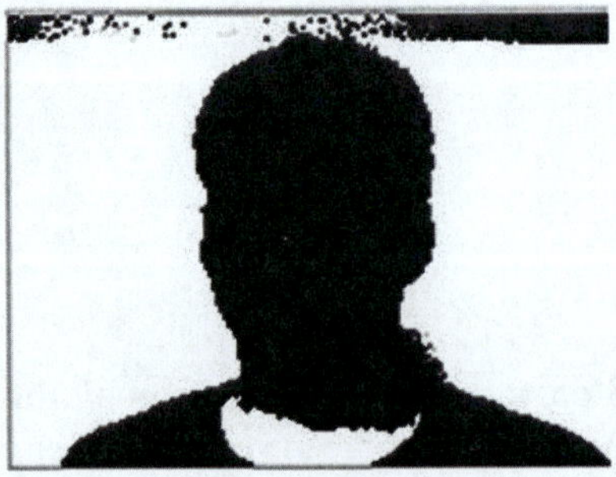

Fig. 8 Shifted image

27.52 28.52 27.52
26.60 26.60 25.60
24.94 22.94 22.94

Here the center value is 26.60. 27.52 is at the (1,1) position, and it is higher than threshold so, 1. And for all the neighbors, count the values. And for 24.94 at (1,3) position, it is less than the threshold value so put 0 at the binary place. Following same for all the neighbors, it will produce the local binary pattern which will be 1 1 1 0 0 0 0 0. Now, 30 is the decimal value of the above binary number. Follow this

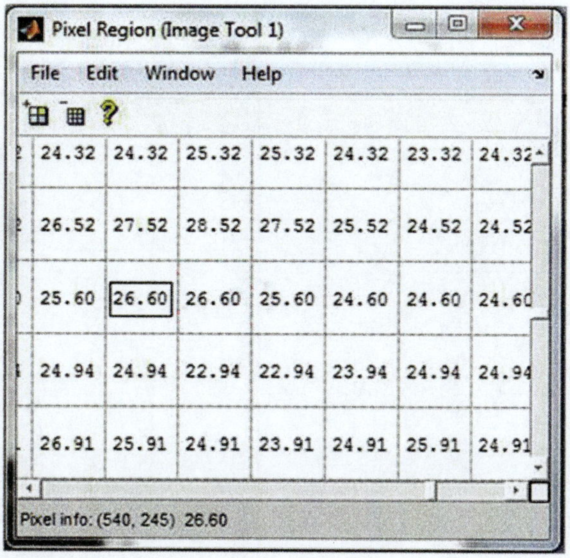

Fig. 9 Pixel value of shifted image using imtool

procedure for all the image pixels, and it will give decimal values for each value [13] (Figs. 10 and 11).

Apply this algorithm on FEI database

Face recognition is using this method to identify the face after this LBP calculation. Using LBP algorithm in this work, we will implement the comparison of proposed algorithm.

Step 6: Eigenvectors and eigenvalues in the matrix of LBP images

It will generate eigenface of the image in matrix form using eigenvectors and eigenvalues (Figs. 12 and 13).

Eig() function is used in MATLAB to find the eigenvectors and eigenvalues. Through this, it will recover the image which has minimum distance from the database. With this LBP, PCA and also proposed algorithm outcomes are matched for the testing and generating the results (Fig. 14).

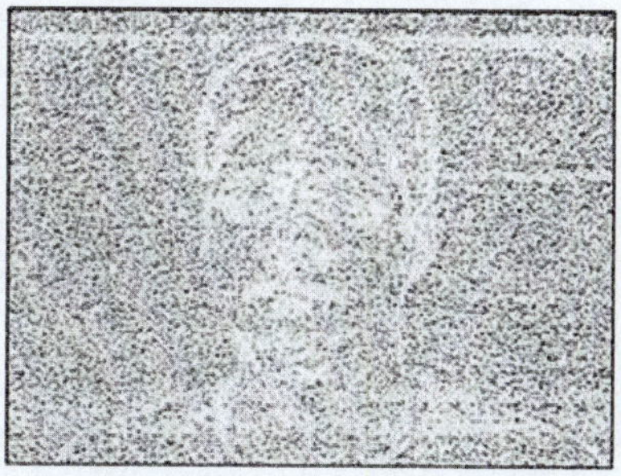

Fig. 10 LBP of planned algorithm

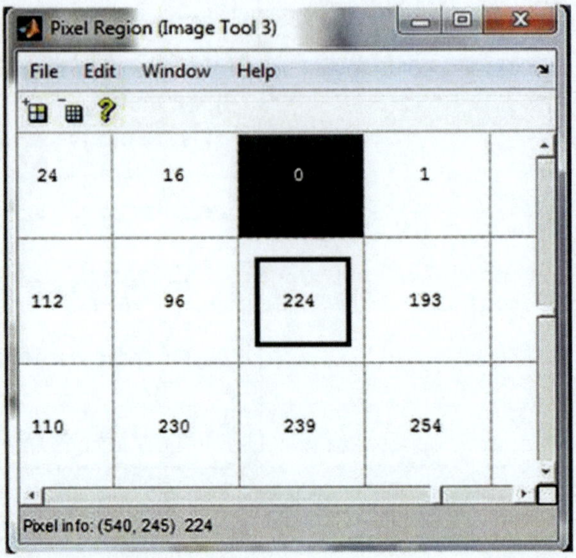

Fig. 11 Pixel value of LBP using imtool

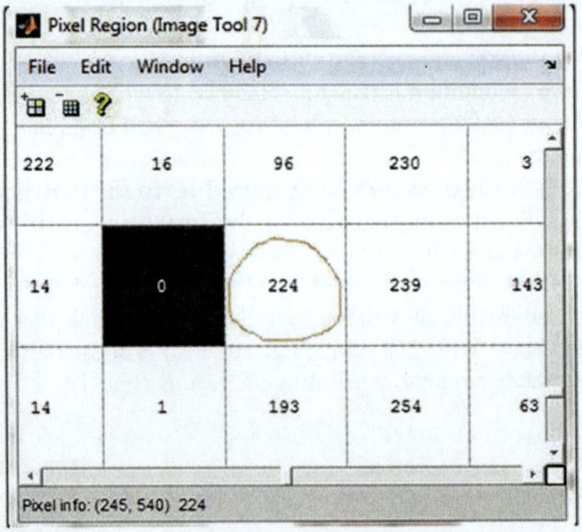

Fig. 12 LBP' pixel values

The present issue of the angle has been experienced in proposed algorithm as it is working with the issue of the angle, whereas LBP and PCA are failing in some detection of the face.

Step 7: Encryption of the image using key

From feature vector, find the mean. And that will consider as key for the encryption and decryption of the image. In below steps, it will create the key and flowchart for the same.

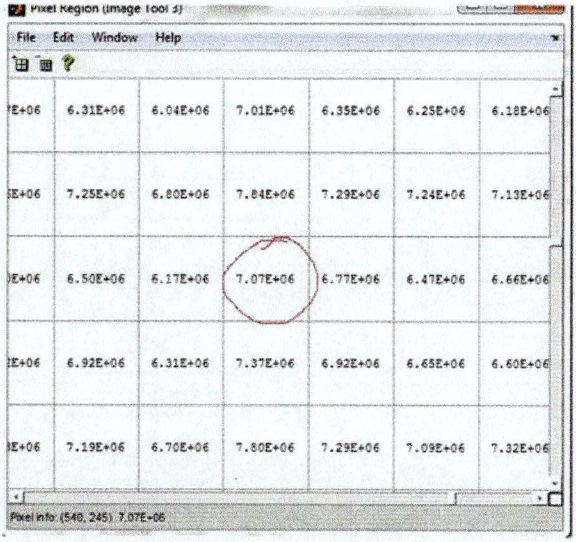

Fig. 13 LBP' * LBP

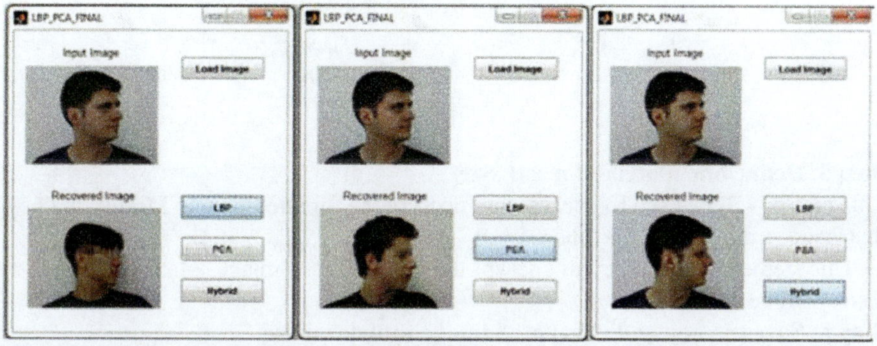

Fig. 14 Experiment with PCA, LBP, and proposed algorithm

5.3.2 Algorithm and Flowchart for Key Generation

Step 1: Calculate 8 * Key
Step 2: Initialize one matrix of size n x **1**
i.e., bin_x = zeros (n, 1, 'unit8 '); also X_N = 0; and L = 0.4 (Any random number between 0 and 1) (Fig. 15).
Step 3: Modify bin_x as given logic in following flowchart
The outcome of this step will be in binary numbers like 0 and 1.
Step 4: Define one matrix of (n/8 x **1) for 0 and 1**
i.e., key = 0 s (n/8, 1, 'unit8'); it will give 65536 * 1 size matrix for all the pixels to save.

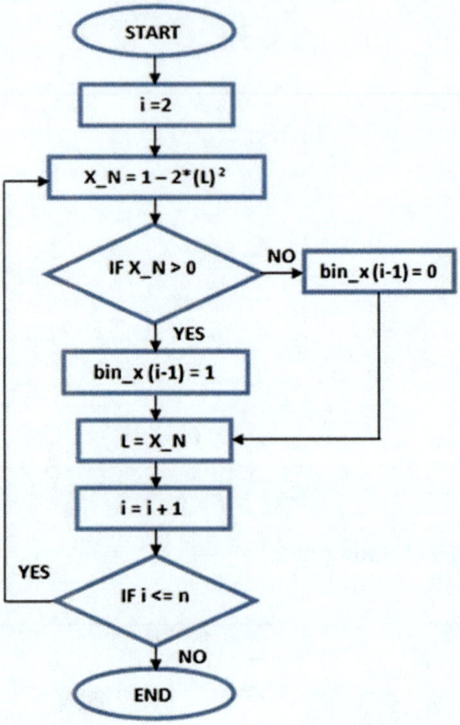

Fig. 15 Convert the key into binary

Step 5: **Define one matrix of n × 1 size**
This matrix will be used to encrypt and decrypt the inputted image. Modify the key matrix according to the flowchart.

This step enables the key to encrypt the inputted face image. Matrix of 65536 * 1 size will be created.

Step 8: **Save the key and encrypted image into database**
Use the key to encrypt and save both into database

Encoding process for image
How to encode the image is shown in below steps, and use the key as generated in earlier steps for the inputted image.

Step 1: **Input the input image to encode**
Find the size of the image by size(img) command, and we will get 256 * 256.

Step 2: **Perform bit-wise XOR between matrix of key and inputted image**
Following is the flowchart for the bit-wise XOR operation and for key generation—first for key generation and second for encryption (Figs. 16, 17, and 18).

Verification stage of proposed algorithm with database
For this step, follow all seven steps similarly like previous method and follow next step. The earlier all the steps are input color image, convert to gray scale, find mean of row, subtract mean from image, find LBP value, and generate eigenvectors a key from the inputted image of face. Now follow next step.

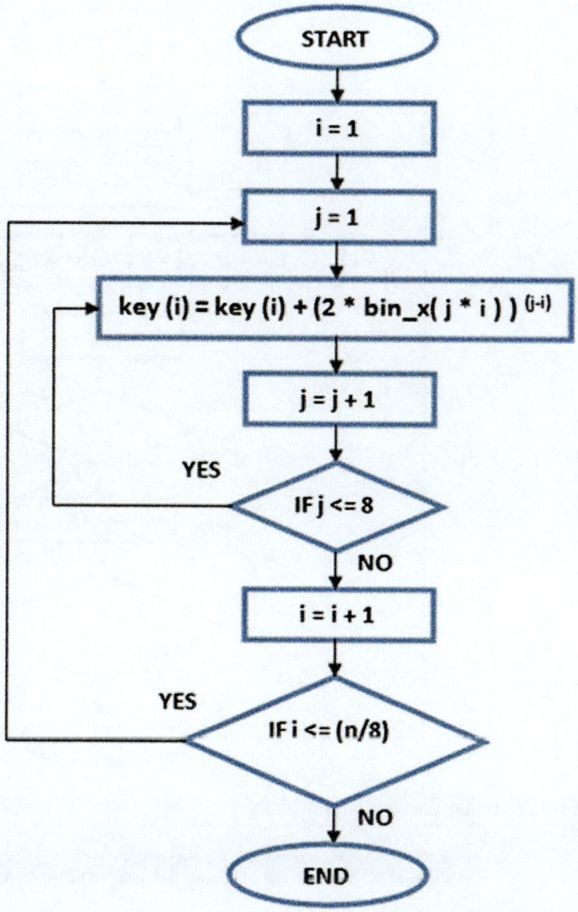

Fig. 16 Actual key generation flowchart

Step 8: **Using that key decrypt all images from database**
Compare with database key with that key, and recover that image which is exactly matching with the database (Fig. 19).

6 Result Analyses of PCA, LBP, and Proposed Algorithms

Explanation: The proposed work proves the results with less false match rate and as it increases the subjects (person) the false match rate is increasing for all the algorithms. And the recognition rate is increasing. And the verification time is average and near to the values of LBP and PCA for the FEI database (Tables 1 and 2).

Explanation: The real-time database has not much affect the results with all the algorithms, but false match rate is decreasing as it increases the number of subjects

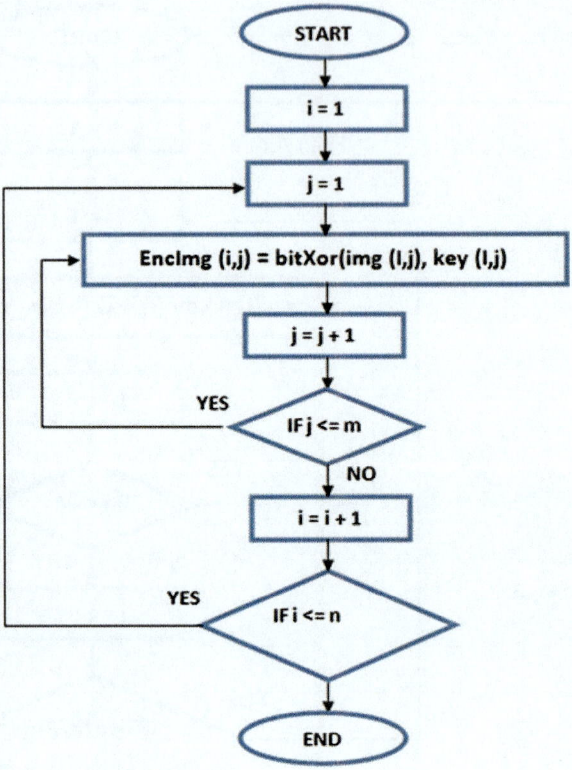

Fig. 17 Image encryption flowchart

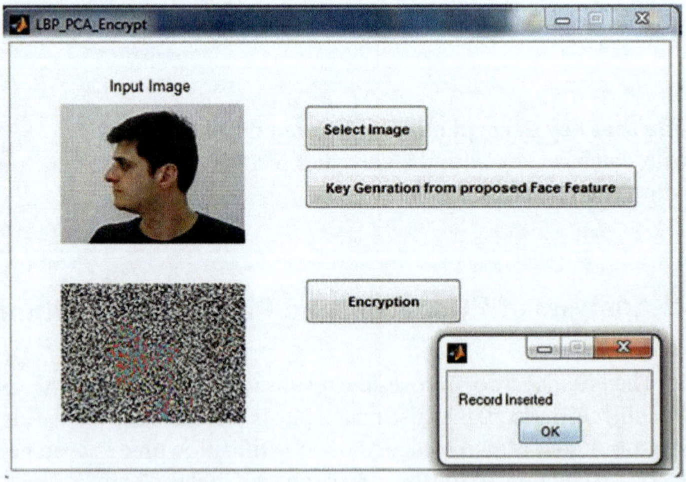

Fig. 18 Encrypted image saved in database

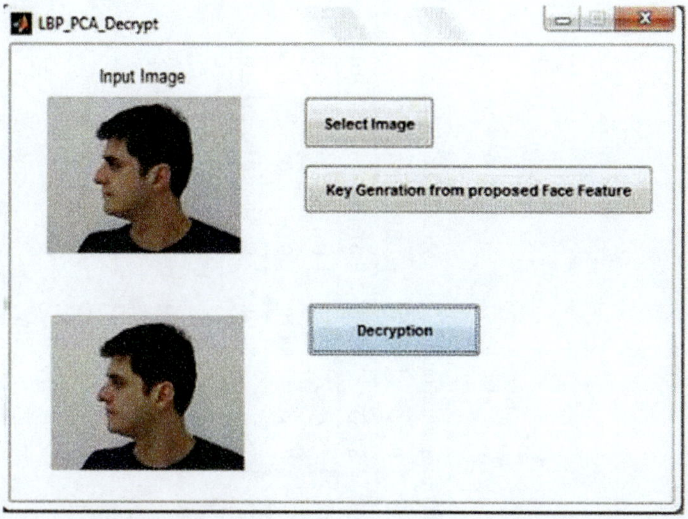

Fig. 19 Decrypted image from database

(person). The recognition rate is little bit higher than existing, but it is accurate and verification time is same as all the algorithms (Table 3).

7 Conclusion

This new method has been offered for face recognition which is using mixture of two popular algorithms, PCA and LBP. To protect the database, the cryptographic method is proposed for the entire face recognition. LBP is having less execution time with real-time database, but it is not feasible with complex face images. It is also having the illumination issue, whereas PCA is having illumination and pose problems, but accuracy is one of the advantages. This combined proposed method is reducing the false match rate and also increasing the recognition rate.

Table 1 Reading of LBP, PCA, and proposed algorithm on FEI database

No. of subjects	Database images	Inputted images	FMR % (false match rate)			RR % (recognition rate)			Average verification time		
			LBP	PCA	PA	LBP	PCA	PA	LBP	PCA	PA
			32.00	17.15	07.15	68.00	82.85	92.85	0.04	0.04	0.04
10	5	14	35.00	24.29	09.29	65.00	75.71	90.71	0.05	0.05	0.06
15	5	14	29.53	23.34	15.72	70.47	76.66	84.28	0.06	0.05	0.04
20	5	14	30.36	24.29	15.36	69.64	75.71	84.64	0.06	0.06	0.07

Table 2 Reading of LBP, PCA, and planned algorithm on real-time database

No. of subjects	Database images	Inputted images	FMR % (false match rate)			RR % (recognition rate)			Average verification time		
			LBP	PCA	PA	LBP	PCA	PA	LBP	PCA	PA
5	3	1	20.00	40.00	40.00	80.00	60.00	60.00	0.39	0.037	0.040
10	3	1	40.00	60.00	30.00	60.00	40.00	70.00	0.044	0.044	0.044
15	3	1	33.34	60.00	26.67	66.66	40.00	73.33	0.048	0.049	0.048
20	3	1	50.00	65.00	40.00	50.00	35.00	60.00	0.052	0.052	0.051

Table 3 Key generation, encryption, and decryption time of existing and planned algorithm

Subject	DB images	Input images	Key generation		Encryption		Decryption	
			Existing	PA	Existing	PA	Existing	PA
5	14	14	0.05	0.00065	0.15	0.12	0.14	0.12
10	14	14	0.05	0.00058	0.14	0.12	0.14	0.12
15	14	14	0.06	0.00059	0.14	0.13	0.14	0.11
20	14	14	0.08	0.00071	0.22	0.13	0.15	0.11

References

1. Solanki K (2017) A novel approach to prevent unauthorized access and to enhance the security of database template
2. Image Encryption Algorithm. Mathsworks.Com
3. Jain AK, Klare B, Park U (2011) Face recognition: some challenges in forensics. In: Face and gesture, pp 726–733. IEEE
4. Upmanyu M, Namboodiri AM, Srinathan K, Jawahar CV (2010) Blind authentication: a secure crypto-biometric verification protocol. IEEE Trans Inf Forensics Secur 5(2):255–268
5. Simoens K, Bringer J, Chabanne H, Seys S (2012) A framework for analyzing template security and privacy in biometric authentication systems. IEEE Trans Inf Forensics Secur 7(2):833–841
6. Patil B, Yardi A (2013) Time face recognition by varing number of eigenvalues. Int J Adv Sci Tech Res 3(1). ISSN 2249-9954
7. Nagar A, Nandakumar K, Jain AK (2011) Multibiometric cryptosystems based on feature-level fusion. IEEE Trans Inf Forensics Secur 7(1):255–268
8. Kekre HB, Thepade SD, Maloo A (2010) Eigenvectors of covariance matrix using row mean and column mean sequences for face recognition. Int J Biometrics Bioinform (IJBB) 4(2):42–50
9. Zou WW, Yuen PC (2011) Very low resolution face recognition problem. IEEE Trans Image Process 21(1):327–340
10. FEI Face Database. http://Fei.Edu.Br/~Cet/Facedatabase.Html
11. Ren J, Jiang X, Yuan J (2013) A complete and fully automated face verification system on mobile devices. Pattern Recogn 46(1):45–56
12. Ojala T, Pietikainen M, Maenpaa T (2002) Multiresolution gray-scale and rotation invariant texture classification with local binary patterns. IEEE Trans Pattern Anal Mach Intell 24(7):971–987
13. Solanki K (2017) Elimination of angular problem in face recognition. Int J Eng Res Technol (IJERT) 06(12)

Inherent Mapping Analysis of Agile Development Methodology Through Design Thinking

Archana Magare, Madonna Lamin, and Prasun Chakrabarti

Abstract Software engineering is a systematic set of activities which leads to the development of software solutions of the user's requirement. In today's information era, there is a need to fulfill the user's requirements quickly in a swift and systematic manner, which is done through agile development methodology. Design thinking facilitates the software development team not only to identify the explicit requirements but also the implicit requirements very keenly. This paper introduces the new framework—Intelli-A, which inherently maps the design thinking approach in agile methodology.

Keywords Agile development · Design thinking · Software development · Story boarding · Intelli-A

1 Introduction

Software development is a classical approach to meet the customer's implicit and explicit requirements within a minimum time span and the end product of the highest quality. Software development in the industry follows the agile approach to have the end user as part of the development team. Moreover, the agile methods [1] always embrace the requirement changes and focus to fulfill the immediate needs of the customers. Agile methodology has various facets such as meeting time constraints, focusing on immediate needs, feature-driven development and changing the team members roles based on the current goal in software development to achieve.

A. Magare (✉) · M. Lamin
Assistant Professor, Dept. of Computer Science and Engineering, ITM(SLS) Baroda University, Vadodara, Gujarat, India
e-mail: archanamagare@gmail.com

P. Chakrabarti
Executive Dean (Research and International Linkage) and Institute Distinguished Senior Chair Professor, Techno India NJR Institute of Technology, Udaipur, Rajasthan, India

Adjunct Distinguished Professor, ITM(SLS) Baroda University, Vadodara, India

K. Kotecha et al. (eds.), *Data Science and Intelligent Applications*,
Lecture Notes on Data Engineering and Communications Technologies 52,
https://doi.org/10.1007/978-981-15-4474-3_57

In teams with a strong design thinking approach, people make ideas tangible. This may involve exploration of complex systems on the whiteboard or working on canvases, or it might be in a customer experience map that shows how people experience a service over time. It could be in the form of a rough prototype prepared to explore how a product works so that it can be evolved and improved through iterative feedback. It goes without saying that these tangible expressions are an excellent improvement against spreadsheets and presentation decks as the primary form of communication. Thus, making ideas tangible allows them to be experienced by the team, the people who might use it and stakeholders.

Buzan and Griffiths in [2] discuss the mind mapping canvas, which is used for better thinking and for brainstorming. During the early stage of design thinking, the team explores ideas, and one of the efficient tools is the mind mapping tool. This tool is employed to explore problem space. It helps us to find the leading causes and factors that impact the problem to address the solution accurately. It is an efficient method to organize ideas and analyze complex situations by decomposing into smaller parts.

Bella and Hanington in [3] discuss the AEIOU canvas, which is an interrelated framework that guides designers in thinking through a problem or scenario from a variety of perspectives: activities, environment, interaction, objects and users. They are useful in organizing thoughts, observations and ideas into distinct categories. It emphasizes strict adherence to these five dimensions of a design phase.

Roman in [4], prior to prototype development, the features and functions of the prioritized solution in the form of product or service identified are listed out along with the user experience in this canvas. The team collaborates to criticize or accept the solution, and suggestions are openly accepted.

Design thinking helps in synthesizing the key elements of what is important, communicates them more clearly, iterates upon them and quickly changes them with what is being learned. The end goal of design thinking is to bring something to life, to bring it to market, to people, to a community. It focuses on making things real. To make things real, the collaborative team includes not only the end users but encompasses people within the business and in other businesses.

The rest of this paper is organized as follows: Sect. 2 describes literature survey of agile methodology and design thinking. Architecture of Intelli-A is presented in Sect. 3. Section 4 describes the inherent mapping analysis of agile development methodology through design thinking. Results and discussion are presented in Sect. 5. Section 6 presents the conclusion of the paper.

2 Literature Survey

2.1 *Agile Methodology-Software Engineering in Today's Era*

Rubin in [5] introduced essential scrum as a framework of software development methodology which guarantees to produce on time, in budget and high quality software by prioritizing the requirements in the product backlog. The scrum framework is made up of scrum values—honesty, openness, focus, respect, principles—commitment to focus on current and not on future requirements and practices—in terms of sprints. These values, principles and practices lead to the formation of current requirements prototype only.

Beck in [6] presented the new software development approach which spent more time on implementation of end software product and little on planning, analyzing and design. This approach helps to accept change in user requirements at any point of software development as the immediate software product only fulfills the immediate needs of the end user.

Highsmith and Cockburn in [7] introduced feature-driven development approach to welcome the changes in user requirements at any time even in the current development phase also by initiating the development which focuses on software features, in which the user is interested in and not on the task to complete.

Ambler in [8] introduced the concept of agile modeling. Agile modeling has three main components as values, principles and practices. The values include communication, feedback and courage. Principles are simplicity, model with purpose, allow multiple models and embrace changes. Practices are keep development light and keep it collaborative. Agile modeling is perfect blend of agile approaches in software development.

Magare and Lamin in [9] introduced the cognitive evolution in software development life cycle through design thinking. Cognitive evolution is described through correlating design thinking and agile methodologies.

2.2 *Design Thinking—A Foundation Backbone*

Sohaib et al. in [10] discussed the integration of design thinking (DT) practices into extreme programming (XP) improves the quality of software product for the end users and enables businesses to achieve creativity and innovation.

Pereira and Russo in [11] concluded that scrum is the agile software design (ASD) methodology most often integrated with the DT approach. Using scrum as an ASD, establish a model of integrated application of DT with agile methods as a practical application.

Dobrigkeit and Paula in [12] discussed and concluded that once trained in design thinking, employees find various ways to implement it not only as a pre-phase to software development but throughout their projects even applying it to their aspects of their surroundings such as the development process, team spaces and team work.

Braz et al. in [13] present and discuss the results of a case study conducted to apply the DT method, adapted through the scrum framework principles, in the software requirements elicitation to model a software product to solve the problem of allocating resources of a Brazilian University.

Brown and Wyatt in [14] discussed that design thinking can lead to hundreds of ideas and ultimately real-world solutions that create better outcomes for organizations and the people they serve.

Hendricks et al. in [15] provide an analysis of the literature on assessing stakeholder participation in health and proposed a framework based on the literature analysis that can be used to assess stakeholder participation during the design thinking process in health innovation.

Cárdenas et al. in [16] illustrated that the main advantages obtained from the implementation of design thinking were related to the consolidation of an innovative culture in the organization, where experimentation, collaboration, empathy with users and discovery of opportunities were motivational for innovation and for the generation of solutions with social impact.

Kinnunen in [17] concluded that DT guides organizations toward being more customer-centered, a positive inference between DT and innovation management by focusing on co-creating value with customers embedded needs and value creation, DT builds a new theory of value creation to society by connecting the company to the customer and end user needs.

Woudhuysen in [18] identifies the role of models in DT to bridge the gap between design and engineering.

Thompson and Cheville in [19] developed a design canvas to minimize redundancy and work toward completeness representations. The canvas classifies representations by actionable questions on two axes—system development and design choices—which in turn are organized hierarchically by scale.

3 Architecture of Intelli-A Framework

Intelli-A framework is the inherent mapping between agile development practices and design thinking paradigm. This framework gives exact mapping between these two major shifts in today's software development methodologies in which design thinking lays strong emphasis on identifying user's needs and aid in identifying innovative ideas to offer solution to the user's needs, whereas agile development focuses on quick software development with user as a part of a team to validate the development is on track. Figure 1 shows the architecture of Intelli-A framework.

Meta layer—The innermost two layers of the architecture are of design thinking. Design thinking helps to empathize and define the requirements of the user/customer through understanding them with the help of mind mapping, AEIOU, Empathy mapping canvases. These canvases state every aspect of the user's problem space clearly

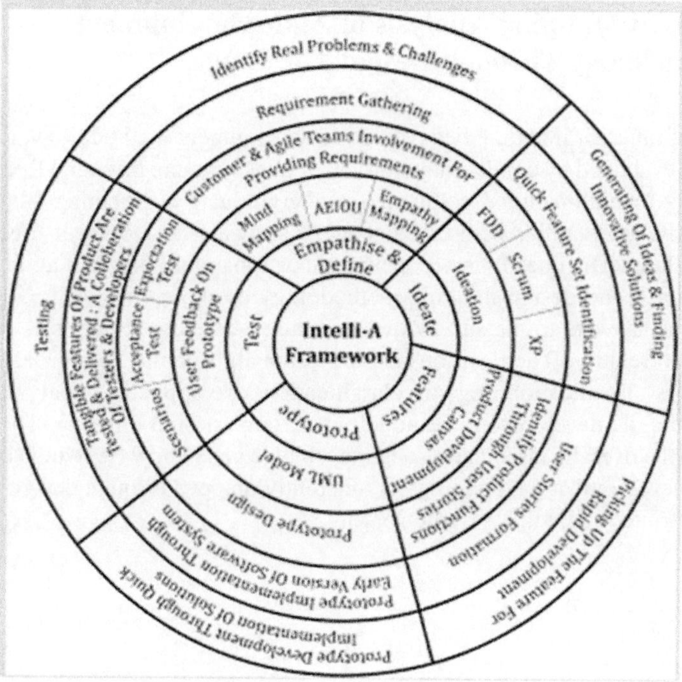

Fig. 1 Intelli-A architecture framework

where the user requirements arise and software solution that is going to be utilized. These three canvases set the foundation for providing requirements and actively participating users/customers in the requirement gathering as in agile development. The inherent fusion of mind mapping, AEIOU, Empathy mapping and user involvement as in agile for requirement gathering enables identifying real and genuine problems and challenges in a very short span of time.

Middle two layers—The middle two layers of the architecture are of agile methodology. Ideation canvas gives number of possible solutions to solve user's/customer's requirement. This serves as backbone for creating set of client valued functions that is feature set as in feature-driven development (FDD) [7] or creating customer's wish list or product backlog in scrum [5] or identifying the simplest solution for the client's problem space. Furthermore, product development canvas helps to define the functions, constraints, design of features which set the basis for user stories formation and completion. Prototype is modeled and implemented through early versions of software. Testing of early releases of the software functions (versions) is a collaborative task which involves users' feedback and user themselves.

Outer layer—The outer layer of the architecture gives a broad overview of software development life cycle phases that are followed cognitively.

4 Inherent Mapping Analysis of Agile Development Methodology Through Design Thinking

Analysis of inherent mapping between agile development methodology and design thinking is evaluated by conducting a survey of questionnaire through IT professionals working in multinational companies, undergraduate engineering students and faculties who are familiar with the concept of software development lifecycle and design thinking. The questionnaire includes exact mapping between agile development approach and design thinking methodology as shown in Fig. 1. The dataset is formed as answers to the survey questionnaire. Attribute selection is done with Pearson's correlation. Then, various classification algorithms (ZeroR classification, Naive Bayes classification, logistic classification) are applied to analyze inherent mapping of agile development methodology through design thinking. The analysis resulted in the overall classification accuracy between 84 and 94%, which concludes that agile development methodology is inherently mapped through design thinking. This results into the Intelli-A framework.

5 Discussion

The quick and rapid solution providing technique to the real-world problems that improve human lives are in huge demand in today's world. The evolution of software development has undergone through many stages and levels. Traditional software development technique has failed to cope up with the ever changing demands and needs of the users/customers. What is the need of the hour is an effective collaboration among the software development team and to bring ideas to life. The team who is well-versed with design thinking along with agile skills naturally leads to Intelli-A which is the requirement of today's software development process. The inherent mapping of the framework is done cognitively without imposing it externally. The feature development can be done in a short span of time, in a matter of hours through Intelli-A, unlike the traditional methods that use to take months. The framework ensures a workable and real entity that is responsive along with compact comprehensive documentation in place of elaborate documentation with a rather longer time until a tangible entity comes to life. Intelli-A follows agile values by welcoming role changing along with innovative solutions through design thinking as a bedrock.

6 Conclusion

The inherent mapping between agile development methodology and design thinking leads to the framework Intelli-A. This framework if followed during software development may lead to serve the customer through innovative solutions within a short span of time and with compact document. Design thinking always serves to find what is actually required by the customer, and Intelli-A serves how fast it can be converted into tangible software product. A software team with design thinking proficiency possesses capability to understand the customer requirements and be equipped with the knowledge of the kind of agile solution that requires to meet those requirements in a quick and rapid manner. This framework utilizes both the methodologies inherently with the aim to fulfill the actual needs of the customer by finding 'what' behind the 'how.'

References

1. Abbas N, Gravell AM, Wills GB (2008) Historical roots of agile methods: where did "Agile thinking" come from?. In: International conference on agile processes and extreme programming in software engineering. Springer, Berlin, Heidelberg, pp 94–103
2. Buzan T, Griffiths C (2013) Mind maps for business. In: Using the ultimate thinking tool to revolutionise how you work, 2nd edn. Pearson, UK
3. Bella M, Hanington B (2012) Universal methods of design. Rockport Publishers, Beverly, p 204
4. Roman P (2013) The product canvas—a lean tool for creating new products. Cambridge Product Management Network
5. Rubin KS (2012) Essential scrum: a practical guide to the most popular agile process. Addison-Wesley
6. Beck K (1999) Embracing change with extreme programming. Computer 32(10):70–77
7. Highsmith J, Cockburn A (2001) Agile software development: the business of innovation. Computer 34(9):120–127
8. Ambler S (2002) Agile modeling: effective practices for extreme programming and the unified process. Wiley
9. Magare A, Lamin M (2017) Cognitive evolution in software development life cycle through design thinking. Comput Model New Technol 21(3):31–34
10. Sohaib O, Solanki H, Dhaliwa N, Hussain W, Asif M (2019) Integrating design thinking into extreme programming. J Ambient Intell Humanized Comput 10(6):2485–2492
11. Pereira JC, de FSM Russo R (2018) Design thinking integrated in agile software development: a systematic literature review. Procedia Comput Sci 138:775–782
12. Dobrigkeit F, de Paula D (2019) Design thinking in practice: understanding manifestations of design thinking in software engineering. In: Proceedings of the 2019 27th ACM joint meeting on European software engineering conference and symposium on the foundations of software engineering, pp 1059–1069
13. dos Santos Braz R, Merlin JR, Trindade DFG, Ribeiro CE, Sgarbi EM, de Sordi Jr F (2019) Design thinking and scrum in software requirements elicitation: a case study. In: International conference on human-computer interaction. Springer, Cham, pp 179–194
14. Brown T, Wyatt J (2010) Design thinking for social innovation. Dev Outreach 12(1):29–43
15. Hendricks S, Conrad N, Douglas TS, Mutsvangwa T (2018) A modified stakeholder participation assessment framework for design thinking in health innovation. In: Healthcare, vol 6(3). Elsevier, pp 191–196
16. Cárdenas LFS, Ibarra ERB, Díaz-Piraquive FN, Rojas IR, González-Crespo R, Calvo A (2018) Design thinking application methodology for pediatric service innovation. In: International conference on knowledge management in organizations. Springer, Cham, pp 215–222
17. Kinnunen S (2018) The inferences between design thinking and innovation management

18. Woudhuysen J (2011) The craze for design thinking: roots, a critique, and toward an alternative. Des Principles Practices 5(6):235–248
19. Thompson MS, Cheville RA (2018) Navigating process-product tensions using a design canvas. In: ASEE annual conference & exposition proceedings

Semantic Enrichment Tool for Implementing Learning Mechanism for Trend Analysis

Pooja Ajwani and Harshal A. Arolkar

Abstract The practice of collecting dataset and attempting to find a pattern for future prediction is known as trend analysis. Due to usage of Internet and accessibility of documents on World Wide Web, the scenario of generating trend analysis has changed. Having the knowledge of trend provides an opportunity to the researchers to know the current popular topics to participate in. To make the trend analysis work better, there arises a need of the tool which can collect the data from the Web and perform the learning mechanism for trend analysis. In this paper, we have developed a system that is capable of identifying keyphrases from text file; classify the keyphrases into a computer science domain. To implement the system, we have developed a crawler that can automatically construct dataset by crawling YouTube videos of computer science playlists. Once the videos and their metadata files are downloaded, we extract the textual data of the videos by transforming the VTT (Video Text Tracks) files of the downloaded videos to the plain text. These plain text files then undergo the keyphrase extraction algorithm, where the keyphrases are filtered and are assigned to specific domains. The domain-related keyphrases files can then be used by the researchers for the purposes like classification, clustering, prediction analysis, trend analysis, recommendation systems and more. We assume that this research will help us further in performing trend analysis of YouTube videos in the domain of computer science.

Keywords Crawler · Semantic analysis · RAKE · Keyphrase extractor · Parser · Trend analysis

P. Ajwani · H. A. Arolkar (✉)
GLS University, Ahmedabad, Gujarat 380006, India
e-mail: harshal.arolkar@glsuniversity.ac.in

P. Ajwani
e-mail: pooja.ajwani@glsuniversity.ac.in

K. Kotecha et al. (eds.), *Data Science and Intelligent Applications*, Lecture Notes on Data Engineering and Communications Technologies 52,
https://doi.org/10.1007/978-981-15-4474-3_58

1 Introduction

Nowadays, there has been a tremendous growth in the usage of social media due to which the user generated content have increased exponentially. This user generated content collection and analysis provides an opportunity to the researchers to search a pattern for the future prediction which is known as trend analysis. The substantial growth in the usage of internet provides an add-on to it. These internet usage scenarios combined with certain learning mechanism can provide a baseline for building a system which is capable of performing analysis like sentiment analysis, recommendation systems, trend analysis, prediction analysis, forecasting, social media analysis, community detection, influence propagation, link analysis and so on [1, 2].

A lot of research has been done on trend analysis using dataset gathered in the form of research paper content, research paper titles, blogs, forums, microblogs, Wikis, news sites, tweets, YouTube videos, podcasts, marketing and more [3–5]. Among many user generated content providers, YouTube has become second most popular social media platform provider. It is also considered as the world's second largest search engine. Every minute more than 400 h of videos are uploaded on YouTube [W1].

With availability of such huge amount of content, researchers have good opportunity of identifying the popular computer science domain through trend analysis. To perform trend analysis, the collected dataset has to be classified in different domains. The classification of the dataset can be done on the basis of domain specific ontology. A set of representational terms which includes concepts and relations is known as ontology [6]. Building ontology of specific domain requires extracting of keyphrases. Keyphrases are the normal key terms of the document which can provide semantic metadata for classifying the document and also shows its relevance to the particular domain. Keyphrases refer to two or more key terms of the document instead of a single keyword. "Cloud computing," "artificial intelligence," "big data" and "management information system" are few examples of keyphrases. Researchers have devised different keyphrase extraction algorithms.

In this paper, we have proposed and implemented a "Semantic Enrichment Tool." This tool will serve as a baseline for developing a domain specific list of keyphrases that can be used for the classification, clustering [7, 8], prediction analysis, recommendation system or trend analysis.

2 Related Work

Rabby et al. [9] proposed a domain-independent tree-based automatic keyphrase extraction technique. The authors first built a tree from the candidate keyphrases. Depending on the similarities of the candidate keyphrases, the tree-level later may grow, collapse or remain. The final keyphrases were extracted from the tree constructed based on the maturity index (MI) value of node present within the tree. The authors compared there algorithm with Rapid Automatic Keyword Extraction (RAKE) algorithm and stated that their algorithm has better accuracy of keyphrase extraction. Gollapalli et al. [10] proposed a keyphrase extraction tool "CiteTextRank," which extracts keyphrases from research papers using citation network. An unsupervised, graph-based model was built to extract information from citation network. The keyphrases extracted were given weightage on parameters like term frequency and inverse document frequency. The authors stated that their algorithm worked around 9–20 % precised manner as compared to other available algorithms. Caragea et al. [11] designed a keyphrase extraction model CeKe (Citation-Enhanced), a supervised algorithm to extract keyphrases from research papers using citation network information. The authors' extracted keyphrases based on the features such as tf-idf, relative, the part of speech, whether the phrase has been cited or occurs in citing contexts along with its citation tf-idf, its distance from the first instance of phrase and threshold. The authors compared their algorithm performance with Hulth-n-grams tags and Keyphrase Extraction Algorithm (KEA) and stated that their model outperforms the other two algorithms. Jain et al. [12] designed a system for extracting the keyphrases from the document in three step. In first step, the documents are scanned and noun phrases are extracted by filtering noise and ignoring certain part of speech (POS) tags. In second step, formatting features of the document are checked and normalized frequency as well as position of the document is calculated. Finally, top-ranked keyphrases are identified. As an outcome, the tool developed exhibits good recall if top 20 keyphrases are taken into consideration. Qureshi et al. [13] extracted domain-specific single key terms, keyphrases and noisy information from websites using Wikipedia articles and Wikipedia categories. Then the Wikipedia articles and categories have been indexed. A subset of intersection of both the indexes has been generated using community detection algorithm. Finally, all n-gram keyphrases have been converted to 1 g keyphrases. The paper stated 77% accuracy in identifying relevant keyphrases. Kim et al. [14] used Wiktionary and FOLDOC, a handcrafted Free On-Line Dictionary Of Computing for allocating keyphrases to a specific domain. It was observed that the system gives its best performance when the semantic features were directly derived from dictionary terms.

3 Our Approach

In this paper, we have proposed a "Semantic Enrichment tool" that allows a user to identify keyphrases of a specific to domain. The Semantic Enrichment Tool consists of crawler, parser and keyphrase extractor. The crawler creates the repository by downloading YouTube videos, its metadata files and Videos Text Track (VTT) files [W2]. The WebVTT files are the timed text tracks included in the video [15]. The metadata file contains attributes which might help the researchers in analysis such as likes, ratings, description comment, tags, title and other such information. To increase the precision of our tool, we downloaded the videos using the computer science playlist.

3.1 Work Flow of Semantic Enrichment Tool

Figure 1 shows the work flow of the Semantic Enrichment Tool. The major components of this tool a crawler, a converter and keyphrase extractor are described briefly in this section.

A. **Crawler**

Dataset collection is major issue in any research. We need huge amount of dataset to create a learning mechanism. We are using YouTube video, metadata and VTT files corresponding to the video as our dataset. The crawler has been designed using

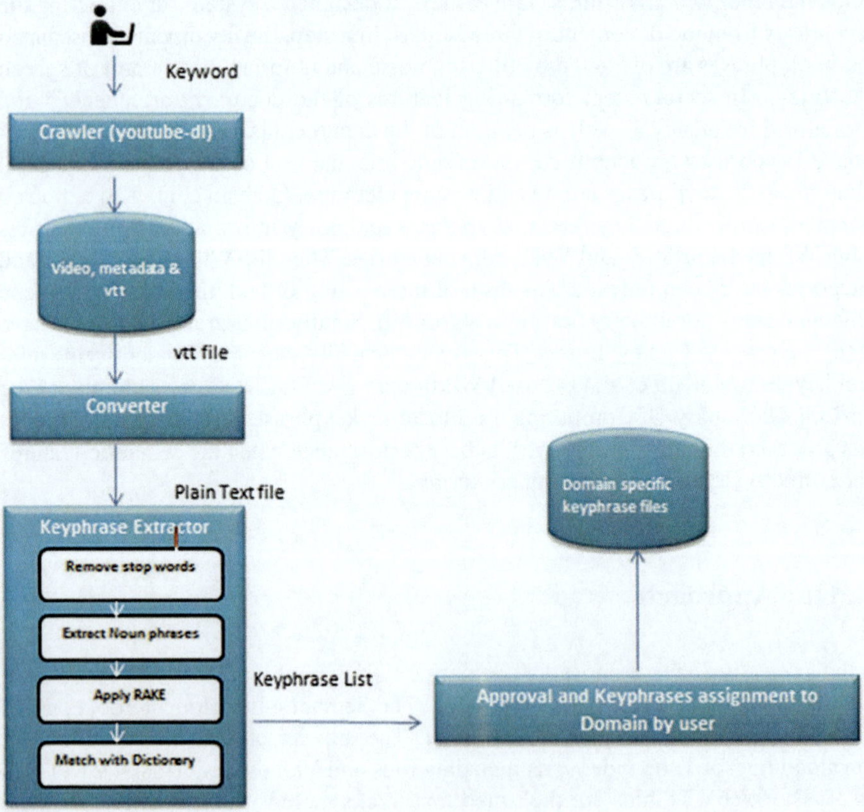

Fig. 1 Work flow of Semantic Enrichment Tool

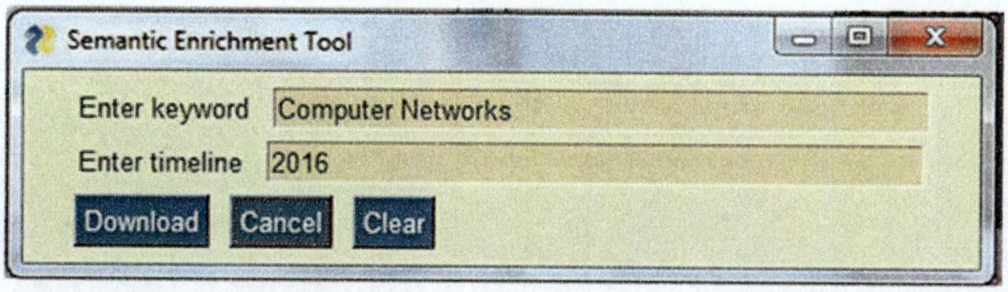

Fig. 2 Parameter selection

an open source tool "youtube-dl" [W3] for the purpose of creating the repository. The user who wants to create repository has been provided an interface as shown in Fig. 2 to enter two specific parameters, keyword and timeline. The keyword here is the topic or domain of computer science and timeline specifies the year from which the videos are to be downloaded till current date.

These parameters help in narrowing the search results for downloading the videos. The videos are downloaded only if they are not located in the local repository. In case the video exist in the repository, their metadata files are updated.

B. **Converter**

This component of the Semantic Enrichment Tool converts the VTT of the downloaded videos found in the repository in to a plain text file. Figure 3a shows a sample VTT file. The converter module is a two-step parser. In the first phase, the converter searches the repository for VTT files. If the VTT file is found, process of extract, transform and load (ETL) is applied on the file to transform it in to a temporary intermediate file. Here, the data cleaning process takes place. In the second phase, the intermediate file is processed to remove any duplication of data if present and a final text file is generated as seen in Fig. 3b.

C. **Keyphrase extractor**

Keyphrases are the sequences of one or more key terms that reflect the relevance of the topic with respect to the document under observation. Researchers have proposed many keyphrase extraction algorithms categorized as supervised and unsupervised algorithms [16]. Few algorithm that need to be mention here are Keyphrase Extraction Algorithm (KEA) [17]; it is an unsupervised algorithm based on the n-gram filtration technique. This algorithm uses a ranking method which is based on position, frequency and χ^2 (chi-square) measure to test the bias of co-occurrence distribution between frequent terms. TextRank algorithm (TRA) [18] is an unsupervised algorithm that uses graph-based ranking model. Keyphrase Extraction Algorithm (KEA) [19] is a supervised algorithm that extracts keyphrases based on the features like the first occurrence of the term, inverse document frequency and term frequency. A modified keyphrase extractor has been implemented that uses Rapid Automatic Keyword Extraction (RAKE) algorithm [20], as its base. RAKE algorithm has been chosen here as it is unsupervised, language independent and domain-independent algorithm. To extract keyphrases, the plain text files within a repository are selected

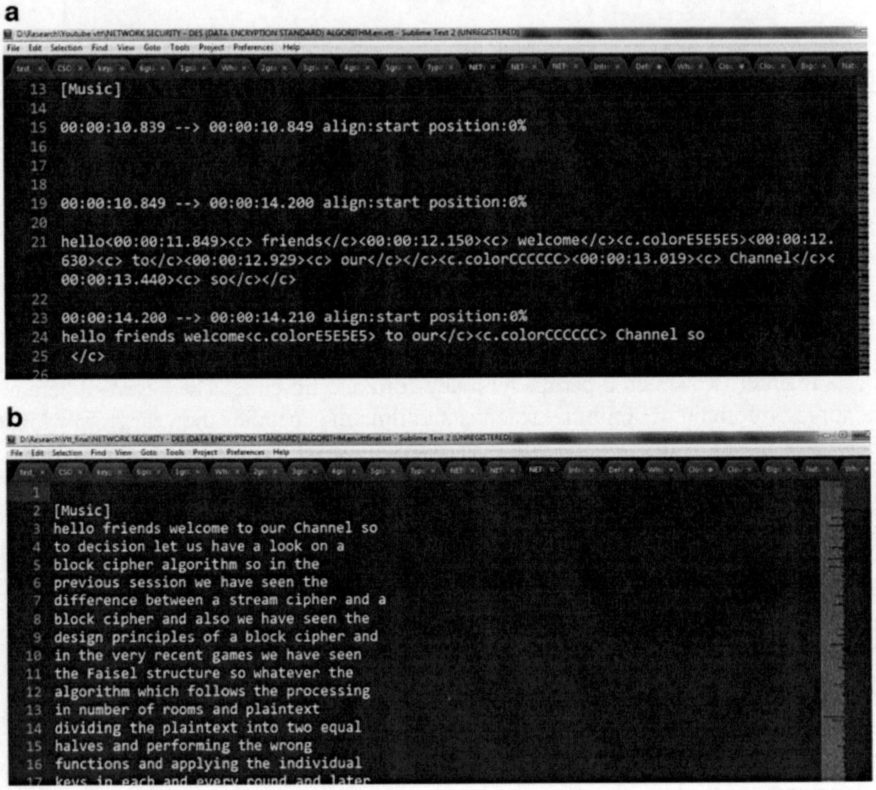

Fig. 3 **a** Sample VTT final, **b** final plain text file

and processed to remove all the stop words. Further, POS tagging [W4] has been applied to extract only noun keyphrases. These noun keyphrases are further refined using RAKE algorithm. To make them more precise, they are compared with an existing computer science ontology (CSO) [21]. The list of keyphrases thus generated is given to the user as shown in Fig. 4. The user, if need, can add, remove, accept or modify the keyphrases and assign the final list to a specific domain(s) of his choice as shown in Fig. 5. The final outcome of this work is a set of files that contain keyphrases belonging to different computer science domains.

4 Experimental Evaluation

The experiment has been performed using 50 downloaded VTT files of computer science videos uploaded on YouTube. The process of conversion from VTT to plain text has been applied on these files. These plain text documents were then provided to the experts of the respective domain for extracting technical keyphrases. The experts

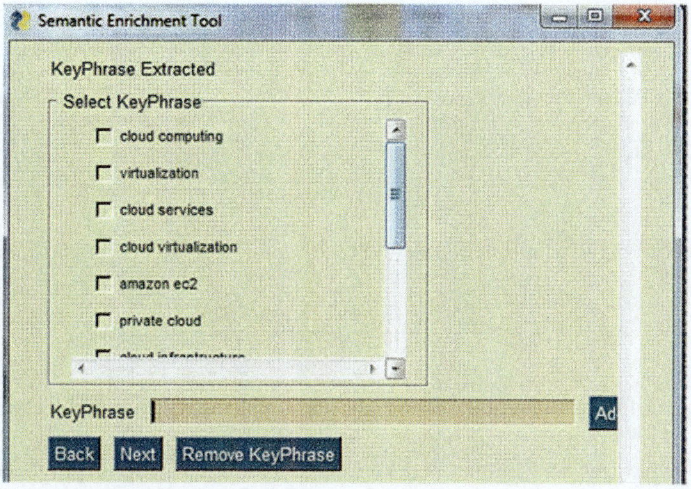

Fig. 4 Extracted keyphrases

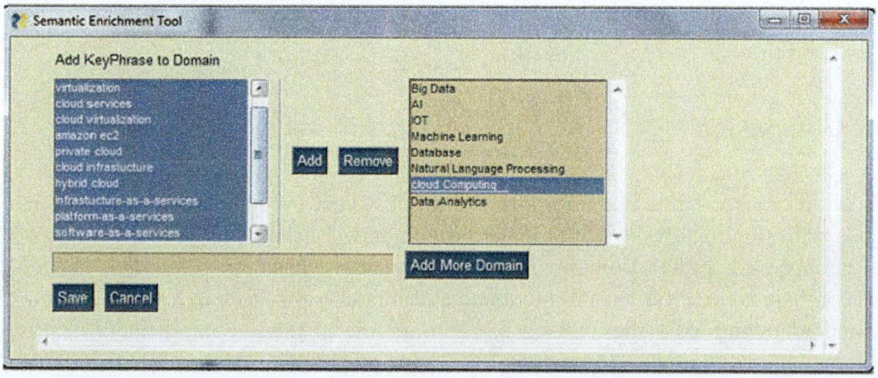

Fig. 5 Keyphrases assigned to domain by user

have identified on an average 15 generic technical keyphrases from each file. The system has extracted on an average 20 technical keyphrases based on a predefined computer science ontology [21]. The authors have then compared the results of keyphrase extracted manually and keyphrases extracted by Semantic Enrichment Tool. The precision and recall have been calculated as:

Precision = Sum of technical keyphrases identified by the system that matches with the author divided by total number of technical keyphrases generated by the system as per CSO.
Recall = Sum of technical keyphrases identified by the system that matches with the author divided by total number of keyphrases identified by the experts.

Table 1 Results of comparison

No. of files evaluated	Average keyphrases matched	Average keyphrases unmatched	Average precision	Average recall	Average F-score
50	5.71	10.88	0.29	0.38	0.24

The F-score is based on precision and recall and calculated as mentioned:

$$F_\beta = \frac{(\beta^2 + 1) * \text{Precision} * \text{Recall}}{\beta^2 * \text{Precision} + \text{Recall}}$$

The value of β has been taken as 1, as the keyphrases generated by the system or identified by the expert has been given equal weightage. The result of average matched keyphrases and unmatched keyphrases along with its average precision, recall and F-score is summarized in Table 1.

From Table 1, it can be observed that out of an average 17 technical keyphrases identified by the expert from each document, the system is able to correctly identify six keyphrases, thus giving an average precision of 0.29.

5 Conclusion

Creation of an accurate and appropriate dataset is very crucial for any research that deals with trend analysis. In this paper, a Semantic Enrichment Tool that allows a researcher to generate a domain specific keyphrase file has been presented. The tool once initiated will allow user to download and create a repository of YouTube videos, its metadata and VTT files. Based on the outcome of the keyphrase identification obtained, it is required to improve the accuracy and precision of the tool. This can be done by either training the CSO file such that it includes generic technical keyphrases or by excluding some of the generic keyphrases (data types and names of programming languages, etc.) identified by the experts. The domain specific keyphrase files generated by processing this repository can help the researcher to perform clustering, prediction analysis, influence propagation, link analysis or trend analysis. Further, this tool can serve as basis in developing computer science ontology.

References

1. Nandi G, Das A (2014) Online social network mining: current trends and research issues. Inter J Res Eng Technol (IJRET). eISSN: 2319-1163
2. Fan W, Gordon MD (2014) The power of social media analytics. Commun ACM 57(6):74–81
3. Nie B, Sun S (2017) Using text mining techniques to identify research trends: a case study of design research. Appl Sci 7(4):401

4. Lu R, Yang Q (2012) Trend analysis of news topics on twitter. Inter J Mach Learn Comput 2(3):327
5. Kondo T, Nanba H, Takezawa T, Okumura M (2009) Technical trend analysis by analyzing research papers' titles. In: Language and technology conference. Springer, Berlin, Heidelberg, pp 512–521
6. Jain S, Pareek J (2010) Automatic topic (s) identification from learning material: an ontological approach. In: 2010 Second international conference on computer engineering and applications, vol 2. IEEE, pp 358–362
7. Jun SP, Yoo HS, Choi S (2018) Ten years of research change using Google trends: from the perspective of big data utilizations and applications. Technol Forecast Soc Chang 130:69–87
8. Antony S, Wagh R (2017) Study on text clustering for topic identification. Inter J Adv Res Comput Sci 8(1)
9. Rabby G, Azad S, Mahmud M, Zamli KZ, Rahman MM (2018) A flexible keyphrase extraction technique for academic literature. Procedia Comput Sci 135:553–563
10. Gollapalli SD, Caragea C (2014) Extracting keyphrases from research papers using citation networks. In: Twenty-eighth AAAI conference on artificial intelligence
11. Caragea C, Bulgarov FA, Godea A, Gollapalli SD (2014) Citation-enhanced keyphrase extraction from research papers: a supervised approach
12. Jain S, Pareek J (2009) Keyphrase extraction tool (ket) for semantic metadata annotation of learning materials. In: 2009 International conference on signal processing systems. IEEE, pp 625–628)
13. Qureshi MA, O'Riordan C, Pasi G (2014) Exploiting wikipedia to identify domain-specific key terms/phrases from a short-text collection. In: IIR, pp 63–74
14. Kim SN, Cavedon L (2011) Classifying domain-specific terms using a dictionary. In: Proceedings of the Australasian language technology association workshop 2011, pp 57–65
15. Steiner T, Mühleisen H, Verborgh R, Champin PA, Encelle B, Prié Y (2014) Weaving the web (VTT) of data
16. Pudota N, Dattolo A, Baruzzo A, Ferrara F, Tasso C (2010) Automatic keyphrase extraction and ontology mining for content-based tag recommendation. Inter J Intell Syst 25(12):1158–1186
17. Youssif AA, Ghalwash AZ, Amer E (2011) KPE: an automatic keyphrase extraction algorithm. In: IEEE proceeding of international conference on information systems and computational intelligence (ICISCI 2011), pp 103–107
18. Mihalcea R, Tarau P (2004) Textrank: bringing order into text. In: Proceedings of the 2004 conference on empirical methods in natural language processing, pp 404–411
19. Witten IH, Paynter GW, Frank E, Gutwin C, Nevill-Manning CG (2005) Kea: practical automated keyphrase extraction. In: Design and usability of digital libraries: case studies in the Asia pacific. IGI global, pp 129–152
20. Rose S, Engel D, Cramer N, Cowley W (2010) Automatic keyword extraction from individual documents. in: Text mining: applications and theory vol 1, pp 1–20
21. Salatino AA, Thanapalasingam T, Mannocci A, Osborne F, Motta E (2018) The computer science ontology: a large-scale taxonomy of research areas. In: International semantic web conference. Springer, Cham, pp 187–205
22. https://www.brandwatch.com/blog/youtube-stats/
23. https://www.youtube.com/
24. https://github.com/rg3/youtube-dl/blob/master/README.md
25. https://www.nltk.org/book/ch05.html

Fingerprint Image Classification

Sudhir Vegad and Zankhana Shah

Abstract Due to the characteristics of human fingerprints like its unique representations, existence, ease to acquire and permanence, they are popular biometrics. The fingerprint texture representation with minutiae can be used to uniquely identify the human fingerprints for number of applications. Universally, arch, left loop, right loop and whorl are the four basic patterns of fingerprint that are considered. As the database of fingerprint is increasing, the classification of fingerprint helps to narrow down the search. Due to the quality of fingerprints and the class ambiguity, classification is still a challenging task. In the process of fingerprint retrieval from a large database, the feature representation of fingerprint images plays key role. In classification phase, for the powerful representation of fingerprint images, to enhance the local information, scale and rotation invariant features are extracted. These features strongly represent the uniqueness of fingerprint images of the similar class. Classification is carried out by applying these features in classification models. The experiments are carried out on FVC2000, FVC2002, FVC2004 and NIST-4 datasets, and results are compared.

Keywords Fingerprint · Image classification · GLCM · LBP · SURF · Bag of features

1 Introduction

With the increasing requirements of authentication of individuals for the developing security needs, it is necessary to find a strong unique representation of individuals. The innovations in biometrics allow the identity of an individual to be robust by using various physical of behavioral attributes. Various biometric representations

S. Vegad (✉)
A. D. Patel Institute of Technology, New V. V. Nagar, Anand, Gujarat, India
e-mail: svegad@gmail.com

Z. Shah
B V M Engineering College, Vallabh Vidyanagar, Anand, Gujarat, India
e-mail: zankhana.shah@bvmengineering.ac.in

K. Kotecha et al. (eds.), *Data Science and Intelligent Applications*, Lecture Notes on Data Engineering and Communications Technologies 52,
https://doi.org/10.1007/978-981-15-4474-3_59

provide the strong uniqueness of an individual which helps in distinguishing the false identity [1]. Due to their uniqueness and comprehensiveness, biometrics has helped researchers to address the challenges of uniquely representing individuals. With the advancements in biometrics, various types of systems and strategies for uniquely identifying the fingerprint images have emerged [2]. Fingerprint impression acknowledgment is proven as one of the strong and widely used among the other known biometrics research advancements due to its financial accessibility. A fingerprint identification system helps to decide whether the given fingerprint image belongs the person claiming. Fingerprints broadly characterized in three classes: arch, whorl and loop. Fingerprint image retrieval tries to search in the vast dataset of fingerprint images to match the query fingerprint image with high accuracy and speed [3, 4].

2 Literature Review

The word "biometric" is based on the Greek words "bio" (life) and "measurements." Due to the significant progress in the area of image handling and computers, automatic biometric systems have become easily accessible since few decades. The human fingerprints have been the first choice for uniquely representing the individuals not only in the current century, but since great many years [5]. To enhance safety efforts and stop terrorism, today, the emphasis is on various recognizing qualities of biometric face, iris, retina, etc.

A grouping framework for fingerprints utilizing particular qualities is built that is being utilized by scientists and educationalists even today. Edward Henry cleared approach to the success of fingerprint recognition by utilizing Galton's hypothesis to recognize prisoners by their fingerprint impressions [6]. A grouping framework that enabled a huge number of fingerprints to be effortlessly documented, looked and followed is devised. As a step before recognition, fingerprint classification is getting attraction due to which the recognition process becomes faster. Various classifications have been tried which include the local and global texture of the fingerprints [7, 8]. Feature extraction methods form images which are crucial for the success of image retrieval and classification.

2.1 *Proposed Classification*

The proposed work shows the classification of fingerprint images using two image classification techniques which use different features as an input. Classification results are compared for the conclusion. In the presented work, the fingerprint image is classified into four basic classes: left loop, right loop, arch and whorl. Feature extraction methods form images which are crucial for the success of image retrieval and classification. Gray-Level covariance metrics (GLCM), local binary pattern (LBP) and speeded-up robust features (SURF) are used for feature extraction and support vector machine (SVM) and bag-of-features (BoG) classifiers which are used for classification. About thirteen different features are extracted from GLCM by Haralick, Shanmugam, Furthermore Dinstein, which are popularly known as Haralick features. These features are utilized even today for few of the fingerprint applications. The GLCM calculates how frequently a pixel for gray-level (grayscale intensity) quality i happens whichever horizontally, vertically or diagonally to contiguous pixels with the quality j.

GLCM direction of analysis is in horizontal, vertical and diagonal, as shown in Fig. 1. As shown in Table 1, four features, energy, contrast, entropy and inverse difference moment, are calculated which form the feature vector.

To describe the texture present in the image with local pixels, local binary pattern (LBP) is well known. This works with 3 × 3 neighborhood of every pixel. The histogram of LBP labels is used as a descriptor of texture present in fingerprint image. 2P different binary patterns are possible with P neighborhood in original LBP. The uniform LBP is the small, commonly found LBP codes, usually observed in natural images. The maximum of two bitwise transitions, ranges from 0 to 1 (or vice versa), is included in uniform LBP when circular bit pattern is considered. Majority of texture microstructures are represented by uniform patterns [9].

A supervised learning model like SVM is used to recognize spatial relationships between groups of date. By focusing on the training cases that lie at the edge of the class distributions, it tries to find the optimal separating hyperplane between classes. It makes it possible to avoid the calculation of the inward result of two

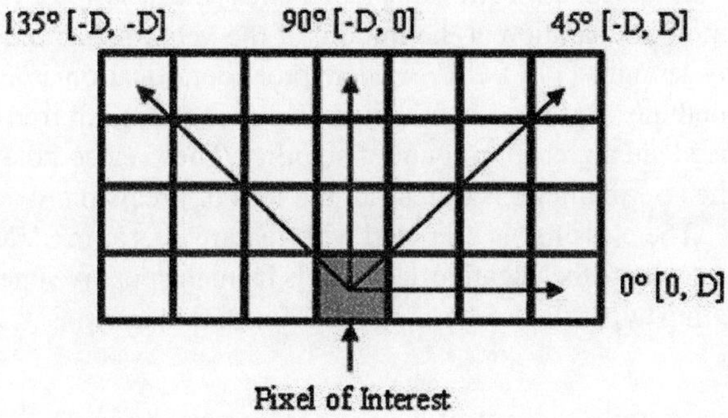

Fig. 1 Directional analysis for GLCM (P_0, P_{45}, P_{90} and P_{135})

Table 1 Table captions should be placed above the tables

1.	Energy	$\mathbf{F}_1 = \sum i \sum j\{p(i,j)\}^2$
2.	Contrast	$\mathbf{F}_2 = \sum_{n=0}^{Ng-1} n^2\{\sum i \sum jp(i,j)\}$
3.	Entropy	$\mathbf{F}_3 = \sum i \sum j\, p(i,j)\log(p(i,j))$
4.	Inverse difference moment	$\mathbf{F}_4 = \sum i \sum j \frac{p(i,j)}{1+(i-j)^2}$

Fig. 2 Calculating LBP features

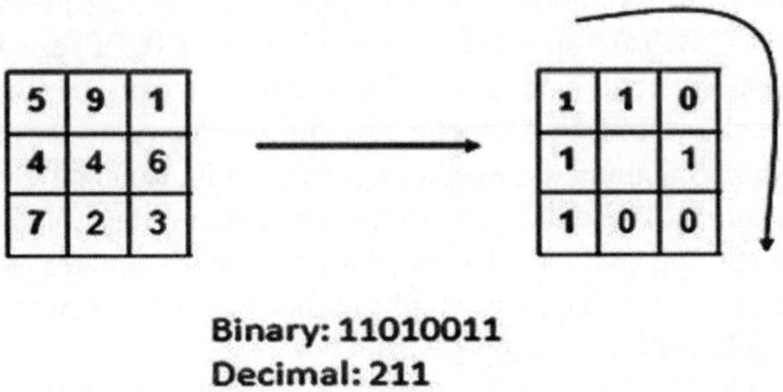

vectors, and by utilizing a specific portion work, it maps the original data space into a highest dimensional component space. The ideal isolating hyperplane with the maximal edge is registered once the examples are mapped to the component space. Along these lines, an upper bound of the normal hazard is limited rather than the observational hazard. With a specific end goal to complete the preparation of the SVMs, we utilize the sequential minimal optimization (SMO) methodology (Fig. 2).

Another robust feature algorithm used for feature extraction is speeded-up robust features (SURF). SURF offers a scale-invariant features for fingerprint verification using Hassian Metrix [10]. The scale and rotation invariant texture points are identified using bifurcations and ridges as shown in Fig. 3.

T-junctions are chosen during dissimilar areas in the picture. Repeatability is an important property from proving the capability of an identifier system. The texture descriptors are generated from the neighborhood contribution. This descriptor need on make dissimilar also at the same time noise, geometric and photometric deformations like identification displacements. The descriptor vectors are matched depending upon a separation at the middle of the vectors, e.g., euclidean separation. As stated by straight [11] SURF's descriptor identification is faster than the other existing methods. Bay, Ess what's more Tinne [11, 12] inferred that Hessian-based detectors need aid a greater amount of stability. They revised the harr-based descriptors using the approximations to restrict the loss of precision.

The classifier is prepared with feature vector in SVM. By Lagrangian multipliers to separate the negative data points from the positive ones, SVM manufactures an isolating hyperplane regarding the information as two sets of points in an n-dimensional

Fig. 3 Detection of SURF feature points (a) input image (b) interest points

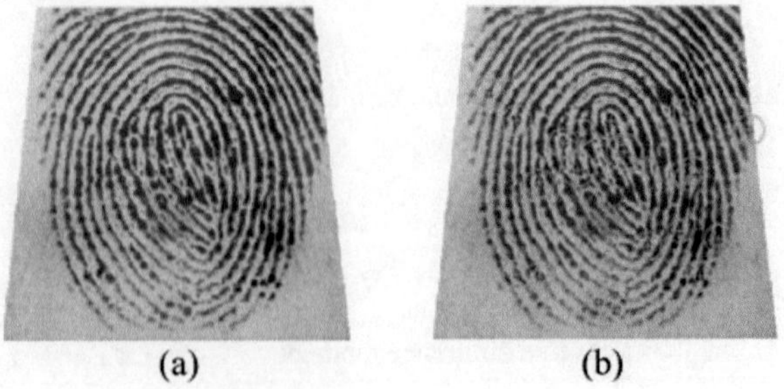

space. When an isolating hyperplane has the largest distance to the limit purposes of the two classes, a great detachment is accomplished. LIBSVM is software for the support vector characterization. It executes four essential bits, among which generally, the radial basis function (RBF) kernel is recommended as the best decision. It additionally gives an instrument named "cross-validation and grid search" to look through the suitable punishment parameter C and bit parameter γ for RBF piece. LIBSVM with RBF bit is utilized to prepare classifiers in the examinations, and to look through the penalty parameter and the kernel parameter, the tools "cross-approval and grid search" are used [13]. The bag-of-features framework is an adaptation of the bag-of-words scheme used for text categorization and text retrieval. A visual vocabulary is constructed by codifying the most representative patterns as code words or visual words in bag of features [14]. Through a simple frequency analysis of each codeword inside the image, the image representation is generated. Csurka et al. describe four steps to classify images using a bag-of-features representation:

1. Feature detection and description,
2. Codebook generation,
3. The bag-of-features construction
4. Training of learning algorithms.

According to the application domain needs, each step of bag-of-features approach may be determined, making BoF approach flexible and adaptable. The main two stages in BoF image classification are: (1) building vocabulary and (2) determine image class.

2.2 *External Results and Analysis*

The GLCM and LBP features extracted from the fingerprint images are fused using concatenation and then applied to SVM classifier. In the other experiment, the SURF features are given to BoG classifier for classification. The effectiveness of the proposed is measured by applying it on standard dataset, NIST-4 which contains 200 images of each fingerprint classes. By using various arbitrary testing images, the experiments have been carried out. Five different test cases are considered while categorizing the testing image which is selected by the system arbitrarily. Testing images are excluded from the training phase in all test cases.

The average results of various test cases and classes by taking GLCM and LBP features with SVM are shown in Figs. 4 and 5. Similarly, the average results of various test cases and classes by taking SURF features with BoG are shown in Figs. 6 and 7.

For various test cases, it is observed that classification rate ranges from 68.75 to 82.50% with SVM classifier. The experiments have also been conducted on various fingerprint classes individually. The results are examined for the fingerprints selected arbitrarily from different classes. Arch class is found most prominent class that have been classified most correctly with 85% classification rate using SVM classifier. The

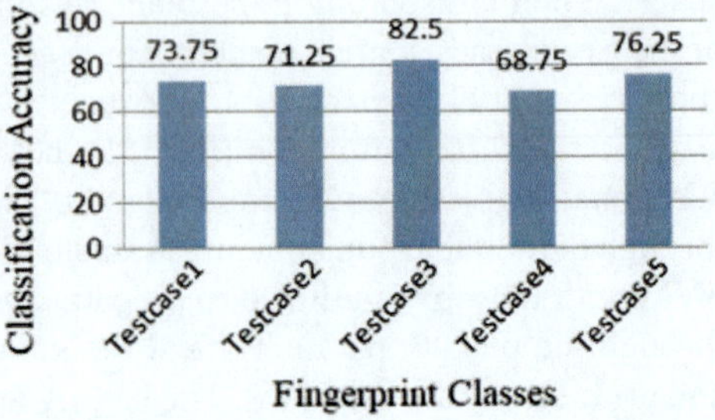

Fig. 4 Analysis of results of various test cases with SVM classifier

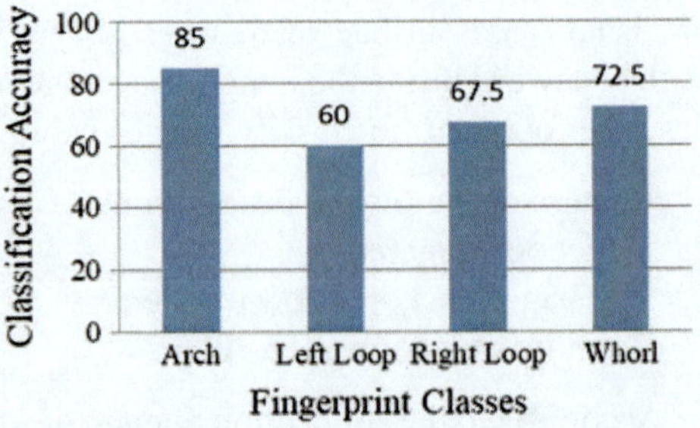

Fig. 5 Analysis of results of various classes with SVM classifier

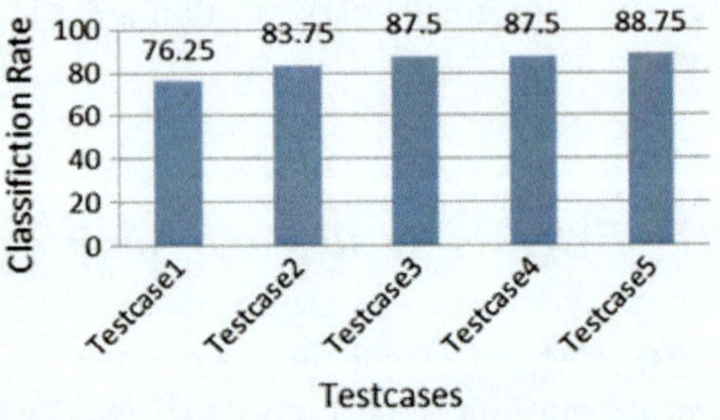

Fig. 6 Analysis of results of various classes with BoG classifier

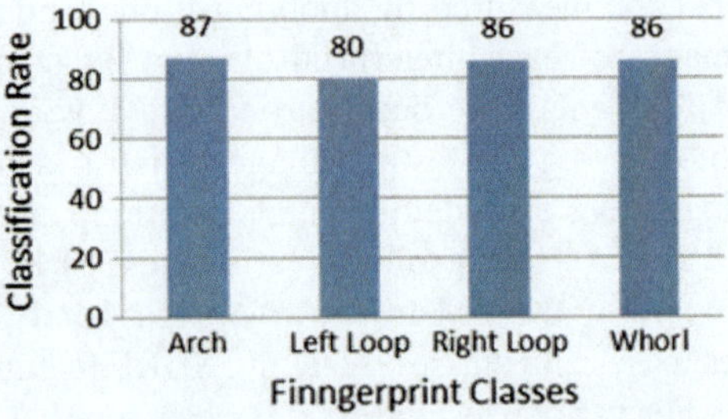

Fig. 7 Analysis of results of various classes with BoF classifier

remaining fingerprints are less correctly recognized due to the limitation of method for proper differentiation of feature points between the classes.

Table 2 Comparison of average test case accuracy and class accuracy of the methods used for classification

S. No.	Description	SVM (%)	BoF (%)
1.	Average test case accuracy	74.50	84.75
2.	Average arch class accuracy	85.00	87.00
3.	Average left loop class accuracy	60.00	80.00
4.	Average right loop class accuracy	67.50	86.00
5.	Average whorl class accuracy	72.50	86.00

It is found that for various test cases, BoF classifier achieves classification accuracy that ranges from 76.25 to 88.75%. The experiments have also been conducted on various fingerprint classes individually. The results are examined for the fingerprints selected arbitrarily from different classes. Observing the accuracy of classification, arch, right loop and whorl are most prominent classes that have been classified most correctly with the classification accuracy of 86–87% . Due to the limitation of the method of differentiating the feature points between classes, the whorl class is less correctly recognized.

3 Conclusion

Table 2 shows the comparison of accuracies of average test case and average class of the methods of classification discussed. The test cases are generated by the system arbitrarily with the bifurcation of fingerprint images as 10% for testing and the rest for training. Hence, the comparison of results of the average accuracies of test cases and classes is considered. It is observed that the BoF classifier is found strong when used with SURF features according to the experiments results.

References

1. Gubbi J, Buyya R, Marusic S, Palaniswami M (2013) Internet of things (IoT): a vision, architectural elements, and future directions. Future Gener Comput syst 29(7):1645–1660
2. Unar JA, Seng WC, Abbasi A (2014) A review of biometric technology along with trends and prospects. Pattern Recogn 47(8):2673–2688
3. Yuan B, Su F, Cai A (2012) Fingerprint retrieval approach based on novel minutiae triplet features. In: 2012 IEEE fifth international conference on biometrics: theory, applications and systems (BTAS). IEEE, pp 170–175
4. Peralta D, Triguero I, García S, Saeys Y, Benitez JM, Herrera F (2018) On the use of convolutional neural networks for robust classification of multiple fingerprint captures. Inter J Intell Syst 33(1):213–230
5. Tiwari S, Chourasia JN, Chourasia VS (2015) A review of advancements in biometric systems. Inter J Innovative Res Adv Eng 2(1):187–204

6. Maniroja M, Sawarkar S (2013) Biometric database protection using public key cryptography. Inter J Comput Sci Netw Secur (IJCSNS) 13(5):20
7. Duvenaud DK, Maclaurin D, Iparraguirre J, Bombarell R, Hirzel T, Aspuru-Guzik A, Adams RP (2015) Convolutional networks on graphs for learning molecular fingerprints. In: Advances in neural information processing systems, pp 2224–2232
8. Zhong Y, Peng X (2016) Multi-level fingerprint continuous classification for large-scale fingerprint database using fractal analysis. Inter J Biometrics 8(2):115–133
9. Zhu C, Bichot CE, Chen L (2013) Image region description using orthogonal combination of local binary patterns enhanced with color information. Pattern Recogn 46(7):1949–1963
10. Meshkini K, Ghassemian H (2017) Texture classification using Shearlet transform and GLCM. In: 2017 Iranian conference on electrical engineering (ICEE). IEEE, pp 1845–1850
11. Tuytelaars T, Mikolajczyk K (2008) Local invariant feature detectors: a survey. Found Trends® Comput Graph Vis 3(3):177–280
12. Zegarra JAM, Leite NJ, da Silva Torres R (2009) Wavelet-based fingerprint image retrieval. J Comput Appl Math 227(2):294–307
13. Yuan C, Sun X, Lv R (2016) Fingerprint liveness detection based on multi-scale LPQ and PCA. China Commun 13(7):60–65
14. Ravindran U, Shakila T (2013) Content based image retrieval for histology image collection using visual pattern mining. Inter J Sci Eng Res 4(4)
15. Paulino AA, Feng J, Jain AK (2012) Latent fingerprint matching using descriptor-based hough transform. IEEE Trans Inf Forensics Secur 8(1):31–45
16. Khodadoust J, Khodadoust AM (2018) Partial fingerprint identification for large databases. Pattern Anal Appl 21(1):19–34
17. Zhou Z, Wang Y, Wu QJ, Yang CN, Sun X (2016) Effective and efficient global context verification for image copy detection. IEEE Trans Inf Forensics Secur 12(1):48–63
18. Patel MS, Patel NM, Holia MS (2015) Feature based multi-view image registration using SURF. In: 2015 International symposium on advanced computing and communication (ISACC). IEEE, pp 213–218
19. Kazerouni MF, Schlemper J, Kuhnert KD (2015) Efficient modern description methods by using surf algorithm for recognition of plant species. Adv Image Video Process 3(2):10–24
20. Penatti OA, Silva FB, Valle E, Gouet-Brunet V, Torres RDS (2014) Visual word spatial arrangement for image retrieval and classification. Pattern Recogn 47(2):705–720
21. Gao Y, Ji R, Liu W, Dai Q, Hua G (2014) Weakly supervised visual dictionary learning by harnessing image attributes. IEEE Trans Image Process 23(12):5400–5411

Automatic Evaluation of Analog Circuit Designs

Poonam Dang and Harshal Arolkar

Abstract The circuit design is part of curriculum of large number of engineering students. The evaluation of circuit design assignments is time consuming as large number of students are enrolled in engineering. It is difficult for teacher to check all permutations and combinations of circuit due to time constraint. Automatic evaluation of circuit designs is thus an important topic in academic research. This paper proposes the automatic evaluation system that evaluates circuit design and allocates marks based on evaluation logic used in an online system. To create circuit design, the researchers are using the eSim EDA tool. Electronic circuit design has been evaluated using logical as well as functional evaluation. The evaluation of circuits can be done according to components list, connectivity of components and generated graph. The online evaluation system provides support to both students and faculties. The system also saves a lot of effort and time of teachers.

Keywords Analog circuits · Automatic evaluation · Electronic design automation · eSim · Half wave rectifier

1 Introduction

One can define ‘e-learning’ as a learning conducted through electronic media or electronic applications. Internet is one of the media of e-learning, and its massive use has created a demand for new methodologies and technologies of e-learning. E-learning provides an environment for the learners to learn at any time and any place. The past decade has been a time of rapid change as e-learning has replaced traditional learning methods, and education in the future is going to bring innovations [1]. LMS is

P. Dang
GLS University, Ahmedabad, Gujarat 380006, India
e-mail: poonam.dang@glsuniversity.ac.in

H. Arolkar (✉)
GLS University, Ahmedabad, Gujarat 380006, India
e-mail: harshal.arolkar@glsuniversity.ac.in

K. Kotecha et al. (eds.), *Data Science and Intelligent Applications*, Lecture Notes on Data Engineering and Communications Technologies 52,
https://doi.org/10.1007/978-981-15-4474-3_60

an e-learning platform used in colleges, universities, communities and schools. It is a platform where students and teachers can work collaboratively [2]. LMS has various features like chats, forums, glossary, questionnaire, scheduler, lesson, assignment uploading, quiz and many more for conducting online tests; the quiz module of an LMS is often used. Quiz module allows users to create a quiz using close-ended questions like multiple choice questions, fill in the blanks, matching, true/false and many more for conducting online tests; the quiz module of an LMS is often used. Quiz module allows users to create a quiz using close-ended questions like multiple choice questions, fill in the blanks, matching, true/false and many more. Open-ended questions still have to be created and evaluated manually.

Mathematical numerical, programming definitions, circuit designs are examples of open-ended questions. Modular Object-Oriented Dynamic Learning Environment (MOODLE) LMS supports various plug-ins for automatic evaluation of programming assignments. Rodríguez-del-Pino [3] Virtual programming lab (VPL) is a plug-in of MOODLE and used to manage automatic evaluation of programming assignments.

The automatic evaluation of programming assignments has been proposed in [4–8]. Like programming languages, engineering students also learn digital and electronic circuit design. VHDL is used for simulation of digital circuit design. In [9, 10], automatic evaluation of VHDL models has been discussed. The automatic evaluation of electronic circuit designs is rarely addressed. M. Nemec. et al. show an automatic evaluation of basic electrical circuits in education [11].

Teachers spend a large amount of their time in evaluating assignments and answer scripts. Automatic evaluation systems can reduce teachers' time and efforts. These systems can give instant feedback to students as well. Automatic evaluation systems provide a flexible and satisfying learning environment for students.

2 Related Work

The automatic evaluation of electrical circuits is shown in [11] by Nemec et al. The circuit has been evaluated in three methods. In the first method, circuit design is given to students by teachers. Students have to calculate current and voltage values across the components which are placed in a given circuit. In the second method, components are given to students. They have to design circuit using given components. Evaluation is done by comparison of students' graph files and teachers' graph file. In the third method, students have to create circuit design with any components, and the circuit is evaluated automatically. Kurmas in [12] designed a JLSCircuitTester tool which is used to test the simulated digital logic circuits. Complex digital circuit designs have multiple inputs and corresponding outputs. Because of time constraints, faculty cannot check all input and output. To solve this problem, the author has created a system. In this system, input file and output file are used with all combinations of inputs and corresponding outputs, respectively. This tool has no grading mechanism. Gutierrez et al. in [9] designed a CTPractical MOODLE module which is used for automatic evaluation of VHDL assignments submitted by students. Students have to submit a zip file of their project. The project source contains VHDL descriptions created using the Xilinx CAD tool by students. Evaluation is done by comparison of the simulated output file of the student and reference file of the teacher. Based on the comparison of both files, assignments are graded. Gutierrez et al. in [13] show an implementation of hardware specifications involving the use of CAD tools with a schematic or VHDL entry. Users of systems are student and teacher. The system provides various services to a student like assignment submission, formal checking of submitted assignments, feedback and evaluation result. The system provides services to the teacher like course configuration, submission period establishment, setting up simulation inputs and feedback mechanisms. Automatic evaluation engine performs formal checking. After this checking, functional tests are performed by the teacher. The system is able to perform two kinds of functional tests based on schematic or VHDL designs. Simulation tool Xilinx is used for schematic design entry. VHDL simulator ModelSim is used for VHDL design. Evaluation is done by comparison between the students' output file and reference file. Jelemenska et al. in [14] discuss the e-assessment of VHDL models. The assessment design is divided into three aspects. The first aspect is the code contents, in which algorithm evaluates keywords and components used by students. The second aspect is the code compilation. Here, the algorithm assigns points if code is successfully compiled. The third aspect is a code simulation. This aspect compares wave log format (WLF) files to calculate the matching percentage of simulation results. The assessment also includes the VHDL source code using language constructs. Source code content analysis must be performed at the beginning of the assessment process of the submitted assignment solution. In source code analysis, properties of respective blocks are recorded and stored. This information is used by other functions to compare student code and reference code. According to simulations of code, points are increased. Evaluation is based on simulation result and comparison between reference code and simulation code. If any student code is not compiled, then the evaluation is done manually.

Rodriguez et al. in [15] discuss the laboratory work management framework to help teachers to manage small projects. It requires in-depth knowledge of the computer I/O subsystem. Students have to develop a set of assembly language programs to ensure that serial line I/O operation is correctly managed under different conditions. The laboratory work environment has three components: delivery agent, project evaluator and plagiarism detection assistant. A delivery agent is used for login, submission of laboratory work and retrieval of result. Appropriate files are sent to automatic project evaluator by delivery agent. A project evaluator is used to inform detailed reports to students. It compares the student file with the reference file provided by the teacher. A plagiarism assistant is used to detect plagiarism. To design laboratory work, BSVC simulator is used. Alajbeg et al. in [16] describe the implementation of EDA tool National Instruments (NI) Multisim and its integration with the existing LMS MOODLE to learn circuit design and evaluation. Authors show examples of circuit design using the Multisim EDA tool. They have also shown the use of various virtual instruments like oscilloscope, power supply, ammeter and voltmeter in the simulation environment. Various types of simulation AC analysis,

DC analysis, transient analysis and frequency analysis are discussed in the paper. LMS provides exercise attendance record, instruction to students for circuit design and simulation and grading of students' assignments. They have created two tests EDA-1 and EDA-2 with three individual tasks. For automatic evaluation, students need to enter simulation values into the LMS exam form. Using these values, LMS performs automatic assessment.

3 Proposed Research Work

In today's scenario, educational electronic laboratories utilize few instruments like DC power supply, an oscilloscope and signal generator. These instruments are costly too. So, a limited number of instruments are available at laboratories. Due to time constraints, students cannot do practice individually. To solve the problems, EDA (electronic design automation) tools can be used even if these instruments are not available at students' home. EDA tools are used for design and simulation of electronic and digital circuit designs. Various open-source and proprietary EDA tools are available in the market. Itagi et al. in [18] show various EDA tools and advantages of simulators used by students to get knowledge of the fundamentals of circuit design. Nehra et al. in [19] show some open-source EDA tools. Few other EDA tools like LTSpice [20], QUCS [21], eSim [22] and many more EDA tools are available. These tools are used to create and simulate circuit designs without an assessment option.

Dang et al. in [23] show the automatic evaluation of electronic circuits using the eSim EDA tool. Evaluation is carried out partially. If components polarities have been changed, then the student will not get full marks. The sequence of choosing components of the circuit on the eSim drawing area of students and faculty should be the same. In a real-life scenario, a student may recreate a similar circuit as that of a faculty, but the way of drawing the circuit can be entirely different. To solve these problems, we have created a circuit building engine that creates all the permutation and combination of circuit designed by faculty. The possible permutations and combinations are created by changing interconnections and polarity of components. Once all the permutations and combinations are generated, the faculty needs to choose the ones that will be used for evaluation. Then, students' output files are compared with the selected test cases and original file of faculty. The research allows automation of exam-related tasks like creation of questions, creation of circuit designs, creation of possible answers of a question, conduction of online examinations and automatic evaluation of circuit designs.

4 System Overview

The researchers have created an online platform integrated with the eSim Web app. The system can be used by an administrator, a teacher and a student. Figure 1 shows the process involved in automatic creation of test cases.

Initially, the teacher needs to login and create questions with all required specifications. Then, the teacher needs to provide a correct answer to this question, so he/she draws the correct circuit design using eSim schematic area. Once the circuit is drawn simulation, details are to be added. If all details and diagram are correct, then the netlist file is created. The netlist file has textual content of circuit design like component details, the interconnection of components and simulation details. Then, teacher has to upload a netlist file for automatic evaluation.

Now, the teacher has to generate and upload a graph file of circuit design. A graph is stored in a text data format. The graph file is also used for automatic evaluation. The system automatically generates different permutation and combinations of entered circuit design and netlist files. Graph files are also generated by eSim for every netlist file in the background. Generated netlist files and graph files are stored and used by the automatic evaluation system. The teacher creates a test and adds created a question with some selected combinations into test and provides evaluation criteria for the question.

Figure 2 shows the process involved in an automatic evaluation of circuit designs. Students need to login and attempt the test. The system will provide eSim interface to draw circuit design and then enter simulation details. After the creation of circuit design, the student needs to generate a netlist file and graph file of created circuit design. Then, both files will be uploaded. Now, the student netlist file is evaluated to search components. If all components are found, then student netlist file is compared with selected netlist which is already stored in system. Selection of netlist files which are going to be compared with student netlist file is done by teacher. If any match found in netlist file comparison, then marks are assigned and further comparison is done. Here, students graph file is compared with selected graph files which are stored. If any match found, then again marks are added. Students need to finish the test and see the marks.

In the case where netlist file comparison fails to generate match, then graph file matching is not done. Marks will be assigned according to components availability and interconnection between them as available in netlist file.

5 Results and Outcomes

To test the proposed system, a half wave rectifier that has capacitor as its part has been created by teacher said with specifications. Use capacitor (100 μF), diode (1n4007), ground, resistor (1 K) and sine voltage source (0, 4, 50, 0, 0). Provide transient simulation (0, 10, 100) and add capacitor parallel with resistor.

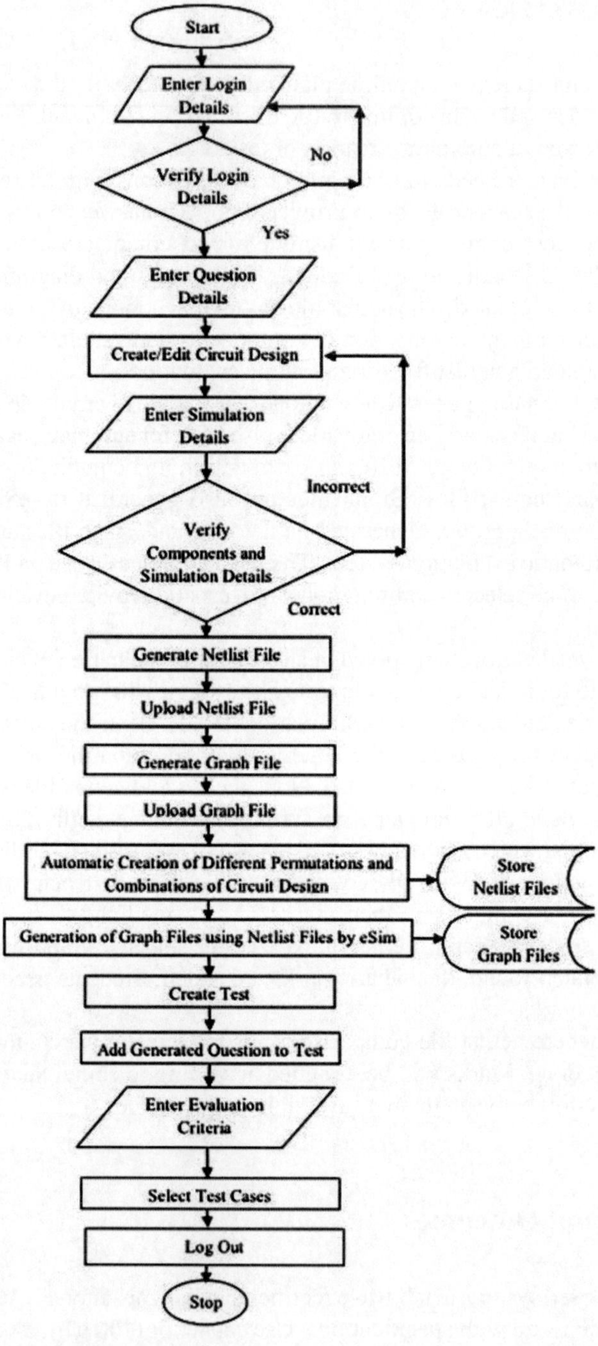

Fig. 1 Automatic creation of circuit design

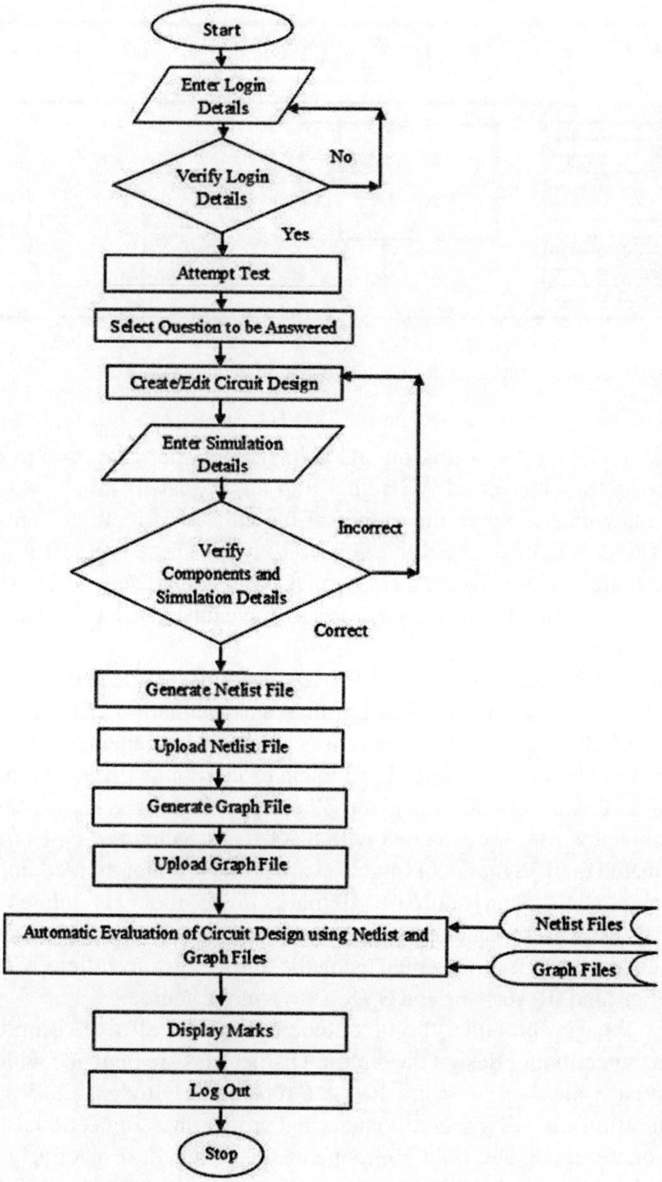

Fig. 2 Automatic evaluation of circuit design

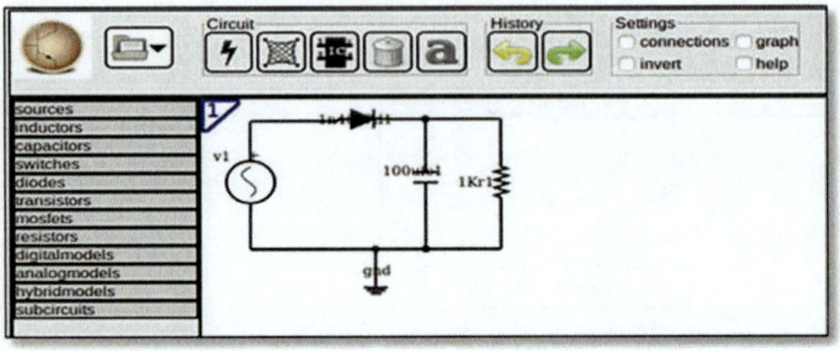

Fig. 3 Half wave rectifier with capacitor circuit design drawn by teacher

Electronic devices are working on DC voltage. Rectifiers are used in electronic devices because they converted AC signal into DC signal. If a half wave rectifier is used without capacitor, then the output of the half wave rectifier is not pure. In that signal, DC is not constant and it varies with time. These types of a signal may damage electronic devices. To get a smooth DC voltage, we need to use some filter. Filter can be implemented with a capacitor. Here, we have used a half wave rectifier with a capacitor as shown in Fig. 3.

Test is created by teacher, and the above question is added to this test. After creating question, teacher needs to assign marks as evaluation criteria to question. Then, teacher needs to select correct test cases from system generated test cases. The evaluation is divided into two parts. The logical evaluation is given weightage of 70 marks, and functional evaluation is given 30 marks weightage. These scenarios are used when students' files are matched with teachers' selected test cases files. If two files do not match then 70 marks of logical evaluation is further divided into 30% (21 marks for components taken) and 70% (49 marks for components' interconnection). Evaluation results of tests as attempted by ten students are shown in Table 1. The component connectivity is represented using the following conventions. C: capacitor, D: diode, G: ground, R: resistor and SVS: sine voltage source.

As can be seen in the table, the first student has used all five components and proper interconnection for design the circuit. His/her files are matched with teachers' selected test case files, so student 1 has got 100 marks. Students 2, 3 and 4 have attempted question with all five components and proper interconnection. But polarity of voltage source, diode and both components are changed, respectively, in circuit designs. All these circuit designs are selected for full marks by teacher, so students will get 100 marks. Student 5 has taken all five components but capacitor has taken in series order with diode and resistor; therefore, interconnection is wrong. This type of circuit does not receive 100 marks; therefore, it is not selected by teacher. This circuit is evaluated by components availability and connectivity between them. According to these two criteria, this student gets 56 marks. Student 6 has taken four components, and capacitor is not available. According to components and their interconnection, this student gets 60 marks. Student 7 has taken four components, and resistor is not

Table 1 Evaluation result of attempt by 10 students

Students	Logical Evaluation		Correct Graph Generated in Functional Evaluation	Marks Obtained		
	Component Used	Component Connectivity		Marks of Logical Evaluation (70 Marks)	Marks of Functional Evaluation (30 Marks)	Total Marks (100 Marks)
Student 1	SVS, R, D, C, G	SVS –> D –> R –> G and C is parallel to R	Yes	70.00	30.00	100.00
Student 2	SVS, R, D, C, G	SVS –> D –> R –> G and C is parallel to R (Changes the polarity of voltage source)	Yes	70.00	30.00	100.00
Student 3	SVS, R, D, C, G	SVS –> D –> R –> G and C is parallel to R (Reverse Diode)	Yes	70.00	30.00	100.00
Student 4	SVS, R, D, C, G	SVS –> D –> R –> G and C is parallel to R (Reverse Diode and changes the polarity of voltage source)	Yes	70.00	30.00	100.00
Student 5	SVS, R, D, C, G	SVS –> D –> C –> R –> G	No	56.00	0.0	56.00
Student 6	SVS, R, D, G	SVS –> D –> R –> G	No	60.00	0.0	60.00
Student 7	SVS, D, C, G	SVS –> D –> C –> G	No	60.00	0.0	60.00

(continued)

available; therefore, circuit is evaluated by component and interconnection between them, and student 7 gets 60 marks. Student 8 has taken all five components, but capacitor is taken parallel with diode. In original circuit, capacitor is connected parallel with resistor. Therefore, circuit interconnection is wrong. Student 8 gets marks according to criteria and gets 56 marks. Student 9 has attempted question with all five components, but capacitor has taken in series order with diode and power

Table 1 (continued)

Students	Logical Evaluation		Correct Graph Generated in Functional Evaluation	Marks Obtained		
	Component Used	Component Connectivity		Marks of Logical Evaluation (70 Marks)	Marks of Functional Evaluation (30 Marks)	Total Marks (100 Marks)
Student 8	R, D, C, G	SVS –> R –> D –> G and C is parallel to D	No	56.00	0.0	56.00
Student 9	SVS, R, D, G, C	SVS–> C –> D –> R –> G	No	49.00	0.0	49.00
Student 10	SVS, R, D, G, C	SVS –> D –> R –> G and R is parallel to C	Yes	70.00	30.0	100.00

source; therefore, interconnection is wrong. This student gets 49 marks. Student 10 has attempted circuit design question with all five components. Here, resistor is taken in parallel with capacitor, and we will get correct interconnection and graph. This student gets 100 marks.

6 Conclusion

The paper has proposed an online circuit creation and automatic evaluation system. The time taken for evaluation of 10 students was approximately 2 m using a client–server architecture. The system thus has potential to reduce the amount of time taken for evaluation of circuits. It further can be used as a virtual laboratory environment to teach and evaluate analog circuit design wherever required.

References

1. Sharma S, Wasim J, Dr Siddiqui J (2014) E-Learning in India. Int J Adv Res Comput Eng Technol (IJARCET) 3(1):113–117
2. Lopes AP (2014) Learning management systems in higher education. In: EDULEARN14 Conference. Proceedings of EDULEARN14 conference-IATED publications, pp 5360–5365
3. Rodríguez-del-Pino JC, Rubio Royo E, Hernández Figueroa Z (2012) A virtual programming lab for moodle with automatic assessment and anti-plagiarism features
4. Douce C, Livingstone D, Orwell J (2005) Automatic test-based assessment of programming: a review. J Educ Resour Comput (JERIC) 5(3):4-es
5. Higgins CA, Gray G, Symeonidis P, Tsintsifas A (2005) Automated assessment and experiences of teaching programming. J Educ Resour Comput (JERIC) 5(3):5-es

6. Caiza JC, Del Alamo JM (2013) Programming assignments automatic grading: review of tools and implementations. In: 7th international technology, education and development conference (INTED2013), p 5691
7. Romli R, Sulaiman S, Zamli KZ (2010) Automatic programming assessment and test data generation a review on its approaches. In: 2010 International symposium on information technology, vol 3. IEEE, pp 1186–1192
8. Queirós R, Leal JP, Gupta S, Dubey S (2012) Programming exercises evaluation systems—an interoperability survey. In: CSEDU vol 1, pp 83–90
9. Gutiérrez E, Trenas MA, Ramos J, Corbera F, Romero S (2010) A new moodle module supporting automatic verification of VHDL-based assignments. Comput Educ 54(2):562–577
10. Mosbeck M, Hauer D, Jantsch A (2018) VELS: VHDL e-learning system for automatic generation and evaluation of per-student randomized assignments. In: 2018 IEEE nordic circuits and systems conference (NORCAS): NORCHIP and international symposium of system-on-chip (SoC). IEEE, pp 1–7
11. Nemec M, Prochazka P, Fasuga R (2013) Automatic evaluation of basic electrical circuits in education. In: 2013 IEEE 11th international conference on emerging elearning technologies and applications (ICETA). IEEE, pp 299–304
12. Kurmas Z (2008) Improving student performance using automated testing of simulated digital logic circuits. In: Proceedings of the 13th annual conference on innovation and technology in computer science education, pp 265–270
13. Gutiérrez E, Ramos J, Romero S, Trenas MA (2007) A learning management system designed for a basic laboratory course on computer architecture. In: IADIS international conference e-Learning, pp 68–74
14. Jelemenska K, Cicak P, Gazik M (2016) VHDL models e-assessment in moodle environment. In: 2016 International conference on emerging elearning technologies and applications (ICETA). IEEE, pp 141–146
15. Rodríguez S, Zamorano J, Rosales F, Dopico AG, Pedraza JL (2007) A framework for lab work management in mass courses. application to low level input/output without hardware. Comput Educ 48(2):153–170
16. Alajbeg T, Sokele M (2019) Implementation of electronic design automation software tool in the learning process. In: 2019 42nd International convention on information and communication technology, electronics and microelectronics (MIPRO). IEEE, pp 532–536
17. Boemo EI (1999) Computer-based tools for electrical engineering education: some informal notes. In: Proceedings computer aided engineering conference, 1999, pp 7–13
18. Itagi M, Verna G, Deshpande A, Udyambag B (2015) Circuit simulators: a revolutionary e-learning platform. In: 2016 The Twelfth International Conference on eLearning for Knowledge-Based Society, 11–12 December 2015, Thailand
19. Nehra V, Tyagi A (2014) Free open source software in electronics engineering education: a survey. Inter J Mod Educ Comput Sci 6(5):15–25
20. Mladenović V (2015) Contemporary electronics with LTSpice and mathematica. In: Synthesis 2015-international scientific conference of IT and business-related research. Singidunum University, pp 134–138
21. Brinson M, Crozier R, Kuznetsov V, Novak C, Roucaries B, Schreuder F, Torri GB (2016) Qucs: an introduction to the new simulation and compact device modelling features implemented in release 0.0. 19/0.0. 19S of the popular GPL circuit simulator. In: Presented at the 13th MOS-AK ESSDERC/ESSCIRC workshop (No. 1/46)
22. Save YD, Rakhi R, Shambhulingayya ND, Srivastava A, Das MR, Choudhary S, Moudgalya KM (2013) Oscad: an open source EDA tool for circuit design, simulation, analysis and PCB design. In: 2013 IEEE 20th international conference on electronics, circuits, and systems (ICECS). IEEE, pp 851–854
23. Dang P, Arolkar H (2019) Automatic evaluation of analog circuits using e-Sim EDA tool. Int J Innovative Technol Exploring Eng (IJITEE) 8(12):1535–1538

A Review on Basic Deep Learning Technologies and Applications

Tejashri Patil, Sweta Pandey, and Kajal Visrani

Abstract Deep learning is a rapidly developing area in data science research. Deep learning is basically a mix of machine learning and artificial intelligence. It proved to be more versatile, inspired by brain neurons, and creates more accurate models compared to machine learning. Yet, due to many aspects, making theoretical designs and conducting necessary experiments are quite difficult. Deep learning methods play an important role in automated systems of perception, falling within the framework of artificial intelligence. Deep learning techniques are used in IOT applications such as smart cities, image recognition, object detection, text recognition, bioinformatics, and pattern recognition. Neural networks are used for decision making in both machine learning and deep learning, but the deep learning framework here is quite different, using several nonlinear layers that generate complexity to obtain more precision, whereas a machine learning system is implemented linearly. In the present paper, those technologies were explored in order to provide researchers with a clear vision in the field of deep learning for future research.

Keywords Deep learning · Neural network · Activation function · Accuracy · Loss function · Weight · Machine learning

1 Introduction

A new field has arisen over the past couple of years and has demonstrated its promise in many existing technologies. Often known as deep neural network, deep learning consists of many layers with a number of neurons in each layer. Such layers may range from a few to thousands, and each layer may contain thousands of neurons (processing unit) in addition. Multiplying the input values with the allocated weight

T. Patil (✉) · S. Pandey · K. Visrani
SSBT's College of Engineering and Technology, Jalgaon, Maharashtra, India
e-mail: ptejashri2293@gmail.com

S. Pandey
e-mail: kajal.visrani@gmail.com

K. Kotecha et al. (eds.), *Data Science and Intelligent Applications*,
Lecture Notes on Data Engineering and Communications Technologies 52,
https://doi.org/10.1007/978-981-15-4474-3_61

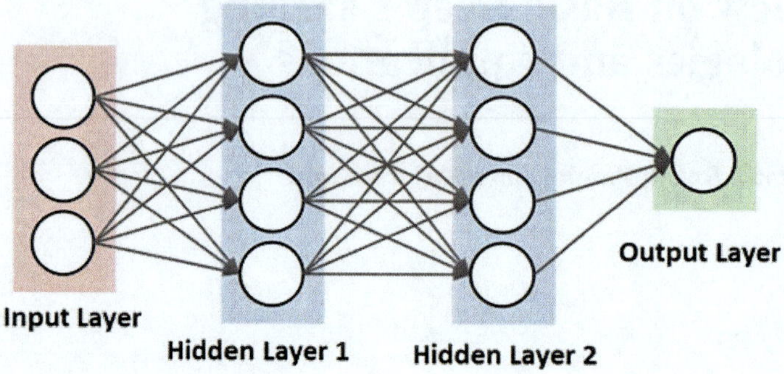

Fig. 1 Structure of deep learning model

to each input and summing up the result are the simplest process in a neuron. This result will be further scrutinized by the activation function. It improves the precision of the deep learning model. Figure 1 illustrates the structure of the deep learning model.

The deep learning model generates results by multiplying the weights of the data, thereby summarizing all the values

$$Y = \sum (\text{weight} * \text{input}) + \text{bias} \tag{1}$$

where Y is the performance of the model and bias is a constant chosen to optimize the model according to the requirement.

Deep learning has applications in numerous areas such as image detection, speech recognition, computer vision, natural language processing, bioinformatics, advertising, e-commerce, digital marketing, robot learning, and many more [1, 2].

2 Literature Survey

Du et al. [3] showed some advanced neural networks of deep learning and their implementation. It also addresses the drawbacks and opportunities of deep learning. Zhou et al. [4] introduced the importance of in-depth learning technology, implementations, and the impact on in-depth learning of the dataset by using new datasets more quickly. This deals with computer vision area, mainly applying deep learning to object detection tasks. Discussed description of the widely used datasets in computer applications and deep learning algorithms. On the other hand, a new dataset is designed according to the widely used datasets and select one of the networks called faster RCNN to operate on this new dataset. Zhou et al. [5] performed a far-reaching research study using in-depth computer well-being learning, including a basic investigation into the relative validity and possible drawbacks of the methodology and, in addition, its future perspective. The paper focuses primarily on major deep learning applications in the fields of translational bioinformatics, medicinal imaging, eventual prediction, restorative computer science, and general well-being. The paper focuses primarily on major deep learning applications in the fields of translational bioinformatics, medicinal imaging, eventual prediction, restorative computer science, and general well-being. Comprehensive empirical evidence of these residual networks being easier to refine and being able to gain accuracy from significantly increased size. Ioffe et al. [6] introduced a new method to dramatically speed up deep device development. It depends on the implementation that covariate movement, known to complicate the preparation of machine learning frameworks, often applies to subsystems and layers, and expelling it from the system's internal activation will aid in the preparation. The suggested approach derives its potential from standardizing initiations and integrating this standardization into the device engineering itself. This means that due care is taken for any programming technique that is used to improve the software. Karpathy et al. [2] showed that CNN models are capable of capturing incredible highlights from weakly labeled knowledge far outperforming well-based execution approaches and surprisingly robust to the subtleties of timely systemic availability. There are interpretable flaws in subjective analysis of system yields and disarray lattices. Lin et al. [7] provided another dataset with the intention of propelling object identification representation by putting the object identification question in relation to the more comprehensive scene understanding inquiry. This is done on social occasions by pictures of complex everyday scenes of common objects in their specific environment. Objects are defined using segmentation, for example, to assist with the exact location of objects. Eventually, categorical specifics are defined by the Gage. Zhou et al. [8] provided millions of scene images for another scene-driven database and proposed new approaches to determine the density and varied variety of image datasets to show that locations are as dense as other scene datasets and are more varied.

2.1 Number of Inputs to be Considered and Finding Noncontributing Columns

Because a dataset can contain a number of attributes, it is basically a good idea to remove the unnecessary attributes when constructing a deep learning model. In addition, removing one class column from the dataset is also necessary. This can be done with the number array in dataset, but choosing the useful attributes is a challenging job.

2.2 Number of Hidden Layers

In a deep learning model, two hidden layers are absolutely essential. A single output layer is used to combine the output of one or more hidden layers. It provides a deeper theoretical model using more hidden layers on the one side, but each additional layer adds computational complexity on the other. In addition, higher numbers of neurons inserted in each layer will also increase the computational cost.

2.3 Gradient Descent Optimizers

The gradient downward tends to minimize the model's expense. The chosen gradient descent shifts weights to reduce the process's cost. Errors are observed based on observations of the input and weight combination. It is also advisable to take reasonable long-term steps when going down as a big step could lead to a situation where global minima could be missed. In deep learning, the main task of a model is to assign weights to different inputs in order to optimize the model based on the inputs given. Weight loss should not be excessive; otherwise, local maxima, depression, and other related issues can arise. In order to optimize the model, various optimizers are programmed to assign weights to the inputs. Choosing an optimizer is a daunting and brainstorming task, however, as upgrading the algorithm with different weights increases the cost of the model and takes more time when training it for large datasets [3, 6].

2.4 Weights

Random weights are randomly choosing eight can be a good choice in which the input values add different weights to achieve good results. In this way, the proper coordination of weight and input can be established. Nonetheless, initializing random weights with low values say 0.1 is a good idea. If the weights of the descent are at zero, in this case, the weights of the corresponding input will never change and the same weight will be repeated, so it is not desirable to choose the weight of the descent as zero. And it has chosen to choose weights with random values. To speed up the model's learning process and overall performance, pick weights very carefully. However, when performing deep learning modeling such as zero, one, random, constant, matrix of defined weights, orthogonal matrix weights, and different possibilities could be available in keras method.

2.5 *Loss Function*

Loss function usually says the discrepancy that is nothing but the error between the actual output and the planned output. The formula calculated is called the function of loss. F (loss) = Expected Output − Actual Output. The real and predicted performance difference can now be calculated in many respects. There are different loss functions to do this. Choosing an acceptable role to lose deep learning is a challenging task.

In deep learning, loss functions are basically convex function to find the n-dimensional interval downward on convex surface to minimize cost while simultaneously finding global minima in learning states. This is because the model will work for the classification of test data with minimal costs associated with it.

Activation Function: There are many activation mechanisms, but they do not produce similar results due to different statistical architecture. Usually, it has been found that the sigmoid function can be used in the output activation function, and the question of binary classification provides the best results. Softmax may be a preferred choice where there is an issue with multi-label classification, but it needs to be avoided for binary classification. There are many activation mechanisms, but they do not produce similar results due to different statistical architecture. Usually, it has been found that the sigmoid function can be used in the output activation function, and the question of binary classification provides the best results.

Type of Network: Dense Network: The network is used to a great extent. In this, each neuron layer is connected to the next neuron layer. Even though it seems to be complicated, it is successful.

LSTM Network: Wide short-term memory network is a technique used over a long period of time to prevent the memory-related issues. In a general neural network, except from the output layer, each layer has the same structure and activation mechanism. Moreover, if different layers have different structures, the LSTM network may be an option as shown in Fig. 1.

3 Methodology

Various numbers of techniques and algorithms are used in deep learning. Some of deep learning techniques are as follows [9].

(i) Recurrent neural networks (RNNs)
(ii) Long short-term memory (LSTM)
(iii) Convolution neural networks (CNNs)
(iv) Deep belief networks (DBNs)
(v) Deep stacking networks (DSNs)

Recurrent Neural Networks: The recurrent neural network is basic network structure; it helps to develop other deep learning structures. Basic multiple layer neural

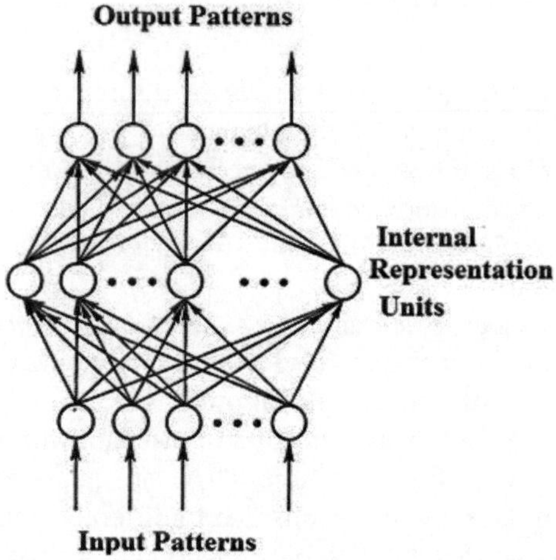

Fig. 2 Recurrent neural networks (RNNs) [15]

network has completely feed-forward links, whereas in case of a recurrent network, it has feedback links connected to the previous layers.

This mechanism enables recurrent neural networks to remember the previous inputs values so that easy process within specific duration.

Figure 2 displays input steps in the current moment showing the example of input, and feedback layers reflect the previous moment's output. Closed-loop feedback related to their past decisions ingests their own outputs as input moment by moment. Inserting memory into neural networks to provide information in the sequence itself and in feed-forward style networks, this is not the case. Back-propagation or back-propagation by time algorithms is used to train recurrent neural networks. Mostly used in speech recognition, recurrent neural networks.

Long Short-term Memory (LSTM): Long short-term memory was earlier structured yet it prominent as of late as a RNN engineering for different applications. In different items will discover LSTMs that utilization consistently, android devices, voice recognition, text captioning etc. rather presented the idea of a memory section. It can hold its value as an element of its information sources for a long or short period of time, allowing the location to recollect what is critical and not just its last registered esteem. In reality, long-term short-term memory network is RNN made of LSTM units. An output input way and a cell for the short term. The cell collects interim values over arbitrary duration and passes the data stream to and from the location of the memory. When a current data is skipped, the short-term cell IS used to handle, allowing the cell to recall new information. In conclusion, the output way handles as data stored in short memory location is utilize in output.

LSTM systems are applicable for preparing, categorizing, and forecasting depending on event sequence information (Fig. 3).

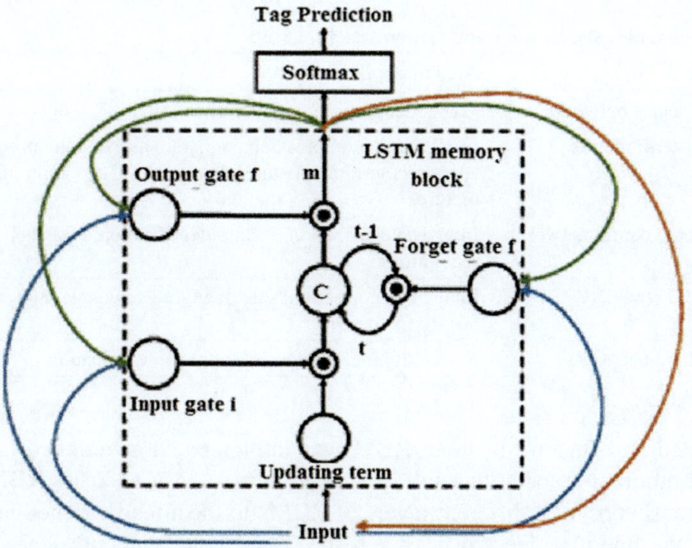

Fig. 3 LSTM networks memory cell arrangement [15]

(i) Convolution neural network (CNN)

Convolutionary neural system is mainly used for image identification, image categorization, and biometric recognition. CNN picture identification accepts a picture as an input and prepares it for groupings of particles (i.e., vehicles, pets, toys) [2, 9].

In Fig. 4, convolution neural network accepts information in the form of an image; in the subsequent stage, convolution layer is used to extract features from an input [9]. Convolution of a picture into extricate features stage performs various activities such as border detection and eliminating noise from a picture by applying filters [10].

Deep Belief Network (DBN): DBN is a class of deep learning system with both directed and undirected edges that involves different layers. It consists of various layers of shrouded units, where each layer is in any case connected to each other, and units are not connected [21]. To learn deep belief networks need to have comprehend two imperative techniques of DBN [9, 21].

Belief Network: It includes stochastic binary unit layers with a certain weight for each associated layer. In belief networks, stochastic binary units have a low (0) or high (1) condition, and the probability and likelihood of becoming 1 are regulated by a tendency and weighted input from other units [21].

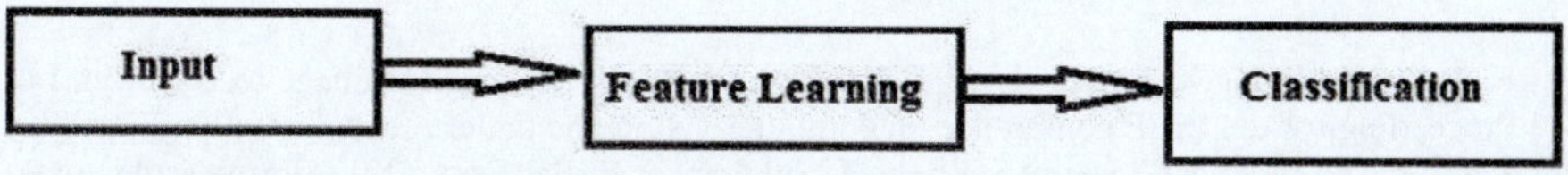

Fig. 4 Basic steps of convolution neural network

Table 1 Techniques and its some application area [4, 15, 16]

Techniques	Area of application
Recurrent neural networks	Voice recognition, text identification
Long short-term memory	In NLP data compression, signature and text identification, voice recognition, posture identification, text captioning for pictures
Convolutional neural networks	Image categorization, visual identification, NLP, behavior recognition
Deep belief networks	Image categorization, search engine, language interpretation, failure forecasting
Deep stacking networks	Search engine, conversation voice identification

Restricted Boltzmann Machine: RBM was implemented as a veiled layer unit that has a minimal association with each hide unit so easy to know RBM. Deep belief systems consist of different layers of RBMs to the already trained layer [21] following the fine-tune feed-forward scheme. Deep belief network learn features from the exposed device in the first level, learn features in a second hide layer in the next step. The entire deep belief network is educated after learning the final layer [11].

Deep Stacking Network (DSN)

The deep stacking network (DSN) is a deep architecture that can be adapted to parallel weight learning [12]. It is prepared in a supervised, block-wise manner, with no back-propagation requirement across all blocks, as it is common in other prominent deep models [13]. The DSN blocks are stacked to form the deep system [14], each consisting of a basic, easy-to-learn module (Table 1).

4 Conclusion

Different techniques and structures available in deep learning help to use in variety of different application areas. Several layers of neural networks in these techniques allow categorization of data by extracting distinct features. Huge amount of dataset is trained in deep learning so that accuracy gets improved.

References

1. Kaiming H, Zhang X, Ren S, Sun J (2016) Deep residual learning for image recognition. In: Proceedings of the IEEE conference on computer vision and pattern recognition, pp 770–778
2. Karpathy A, Toderici G, Shetty S, Leung T, Sukthankar R, Fei-Fei L (2014) Large-scale video classification with convolutional neural networks. In: Proceedings of the IEEE conference on computer vision and pattern recognition, pp 1725–1732

3. Du X, Cai Y, Wang S, Zhang L (2016) Overview of deep learning. In: 31st Youth academic annual conference of chinese association of automation (YAC). IEEE, pp 159–164
4. Zhou X, Gong W, Fu W, Du F (2017) Application of deep learning in object detection. In: IEEE/ACIS 16th international conference on computer and information science. IEEE, pp 631–634
5. Ravì D, Wong C, Deligianni F, Berthelot B, Andreu-Perez J, Lo B, Yang G-Z (2016) Deep learning for health informatics. IEEE J Biomed Health Inform 21(1):4–21
6. Ioffe S, Szegedy C (2015) Batch normalization: accelerating deep network training by reducing internal covariate shift
7. Lin T-Y, Maire M, Belongie S, Bourdev L, Girshick R, Hays J, Perona P, Ramanan D, Lawrence Zitnick C, Dollar P (2015) Microsoft COCO: common objects in context
8. Zhou B, Lapedriza A, Xiao J, Torralba A, Oliva A (2014) Learning deep features for scene recognition using places database. In: Advances in neural information processing systems, pp 487–495
9. LeCun Y, Bengio Y, Hinton G (2015) Deep learning. Nature 14539, 52128:436–444
10. Hariharan B, Arbeláez P, Girshick R, Malik J (2014) Simultaneous detection and segmentation
11. Rubi CR (2015) A review: speech recognition with deep learning methods. Inter J Comput Sci Mob Comput (IJCSMC) 4(5):1017–1024
12. Bengio Y (2009) Learning deep architectures for AI. Found Trends Mach Learn 2(1):1–127
13. Hutchinson B, Deng L, Yu D (2012) Tensor deep stacking networks. IEEE Trans Pattern Analysis Mach Intell (Special issue in learning deep architectures), 1–14
14. Makhlysheva A, Budrionis A, Chomutare T, Nordsletta AT, Bakkevoll PA, Henriksen TD, Hurley JS (2018) Health analytics. Norwegian Center for E-health Research
15. Understanding of Convolutional Neural Network (CNN)—Deep learning. https://medium.com/@RaghavPrabhu/understanding-of-convolutional-neural-network-cnn-deep-learning-99760835f148
16. Deng L, Yu D (2011) Deep convex net: a scalable architecture for speech pattern classification. In: Twelth annual conference of the international speech communication association
17. Gong Y, Wang L, Guo R, Lazebnik S Multi-scale orderless pooling of deep convolutional activation features (published in Illinois)
18. Liu Y, Liu S, Zhao X (2017) Intrusion detection algorithm based on convolutional neural network. In: 4th International conference on engineering technology and application, pp 9–13
19. Tim Jones M (2017) Deep learning architectures. Artificial Intelligence
20. Mahmud M, Kaiser MS, Hussain A, Vassanelli S (2018) Applications of deep learning and reinforcement learning to biological data. IEEE Trans Neural Netw Learn Syst, 1–17

Author Index

K. Kotecha et al. (eds.), *Data Science and Intelligent Applications*,
Lecture Notes on Data Engineering and Communications Technologies 52,
https://doi.org/10.1007/978-981-15-4474-3